S0-AKX-379

Northeast Lakeview College
33784 0001 2834 8

Receptor Signal Transduction Protocols

METHODS IN MOLECULAR BIOLOGY™

John M. Walker, SERIES EDITOR

METHODS IN MOLECULAR BIOLOGY™

Receptor Signal Transduction Protocols

SECOND EDITION

Edited by

Gary B. Willars

R. A. John Challiss

Department of Cell Physiology and Pharmacology
University of Leicester
Leicester, UK

Humana Press **Totowa, New Jersey**

999 Riverview Drive, Suite 208
Totowa, New Jersey 07512

www.humanapress.com

This publication is printed on acid-free paper. ∞
ANSI Z39.48-1984 (American Standards Institute) Permanence of Paper for Printed Library Materials.

Cover design by Patricia F. Cleary.
Cover: Confocal image showing membrane localization of fluorescently labelled neuromedin U-8 (NmU-8) bound to HEK293 cells expressing recombinant human NmU type 1 receptors (NmU-R1). NmU-8, N-terminally labelled with Cy3B (Amersham, Bucks, UK), was kindly provided by J. Scott and M. Ruediger (GlaxoSmithKline, Harlow, UK). Image: P. J. Brighton and G. B. Willars (University of Leicester, UK). Inset: Cartoon depicting GPCR structure.

For additional copies, pricing for bulk purchases, and/or information about other Humana titles, contact Humana at the above address or at any of the following numbers: Tel.: 973-256-1699; Fax: 973-256-8341; E-mail: humana@humanapr.com; or visit our Website: www.humanapress.com

Printed in the United States of America. 10 9 8 7 6 5 4 3 2 1

ISSN 1064-3745

E-ISBN 1-59259-754-8

Library of Congress Cataloging in Publication Data

Receptor signal transduction protocols / edited by Gary B. Willars,
R. A. John Challiss.-- 2nd ed.
p. ; cm. -- (Methods in molecular biology ; 259)
Includes bibliographical references and index.
ISBN 1-58829-329-7 (alk. paper)
1. Cell receptors--Research--Methodology. 2. Cellular signal transduction--Research--Methodology. [DNLM: 1. Signal Transduction--physiology--Laboratory Manuals. QH 601 R2954 2004] I. Willars, Gary B. II. Challiss, R. A. J. (R. A. John) III. Series: Methods in molecular biology (Clifton, N.J.) ; v. 259.
QH603.C43R42 2004
572'.69--dc22

2003027534

Preface

This second edition of *Receptor Signal Transduction Protocols* not only has a new editor, but also a greater focus on G-protein-coupled receptors, their properties *per se,* and their coupling to immediate downstream binding partners—principally, although not exclusively, the heterotrimeric G-proteins. The new edition combines updates of key chapters from the first edition, as well as a large number of new contributions covering key methodologies that have emerged, or been extended to receptor/G-protein research, in the past 5–6 years.

In common with many fields, the range of methods used to assess the first steps in signal transduction are continually expanding and methods that might have been considered too specialized five years ago are now sufficiently routine to be included here. Unlike many research areas, where off-the-shelf kits have made research basically foolproof, signal transduction research still requires considerable expertise, and the methods included here are provided by internationally recognized experts in their fields who have many years of experience using the methods they describe. This not only allows each chapter to impart a clear description of the method, but also to furnish invaluable troubleshooting advice for when things do not go entirely according to plan.

Once again we would like to thank the Series Editor, John Walker, for the invitation to compile this second edition, and to express our gratitude to all of the authors who have enthusiastically agreed to provide the uniformly excellent contributions.

Gary B. Willars
R. A. John Challiss

Preface to the First Edition

Gaining an understanding of the mechanisms by which cells process and respond to extracellular cues has become a major goal in many areas of biology and has attracted the attentions of almost every traditional discipline within the biological sciences. At the heart of these divergent endeavors are common methods that can aid biochemists, physiologists, and pharmacologists in tackling the specific questions addressed by their research.

In *Receptor Signal Transduction Protocols,* a diverse array of methodologies employed to interrogate ligand–receptor and receptor–effector interactions are described by authors who have devised and successfully applied them. The authors blend excellent descriptions and applications of fairly well established methodologies with new technologies at the cutting-edge of signal transduction research and as such I hope the present volume will complement and extend a previous excellent volume in this series edited by David Kendall and Stephen Hill (*Methods in Molecular Biology,* vol. 41, *Signal Transduction Protocols*).

The objective of the *Methods in Molecular Biology* series is to present easy-to-follow protocols that can be used at the bench. Although a laudable aim, some of the methodologies presented here are complex and require considerable preparation and initial experimental validation in the reader's own hands. Indeed, many chapters present in a concise form hard-won methodologies refined over months or years in the author's laboratory; as such it would be naive (and probably misguided) to assume that they are "recipes" for instant research success. Therefore, I hope that the success of the chapters in this volume will not necessarily be judged by the ease with which they can be directly applied experimentally, but that they will be considered successful if they provide other workers with sound bases to be adopted and adapted to their own experimental objectives.

Finally, I would like to take this opportunity to thank the series editor, John Walker, for his invitation to put this volume together, and to express my gratitude to all of the authors for their uniformly excellent contributions.

R. A. John Challiss

Contents

Contributors

ATSU AIBA • *Department of Molecular and Cellular Biology, Kobe University Graduate School of Medicine, Kobe, Japan*
JÜRGEN E. BADER • *Institute of Biochemistry, University of Leipzig, Leipzig, Germany*
ANNETTE G. BECK-SICKINGER • *Institute of Biochemistry, University of Leipzig, Leipzig, Germany*
JEFFREY L. BENOVIC • *Department of Microbiology and Immunology, Kimmel Cancer Center, Thomas Jefferson University, Philadelphia, PA*
ANDREE BLAUKAT • *Oncology Research Darmstadt, Merck KGaA, Darmstadt, Germany*
MARION BLOMENRÖHR • *Physiological Chemistry, Utrecht University, Utrecht, The Netherlands*
JAN BOGERD • *Department of Endocrinology, Utrecht University, Utrecht, The Netherlands*
ASHLEY E. BRADY • *Vanderbilt University Medical Center, Nashville, TN*
DAVID B. BYLUND • *Department of Pharmacology, University of Nebraska Medical Center, Omaha, NE*
R. A. JOHN CHALLISS • *Department of Cell Physiology and Pharmacology, University of Leicester, Leicester, United Kingdom*
ANTONIO DE BLASI • *Laboratory of Cellular and Molecular Neurobiology, I.N.M. Neuromed, I.R.C.C.S., Pozzilli, Italy*
JEAN D. DEUPREE • *Department of Pharmacology, University Nebraska Medical Center, Omaha, NE*
LUCY F. DONALDSON • *Department of Physiology, University of Bristol School of Medical Sciences, Bristol, United Kingdom*
MARK R. DOWLING • *Novartis Horsham Research Center, Horsham, West Sussex, United Kingdom*
KARIN A. EIDNE • *Molecular Endocrinology Group/7TM Receptor Laboratory, Western Australian Institute for Medical Research, Centre for Medical Research, The University of Western Australia, Perth, Australia*
PAT FREEMAN • *School of Health and Bioscience, University of East London, London, United Kingdom*

Michael Freissmuth • *Institute of Pharmacology, Vienna University, Vienna, Austria*
Blair D. Grubb • *Department of Cell Physiology and Pharmacology, University of Leicester, Leicester, United Kingdom*
Emmanuel Hermans • *Laboratory of Experimental Pharmacology, Catholic University of Louvain, Brussels, Belgium*
Martin Hohenegger • *Institute of Pharmacology, Vienna University, Vienna, Austria*
Stephen R. Ikeda • *Laboratory of Molecular Physiology, National Institute on Alcohol Abuse and Alcoholism, National Institutes of Health, Bethesda, MD*
Jennifer A. Koenig • *Department of Pharmacology, University of Cambridge, Cambridge, United Kingdom*
Karen M. Kroeger • *Molecular Endocrinology Group/7TM Receptor Laboratory, Western Australian Institute for Medical Research, Centre for Medical Research, The University of Western Australia, Perth, Australia*
Sebastian Lazareno • *MRC Technology, London, United Kingdom*
Lee E. Limbird • *Vanderbilt University Medical Center, Nashville, TN*
Rafael Luján • *Centro Regional de Investigaciones Biomédicas, Universidad de Castilla-La Mancha, Albacete, Spain*
John J. Mackrill • *Department of Biochemistry, University College Cork, Cork, Ireland*
Adriano Marchese • *Department of Microbiology and Immunology, Kimmel Cancer Center, Thomas Jefferson University, Philadelphia, PA*
Ikuo Matsuda • *Department of Molecular and Cellular Biology, Kobe University Graduate School of Medicine, Kobe, Japan*
R. A. Jeffrey McIlhinney • *Medical Research Council Anatomical Neuropharmacology Unit, Oxford, United Kingdom*
Graeme Milligan • *Molecular Pharmacology Group, University of Glasgow, Glasgow, Scotland, United Kingdom*
Laura C. Mongan • *Department of Cell Physiology and Pharmacology, University of Leicester, Leicester, United Kingdom*
Ian Mullaney • *Molecular Pharmacology Group, University of Glasgow, Glasgow, Scotland, United Kingdom*
Stefan R. Nahorski • *Department of Cell Physiology and Pharmacology, University of Leicester, Leicester, United Kingdom*
Christian Nanoff • *Institute of Pharmacology, Vienna University, Austria*

ANTONIO PORCELLINI • *Department of Molecular Pathology, I.N.M. Neuromed, I.R.C.C.S., Pozzilli, Italy; Department of Experimental Medicine and Pathology, University of Rome, Rome, Italy*
DOMENICO SPINA • *The Sackler Institute of Pulmonary Pharmacology, King's College London, London, United Kingdom*
MICHAEL TANOWITZ • *Department of Psychiatry, University of California, San Francisco, CA*
ANDREW B. TOBIN • *Department of Cell Physiology and Pharmacology, University of Leicester, Leicester, United Kingdom*
MYRON L. TOEWS • *Department of Pharmacology, University of Nebraska Medical Center, Omaha, NE*
HENRY F. VISCHER • *Department of Endocrinology, Utrecht University, Utrecht, The Netherlands*
QIN WANG • *Vanderbilt University Medical Center, Nashville, TN*
RICHARD J. WARD • *Molecular Pharmacology Group, University of Glasgow, Glasgow, Scotland, United Kingdom*
MARK WHEATLEY • *School of Biosciences, University of Birmingham, Brimingham, United Kingdom*
GARY B. WILLARS • *Department of Cell Physiology and Pharmacology, University of Leicester, Leicester, United Kingdom*
MARK VON ZASTROW • *Department of Psychiatry, University of California, San Francisco, CA*

1

Radioligand-Binding Methods for Membrane Preparations and Intact Cells

David B. Bylund, Jean D. Deupree, and Myron L. Toews

Summary

The radioligand-binding assay is a relatively simple but powerful tool for studying G-protein-coupled receptors. There are three basic types of radioligand-binding experiments: (1) saturation experiments from which the affinity of the radioligand for the receptor and the binding site density can be determined; (2) inhibition experiments from which the affinity of a competing, unlabeled compound for the receptor can be determined; and (3) kinetic experiments from which the forward and reverse rate constants for radioligand binding can be determined. Detailed methods for typical radioligand-binding assays for G-protein-coupled receptors in membranes and intact cells are presented for these types of experiments. Detailed procedures for analysis of the data obtained from these experiments are also given.

Key Words

Affinity, assay, binding, competition, G-protein-coupled receptor, inhibition, intact cell, kinetic, nonspecific binding, radioligand, radioreceptor, rate constant, receptor, saturation.

1. Introduction

The radioligand-binding assay is a relatively simple but powerful tool for studying G-protein-coupled receptors. It can be used to determine the affinity of numerous drugs for these receptors, and to characterize regulatory changes in receptor number and in subcellular localization. As a result, this assay is widely used (and often misused) by investigators in a variety of disciplines. Our focus in this chapter is on radioligand-binding assays in membrane preparations from tissues and cell lines, and in intact cells. Similar techniques, however, can be used to study solubilized receptors, receptors in tissue slices (receptor autoradiography), or receptors in intact animals.

From: *Methods in Molecular Biology, vol. 259, Receptor Signal Transduction Protocols, 2nd ed.*
Edited by: G. B. Willars and R. A. J. Challiss © Humana Press Inc., Totowa, NJ

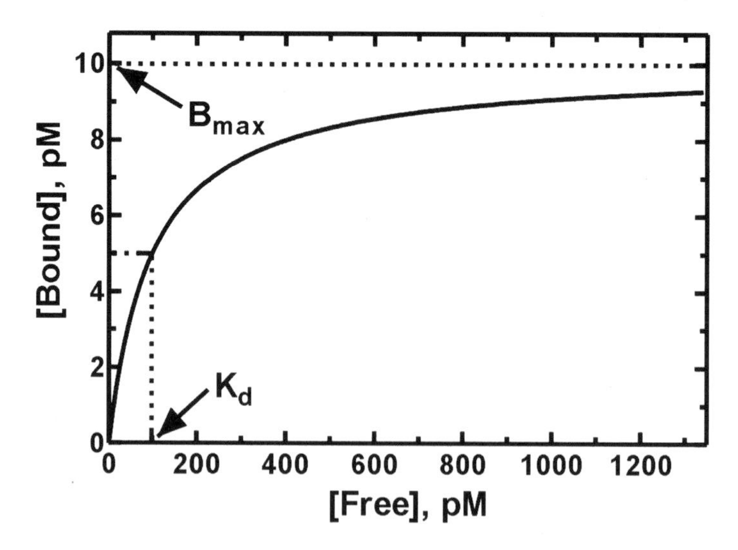

Fig. 1. Typical saturation experiment. In this simulation the B_{max} (receptor density) is 10 p*M* and the K_d (the dissociation constant or the free concentration that gives half-maximal binding) is 100 p*M*.

There are three basic types of radioligand-binding experiments: (1) saturation experiments from which the affinity (K_d) of the radioligand for the receptor and the binding site density (B_{max}) can be determined; (2) inhibition experiments from which the affinity (K_i) of a competing, unlabeled compound for the receptor can be determined; and (3) kinetic experiments from which the forward (k_{+1}) and reverse (k_{-1}) rate constants for radioligand binding can be determined. This chapter presents methods for typical radioligand-binding assays for G-protein-coupled receptors.

1.1. Saturation Experiment

Saturation experiments are frequently used to determine the change in receptor density (number of receptors) during development or following some experimental intervention, such as treatment with a drug. A saturation curve is generated by holding the amount of receptor constant and varying the concentration of radioligand. From this type of experiment the receptor density (B_{max}) and the dissociation constant (K_d) of the receptor for the radioligand can be estimated. The results of the saturation experiment can be plotted with bound (the amount of radioactive ligand that is bound to the receptor) on the *y*-axis and free (the free concentration of radioactive ligand) on the *x*-axis. As shown in **Fig. 1**, as the concentration of radioligand increases the amount bound

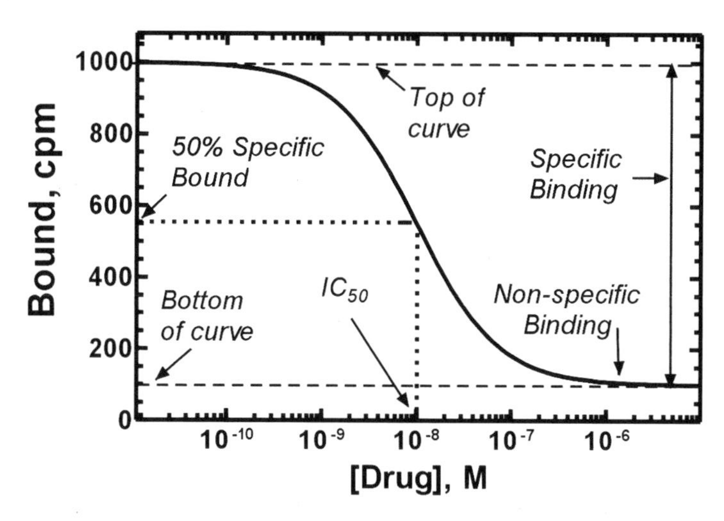

Fig. 2. Typical inhibition experiment. In this simulation the specific binding is 900 cpm and the IC_{50} (the concentration of drug that inhibits 50% of the specific binding) is 10 n*M*.

increases until a point is reached at which more radioactive ligand does not significantly increase the amount bound. The resulting graph is a rectangular hyperbola and is called a saturation curve. B_{max} is the maximal binding which is approached asymptotically as radioligand concentration is increased. B_{max} is the density of the receptor in the tissue being studied. K_d is the concentration of ligand that occupies 50% of the binding sites.

1.2. Inhibition Experiment

The great utility of inhibition experiments is that the affinity of any (soluble) compound for the receptor can be determined. Thus these assays are heavily used both for determining the pharmacological characteristics of the receptor and for discovering new drugs using high-throughput screening techniques. In an inhibition experiment, the amount of an inhibitor (nonradioactive) drug included in the incubation is the only variable, and the dissociation constant (K_i) of that drug for the receptor identified by the radioligand is determined. A graph of the data from a typical inhibition experiment is shown in **Fig. 2**. The amount of radioligand bound is plotted vs the concentration of the unlabeled ligand (on a logarithmic scale). The bottom of the curve defines the amount of nonspecific binding. The IC_{50} value is defined as the concentration of an unlabeled drug required to inhibit specific binding of the radioligand by 50%. The K_i is then calculated from the IC_{50}.

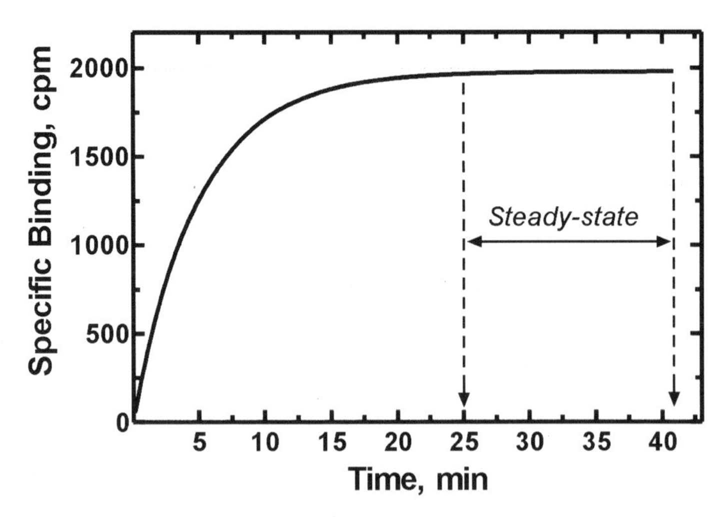

Fig. 3. Typical association experiment. In this simulation steady state is reached after approx 25 min and lasts until the end of the experiment (42 min).

1.3. Kinetic Experiments

Kinetic experiments have two main purposes. The first is to establish an incubation time that is sufficient to ensure that steady state (commonly called equilibrium) has been reached. The second is to determine the forward (k_{+1}) and reverse (k_{-1}) rate constants. The ratio of these constants provides an independent estimate of the K_d (k_{-1}/k_{+1}). If the amounts of receptor and radioligand are held constant and the time varied, then kinetic data are obtained from which forward and reverse rate constants can be estimated. A graph of the data from a typical association kinetic experiment is shown in **Fig. 3**. Initially the rate of the forward reaction exceeds the rate of the reverse reaction. After approx 25 min the amount of specific binding no longer increases and thus steady state has been reached. From these data, the k_{+1} can be calculated.

For a dissociation experiment, the radioligand is first allowed to bind to the receptor and then the dissociation of the radioligand from the receptor is monitored by the decrease in specific binding (**Fig. 4**). The rebinding of the radioligand to the receptor is prevented by the addition of a high concentration of a nonradioactive drug that binds to the receptor and thus blocks the receptor binding site, or by "infinite" dilution which reduces the free concentration of the radioligand. Dissociation follows first-order kinetics and thus k_{-1} is equal to the $t½$ for dissociation divided by 0.693 (natural logarithm of 2).

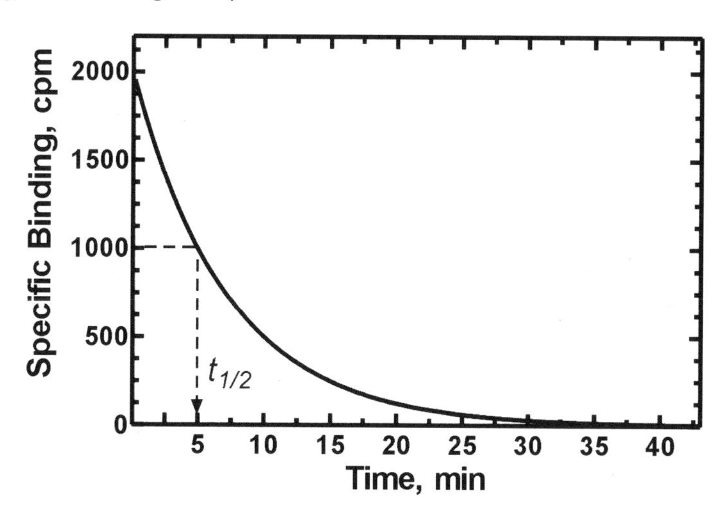

Fig. 4. Typical dissociation experiment. In this simulation the $t_{1/2}$ (the time at which the specific binding has decreased by 50%) is 5 min.

1.4. Assays in Intact Cells

Although isolated membranes are by far the most common preparation used for radioligand-binding assays, for some purposes it is preferable to use intact cells. The most obvious advantage of assays with intact cells is that the receptor is being studied in its native environment in the cell. A related advantage of intact cell assays is that the binding properties of the receptor can be assessed in the same preparation and under essentially the same conditions as the functional responses mediated by the receptor are measured. This allows a more direct comparison of the receptor binding properties with a wide variety of physiological responses following activation or inhibition of the receptor. Intact cell assays may also be advantageous when a large number of different cell samples need to be studied, because intact cell assays eliminate the need to lyse cells and isolate membranes prior to assay. For example, intact cell assays have proven very useful for preliminary screening of cell colonies following transfection with cDNA for various G-protein-coupled receptors, thus allowing rapid identification of clones for amplification and further analysis.

Most of the considerations that make intact cell assays advantageous in certain cases also represent limitations of intact cell assays in other cases. For example, intact cell assays allow studies under physiological conditions, but they make it much more difficult to vary or control the assay conditions to identify factors that modulate receptor binding. Radioligand uptake into cells by various transport processes can occur with intact cells, and care must be taken to ensure that radioligand association with intact cells is due to binding rather than uptake. The occurrence of adaptive regulatory changes in receptor number, localization, and binding properties during the course of binding assays with intact cells can also present a serious complication *(1)*. Finally, intact cells have membrane permeability barriers that are not present in isolated membrane preparations, and therefore the lipid solubility and membrane permeability of both the radioligand and the competing ligands must be considered in assays with intact cells. Lipophilic ("lipid-loving") ligands generally cross all cell membranes easily and thus have access to both cell surface receptors and those in intracellular compartments such as endosomes. In contrast, hydrophilic ("water-loving") ligands are relatively impermeable to the plasma membrane, and thus these ligands label only cell surface receptors. Although these properties can complicate assays with intact cells, they also provide the basis for important radioligand-binding-based assays for receptor internalization, as discussed previously *(1)*.

2. Materials

The information given in this section is specifically for assays with membrane preparations. Additional information for intact cell assays is given in **Subheading 3.4.**

1. A radioligand appropriate for the receptor being studied (*see* **Note 1**). For membrane saturation experiments, add the appropriate volume of radioligand into 550 µL of 5 m*M* HCl in a glass test tube. Thoroughly mix and add 200 µL of this solution to 300 µL of 5 m*M* HCl. Prepare successive dilutions in the same manner by adding 200 µL of each dilution to 300 µL of 5 m*M* HCl to obtain the next lower dilution until six concentrations of radioligand have been prepared. This dilution strategy gives a 100-fold range of radioligand concentrations. Other dilution strategies will give different ranges as indicated in **Table 1** (*see* **Note 2**). For membrane inhibition and kinetic experiments, only a single concentration of radioligand is needed (*see* **Note 3**).
2. A source of receptor, either membranes or intact cells. The standard procedure for a membrane assay is to homogenize the tissue or cells of interest in a hypotonic buffer using either a Polytron (Brinkman) or similar homogenizer. Remarkably, most receptors are stable at room temperature (generally for hours), although it is wise to put the tissue on ice quickly. Homogenize about 500 mg of tissue in approx 35 mL of wash buffer (50 m*M* Tris-HCl or similar buffer at pH 7.0–8.0) using a Polytron (PT10-35 generator with PT10/TS probe) at setting 7 for 20 s (*see* **Note 4**). The actual weight of tissue used should be recorded. Centrifuge at 20,000 rpm (48,000*g*) in a Sorvall RC5-B using an SS34 rotor (or similar centrifuge and rotor) for 10 min at 4° C (*see* **Note 5**). Decant the supernatant, and

Table 1
Dilution of Radioligand for Saturation Experiments

μL of radioligand	250	200	150
μL of diluent	250	300	350
Dilution number	Relative concentration		
1	100	100	100
2	50	40	30
3	25	16	9.0
4	12.5	6.4	2.7
5	6.2	2.6	0.81
6	3.1	1.0	0.24
7	1.6	0.41	
8	0.78		
9	0.39		

repeat the homogenization and centrifugation. The tissue preparation can either be used immediately or stored frozen as a pellet until needed (*see* **Note 6**). Generally protease inhibitors are not needed, but could be important in the case of certain tissues or with certain receptors.

3. Membrane assay buffer, 25 m*M* at pH 7.4, such as sodium phosphate or Tris. For a few receptors the choice of buffer is important, but for most it is not.
4. Wash buffer such as 25 m*M* Tris, pH 7.4. Almost any buffer at neutral pH will serve the purpose.
5. 5 m*M* HCl for diluting labeled and unlabeled ligands. For many ligands, using a slightly acidic diluent will increase stability and decrease binding to test tubes.
6. Appropriate unlabeled ligands in solution.
7. 0.1 *M* NaOH for samples to be used to assay protein.
8. Polypropylene test tubes, 12 × 75 mm (assay tubes).
9. Borosilicate glass test tubes, 12 × 75 mm (dilution tubes).
10. Glass fiber filters (GF/A circles and GF/B strips).
11. Filtration manifold.
12. Scintillation vials if using a ^{3}H-radioligand, or test tubes if using a ^{125}I-radioligand.
13. Scintillation cocktail (if using a ^{3}H-radioligand).

3. Methods

3.1. Saturation Experiment (Membrane Assay)

1. Resuspend washed membrane preparation in distilled water by homogenization.
2. Add three 20-μL aliquots of the tissue suspension to 80 μL of 0.1 *M* NaOH for estimating protein concentration.

3. Add sufficient ice-cold assay buffer to the membrane suspension to give the appropriate final concentration (*see* **Note 7**).
4. Set up a rack of 24 polypropylene incubation tubes, 6 tubes across and 4 tubes deep. If using a ^{125}I-ligand add two additional test tubes to each of the 6 sets of tubes (for the determination of total added radioactivity). If using a ^{3}H-ligand, prepare a set of 12 uncapped scintillation vials with GF/A glass fiber filter discs (*see* **Note 8**).
5. To the 12 tubes on the last two rows, add 10 μL of a high concentration of an unlabeled ligand to determine nonspecific binding (*see* **Note 9**).
6. To all 24 tubes add 970 μL of the membrane preparation. Because this is a particulate suspension, it should be stirred slowly while aliquots are being removed.
7. Starting with the most dilute radioligand solution, add 20 μL to the columns of four tubes, and mix each tube. Also add 20 μL of the radioligand solution to the two filter papers on the scintillation vials (if using a ^{3}H-radioligand) or two test tubes (if using a ^{125}I-ligand) for the determination of total added radioactivity.
8. Mix all the tubes again and incubate (usually at room temperature) for 45 min. Assuming that the system is at steady state, the exact time is not critical. The tubes may need to be rearranged to be compatible with the specific style of filtration manifold used.
9. Filter the contents of the tubes and wash the filters twice with 5 mL of wash buffer. Depending on the rate of dissociation of the radioligand from the receptor, it may be important to use ice-cold wash buffer.
10. Place the filters into scintillation vials, add 5 mL of scintillation cocktail and cap if using a ^{3}H-radioligand; or into test tubes if a using ^{125}I-ligand.
11. Shake the scintillation vials gently for 1 h (or let stand at room temperature overnight) and then count in a liquid scintillation counter (if using a ^{3}H-radioligand); or count in a gamma counter (if using a ^{125}I-ligand) (*see* **Note 10**).

3.1.1. Calculation of Results from a Saturation Experiment

Data from a sample saturation experiment are shown in **Table 2**. (The methods and calculations for the sample competition and inhibition experiments are also available in an interactive format at http://www.unmc.edu/Pharmacology/receptortutorial/.)

1. Total binding and nonspecific binding can be plotted vs total added as shown in **Fig. 5**. for the sample experiment. This plot allows one to detect data points that may be problematic. Note that the nonspecific binding is linear (except possibly at the lowest concentrations), and that the specific binding saturates (is relatively constant) at high radioligand concentrations.
2. Specific binding is determined by subtracting nonspecific binding from total binding at each concentration of radioligand (*see* **Table 2**).
3. The cpm values are converted to picomolar values using a conversion factor that accounts for specific activity for the radioligand, the counting efficiency of the particular scintillation counter used, and the conversion factor 2.2×10^{12} dpm/Ci. For this experiment the counting efficiency was 0.36 and the specific radioactiv-

Table 2
Results of a Sample Saturation Experiment[a] in cpm

Total added (cpm)	Total bound (cpm)	Nonspecifically bound (cpm)	Specifically bound[b] (cpm)
2360	208	18	190
5601	394	25	369
14,491	597	46	551
32,011	782	88	694
82,520	984	189	795
199,248	1210	416	794

[a] $[^3H]$RX821002 binding to human α_{2A}-adrenergic receptors in HT-29 cells.
[b] The amount of radioligand specifically bound was also determined by subtracting the nonspecifically bound from the total bound.

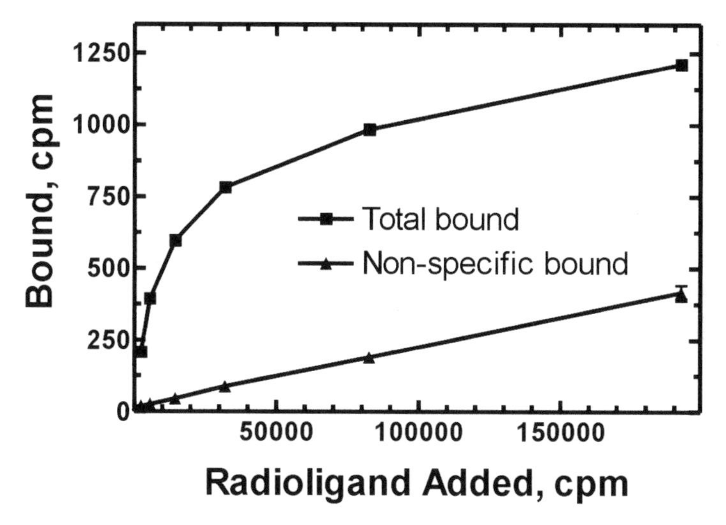

Fig. 5. Total binding and nonspecific binding vs total added for a sample experiment from **Table 2**.

ity of the radioligand was 60 Ci/mmol, and the factor for converting cpm to dpm is 0.0210 as shown:

$$\frac{\text{cpm}}{\text{mL}} \times \frac{\text{dpm}}{0.36\ \text{cpm}} \times \frac{\text{Ci}}{2.2 \times 10^{12}\ \text{dpm}} \times \frac{\text{mmol}}{60\ \text{Ci}} \times \frac{1000\ \text{mL}}{\text{L}} \times \frac{1\ \text{mol}}{1000\ \text{mmol}} \times \frac{10^{12}\ \text{pmoL}}{\text{mol}} = \frac{2.10 \times 10^{-2}\ \text{pmol}}{\text{L}}$$

The results of this conversion for the sample experiment are shown in **Table 3**.

Table 3
Results of Sample Saturation Experiment Given in Table 2 Converted to p*M* Units

Total added (p*M*)	Specifically bound (p*M*)	Free (p*M*)	Bound/free
4.37	3.99	46	0.0867
8.27	7.74	110	0.0704
12.5	11.5	292	0.0394
16.4	14.5	658	0.0220
20.7	16.7	1720	0.00971
25.4	16.7	4020	0.00415

4. The free concentration of radioligand is calculated by subtracting specifically bound from total added as shown in **Table 3**.
5. The data are then plotted as bound vs free as shown in **Fig. 6** for the typical saturation experiment (*see* **Note 11**). The K_d and B_{max} values are generally calculated by nonlinear regression of the specific binding vs the concentration of radioligand using a computer program as such Prism (GraphPad, San Diego, CA) or a variety of other software packages using the following equation:

$$B = \frac{B_{max} \times F}{K_d + F}$$

where B is the amount of radioligand specifically bound, F is the free radioligand concentration, B_{max} is the radioligand concentration required to saturate all of the binding sites, and K_d is the dissociation constant for the radioligand at these receptors.
6. The B_{max} values are dependent on the concentration of protein in the assay. Note that the results are given in pM units. To convert the B_{max} values to pmol/mg of protein, the p*M* values are converted to pmol/mL and divided by the protein concentration (mg/mL) used in the assay. In this example the protein concentration is 0.072 mg/assay tube and the B_{max} value is calculated as shown:

$$\frac{17.7 \text{ pmol}}{\text{L}} \times \frac{1 \text{ L}}{1000 \text{ mL}} \times 1 \text{ mL} \times \frac{1 \text{ mL}}{0.072 \text{ mg}} = \frac{0.246 \text{ pmol}}{\text{mg}}$$

7. To visualize the results better and to detect potential problems, the data are frequently transformed (as shown in **Table 3**) and viewed as a Rosenthal plot *(2)* in

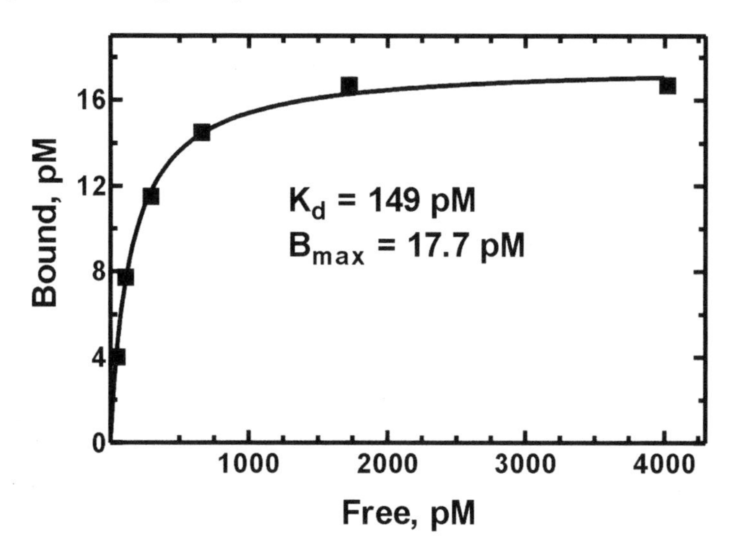

Fig. 6. Saturation curve for data from a sample saturation experiment. Specifically bound (p*M* units) from **Table 3** is plotted vs the free (p*M*) concentration of the radioligand. The line was drawn using nonlinear regression analysis for one-site binding using the Prism computer program (GraphPad, San Diego, CA).

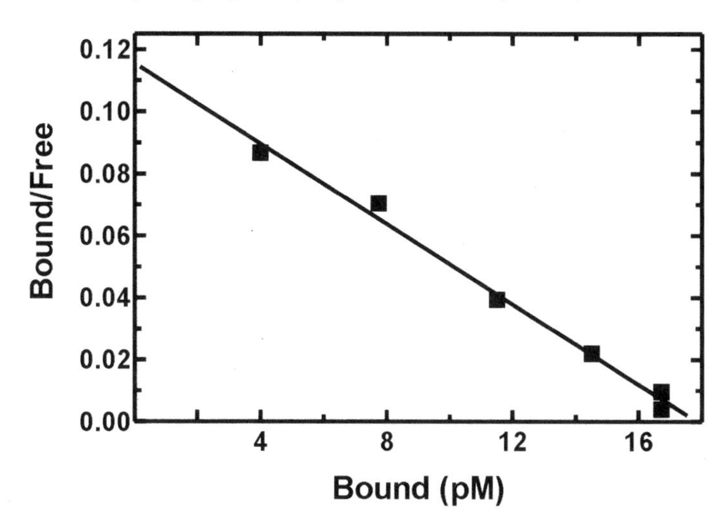

Fig. 7. Rosenthal plot for sample saturation experiment. The bound/free vs bound data from **Table 3** are plotted to obtain the Rosenthal plot.

the form of bound/free vs bound as shown in **Fig. 7** (*see* **Note 12**). The equation for the line is:

$$\frac{\text{Bound}}{\text{Free}} = -\frac{1}{K_d}\text{Bound} + \frac{B_{max}}{K_d}$$

In this plot, the intercept with the *x*-axis (abscissa) is the B_{max} and the K_d is the negative reciprocal of the slope (*see* **Note 13**). The data points fall close to a straight line, indicating a single class of binding sites.

3.2. Inhibition Experiment

1. Prepare a 1 m*M* solution of the inhibitor(s) in 5 m*M* HCl (or other diluent as appropriate). Dilute 0.3 mL of this solution with 0.7 mL of 5 m*M* HCl to give a 0.3 m*M* solution. Prepare 100 μ*M*, 10 μ*M*, 1 μ*M*, 100 n*M*, 10 n*M*, 1 n*M*, and 0.1 n*M* solutions by sequentially diluting 100 μL of the previous solution (i.e., 10-fold higher concentration) with 900 μL of 5 m*M* HCl. Similarly, prepare 30 μ*M*, 3 μ*M*, 300 n*M*, and 30 n*M* solutions from the 0.3 m*M* solution (*see* **Note 14**).
2. If using a ^{3}H-ligand, prepare two uncapped scintillation vials with GF/A glass fiber filter discs (*see* **Note 8**).
3. Set up 24 assay tubes in two rows of 12. Add 10 μL of 5 m*M* HCl (or other diluent) to the first pair of tubes. Add 10 μL of the appropriate dilution (concentration) of the inhibitor to the other pairs of tubes, starting with the lowest concentration (*see* **Note 15**).
4. Add 970 μL of the membrane preparation to each of the 24 tubes.
5. Add 20 μL of radioligand to each of the tubes and mix to start the incubation. Pipet 20 μL of the radioligand solution directly onto duplicate GF/A glass fiber filter discs (if using a ^{3}H-radioligand) or into two test tubes (if using a ^{125}I-ligand) to determine the total added radioactivity.
6. Mix all of the tubes again and incubate at room temperature for 45 min. Assuming that the system is at steady state, the exact time is not critical.
7. Filter the contents of the tubes and wash the filters twice with 5 mL of ice-cold wash buffer. The tubes may need to be rearranged to be compatible with the specific style of filtration manifold used.
8. Place the filters into scintillation vials, add 5 mL of scintillation cocktail, and cap (if using a ^{3}H-radioligand); or place the filters into test tubes (if using a ^{125}I-ligand).
9. Shake the scintillation vials gently for 1 h (or let stand at room temperature overnight) and then count in a liquid scintillation counter (if using a ^{3}H-radioligand) or in a gamma counter (if using a ^{125}I-ligand) (*see* **Note 10**).

3.2.1. Calculation of Results from an Inhibition Experiment

The calculation of K_i values from inhibition experiments is relatively straightforward. The inhibition data are simply fit to a sigmoidal curve with the logarithm of concentration of the inhibitor on the abscissa, and the IC_{50} value (the concentration of the inhibitor that inhibits 50% of the specific binding) determined using a Hill slope of 1.

$$Y = \text{Bottom} + \frac{\text{Top} - \text{Bottomn}}{1 + 10^{(X - \text{LogIC}_{50})(\text{Hill slope})}}$$

Top and bottom refer to the concentration of bound radioligand at the top and bottom of the curve. *Y* is the amount of radioactive ligand bound at each concentration of inhibitor *X*. The K_i value is calculated from the IC_{50} value using the equation

$$K_i = \frac{IC_{50}}{1 + \frac{F}{K_d}}$$

where *F* is the free radioligand concentration and K_d is the affinity of the radioligand. This is often called the Cheng–Prusoff equation *(3)* (*see* **Note 16**). Thus, if the radioligand is present at its K_d concentration, then the K_i is one half of the IC_{50}.

If more than one binding site is suspected, the Hill slope can be allowed to vary or the equation for two-site binding can be used.

$$Y = \frac{\text{Span} \times \text{Fraction 1}}{1 + 10^{X - \text{Log IC}_{50_1}}} + \frac{\text{Span} \times (1 - \text{Fraction 1})}{1 + 10^{X - \text{LogIC}_{50_2}}}$$

In this equation span refers to the difference between the top and bottom of the curve, fraction 1 is the amount of radioligand bound to the high-affinity site, and log IC_{50_1} and IC_{50_2} refer to the inhibition of binding to high- and low-affinity sites, respectively. An *F*-test can be used to determine whether the data better fit a one-site or a two-site model (*see* **Example 1**).

These analyses are illustrated with the data from a sample experiment given in **Table 4**.

1. The data are plotted as bound (cpm or p*M* units) vs logarithm of the inhibitor concentration as shown for the sample experiment in **Fig. 8**. The bottom of the curve plateaus at the same bound value as obtained with norepinephrine, indicating that prazosin is likely binding to the same sites as norepinephrine.
2. The solid curve was obtained using a nonlinear regression analysis of a one-site competition equation using the Prism computer program (GraphPad, San Diego, CA). The data were also fit to a sigmoid curve using nonlinear regression analysis with a variable Hill slope (n_H) as is indicated by the dashed line. As is shown in **Table 5**, the results of the two analyses are essentially identical because the n_H is not different from unity. As a rough approximation, n_H values need to be < 0.8 to be significantly different from 1.0 and suggest more complex binding.
3. Similarly the data can also be fit to a two-site competition equation. The results of this fit are shown in **Fig. 8** and **Table 5** (*see* **Note 17**).
4. An *F*-test is used to determine whether the data fit a one- or two-site equation better. The *F*-test for the sample experiment is shown in **Example 1**. The two-site fit was not significantly better than the one-site fit; thus the curve for the one-site fit was chosen, and the IC_{50} values from the one-site fit were used to calculate the K_i values.

Example 1.

***F*-test for comparison of fit of data from sample competition experiment to a one- vs two-site fit.**

The *F*-test is used to compare the one-site and the two-site models. The basic steps are:

1. Analyze the data for a one- and two-site fit using nonlinear regression analysis.
2. Apply the sum of the squares and degrees of freedom to the equation for an *F*-test:

$$F = \frac{(SS1 - SS2) / (DF1 - DF2)}{SS2 / DF2}$$

where SS1 = sum of squares for one-site fit
SS2 = sum of squares for two-site fit
DF1 = degrees of freedom for one-site fit
DF2 = degrees of freedom for two-site fit

3. Determine the *p* value from an *F*-table of statistics.
4. The *p* value answers the question: if model 1 (one-site fit) is correct, what is the chance that you would randomly obtain data that fits model 2 (two-site fit) much better?
5. If *p* is low, you conclude that model 2 (two-site fit) is significantly better than model 1.
6. The calculation for the sample saturation experiment:

$$F = \frac{(25{,}467 - 14{,}394) / (8 - 6)}{14{,}394 / 6}$$

$$F = 2.308$$

$$p = 0.1806$$

7. Because $p > 0.05$, the two-site model does not give a significantly better fit and the one-site model is accepted.

5. The K_i is calculated using the Cheng–Prusoff equation *(3)*. For the sample experiment, the concentration of radioligand was 0.75 n*M*, the K_d was 0.89 n*M*, and the log of IC_{50} was –8.04 (9.08 n*M*). Putting these numbers into the Cheng–Prusoff equation gives a K_i of 4.9 n*M*).

$$K_i = \frac{IC_{50}}{1 + \frac{F}{K_d}} = \frac{9.08 \text{n}M}{1 + \frac{0.75 \text{ n}M}{0.89 \text{n}M}} = 4.9 \text{ n}M$$

Table 4
Data from a Sample Inhibition Experiment[a]

Prazosin concentration (nM)	– Log of prazosin concentration (M)	Average bound (cpm)	Specifically bound[b] (cpm)	% Specifically bound[c]
(0)[d]	–12	1395	1314	100
0.3	–9.52	1292	1211	92.2
1	–9	1256	1175	89.4
3	–8.52	958	877	66.7
10	–8	730	649	19.4
30	–7.52	417	336	25.6
100	–7	198	117	8.90
300	–6.52	126	45	3.43
1000	–6	116	35	2.67
3000	–5.52	82	1	0.076
10,000	–5	81	0	0
NE[e]		92		

[a]Prazosin inhibition of [^{3}H]RX821002 binding to α_{2B}-adrenergic receptors transfected into CHO cells with K_d = 0.89 nM. The concentration of radioligand used in all the tubes was 0.75 nM.

[b]The amount of radioligand specifically bound was also determined by subtracting the amount bound in the presence of the highest concentration of prazosin.

[c]The data were normalized by dividing specifically bound by amount of radioligand bound in the absence of prazosin.

[d]Although the first inhibitor concentration is zero, to run the nonlinear regression program a number needs to be used. Routinely a concentration that is at least one log unit lower than the lowest unlabeled drug concentration is used.

[e]Norepinephrine (NE) was used at a concentration of 0.3 mM to determine the extent of nonspecific binding in this experiment.

6. Frequently the total binding for different receptor preparations or different unlabeled ligands is different, so inhibition curves are often normalized so that the percent inhibition can be more easily compared as shown in **Fig. 9**.

3.3. Kinetic Experiments

3.3.1. Dissociation Experiment

1. If using a ^{3}H-ligand, prepare two uncapped scintillation vials with GF/A glass fiber filter discs (*see* **Note 8**).
2. Set up 48 assay tubes in four rows of 12. To the 12 tubes on the last two rows, add 10 µL of a high concentration of an unlabeled ligand to determine nonspecific binding (*see* **Note 9**).
3. Add 970 µL of the membrane preparation to each of the 48 tubes.
4. Add 20 µL of radioligand to all tubes and mix to start the incubation. The reaction is allowed to proceed until steady-state conditions are reached (45 min). At

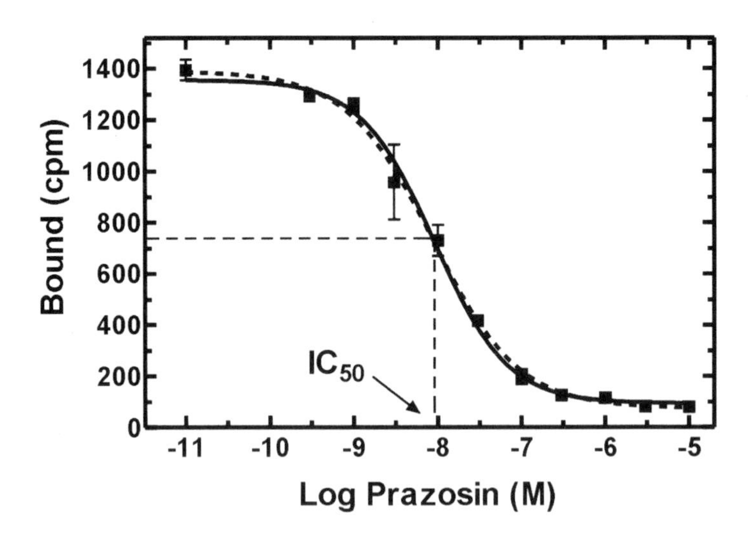

Fig. 8. Plot of inhibition data for a sample experiment. The data from **Table 4** are plotted as bound (cpm) vs the prazosin concentration (in log molar units). The *solid curve* was obtained using a one-site model. Determination of the IC_{50} is based on the middle of the curve (725 cpm bound) and not half of the binding in the absence of prazosin (697 cpm bound). The *dashed curve* is from both a two-site analysis and an analysis of a sigmoid equation with a variable Hill slope (the curves are essentially identical for these data).

Table 5
Results of Analysis of the Sample Inhibition Experiment Presented in Table 4

Parameter	Single-site fit	Variable Hill slope fit	Two-site fit
Bottom of curve	93 cpm	74 cpm	80 cpm
Top of curve	1356 cpm	1390 cpm	1389 cpm
Log IC_{50}	–8.04	–8.06	–8.62
IC_{50}	–9.08 n*M*	–8.72 n*M*	–2.39 n*M*
Log IC_{50}, second site			–7.76
IC_{50}, second site			17.2 n*M*
Fraction, second site			0.64

The curves for these data are shown in **Fig. 8**.

appropriate time intervals, add a high concentration (50 times the IC_{50}) of unlabeled ligand to tubes 2–12 in each row to start the dissociation reaction. The first tube in each row is used to determine binding at zero time at the start of the dissociation reaction. All incubations will be terminated at the same time by filtration. Thus, the unlabeled ligand is added at various times, for example 1, 2, 4, 7, 10, 15, 20, 25, 30, 40, and 60 min before filtration. Also pipet 20 µL of the radi-

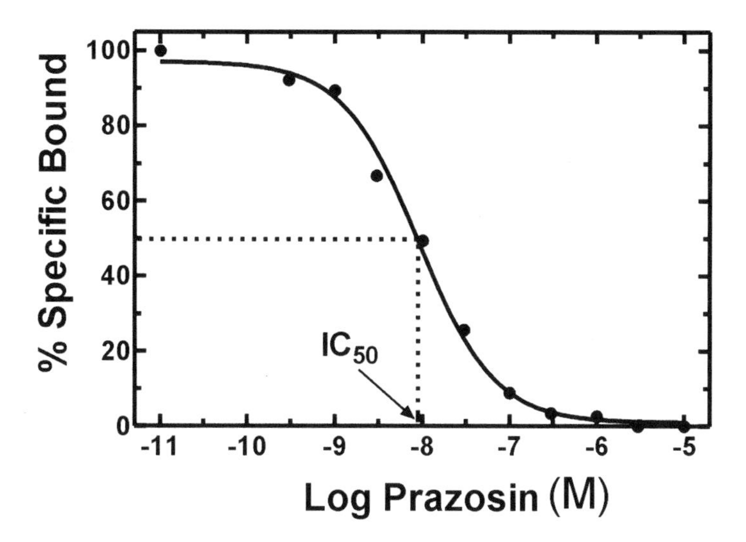

Fig. 9. Plot of normalized data for the sample inhibition experiment. The normalized data from **Table 4** are plotted as a function of the log of the inhibitor concentration. This plot is more useful when multiple data sets are being compared.

oligand solution directly onto duplicate GF/A glass fiber filter discs (if using a ^{3}H-radioligand) or two test tubes (if using a ^{125}I-ligand) to estimate the total added radioactivity (*see* **Note 18**).

5. Filter the contents of the tubes and wash the filters twice with 5 mL of ice-cold wash buffer. The tubes may need to be rearranged to be compatible with the specific style of filtration manifold used.
6. Place the filters into scintillation vials, add 5 mL of scintillation cocktail, and cap (if using a ^{3}H-radioligand); or place the filters into test tubes (if a using ^{125}I-ligand).
7. Shake the scintillation vials gently for 1 h (or let stand at room temperature overnight) and then count in a liquid scintillation counter (if using a ^{3}H-radioligand), or count in a gamma counter (if using a ^{125}I-ligand) (*see* **Note 10**).

3.3.2. Calculation of Results from a Dissociation Experiment

Data from a sample experiment are given in **Table 6**. Note that in a dissociation experiment time zero is the time at which the unlabeled ligand is added to the assay tube. The time course for the dissociation experiment then becomes the time between when the unlabeled ligand is added to the assay tube and the time when the samples are filtered.

1. Nonspecific binding is subtracted from total binding at each time point. Specific binding from a sample experiment is presented in **Table 6**.
2. The data are plotted as bound vs dissociation time. The data from the sample experiment are plotted in **Fig. 10**.

Table 6
Data from a Sample Dissociation Experiment[a]

Time (min)	Specifically bound (cpm)
0	577
1	490
2	460
3	400
4	360
6	280
8	215
10	165
12	140

[a] $[^3H]$DHA was incubated with guinea pig cerebral cortex membranes for 20 min to label β-adrenergic receptors. Isoproterenol (2 μ*M*) was added at time zero. Samples were filtered at the times indicated.

3. The data are analyzed using a nonlinear regression program using the equation for exponential decay:

$$Y = \text{Span} * e^{-k*X} + \text{Nonspecific binding}$$

In a dissociation experiment span refers to specific binding and k is k_{-1}. Analysis of the data for the sample experiment using a one-site exponential decay analysis and the Prism computer program (GraphPad, San Diego, CA) gives a k_{-1} of 0.117 min^{-1}.

3.3.3. Association Experiments

1. If using a ^3H-ligand, prepare two uncapped scintillation vials with GF/A glass fiber filter discs (*see* **Note 8**).
2. Set up 48 assay tubes in four rows of 12. To the 12 tubes on the last two rows add 10 μL of a high concentration of an unlabeled ligand to determine nonspecific binding (*see* **Note 9**).
3. Add 970 μL of the membrane preparation to each of the 48 tubes.
4. At appropriate time intervals, add 20 μL radioligand to all of the tubes and mix. All incubations will be terminated at the same time by filtration. Thus, the radioactivity is added at, for example, 1, 2, 4, 7, 10, 15, 20, 25, 30, 40, 50, and 60 min before filtration. Also pipet 20 μL of the radioligand solution directly onto duplicate GF/A glass fiber filter discs (if using a ^3H-radioligand) or two test tubes (if using a ^{125}I-ligand) to estimate the total added radioactivity.

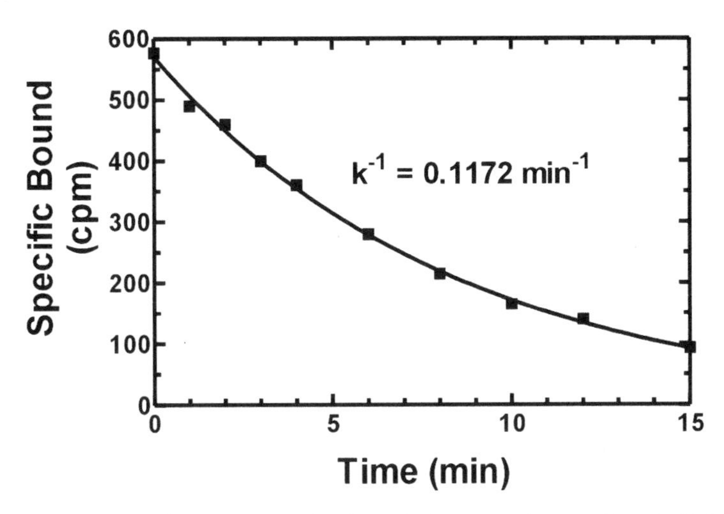

Fig. 10. Plot of data from the sample dissociation experiment given in **Table 6**. The line was drawn using the nonlinear regression equation for one-phase exponential decay using the Prism computer program (GraphPad, San Diego, CA).

5. Filter the contents of the tubes and wash the filters twice with 5 mL of ice-cold wash buffer. The tubes may need to be rearranged to be compatible with the specific style of filtration manifold used.
6. Place the filters into scintillation vials, add 5 mL of scintillation cocktail, and cap (if using a ^{3}H-radioligand); or place the filters into test tubes (if using a ^{125}I-ligand).
7. Shake the scintillation vials gently for 1 h (or let stand at room temperature overnight) and then count in a liquid scintillation counter (if using a ^{3}H-radioligand); or count in a gamma counter (if using a ^{125}I-ligand) (*see* **Note 10**).

3.3.4. Calculation of Results from an Association Experiment

1. Nonspecific binding is subtracted from total binding at each time point to give specific binding. The data from a sample experiment are given in **Table 7**.
2. Amount bound is plotted vs time as shown in **Fig. 11** for the sample association experiment.
3. The data are analyzed using a nonlinear regression analysis of the equation for an exponential association curve:

$$Y = Y_{max}(1 - e^{-K_{ob}t})$$

The rate constant (k_{ob}) obtained is a combination of k_1 and k_{-1} and will vary with the concentration of radioligand (F) added to the assay according to the following equation:

$$k_{ob} = k_1F + k_{-1}$$

Table 7
Data from a Sample Association Experiment[a]

Time (min)	Specifically bound (cpm)
1.5	0.0
0.0	170
2.5	285
4.0	380
6.0	475
8.0	560
10.0	610
13.0	655
16.0	680
20.0	710

[a]Radioligand binding of 0.36 n*M* [^{3}H]DHA to guinea pig cerebral cortex.

k_{-1} can be determined using a separate dissociation experiment as described above. Then, k_1 can be determined using the following rearrangement of the above equation:

$$k_1 = \frac{k_{ob} - k_{-1}}{F}$$

Nonlinear regression analysis of the sample association data (**Table 7**; **Fig. 11**) gave a $k_{ob} = 0.187 \text{ min}^{-1}$ at a radioligand concentration of 0.36 n*M*. The dissociation rate constant determined from the experiment described above was 0.117 min^{-1}. The association rate constant for the sample experiment thus becomes:

$$k_1 = \frac{0.187 \text{ min}^{-1} - 0.117 \text{ min}^{-1}}{0.36 \text{ n}M} = 0.194 \text{ min}^{-1} \text{ n}M^{-1}$$

An alternate way to determine k_1 is to perform the association experiment at various concentrations of radioactive ligand. The k_{ob} determined from these association experiments can then be plotted vs the concentration of radioligand (F). The y-intercept is the k_{-1} and the slope of the line is k_1.

4. K_d is determined by dividing k_{-1}/k_1. For the sample experiment:

$$K_d = \frac{0.117 \text{ min}^{-1}}{0.194 \text{ min}^{-1} \text{ n}M^{-1}} = 0.60 \text{ n}M$$

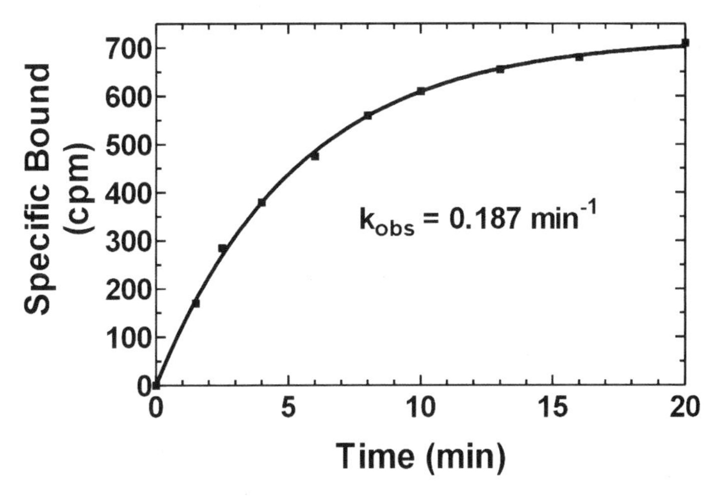

Fig. 11. Plot of data for sample association experiment presented in **Table 7**. The line was drawn using the nonlinear regression equation for exponential association using the Prism computer program (GraphPad, San Diego, CA).

3.4. Sample Protocol for a Saturation Experiment With Monolayer Cells

Assays with intact cells in suspension are quite similar to assays with membranes. The manipulations required for assays with monolayer cells are more involved, however, and thus a detailed protocol for monolayer cells is presented below (*see* **Note 19**). The protocol described is for an eight-point saturation experiment with triplicate determinations for both total and nonspecific binding. The protocol assays the cells in sets of six (three total binding and three nonspecific binding), because this is a convenient number to manipulate within a 1-min time frame (1 every 10 s). Accordingly the cells are plated on 6-well plates, and the plates are treated at 5-min intervals.

1. Grow cells to near confluence on eight 6-well plates in 2 mL of growth medium per well (*see* **Note 20**).
2. Prepare 7.5 mL of 4-(2-hydroxyethyl)piperazine-1-ethanesulfonic acid (HEPES)-buffered serum-free growth medium (*see* **Note 21**) containing each of the eight concentrations of radioligand, labeled as A–H. Transfer 3.5 mL of these solutions to each of two polypropylene tubes with a large enough diameter to allow easy use of a 1-mL pipettor tip (*see* **Note 22**). To one of each pair of tubes (for total binding), add 35 µL of the vehicle for the agent used to define nonspecific binding and label as "AT-HT." To the other (for nonspecific binding), add 35 µL of 100×-concentrated solution of the agent used to define nonspecific binding and

label as "AN-HN." Place these tubes in a 37°C water bath to reach physiological temperature (*see* **Note 23**).

3. Place a beaker with approx 300 mL of HEPES-buffered serum-free growth medium in the 37°C water bath as well, to be used as preincubation wash medium. This same medium can be used as postincubation wash buffer, although in some cases it is beneficial for the postincubation wash buffer to contain a drug to reduce nonspecific binding (*see* **Note 23**).
4. Prepare a Pasteur pipet connected by Tygon tubing to a vacuum flask connected to a vacuum pump or vacuum line to use for aspirating medium from dishes.
5. A detailed time course for a 60-min assay in which a single investigator can conduct the entire assay is presented below, with times presented in minutes. All solution additions are done with a Pipetman or similar hand-held adjustable pipettor. These solutions should be gently added against the inside wall of the dish, not directly onto the monolayers, to avoid loss of cells from the dish due to the multiple medium changes.

a. Initiation of binding:

t = 0 min: At 10-s intervals, aspirate the growth medium and add 2 mL of 37°C preincubation wash buffer to the six wells of the first plate. This step is to wash away serum and bicarbonate and switch the cells to the medium used as assay buffer.

t = 1 min: At 10-s intervals, aspirate the wash buffer from the plate and add 1 mL of AT solution to the top three wells and 1 mL of AN solution to the bottom three wells. By starting the T wells before the N wells, it is not necessary to change pipet tips between the total and nonspecific binding solutions. However, the tip must be changed before the next concentration of total binding solution is added to the next set of dishes.

t = 2 min: Transfer the plate to a 37°C non-CO_2 incubator for the 60-min binding time.

t = 5, 6, 7 min: Repeat the above steps for the second plate, using the BT and BN solutions.

t = 10, 11, 12, 15, 16, 17 min, etc.: Repeat the above steps for each of the remaining six concentrations of radioligand.

b. Termination of binding:

t = 60 min: At 10-s intervals, aspirate the binding medium and add 2 mL of 37°C postincubation wash buffer (*see* **Note 24**) to the three AT wells and then the 3 AN wells. This step is to stop the binding reaction and wash away unbound radioligand.

t = 61 min: Repeat the preceding step for a second wash of the first plate.

t = 62 min: At 10-s intervals, aspirate the wash buffer and add 1 mL of 0.2 *M* NaOH to each of the six wells. Set the plate aside (*see* **Note 25**).

t = 65, 66, 67 min: Repeat the above steps for the BT and BN plates.

t = 70, 71, 72, 75, 76, 77 min, etc.: Repeat the above steps for each of the remaining six plates.

Transfer and quantitation of bound radioactivity: after the binding and wash steps are completed for all eight plates, transfer the dissolved cells and the associated radioactivity to vials for scintillation counting or to tubes for gamma counting, depending on the radioligand used (*see* **Note 26**).

6. Only minimal modifications to this protocol are needed for competition rather than saturation assays.
7. Kinetic assays of association and dissociation become somewhat more complicated with intact cells grown in monolayer culture. Because each well must be started and stopped individually, careful planning of the time points is required. In general, the longer time points are started first and stopped last to complete the entire experiment in the shortest possible time. An example of sample timing that allows a single investigator to conduct a time course experiment with time points at 2, 5, 10, 20, 40, and 60 min is presented below.

t = 0, 1, 2 min:	Start the binding reaction for the 60-min plate.
t = 5, 6, 7 min:	Start the binding reaction for the 40-min plate.
t = 10, 11, 12 min:	Start the binding reaction for the 20-min plate.
t = 15, 16, 17 min:	Start the binding reaction for the 10-min plate.
t = 25, 26, 27 min:	Stop the reaction for the 10-min plate.
t = 30, 31, 32 min:	Stop the reaction for the 20-min plate.
t = 35, 36, 37 min:	Start the binding reaction for the 5-min plate.
t = 40, 41, 42 min:	Stop the reaction for the 5-min plate.
t = 45, 46, 47 min:	Stop the reaction for the 40-min plate.
t = 50, 51 min:	Start the reaction for the 2-min plate; this plate will be stopped immediately.
t = 52, 53, 54 min:	Stop the reaction for the 2-min plate.
t = 65, 66, 67 min:	Stop the reaction for the 60-min plate.

4. Notes

1. The decision as to which radioligand to use is based both on the characteristics of the radioligand and on the specific scientific questions being asked. The important characteristics to be considered include the radioisotope (^3H or ^{125}I), the extent of nonspecific binding, the selectivity and affinity of the radioligand for the receptor, and whether the radioligand is an agonist or an antagonist.
   The advantages of ^3H over ^{125}I as a radioisotope include that the radioligand is chemically unaltered and thus biologically indistinguishable from the unlabeled compound, and that it has a longer half-life (12 yr vs 60 d). Because of their short half-lives, iodinated radioligands are usually purchased or prepared every 4–6 wk, whereas ^3H-ligands can often be used for several months or even longer. An advantage of the iodinated radioligands is their higher specific activity (up to 2200 Ci/mmol vs 30–100 Ci/mmol for ^3H-ligands), which makes them particularly useful if the density of receptors is low or if the amount of tissue is small. It is easier and less expensive to use iodinated ligands, because scintillation cocktail is not needed, thus eliminating purchasing and disposal costs associated with scintillation cocktail.

Each radioligand has a unique pharmacological profile. The radioligand should bind selectively to the receptor type or subtypes of interest under the assay conditions used. Although no radioligand is completely selective for any given receptor or receptor subtype, some are better than others. If, for example, several subtypes of a receptor are present in a given tissue, and if the intent is to label all the subtypes, then a subtype nonselective radioligand that has nearly equal affinity for all three of the subtypes would be chosen. By contrast, if only a single subtype is of primary interest, then a radioligand having higher affinity for that particular subtype as compared to the other subtypes would be the preferred radioligand.

Usually the higher the affinity the better, because a lower concentration of the radioligand can be used in the assay, which results in a lower level of nonspecific binding. Furthermore, a higher affinity usually means a slower rate of dissociation, which provides for a more convenient assay.

Agonist radioligands may label only a portion of the total receptor population (the high-affinity state for G-protein-coupled receptors), whereas antagonist ligands generally label all receptors. On the other hand, an agonist radioligand may more accurately reflect receptor alterations of biological significance, because it is agonists that activate the receptor.

Usually the radioligand with the lowest nonspecific (or nonreceptor) binding is best. An assay is considered barely adequate if 50% of the total binding is specific; 70% is good and 90% is excellent.

Most radioligands are stored in an aqueous solution that often contains an organic solvent such as ethanol. These solutions should be stored cold but not frozen, because freezing of the solution tends to concentrate the radioligand locally and increase its radiolytic destruction.

2. At least six concentrations of radioligand should be used with an equal number above and below the anticipated K_d value. Thus, the amount of stock radioligand used will depend on the K_d of the radioligand for the receptor type and subtype assayed. The amount of radioligand prepared in this manner is sufficient for three saturation experiments. We routinely use 5 m*M* HCl to dilute the radioactivity because we have found that it reduces the nonspecific binding of many radioligands to the dilution tubes and help to ensure ligand stability. This may not be necessary for all radioligands. The small amount (20 µL) of dilute HCl (5 m*M*) carried over to the 1.0-mL assay in 25 m*M* buffer does not alter the pH of the assay. Because the amount of specific binding approaches the B_{max} asymptotically, the specific binding will never actually reach the B_{max} and thus true saturation will be never be achieved. Furthermore, the use of very high radioligand concentration is usually limited by the associated high level of nonspecific binding and by cost. In practice, for an assay that conforms to a single site, it is sufficient if the highest concentration gives specific binding of >90% of the B_{max} and that the Rosenthal line is linear.
3. The radioligand concentration used in inhibition experiment should be less than its K_d value. For kinetic experiments to establish steady state, the lowest practical concentration should be used. For experiments to calculate K_d from k_{+1} and k_{-1}, a concentration near the K_d usually works well.

4. Fibrous tissues, such as the lung, should be filtered through a nylon mesh (about 50 µm). A slow-speed centrifugation step (500*g*) is also helpful with some tissues to remove unwanted pieces of tissue.
5. This centrifugation step is generally done at the highest speed possible (without using an ultracentrifuge) for 5–10 min. Centrifugations are routinely carried out at 4°C, although for many receptors this may not be necessary.
6. This tissue preparation is variously called a crude particulate fraction or a membrane fraction. The purpose of the two centrifugation steps is to remove any soluble substances, such as endogenous neurotransmitters and guanine nucleotides, that may interfere with the radioligand binding assay. The choice of buffer for the homogenization is generally not critical; any buffer at neutral pH appears to be sufficient for most receptor preparations. For some receptor assays it is recommended that EDTA be added to the homogenization buffer and/or an incubation (20 min at 37°C) be included after the second homogenization (but before the second centrifugation) to remove more completely various endogenous substances. Most receptor preparations are stable to freezing and can be stored at –20°C or –80°C for extended periods of time, either as the original tissue or as a pellet after the first homogenization/centrifugation. Experience indicates that some receptors and small pieces of tissue (10 mg) do not store well and thus in these particular cases the assay should be run on fresh tissue.
7. A membrane concentration in the range of 2–10 mg original wet weight of tissue per milliliter (i.e., a dilution of 500–100 volumes) is usually appropriate. This gives a concentration of about 100–500 µg of membrane protein per milliliter. For transfected cells that overexpress the receptor, the protein concentration in the assay will be lower. Within reason, the higher the membrane (receptor) concentration, the better. Increasing the receptor concentration generally increases the ratio of specific to nonspecific binding, as a large portion of the nonspecific binding is binding to the glass fiber filter. A rule of thumb is that if more than 10% of the added radioligand is bound, then the tissue concentration is too high.
8. Aliquots of the diluted ^{3}H-ligand will be added directly onto the filter paper to determine the total added radioactivity. GF/B filters can be used for this purpose, but GF/A filters are less expensive, because they are only half as thick.
9. The choice of the ligand and the concentration used to determine nonspecific binding are critical to the success of the experiment. It is best to use a ligand that is chemically dissimilar to the radioligand to prevent the labeling of specific nonreceptor sites. If at all possible, avoid the use of the unlabeled form of the radioligand. The concentration of the ligand should be sufficiently high to inhibit all specific binding, but none of the nonspecific binding. This can be checked by doing inhibition experiments with several ligands and ensuring that they all give the same level of nonspecific binding.
10. The time required for shaking and/or waiting before counting will depend on the scintillation cocktail used. The number of cpm in the samples should be identical (within counting error) when recounted 5–10 h later. Many radioligands are

lipophilic and for such ligands a nonaqueous scintillation cocktail can be used. For ligands that are less lipophilic, a more expensive aqueous cocktail must be used. If an aqueous cocktail is used, then the 20-μL aliquots of radioligand solution do not need to be spotted onto filter paper but can be added directly to the cocktail. When used with a nonaqueous cocktail the filter paper increases the surface area dramatically so that the radioligand can better partition into the cocktail.

11. In a well-designed experiment there should be an equal number of points above and below the K_d and the highest ligand concentrations should be approx 10 times the K_d. This sample experiment could be improved by using one half to one fourth as much radioligand so as to give a better spread of the data points.
12. The term *Scatchard analysis* is frequently used to describe this linear transformation of saturation data. However, Scatchard's article ***(4)*** is often not referenced. Even when it is referenced, it seems that the authors either have not read it or do not understand it. The Scatchard derivation assumes a single species of binding macromolecule of known molecular weight and concentration, and the intercept at the abscissa is the number of ligand binding sites per macromolecule. The bound/free vs bound plot was first used by Rosenthal ***(2)***, and for most receptor binding studies this is the more appropriate reference. One unique feature of this plot is that radial lines through the origin represent the free radioligand concentration.
13. Ideally the data points should be equally spaced along the line and randomly distributed about the line. In addition the lowest bound point should have a bound/free ratio of < 0.1. At ratios > 0.1, $>10\%$ of the free ligand is depleted, and the equations used for analyzing the data are no longer valid. If ratios > 0.1 are obtained, the tissue concentration should be reduced or the volume increased.
14. An equal number of concentrations above and below the anticipated IC_{50} value should be used. Typically, nine or ten concentrations of the inhibitor are used. A concentration spacing of half-log units is frequently appropriate. Because the inhibitor will be diluted 100-fold in the assay, the stock solutions are made up at a 100-fold higher concentration.
15. If desired, add 10 μL of the compound normally used to determine nonspecific binding (rather than the inhibitor) to the 12th pair of tubes to confirm that the inhibition caused by highest concentration of the inhibitor is consistent with that caused by the "standard" compound.
16. The equation published by Cheng and Prusoff was independently derived by Jacobs et al. ***(5)***, but is valid only if the Hill slope is unity. For cases when it is not unity, a revised equation has been developed ***(6)***.
17. Data from inhibition experiments are always best analyzed using nonlinear regression techniques. If the data are consistent with a single binding site interaction, then the data can be visualized as a simple sigmoidal inhibition curve of bound radioligand (as percentage of maximum) vs the logarithm of the inhibitor concentration. If there are multiple sites, however, the data are best visualized using a plot of bound vs (bound × inhibitor) concentration ***(7)***. An interesting special case is when the radioligand and the competing ligand have the same affinity for

the receptor, such as if the inhibitor is the unlabeled form of the radioligand. In this case, for a plot of bound vs (bound × total free ligand) (i.e., radioligand plus inhibitor), the negative reciprocal of the slope is the K_d rather than the IC_{50}.

18. Total added is not needed to calculate *k*–1, but it is useful to know that the anticipated amount of radioactivity was actually added to the assay tubes.
19. Several complications of intact cell binding assays are unique to the case of cells in monolayer culture. Whereas the receptor concentration in the assay can be easily varied for suspension cells, for monolayer cells it is difficult to vary the receptor concentration except by varying the number of cells plated per culture dish or by varying the extent of confluence at which the cells are assayed. Furthermore, the monolayer cells are not "in solution" in the assay medium and their concentration is not uniform throughout the medium, and this may complicate some of the theoretical aspects of receptor analysis.
20. The cells must first be grown in the appropriate number and type of vessels for the assays to be performed, with either monolayer cells or cells in suspension. Cells grown in monolayer culture can also be released from the monolayers and assayed in suspension if this is more convenient. An advantage of monolayer culture is that the various medium changes and washes required for the assays can be accomplished by simply aspirating with vacuum and replacing with the next solution. In contrast, cells in suspension culture require centrifugation for medium changes and washing, which may take longer.

    Suspension culture cells can be grown in a single vessel, harvested and washed, and then used in assays with individual tubes essentially identical to assays with isolated membrane preparations. The assay tubes can generally be prepared on ice, all started simultaneously by placing the rack of tubes in a water bath, and all terminated simultaneously by filtration with a cell harvester. In contrast, for monolayer culture the cells must be grown in as many separate vessels as the number of assays that are to be performed. For a typical saturation experiment with eight concentrations assayed in triplicate for total and nonspecific binding, 48 separate dishes or wells need to be plated in advance. Thus, details of the specific experiment to be conducted need to be known at the time of plating, so that the proper number of dishes or wells are available. Monolayer cells can be plated on either 35-mm dishes and processed individually, or they can be plated on multiwell plates, in which case multiwell plate washers and cell harvesters can be used for washing and collecting the samples. For saturation and competition assays, where the binding time is constant for all samples, multiwell plates are most convenient, as all of the samples can be washed and harvested simultaneously. In contrast, for kinetic experiments where the time of association or dissociation is varied, each set of dishes must be assayed separately and manually; nonetheless, 4- or 6-well plates are still more conveniently handled than the same number of individual 35-mm dishes. At the end of the assays, the final collection of samples for counting is by filtration for cells assayed in suspension or by dissolving the cell sheet with its bound radioligand in NaOH or detergent for cells assayed as monolayers.

21. The assay buffer used for intact cells is also critical. To maintain the cells intact and viable, the assay buffer should be isotonic and should contain an adequate energy source. This is most easily accomplished by utilizing serum-free growth medium as the assay buffer, but a balanced salt solution supplemented with glucose as energy source can also be used. Because various portions of the assays are done outside of the culture incubator, a nonvolatile buffer such as Tris or HEPES should be used rather than the CO_2–HCO_3^- buffer system used for growing the cells, and the binding incubations should be done in a cell culture incubator without CO_2.
22. Sterile technique is no longer needed.
23. Slightly more solution is prepared at each step than is needed, to ensure that the appropriate number of identical aliquots can be recovered from each tube.
24. Including 100 μM propranolol in the postincubation wash buffer has been shown to reduce nonspecific binding by unknown mechanisms, not only for the β-adrenergic receptors for which propranolol is a high-affinity antagonist, but also for α_1-adrenergic receptors and muscarinic receptors *(1)*. Phentolamine has also proven useful for some receptors. Thus, testing a variety of drugs for inclusion in the postincubation wash buffer to reduce nonspecific binding may prove beneficial.
25. Pull the NaOH solution into the pipettor tip, and while holding the plate at a 45° angle, rinse the plate from top to bottom twice with the same solution before transferring; we have not found it important to rinse the plate with a second aliquot to obtain reproducible and quantitative transfer.
26. The NaOH may cause problems with chemiluminescence in some scintillation cocktails; this can be avoided by neutralizing the NaOH after transfer or by choosing a different scintillation cocktail.

References

1. Toews, M. L. (2000) Radioligand-binding assays for G protein-coupled receptor internalization, in *Regulation of G protein-coupled receptor function and expression* (Benovic, J. L., ed.), John Wiley & Sons, New York, pp. 199–230.
2. Rosenthal, H. E. (1967) Graphical method for the determination and presentation of binding parameters in a complex system. *Anal. Biochem.* **20,** 525–532.
3. Cheng, Y-C. and Prusoff, W. H. (1973) Relationship between the inhibition constant (K_i) and the concentration of inhibitor which causes 50 per cent inhibition (IC_{50}) of an enzymatic reaction. *Biochem. Pharmacol.* **22,** 3099–3108.
4. Scatchard, G. (1949) The attractions of proteins for small molecules and ions. *Ann. NY Acad. Sci.* **51,** 660–672.
5. Jacobs, S., Chang, K. J., and Cuatrecasas, P. (1975) Estimation of hormone-receptor affinity by competitive displacement of labeled ligand: effect of concentration of receptor and labeled ligand. *Biochem. Biophys. Res. Commun.* **66,** 687–695.
6. Cheng, H. C. (2002) The power issue: determination of K_B or K_i from IC_{50}: a closer look at the Cheng–Prusoff equation, the Schild plot and related power equations. *J. Pharmacol. Toxicol. Methods* **46,** 61–71.
7. Bylund, D. B. (1986) Graphic presentation and analysis of inhibition data from ligand-binding experiments. *Anal. Biochem.* **159,** 50–57.

2

Determination of Allosteric Interactions Using Radioligand-Binding Techniques

Sebastian Lazareno

Summary

Methods are presented for identifying and quantifying allosteric interactions of G-protein-coupled receptors with labeled and unlabeled ligands using radioligand-binding assays. The experimental designs and analyses are based on the simplest ternary complex allosteric model.

Key Words

Affinity ratio, allosteric, radioligand binding, screening, ternary complex model.

1. Introduction

This chapter describes radioligand-binding methods for detecting and quantifying a certain type of allosteric interaction at G-protein-coupled receptors (GPCRs). An allosteric ligand binds reversibly to a site on the receptor that is spatially distinct from the primary site (also called the orthosteric site) to which the endogenous ligand (or another primary ligand) binds, allowing the primary and allosteric ligands to bind simultaneously and form a ternary complex with the receptor *(1,2)*. Any mechanism that is involved with the interaction of the primary ligand and the receptor may be modified by the presence of the allosteric ligand, so complex models may be required to account for some allosteric interactions.

Muscarinic receptors for acetylcholine (ACh) were the first GPCRs at which allosterically acting ligands were identified, and they are the most extensively characterized GPCRs with regard to allosteric ligands *(3–5)*. In general, the effects of allosteric ligands on muscarinic receptors are consistent with the allosteric mechanism depicted in **Fig. 1**, and the methods described here relate to this model, which is the simplest possible allosteric model *(6,7)*. This model

From: *Methods in Molecular Biology, vol. 259, Receptor Signal Transduction Protocols, 2nd ed.*
Edited by: G. B. Willars and R. A. J. Challiss © Humana Press Inc., Totowa, NJ

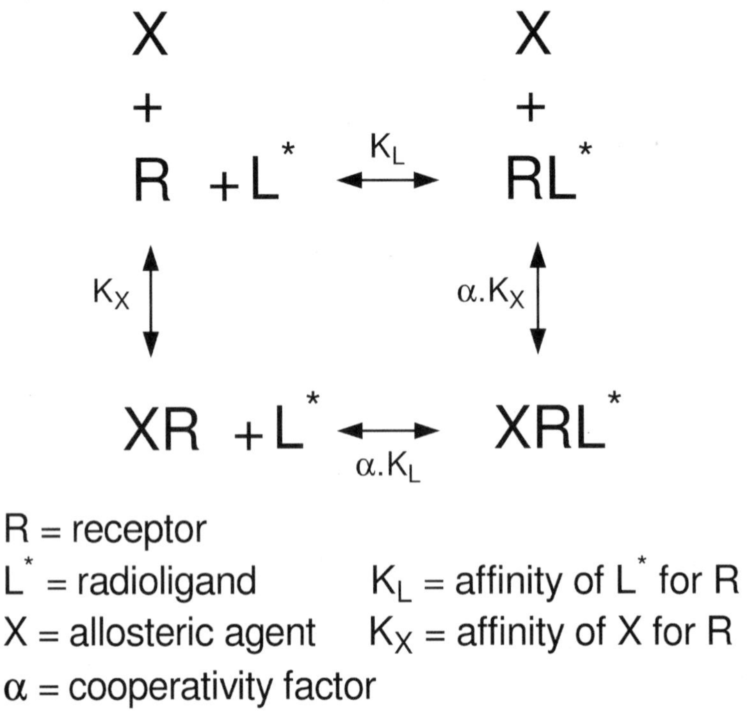

$$
\begin{array}{ccc}
X & & X \\
+ & & + \\
R + L^{*} & \overset{K_L}{\longleftrightarrow} & RL^{*} \\
K_X \updownarrow & & \updownarrow \alpha.K_X \\
XR + L^{*} & \underset{\alpha.K_L}{\longleftrightarrow} & XRL^{*}
\end{array}
$$

R = receptor
L^{*} = radioligand K_L = affinity of L^{*} for R
X = allosteric agent K_X = affinity of X for R
α = cooperativity factor

Fig. 1. Receptor R has two binding pockets: one for radioligand L* and the other for allosteric agent X. The two types of ligand can bind simultaneously to R to form a ternary complex. The affinity ($1/K_d$) of L* for binding to free R, K_L, is different from the affinity with which L* binds to XR, $\alpha \cdot K_L$, and the ratio of the two affinities, α, is the cooperativity of the system. The same cooperativity factor changes the affinity of X from K_X at free R to $\alpha \cdot K_X$ at RL.

consists of a receptor with two binding sites (primary and allosteric sites), and the only effect of the bound allosteric ligand is to modify the affinity and binding kinetics of the primary ligand. The model does not consider different activation or desensitization states of the receptor, the binding of the receptor to the G-protein or other accessory proteins, or the possibility of receptor dimers or oligomers, and it cannot account for changes in the functional effects of agonists that are not reflected in changes in their binding characteristics, as is seen with allosteric ligands at class 3 GPCRs (γ-aminobutyric acid B [$GABA_B$], metabotropic glutamate and calcium sensing receptors). Nevertheless, the model is completely defined and, where studied, accounts for most data observed with allosteric ligands at muscarinic receptors ***(3–5)***, adrenoceptors

(8–11), and dopamine receptors ***(12)***, and it provides a benchmark for assessing more complex allosteric interactions.

The model defined in **Fig. 1** has the following characteristics. L is the primary ligand and X is the allosteric ligand. L binds with different affinities to the X-liganded and free receptor, and the ratio of these affinities, α, is the cooperativity of the system. Values of $\alpha > 1$ indicate positive cooperativity, and values < 1 indicate negative cooperativity, with a limiting value of zero which is equivalent to a competitive interaction. A value of 1 indicates neutral cooperativity, where the allosteric ligand binds to the receptor but does not modify the affinity of the primary ligand.

It is important to note that the cooperativity of the system depends on all three components of the ternary complex, so an allosteric agent may well have different effects on different primary ligands, or on the same primary ligand at different receptors. In a therapeutic setting an allosteric agent will be used to modulate the effect of the endogenous ligand or another primary ligand, and in many cases these ligands will not be available as useful radioligands. For example, an allosteric agent that is positively cooperative with ACh at muscarinic M_1 receptors could be of use in treatment of Alzheimer's disease, but $[^3H]$ACh does not bind usefully to M_1 receptors. It is therefore important to be able to screen for the desired interaction using the unlabeled primary ligand in an indirect screening assay (*see* **Notes 1** and **2**). **Figure 2** is an extension of **Fig. 1** to include a second, unlabeled, primary ligand, and the equations derived from this figure underlie the medium-throughput assay described in **Subheadings 3.4.** and **3.5.** and **Fig. 3**.

Allosteric effects may be small and hard to quantify, or may occur through a mechanism that is different from that shown in **Fig. 1**. It is therefore useful to have a check on the results. In many cases allosteric agents modify the dissociation rate of the radioligand. In theory the reciprocal of the IC_{50} or EC_{50} of the allosteric agent for causing this effect corresponds to the affinity ($1/K_d$) of the agent for the radioligand-occupied receptor ($\alpha \cdot K_X$ in **Fig. 1**). The same quantity is also estimated from equilibrium assays (the product of α and K_X), so the results from these two types of assay may be compared to determine whether they are internally consistent with the allosteric model.

The assays described here should be applicable to any GPCR for which both an antagonist radioligand and the unlabeled ligand of interest are available, but there are many other ways in which allosteric effects may be detected. Allosteric effects with an agonist radioligand may predict similar effects with the endogenous ligand, for example. In particular, functional assays may be crucial for detecting allosteric agents that do not affect radioligand binding ***(13)*** or that alter the efficacy of an agonist ligand ***(14,15)***. Models that are more

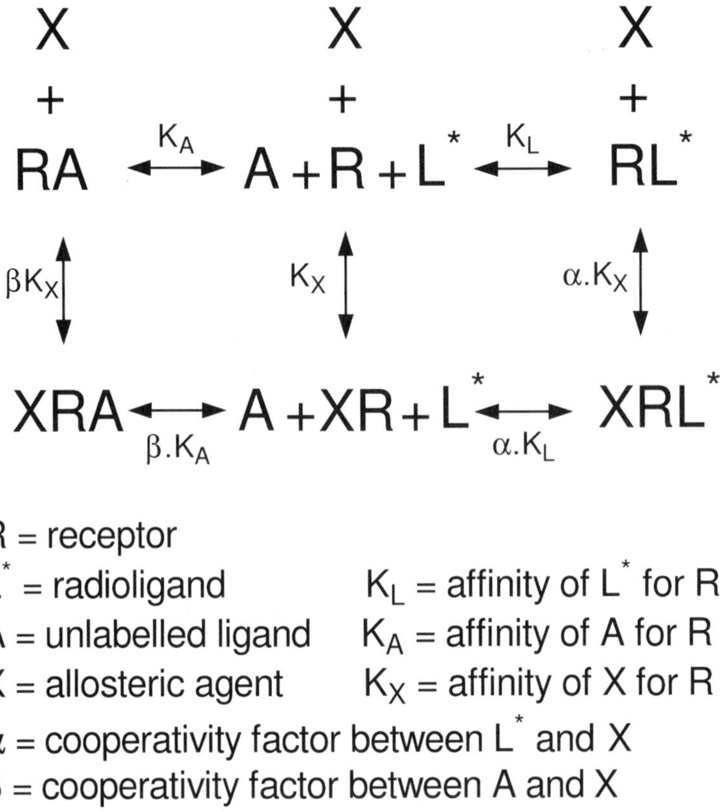

$$
\begin{array}{ccccc}
X & & X & & X \\
+ & & + & & + \\
RA & \xleftrightarrow{K_A} & A + R + L^* & \xleftrightarrow{K_L} & RL^* \\
\beta K_X \updownarrow & & K_X \updownarrow & & \alpha . K_X \updownarrow \\
XRA & \xleftrightarrow[\beta . K_A]{} & A + XR + L^* & \xleftrightarrow[\alpha . K_L]{} & XRL^*
\end{array}
$$

R = receptor

L^* = radioligand K_L = affinity of L^* for R

A = unlabelled ligand K_A = affinity of A for R

X = allosteric agent K_X = affinity of X for R

α = cooperativity factor between L^* and X

β = cooperativity factor between A and X

Fig. 2. As **Fig. 1** but including unlabeled ligand A with affinity K_A that competes with radioligand L and has a cooperativity of β with allosteric agent X.

complex than the one shown in **Fig. 1** will be required to account for such findings ***(16,17)***.

2. Materials

1. An antagonist radioligand (*see* **Note 2**).
2. Allosteric ligand(s) of interest.
3. Unlabeled primary ligand(s) of interest.
4. A source of membranes containing the receptor of interest.
5. Assay buffer, e.g., 20 m*M* 4-(2-hydroxyethyl)piperazine-1-ethanesulfonic acid (HEPES)/Na HEPES, pH 7.4, 100 m*M* NaCl, 10 m*M* $MgCl_2$.
6. 5-mL polystyrene tubes or 96-well 1 mL deep well plates (*see* **Note 3**).
7. Filtration apparatus.
8. Liquid scintillation counter.

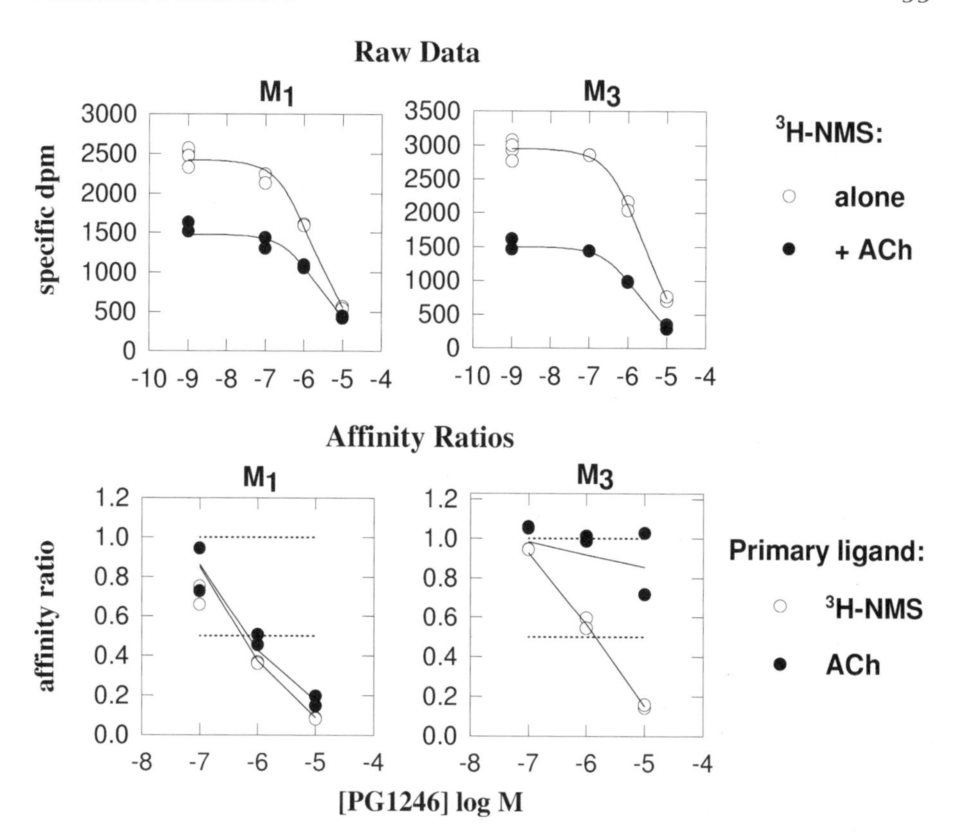

Fig. 3. Effect of an allosteric agent on [^{3}H]NMS binding at M_1 and M_3 receptors, alone and in the presence of a single concentration of ACh, all in the presence of 0.2 m*M* GTP. The **top panel** shows raw data; the **bottom panel** shows affinity ratios calculated from the raw data. The lines are from nonlinear regression analysis of the raw data. From inspection of the affinity ratio plots it is clear that the compound has a log affinity of approx 6 at both subtypes, and shows strong negative cooperativity with [^{3}H]NMS at M_1 and M_3 receptors and with ACh at M_1 receptors, but shows neutral cooperativity with ACh at M_3 receptors.

3. Methods

3.1. Effect of an Allosteric Agent on Radioligand Dissociation

The ability of a compound to modify the rate of dissociation of an antagonist radioligand is a clear indication of an allosteric action. Radioligand dissociation is measured by allowing the radioligand to bind and then preventing its reassociation by diluting the sample or adding an excess of an unlabeled competing compound (1 μ*M* quinuclidinyl benzilate [QNB] or 10 μ*M* atropine for muscarinic receptors: note that, if possible, the unlabeled compound should be chemically different from the radioligand). Dissociation may be measured at a number of time points in the presence of a few concentrations of compound, which is useful for studying the pattern of dissociation, or at a single time point, which is quicker and useful for studying a broader range of compound concentrations. The nonspecific binding characteristics of the radioligand determine whether it can be used at a high concentration in a diluting plus blocking protocol (*see* **Note 4**).

3.1.1. Single Time Point, Dilution and Blocking

1. Dilute radioligand to a concentration of 100–500X K_d.
2. Prepare unlabeled blocking agent to twice the final concentration (e.g., 2 μ*M* QNB) (0.5 mL/sample).
3. Prepare 10 μL of unlabeled blocking agent to 10 times the final concentration (e.g., 10 μ*M* QNB).
4. Prepare test agents at double the final concentration (0.5 mL/sample) using the same vehicle if possible for all samples (e.g., 1% dimethyl sulfoxide [DMSO] in buffer).
5. Dilute membranes in buffer to give a total binding capacity of 500,000 dpm/mL (*see* **Note 5**).
6. Retain 30 μL of dilute membranes for measurement of nonspecific binding (nsb).
7. Prepare nsb mixture: mix 5 μL of radioligand with 5 μL of 10 μ*M* QNB, and add 6 μL of this mixture to 30 μL of membranes.
8. Prepare test tubes (or wells) in duplicate (*see* **Note 6**) with 0.5-mL volumes of a high concentration of unlabeled ligand (from **step 2**) and 0.5 mL of test compounds (from **step 4**) as follows:
   a. No additions, to measure binding before dissociation.
   b. 11 μL of nsb mixture only, to measure nsb before dissociation.
   c. QNB alone plus vehicle, to measure the control rate of dissociation.
   d. QNB plus test agent, to measure the rate of dissociation in the presence of test agent.
9. Add radioligand to the remaining membranes in a 1:10 ratio, to give a final radioligand concentration of 10–50X K_d, and incubate for a few minutes (*see* **Note 7**).
10. Initiate dissociation by adding 10 μL of labeled membranes to all tubes except those containing the nsb mixture.
11. Terminate the reaction by filtration after about two to three dissociation half-lives (*see* **Note 8**).
12. *See* **Note 9** for data analysis.

3.1.2. Single Time Point, Blocking Only

1. Dilute radioligand to 100 times the final concentration (e.g., 100X K_d).
2. Prepare unlabeled blocking agent to 200 times the final concentration (e.g., 200 μ*M* QNB) (5 μL/sample).

3. Prepare test agents at 200 times the final concentration (5 µL/sample) using the same vehicle if possible for all samples (e.g., DMSO).
4. Mix each test agent concentration, and vehicle alone, with an equal volume of unlabeled blocking agent.
5. Dilute membranes in buffer to give a total binding capacity of 5000 dpm/mL (*see* **Note 5**).
6. Prepare tubes or wells with 10 µL of radioligand.
7. Prepare 30 µL of unlabeled blocking agent to 100 times the final concentration (e.g., 100 µ*M* QNB).
8. Add 10 µL of this to two tubes (for measurement of nsb, *see* **Note 6**).
9. Add 1 mL of membranes to each tube and incubate for some time, ideally (although not necessarily) until binding equilibrium is reached.
10. Initiate dissociation by adding 10 µL of blocker–vehicle mixture (control dissociation) or 10 µL of blocker–allosteric agent to the appropriate tubes. Do NOT add blocker to tubes measuring total binding (no dissociation) or nsb.
11. Terminate the reaction by filtration after about two to three dissociation half-lives (*see* **Note 8**).
12. *See* **Note 9** for data analysis.

3.1.3. Multiple Time Points

The design of dissociation experiments depends on whether the radioligand binds relatively rapidly and reaches an asymptotic level of binding that is stable over the complete time course of the assay. If the radioligand has these characteristics, then, depending on the nsb characteristics of the radioligand, either the dilution + blocking protocol (**Subheading 3.1.1.**) or the blocking only protocol (**Subheading 3.1.2.**) may be used. In either case dissociation is initiated at different times before the assay is terminated by filtration.

If the radioligand does not reach a stable asymptotic level of binding then the blocking only protocol (**Subheading 3.1.2.**) must be used, and both the initiation and termination of the membrane labeling must be timed so that each sample associates for the same time and dissociates for the appropriate time.

See **Note 10** for data analysis.

3.2. Effect of an Allosteric Agent on Radioligand Affinity and B_{max}

This assay contains radioligand saturation curves alone and in the presence of one or more concentrations of allosteric agent.

1. Dilute membranes to a total binding capacity of 5000 dpm/mL (*see* **Note 3**).
2. Prepare radioligand solutions at 100 times the final concentrations, of which there should be at least three and which ideally should span at least 0.1X K_d to 10X K_d.
3. Prepare test agent at 100 times the final concentration(s).
4. Prepare unlabeled blocking agent to define nsb at 100 times the final concentration (e.g., 100 µ*M* QNB).

5. Add 10 µL of each radioligand concentration to the appropriate tubes to measure total and nonspecific binding.
6. Add 10 µL of vehicle or test agent to the appropriate tubes (*see* **Note 11**).
7. Add 10 µL of buffer to "total" tubes and 10 µL of unlabeled blocking agent to "nsb" tubes (*see* **Note 11**).
8. Add 1 mL of membranes to each tube.
9. Incubate for at least five dissociation half-lives (*see* **Note 7**) before filtration.
10. In addition to counting bound radioligand, count samples of each radioligand dilution.
11. *See* **Note 12** for analysis and interpretation.

3.3. Effect of an Allosteric Agent on Radioligand Affinity

If it is assumed that a test agent does not affect B_{max}, then its effect on the affinity of the radioligand (i.e., its cooperativity with the radioligand) can be assessed using a single concentration of radioligand.

1. Dilute membranes to a total binding capacity of 5000 dpm/mL (*see* **Note 3**).
2. Prepare radioligand at 100 times the final concentration (*see* **Note 13**).
3. Prepare 100 µL of a second, higher concentration of radioligand (optional).
4. Prepare unlabeled blocking agent to define nsb at 100 times the final concentration (e.g., 100 µ*M* QNB).
5. Prepare a range of concentrations of test agent(s) at 100 times the final concentration using the same vehicle if possible (e.g., DMSO).
6. Prepare tubes/wells with 10 µL of each of radioligand, buffer, and test agent or vehicle: also prepare tubes containing 10 µL of each of radioligand, unlabeled blocker, and vehicle to measure nsb. The assay should contain tubes to measure:
   a. Total binding of the high [radioligand] (optional).
   b. Nonspecific binding of the high [radioligand] (optional).
   c. Total binding of the low [radioligand] alone.
   d. Nonspecific binding of the low [radioligand].
   e. Binding of the low [radioligand] in the presence of each concentration of allosteric agent.
7. Add 1 mL of membranes to each tube.
8. Incubate for at least five dissociation half-lives (*see* **Note 7**) before filtration.
9. In addition to counting bound radioligand, count samples of each radioligand dilution.
10. *See* **Note 14** for analysis and interpretation.

3.4. Effect of an Allosteric Agent on the Affinity of an Unlabeled Ligand

In this assay the effects of each concentration of allosteric agent and each concentration of unlabeled ligand are measured alone and in combination. The cooperative effect on the unlabeled ligand is disentangled from the cooperative effect on the radioligand using nonlinear regression. The assay would typically have the form of a series of inhibition curves of the unlabeled ligand, alone and in the presence of each concentration of allosteric agent.

1. Dilute membranes to a total binding capacity of 5000 dpm/mL (*see* **Note 3**).
2. If the unlabeled ligand is an agonist, add GTP to the membranes to a concentration of 0.2 m*M* (*see* **Note 1**).
3. Prepare radioligand at 100 times the final concentration (*see* **Note 13**).
4. Prepare 100 μL of a second, higher concentration of radioligand (optional).
5. Prepare unlabeled blocking agent to define nsb at 100 times the final concentration (e.g., 100 μ*M* QNB).
6. Prepare a range of concentrations of test agent(s) at 100 times the final concentration using the same vehicle if possible (e.g., DMSO).
7. Prepare unlabeled ligand at 100 times the final concentration(s).
8. Prepare tubes/wells with 10 μL each of radioligand, unlabeled ligand or buffer, and test agent or vehicle: also prepare tubes containing radioligand, solvent, and unlabeled blocker to measure nsb. The assay should contain tubes to measure:
   a. Total binding of the high [radioligand] (optional).
   b. Nonspecific binding of the high [radioligand] (optional).
   c. Total binding of the low [radioligand] alone.
   d. Nonspecific binding of the low [radioligand].
   e. Binding of the low [radioligand] in the presence of each concentration of unlabeled ligand.
   f. Binding of the low [radioligand] in the presence of each concentration of test agent.
   g. Binding of the low [radioligand] in the presence of each concentration of unlabeled ligand and test agent.
9. Add 1 mL of membranes to each tube.
10. Incubate for at least five dissociation half-lives (*see* **Note 7**) before filtration.
11. In addition to counting bound radioligand, count samples of each radioligand dilution.
12. *See* **Note 15** for analysis and interpretation.

3.5. Effect of an Allosteric Agent on the Affinity of an Unlabeled Ligand—Single Concentration Screening Assay

This is a special case of the protocol described in **Subheading 3.4.**, in which a single concentration of unlabeled ligand (usually the IC_{50}) is used. In addition to analysis using nonlinear regression (*see* **Note 15**), this design can be analyzed semiquantitatively by visual inspection of plotted transformed data (*see* **Note 16** and **Fig. 3**).

4. Notes

1. The allosteric model is concerned only with ligand interactions at the free receptor. In many GPCR systems, however, agonists bind with higher affinity to the G-protein-coupled form of the receptor than to the free receptor, so agents that

modify G-protein–receptor coupling will also affect agonist potency and may be misinterpreted as allosteric agents. High-affinity agonist binding is largely or completely removed by high concentrations of guanine nucleotides, so 0.2 m*M* GTP is included in assays with unlabeled agonists to minimize the confounding effects related to G-protein coupling.

2. The techniques described here may be used with any radioligand, but agonists will often not label the entire receptor population, multiple affinity states may occur, and high-affinity agonist binding often involves some degree of receptor–G-protein interaction, so the assumptions underlying the experimental designs and analyses described here will generally not be applicable to agonist radioligands.
3. If possible, in an equilibrium assay, bound radioligand should not be more than about 15% of added radioligand, and there must be sufficient bound radioligand to measure accurately changes in binding caused by other compounds. For potent radioligands (e.g., K_d = 0.1 n*M*) this will be difficult to achieve in a 250-µL volume, so compromises must be made, usually leading to binding of >15% added radioligand. This was not a serious problem in a 250-µL assay using $[^3H]$*N*-methyl scopolamine ($[^3H]$NMS) and muscarinic M_3 receptors (K_d ~0.2 n*M*) but might be a problem in other systems.
4. In the dilution + blocking protocol, radioligand binding is prevented both by a 100-fold dilution and by exposing the receptors to a high level of unlabeled blocker. Features of this protocol are:
   a. Security: it is conceivable that an allosteric agent could reduce the affinity of the unlabeled blocking agent so much that radioligand binding was not completely blocked, which would give the appearance of a reduction in the rate of radioligand dissociation. With the additional dilution component of the assay this artefact should not occur.
   b. Efficient use of receptor: with a high concentration of both receptor and radioligand during the labeling phase, almost all the receptors will be bound.
   c. Good nsb: this does depend on the characteristics of the radioligand and its affinity for the receptor. If the affinity of the radioligand is sufficiently high to allow radioligand and receptor to be present at the same high concentration, then most of the receptors will be labeled but the free concentration of radioligand, and hence nsb, will be low because of radioligand depletion. Also, the radioligand concentration after dilution is low, also leading to low nsb.
   d. Bad nsb: if the radioligand is not potent enough it will not be possible to use equal concentrations of radioligand and receptor, so there will be no depletion of radioligand and nsb will be high during the labeling phase. The dilution will result in lower nsb eventually, but this may not happen rapidly. Also, if the radioligand has high affinity but also high nsb it will not be possible to measure accurately the initial level of specific binding before dilution, because the samples measuring nsb will have a much higher free concentration of radioligand than the samples measuring total binding (because of radioligand depletion in the latter samples). If the radioligand–receptor system has either of these characteristics then the blocking-only protocol is preferable.

5. The concentration of binding sites should, for convenience, result in at least 2500 dpm bound to the control samples, but if there are fewer bound dpm (e.g., because of a low specific activity radioligand or a shortage of receptors) this is acceptable as long as accurate data can be obtained.
6. The number of replicates for each point will depend on the specific/nonspecific binding ratio of the radioligand, other sources of variability in the assay, and possibly the cost or availability of materials. I will assume that duplicates are used.
7. The observed association rate constant, k_{obs}, defines the rate of radioligand association (time for 50% association is $\ln(2) / k_{obs}$). k_{obs} is related to the dissociation rate constant, k_{off}, and association rate constant, k_{on}:

$$k_{obs} = L \cdot k_{on} + k_{off} \tag{1}$$

where L is the radioligand concentration.

The K_d of the radioligand is also related to k_{on} and k_{off}:

$$K_d = k_{off} / k_{on} \tag{2}$$

It therefore follows that

$$k_{obs} = k_{off} \cdot (1 + L / K_d) \tag{3}$$

so if $L << K_d$ then the observed association rate is the same as the dissociation rate, and if $L >> K_d$ then $k_{obs} \approx k_{off} \cdot L / K_d$.
8. The particular dissociation time should not affect the estimate of the affinity of the allosteric agent for the radioligand-bound receptor. Speeding of dissociation is more accurately measured with a short dissociation time, whereas slowing of dissociation is more accurately measured at long dissociation times. We find that a dissociation time of two to three half-lives is a suitable compromise.
9. This analysis contains a number of assumptions:
   a. Radioligand dissociation is monoexponential.
   b. The allosteric agent affects only the rate constant, not the shape of the curve.
   c. The allosteric agent has rapid kinetics so it is always in equilibrium.

   If these assumptions are correct then each experimental point can be converted into an observed rate constant, and the plot of observed rate constant vs log (allosteric agent) has an IC_{50} or EC_{50} corresponding to the K_d of the allosteric agent for the radioligand-occupied receptor and an asymptotic value corresponding to the dissociation rate of the radioligand from the allosteric agent-occupied receptor.

   Subtract nonspecific binding from all the data points.

   Transform the data to observed rate constants, k_{offobs}, using the formula

$$k_{offobs} = \ln(B_0 / B_t) / t \tag{4}$$

where B_0 is initially bound radioligand and B_t is bound radioligand remaining after t min of dissociation.

If comparing curves obtained using different receptors, normalize the data, that is, express each individual k_{offobs} value as a fraction of the mean control k_{offobs} value that was obtained in the absence of allosteric agent.

Fit the k_{offobs} or normalized values to a logistic function to yield the K_d of the allosteric agent for the radioligand-occupied receptor and dissociation rate constant of the radioligand from the allosteric agent-occupied receptor. In theory the slope factor should be 1—if it is different from 1 then the interpretation of the IC_{50} or EC_{50} is unclear.

Inaccuracies in the timing of the assay (*see* **Note 17**) may lead to the initial binding of some curves to be >100% and/or the asymptotic binding to be <0% (if the off rate is inhibited). This is not a serious problem with respect to measuring the potency of the allosteric agent as long as the top and bottom of the curve are defined by the curve itself and not by inappropriate constraints in the fitting process.

10. Subtract nonspecific binding from all the data points.

If the fitting routine can handle only one independent variable then fit the specific binding data from each curve to the equation:

$$B_t = B_0 \cdot \exp(-t \cdot k_{offobs}) \quad (5)$$

where B_0 and B_t are the observed specific binding at times 0 and t respectively, and k_{offobs} is the observed dissociation rate constant. Alternatively, plot $\ln(B_t/B_0)$ vs time: if dissociation is monoexponential then this plot will be a straight line with a slope of $-k_{offobs}$. A plot of k_{offobs} vs log[allosteric agent] will reveal the potency of the compound for the radioligand-bound receptor (*see* **Note 9**).

If the fitting routine can handle more than one independent variable (e.g., SigmaPlot), then fit the data to:

$$B_t = B_0 \cdot \exp\left(-t \cdot \frac{X \cdot \alpha \cdot K_X \cdot k_{offX} + k_{off}}{1 + X \cdot \alpha \cdot K_X}\right) \quad (6)$$

where $\alpha \cdot K_x$ is the affinity of the allosteric agent for the radioligand-occupied receptor and k_{offX} is the dissociation rate of the radioligand from the allosteric agent-occupied receptor.

If $k_{offX} << k_{off}$ then a high concentration of X almost completely inhibits radioligand dissociation and **Eq. 6** can be simplified to:

$$B_t = B_0 \cdot \exp\left(-t \cdot \frac{k_{off}}{1 + X \cdot \alpha \cdot K_X}\right) \quad (7)$$

11. If radioligand nsb is low, small molecules usually have little or no effect on it, so it is not normally necessary to measure nsb in the presence of the allosteric agent—if nsb is changed then this will be reflected in changes in apparent B_{max}. If, however, nsb is high then it would be prudent to measure it in the presence of each concentration of allosteric agent.
12. Subtract the appropriate value of nonspecific binding. The following equations refer to specific binding.

Radioligand binding in the presence of an allosteric agent alone is given by:

$$B_{LX} = \frac{B_{max} \cdot L \cdot K_L \cdot (1 + X \cdot K_X \cdot \alpha)}{1 + X \cdot K_X + L \cdot K_L \cdot (1 + X \cdot K_X \cdot \alpha)} \quad (8)$$

where B_{LX} (bound L in the presence of X) is specific binding at equilibrium; L and X are the concentrations of radioligand and allosteric agent respectively; K_L and K_X are the affinities of radioligand and allosteric agent respectively for the free receptor; α is the cooperativity factor for the interaction of the allosteric agent with the radioligand.

If the concentration dependence of the effect of the allosteric agent appears to be steep or shallow then a Schild slope factor, s, can be introduced by substituting $(X \cdot K_X)^s$ for $X \cdot K_X$ in **Eq. 8**.

If the allosteric agent inhibits radioligand dissociation then radioligand association may also be reduced (*see* **Note 7**) over the same concentration range. This may be a problem for agents that show positive, neutral, or low negative cooperativity with the radioligand because such agents will have similar affinities for the free and liganded receptor. Partial inhibition of dissociation is hard to adjust for, but if the allosteric agent is known to inhibit almost completely the dissociation of the radioligand, and if some binding in the presence of allosteric agent may not have been at equilibrium, then nonequilibrium radioligand binding is given by:

$$B_{LXt} = B_{LX} + (B_0 - B_{LX}) \cdot \exp\left(\frac{-t \cdot k_{off}}{1 + X \cdot K_X \cdot \alpha} + \frac{-t \cdot k_{off} \cdot L \cdot K_L}{1 + X \cdot K_X}\right) \quad (9)$$

where B_{LXt} is nonequilibrium binding after t min incubation; B_{LX} is the predicted binding at equilibrium, **Eq. 8**; B_0 is the initial amount of radioligand binding, set to 0 in this case; and k_{off} is the known dissociation rate constant of the radioligand. Data from a family of saturation curves may be fitted simultaneously to one of the above equations. Alternatively:

a. Fit each curve individually to the equation,

$$\text{bound} = B_{max} \cdot \frac{L}{L + K_d} \quad (10)$$

b. For each concentration of allosteric agent, divide the control radioligand K_d by the radioligand K_d obtained in the presence of allosteric agent to obtain affinity ratios,
c. Plot the affinity ratio vs the corresponding log concentration of allosteric agent and fit to a logistic function: the slope should be 1, the EC_{50} or IC_{50} corresponds to the K_d of the allosteric agent for the free receptor, and the E_{max} corresponds to the cooperativity of the system (*see* **Note 16**).

   Note that an allosteric agent acting in accordance with the model should not affect B_{max}.

13. Whereas agents with weak negative cooperativity (e.g., 0.5, or twofold negative) will show a clear nonzero asymptotic level in an inhibition curve type of binding assay, agents with strong negative cooperativity (e.g., 0.001, or 1000-fold negative) may be difficult to distinguish from competitive ligands (cooperativity = 0). The asymptotic level of binding will increase with increasing radioligand concentration, so a high concentration of radioligand will be needed to distinguish

strong negative cooperativity from a competitive interaction. Conversely, agents that are positively cooperative with the radioligand are best detected at low radioligand concentrations ($<<K_d$). A general-purpose assay should therefore use the radioligand at around its K_d value, bearing in mind the difficulties this imposes on the detection of weak positive or strong negative cooperativity.

14. Subtract nonspecific binding from all points.

    If a high radioligand concentration was not included then the K_d of the radioligand is assumed.

    Fit the data to **Eq. 8** (*see* **Note 12**).

    Alternatively:

    a. Fit the curve to a logistic function (the slope should be 1) to obtain an IC_{50} or EC_{50} value and an asymptotic level of binding.
    b. If a high radioligand concentration was included then calculate the radioligand K_d with Eq. (18) (*see* **Note 16**); otherwise use the known value.
    c. E_{max} = asymptotic bound / control bound (11)
    d. $\alpha = E_{max} / (1 + L \cdot (1 - E_{max})/K_d)$ (12)
    e. $K_X = (1 + L / K_d) / (1 + \alpha \cdot L / K_d) / IC_{50}$ (13)

    where L is the radioligand concentration, α is the cooperativity of the system, and K_X is the affinity of the allosteric agent for the free receptor.

15. Equations describing radioligand binding in the presence of an unlabeled ligand and an allosteric agent are presented below. Nonlinear regression analysis requires a program such as SigmaPlot or Prism 4 that can handle more than one independent variable.

    Subtract the appropriate value of nonspecific binding. The following equations refer to specific binding. For each data point, make available to the fitting procedure the specific dpm bound, the free concentration of radioligand, the concentration of unlabeled ligand, and the concentration of allosteric agent. Fitting the data to the following equation provides estimates of the pharmacological parameters of all three ligands, unless binding of a high radioligand concentration was not measured in which case the affinity of the radioligand (K_L, or $1/K_d$) must be provided.

$$B_{LAX} = \frac{B_{max} \cdot L \cdot K_L \cdot (1 + X \cdot K_X \cdot \alpha)}{1 + X \cdot K_X + (A \cdot K_A)^n \cdot (1 + X \cdot K_X \cdot \beta) + L \cdot K_L \cdot (1 + X \cdot K_X \cdot \alpha)} \quad (14)$$

where B_{LAX} (bound L in the presence of A and X) is specific binding at equilibrium; L, A, and X are the concentrations of radioligand, unlabeled ligand and allosteric agent respectively; K_L, K_A, and K_X are the affinities of radioligand, unlabeled ligand, and allosteric agent respectively for the free receptor; α and β are the cooperativity factors for the interactions of the allosteric agent with the radioligand and unlabeled ligand respectively; n is a slope factor to account for the shape of the inhibition curves of the unlabeled ligand. If only one concentration of unlabeled ligand was used then n should be set to 1. If the concentration dependence of the effect of the allosteric agent appears to be steep or shallow then a Schild slope factor, *s*, can be introduced by substituting $(X \cdot K_X)^s$ for $X \cdot K_X$ in the above equation.

If the allosteric agent is known to inhibit completely the dissociation of the radioligand then it is possible that binding in the presence of high concentrations of allosteric agent was not at equilibrium (*see* **Note 7**). In this case the general equation for nonequilibrium radioligand binding, is given by

$$B_{LAXt} =$$

$$B_{LAX} + (B_0 - B_{LAX}) \cdot \exp\left(\frac{-t \cdot k_{off}}{1+X \cdot K_X \cdot \alpha} + \frac{-t \cdot k_{off} \cdot L \cdot K_L}{1+X \cdot K_X + (A \cdot K_A)^n \cdot (1+X \cdot K_X \cdot \beta)}\right) \quad (15)$$

where: B_{LAXt} is nonequilibrium binding at time *t*; B_{LAX} is the predicted binding at equilibrium, **Eq. 14**; B_0 is the initial amount of radioligand binding, set to 0 in this case; *t* is the incubation time; and k_{off} is the known dissociation rate constant of the radioligand. As above, a slope factor, *s*, can be introduced to account for steep or shallow binding curves for the allosteric agent if required.

Note that if it is necessary to introduce slope factors (*n* for the unlabeled ligand *A*, and *s* for the allosteric agent *X*) then the model changes from a mechanistic model to a mixed mechanistic/empirical model, and the mechanistic interpretation of the respective affinity estimates (K_A and K_X) is unclear.

Allosteric effects on the binding of the radioligand are easy to see from the raw data, but effects on the binding of the unlabeled ligand may be unclear. To visualize the allosteric effects of the agent with both primary ligands on the same scale, calculate affinity ratios from the parameters of the fit:

$$r_L = \frac{1 + \alpha \cdot X \cdot K_X}{1 + X \cdot K_X} \qquad r_A = \frac{1 + \beta \cdot X \cdot K_X}{1 + X \cdot K_X} \quad (16, 17)$$

where r_L and r_A are the affinity ratios of the labeled and unlabeled ligand respectively; α and β are the cooperativity factors for the interactions of the allosteric age with the labelled and unlabeled ligand respectively; *X* is the concentration of allosteric agent and K_X is its affinity. Plot r_L and r_A vs the log concentration of allosteric agent. *See* **Note 16** for an explanation of affinity ratios and the affinity ratio plot.

16. This assay provides an efficient screening tool for detecting effects of allosteric agents on the binding of a labeled and unlabeled ligand, and measuring them semiquantitatively by visual inspection. The specific binding data will be transformed into affinity ratios, each of which is the apparent affinity of the primary ligand (labeled or unlabeled) in the presence of a particular concentration of allosteric agent divided by the apparent affinity of the ligand in the absence of allosteric agent. The affinity ratios are plotted against the log concentration of allosteric agent. In theory, the EC_{50} or IC_{50} of this affinity ratio plot corresponds to the K_d of the allosteric agent at the free receptor and the asymptotic value of the plot corresponds to the cooperativity with the primary ligand: these parameters can be estimated from the plot by visual inspection.

Subtract the appropriate values of nonspecific binding from all points to give specific binding.

a. The following quantities are measured directly in the assay:
B_L, binding in the presence of the low [radioligand] alone,
B_{L1}, binding in the presence of the high [radioligand] (optional),
B_{LA}, binding in the presence of the low [radioligand] and unlabeled ligand,
B_{LX}, binding in the presence of the low [radioligand] and a particular concentration of allosteric agent,
B_{LAX}, binding in the presence of the low [radioligand], unlabeled ligand and the same concentration of allosteric agent,
L, the low radioligand concentration,
L_1, the high radioligand concentration (optional),
A, the concentration of unlabeled ligand,
q, the ratio of low and high radioligand concentrations, L/L_1,
(p, the ratio of low [radioligand] and assumed K_d, L/K_d, if a high [radioligand] was not used).

b. The following quantities are calculated (equations contain q if a high radioligand concentration was used or p if the K_d of the radioligand is assumed):

radioligand K_d,

$$K_d = \frac{(B_{L1} - B_L) \cdot L \cdot L_1}{B_L \cdot L_1 - B_{L1} \cdot L} \tag{18}$$

the receptor concentration

$$B_{max} = \frac{B_L \cdot B_{L1} \cdot (1 - q)}{B_L - q \cdot B_{L1}} \quad \text{or} \quad B_{max} = \frac{B_L \cdot (p + 1)}{p} \tag{19}$$

unlabeled ligand apparent affinity,

$$K_A = \frac{B_{max} \cdot (B_L - B_{LA})}{A \cdot B_{LA} \cdot (B_{max} - B_L)} \quad \text{or} \quad K_A = \frac{B_{max} \cdot (B_L - B_{LA})}{A \cdot B_{LA} \cdot (B_{max} - B_L)} \tag{20}$$

The affinity ratio of radioligand in the presence of a single concentration of allosteric agent,

$$r_L = \frac{B_{LX} \cdot (B_{L1} - B_L)}{B_{L1} \cdot B_L \cdot (1 - q) - B_{LX} \cdot (B_L - q \cdot B_{L1})} \quad \text{or} \quad r_L = \frac{B_{LX}}{B_L \cdot (p + 1) - p \cdot B_{LX}} \tag{21}$$

The affinity ratio of unlabeled ligand the presence of a single concentration of allosteric agent,

$$r_A = \frac{B_L \cdot B_{LA} \cdot (B_{L1} - B_L) \cdot (B_{LX} - B_{LAX})}{B_{LAX} \cdot (B_L - B_{LA}) \cdot [B_{L1} \cdot B_L \cdot (1 - q) - B_{LX} \cdot (B_L - q \cdot B_{L1})]} \quad \text{or} \tag{22a}$$

$$r_A = \frac{B_L \cdot B_{LA} \cdot (B_{LX} - B_{LAX})}{B_{LAX} \cdot (B_L - B_{LA}) \cdot [B_L \cdot (p + 1) - p \cdot B_{LX}]} \tag{22b}$$

Note that these calculations cannot take account of possible kinetic artefacts (*see* **Notes 12** and **15**).

17. In a single-time-point assay there will inevitably be some inaccuracy owing to the time needed to initiate the dissociation in each tube, but this should not be a problem if the time needed to initiate dissociation is 10% or less of the dissociation time used. This imposes a practical limit on the size of the assay if short dissociation times are used. We find that it is just possible to run a 30-tube assay with a 6-min dissociation without major inaccuracies.

References

1. Christopoulos, A. (2002) Allosteric binding sites on cell-surface receptors: novel targets for drug discovery. *Na.t Rev. Drug Discov.* **1,** 198–210.
2. Christopoulos, A. and Kenakin, T. (2002) G protein-coupled receptor allosterism and complexing. *Pharmacol. Rev.* **54,** 323–374.
3. Christopoulos, A., Lanzafame, A., and Mitchelson, F. (1998) Allosteric interactions at muscarinic cholinoceptors. *Clin. Exp. Pharmacol. Physiol.* **25,** 185–194.
4. Ellis, J. (1997) Allosteric binding sites on muscarinic receptors. *Drug Dev. Res.* **40,** 193–204.
5. Holzgrabe, U. and Mohr, K. (1998) Allosteric modulators of ligand binding to muscarinic acetylcholine receptors. *Drug Discov. Today* **3,** 214–222.
6. Lazareno, S. and Birdsall, N. J. M. (1995) Detection, quantitation, and verification of allosteric interactions of agents with labeled and unlabeled ligands at G protein-coupled receptors: interactions of strychnine and acetylcholine at muscarinic receptors. *Mol. Pharmacol.* **48,** 362–378.
7. Lazareno, S., Gharagozloo, P., Kuonen, D., Popham, A., and Birdsall, N. J. M. (1998) Subtype-selective positive cooperative interactions between brucine analogues and acetylcholine at muscarinic receptors: radioligand binding studies. *Mol. Pharmacol.* **53,** 573–589.
8. Leppik, R. A., Lazareno, S., Mynett, A., and Birdsall, N. J. (1998) Characterization of the allosteric interactions between antagonists and amiloride analogues at the human α2A-adrenergic receptor. *Mol. Pharmacol.* **53,** 916–925.
9. Leppik, R. A. and Birdsall, N. J. (2000) Agonist binding and function at the human α(2A)-adrenoceptor: allosteric modulation by amilorides. *Mol. Pharmacol.* **58,** 1091–1099.
10. Leppik, R. A., Mynett, A., Lazareno, S., and Birdsall, N. J. (2000) Allosteric interactions between the antagonist prazosin and amiloride analogs at the human α(1A)-adrenergic receptor. *Mol. Pharmacol.* **57,** 436–445.
11. Swaminath, G., Steenhuis, J., Kobilka, B., and Lee, T. W. (2002) Allosteric modulation of β2-adrenergic receptor by Zn(2+) *Mol. Pharmacol.* **61,** 65–72.
12. Hoare, S. R., Coldwell, M. C., Armstrong, D., and Strange, P. G. (2000) Regulation of human D(1), d(2(long)), d(2(short)), D(3) and D(4) dopamine receptors by amiloride and amiloride analogues. *Br. J. Pharmacol.* **130,** 1045–1059.

13. Litschig, S., Gasparini, F., Rueegg, D., et al. (1999) CPCCOEt, a noncompetitive metabotropic glutamate receptor 1 antagonist, inhibits receptor signaling without affecting glutamate binding. *Mol. Pharmacol.* **55,** 453–461.
14. Urwyler, S., Mosbacher, J., Lingenhoehl, K., et al. (2001) Positive allosteric modulation of native and recombinant γ-aminobutyric acid(B) receptors by 2,6-Di-tert-butyl-4-(3-hydroxy-2,2-dimethyl-propyl)-phenol (CGP7930) and its aldehyde analog CGP13501. *Mol. Pharmacol.* **60,** 963–971.
15. Zahn, K., Eckstein, N., Trankle, C., Sadee, W., and Mohr, K. (2002) Allosteric modulation of muscarinic receptor signaling: alcuronium-induced conversion of pilocarpine from an agonist into an antagonist. *J. Pharmacol. Exp. Ther.* **301,** 720–728.
16. Hall, D. A. (2000) Modeling the functional effects of allosteric modulators at pharmacological receptors: an extension of the two-state model of receptor activation. *Mol. Pharmacol.* **58,** 1412–1423.
17. Parmentier, M. L., Prezeau, L., Bockaert, J., and Pin, J. P. (2002) A model for the functioning of family 3 GPCRs. *Trends Pharmacol. Sci.* **23,** 268–274.

3

Generation, Use, and Validation of Receptor-Selective Antibodies

John J. Mackrill

Summary

Antibodies have proved invaluable in the study of G-protein-coupled receptors (GPCRs). The utility of these immunoglobulin probes for investigation of protein structures and functions arises from their selectivity as well as their versatility. Antibodies can be used to analyze GPCR size, abundance, distribution, turnover, modification, interaction with other proteins, and functional properties. In this chapter, techniques for the generation and characterization of receptor-selective antibodies are described. Two protocols are given for the generation of antibodies: (1) development of polyclonal antibodies (PAbs) against synthetic peptides corresponding to a specific site within a GPCR and (2) selection of synthetic single-chain fragment variable (scFv) monoclonal antibodies (MAbs) from libraries expressed on the surface of bacteriophage. Immunoblot and enzyme-linked immunosorbent assays for characterization of the selectivity and affinity of such antibodies are described. Finally, methods are given for improvement of the titer and specificity of PAbs.

Key Words

Enzyme-linked immunosorbent assay, immunoblot, monoclonal antibody, 'phage-display, polyclonal antibody, single-chain fragment variable.

1. Introduction

Few tools for the study of G-protein-coupled receptors (GPCRs) have proved as versatile as antibodies. These immunoglobulin probes have been employed to characterize the molecular identity, subcellular localization (Chaps. 4, 7, and 15), posttranslational modifications (Chaps. 16–18), and interaction with accessory proteins (Chap. 22) of GPCRs. Antibodies are unique in their selectivity, frequently being able to distinguish isoforms of a receptor family for which subtype-specific pharmacological reagents are unavailable. Antibodies occasionally influence receptor activity, being utilized as functional probes. Further-

From: *Methods in Molecular Biology, vol. 259, Receptor Signal Transduction Protocols, 2nd ed.*
Edited by: G. B. Willars and R. A. J. Challiss © Humana Press Inc., Totowa, NJ

more, these immunological reagents are invaluable for the characterization of "orphan receptors" identified in genome sequencing and proteomic projects, for which pharmacological tools are unidentified.

Antibodies can be broadly divided into two types: polyclonal antibodies (PAbs) consist of a population of antibody molecules of various affinities and specificities for the target molecule, the antigen; monoclonal antibodies (MAbs) interact with their antigen with single affinity and specifity. PAbs are often used in the form of antisera, prepared from the blood of immunized animals, typically rabbits or guinea pigs. MAbs are usually generated from immortalized B-lymphocyte hydridoma cell lines from immunized animals, typically mice or rats *(1)*. Each type of antibody has advantages and disadvantages for the study of GPCRs. Once a clonal hybridoma cell line has been established, it provides a potentially unlimited supply of a MAb, whereas the availability of PAbs is dependent on the volume of serum obtained from the immunized animal. MAbs can be more useful in studying the structure–function of GPCRs, as they can be purified and used as a probe of single-affinity binding to a single site in the target antigen. However, PAbs often give stronger signals in immunoassays, are less expensive and easier to produce, and can be more effective at immunoprecipitation owing to their interaction with the target molecule potentially at multiple sites. Furthermore, production and screening of rodent MAbs is labor intensive and requires tissue-culture expertise. Generation of anti-GPCR PAbs against synthetic peptides is a suitable option for most laboratories involved in receptor research; commercial services are available for synthesis of antigens, immunization of host animals, and preparation of antisera. Over the past decade, techniques have been developed for the generation of synthetic MAb reagents, such as selection of single-chain fragment variable (scFv) antibodies *(2)* from bacteriophage-displayed ('phage-displayed) libraries. Such methods offer numerous advantages over the production of rodent MAbs, including ease of selection of desired antibody specificities, no requirement for tissue culture, the ethical advantage of no animal usage, as well as the ability to select probes against self- or cytotoxic antigens. The main disadvantage of scFv antibodies is that they consist of a single heavy and light variable region, vs a pair of these sequences in natural immunoglobulin molecules, and so tend to be of lower affinity (*see* **Fig. 1**).

Numerous strategies are available for obtaining antibodies against a particular GPCR. The simplest of these is to purchase such an antibody from one of numerous biotechnology companies that supply such probes, or to obtain them from a collaborating research laboratory. However, such reagents might not be available, are often very expensive, are usually supplied in limiting quantities, and still have to be validated for the desired application. In this chapter, the generation of antisynthetic peptide PAbs and the selection of 'phage-displayed

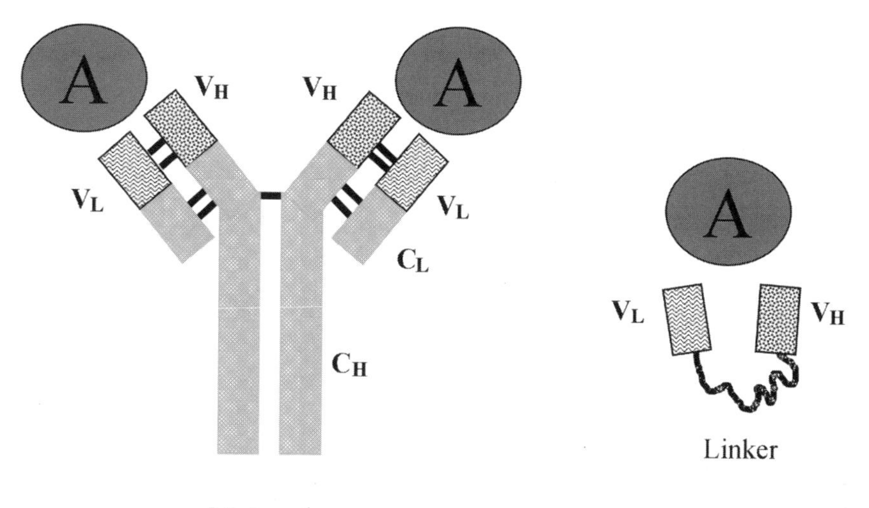

Fig. 1. Structures of natural and scFv antibodies. Natural antibodies, or immunoglobulins, consist of two pairs of polypeptides, called the heavy (H) and light (L) chains, joined by disulfide bridges. Constant regions of these polypeptides, C_H and C_L, are involved in antibody effector functions. The variable regions, V_H and V_L, interact with the target molecule that the antibody recognizes, known as the antigen (A). Synthetic scFv antibodies consist of single, rather than paired, V_H and V_L fragments joined by a flexible linker region.

scFvs are outlined. General procedures for the validation and use of such antibodies are also described.

2. Materials

2.1. Reagents

All reagents were from Sigma-Aldrich (Dorset, UK), except where noted.

1. Phosphate-buffered saline (PBS): 8 g/L of NaCl, 0.2 g/L of KCl, 1.44 g/L of Na_2HPO_4, 0.18 g/L of KH_2PO_4, adjusted to pH 7.2–7.4 with HCl. PBS-T: 0.2% (v/v) Tween-20 in PBS; 2%/4% /5% mPBS: 2%/4%/5% (w/v) nonfat milk powder (Marvel or equivalent) in PBS; mPBS-T: 5% (w/v) nonfat milk in PBS-T.
2. 4X Separating gel buffer: 1.5 *M* Tris-HCl, pH 8.3.
3. 4X Stacking buffer: 0.5 *M* Tris-HCl, pH 6.8.
4. 30% Acrylamide stock (Ultrapure ProtoGel®, 30% (w/v) acrylamide, 0.8% (w/v) bisacrylamide (37.5:1), from National Diagnostics, GA, or equivalent). **Caution:** acrylamide is toxic, so it is essential to wear gloves whenever handling this reagent.

5. Separating gel mix (8%): 4266.7 μL of 30% acrylamide stock, 4000 μL of 4X separating buffer, 7517.3 μL of distilled deionized water, 160 μL of 10% (w/v) sodium dodecyl sulfate (SDS). Add 16 μL of *N,N,N'*,N'-tetramethylethylene diamine (TEMED) and 40 μL of 10% (w/v) ammonium persulfate (APS) just prior to pouring the gel.
6. Stacking gel mix (5%): 1666.7 μL of 30% acrylamide stock, 2500 μL of 4X stacking buffer, 5683.3 μL of distilled deionized water, 100 μL of 10% SDS. Add 10 μL of TEMED and 40 μL of APS just prior to pouring.
7. 5X SDS-polyacrylamide gel electrophoresis (PAGE) sample buffer (10 mL): 1250 μL of 1 *M* Tris-HCl, pH 6.8, 4000 μL of 10% SDS, 1000 μL of 1% (w/v) bromophenol blue, 2000 μL of glycerol, 10 m*M* dithiothreitol. Dissolve, warming in a 37°C water bath if necessary, then make up to 10 mL with distilled deionized water. Aliquot 1-mL volumes and store at –20°C.
8. 10X SDS-PAGE running buffer: 30.2 g of Tris, 188 g of glycine, 10 g of SDS in 1 L of distilled, deionized water.
9. 10X Blotting buffer: 1.5 *M* glycine, 200 m*M* Tris, 0.37% (w/v) SDS.
10. Nitrocellulose membranes, such as Protran nitrocellulose (Schleider and Schuell, Dassel, Germany).
11. 10X Ponceau S: 2% (w/v) Ponceau S, 5% (v/v) acetic acid.
12. Coomassie stain: 0.25% (w/v) Coomassie brilliant blue R250 in 50% (v/v) methanol, 10% (v/v) acetic acid.
13. Destain: 25% (v/v) methanol, 7% (v/v) acetic acid.
14. Horseradish peroxidase (HRP)-conjugated anti-rabbit, anti-sheep, and anti-guinea pig IgG antisera (Sigma or equivalent).
15. Pierce Supersignal chemiluminescence reagents (from Pierce and Warriner, Chesire, UK) or equivalent.
16. Coating buffer: 10 m*M* $NaHCO_3$, pH 9.
17. Protein A conjugated to agarose (such as Protein A Superose from Pharmacia Bioscience, Uppsala, Sweden).
18. Column buffer A: 100 m*M* Tris-HCl, pH 8.0.
19. Column buffer B: 1 *M* Tris-HCl, pH 8.0.
20. Column buffer C: 100 m*M* triethylamine, pH 11.5, freshly prepared by addition of 70 μL of stock triethylamine to 5 mL of distilled, deionized water.
21. Column buffer D: 0.5 *M* NaCl, 20% ethanol.
22. M13/Fd bacteriophage scFv library, such as the GRIFFIN.1 $V_H + V_L$ scFv library, HB2151, and TG1 host *Escherichia coli* host strains from Department of Protein Engineering, Laboratory of Molecular Biology, Hills Road, Cambridge (*see* http://www.mrc-cpe.cam.ac.uk/g1p.php?menu=1808).
23. Glucose (20%): 40 g of glucose in a total volume of 200 mL of distilled, deionized water. Autoclave for 20 min at 121°C and 15 psi. Store at room temperature for up to 1 mo.
24. Ampicillin stock: 1 g of ampicillin in 10 mL of distilled, deionized water. Sterilize by filtration through a 0.45-μm sterile syringe filter unit and store in 1-mL aliquots at –20°C.

25. Kanamycin stock: 0.25 g of kanamycin in 10 mL of distilled, deionized water. Sterilize and store as for the ampicillin stock.
26. 2X TY: 16 g/L of bactotryptone, 10 g/L of yeast extract (both from DIFCO Products, Unitech, Dublin), 5 g/L of NaCl in distilled, deionized water. Autoclave, add glucose and ampicillin as required, and then store at 4°C for up to 2 wk.
27. TYE: 15 g/L of Bacto-Agar, 8 g/L of NaCl, 10 g/L of tryptone, 5 g/L of yeast extract. Autoclave, cool to approx 55°C, add ampicillin and glucose as required, then pour into-100 mm Petri dishes or Nunc Bio-Assay dishes (~20 mL/petri dish, ~100 mL/Bio-Assay dish) under sterile conditions. Allow to set, invert, and dry by overnight incubation at 37°C. Store at 4°C for up 2 wk.
28. VCS-M13 Helper Phage from Stratagene Cloning Systems, La Jolla, CA, USA.
29. PEG buffer: 20% (w/v) polyethylene glycol 6000, 2.5 *M* NaCl. Autoclave.

2.2. Apparatus

1. For immunoblotting: Bio-Rad Protean SDS-PAGE minigel apparatus or similar; Bio-Rad miniwet blotter or equivalent; power supply capable of generating an output of at least 400 mA and 300 V; 3MM filter paper (Whatman Kent, UK).
2. For enzyme-linked immunosorbent assays (ELISA): 96-well polyvinyl chloride plates (Nunc A/S., Rosklide, Denmark) and microtiter plate absorbance reader, such as Bio-Rad Model 450.
3. For chromatography: 10-mL disposable plastic columns, such as Bio-Rad Econo-Pac columns or Pierce Disposable Plastic columns (Pierce Biotechnology, Rockford, IL, USA). Exocellulose GF-5 Desalting Columns (5 mL) from Pierce.
4. For selection of 'phage-scFv: Dynal magnetic separators (MPC-1 and MPC-S) and Dynabeads® M-280 Streptavidin-coated paramagnetic beads (Dynal Biotech UK, Wirral, UK); autoclave; shaking incubators at 30°C and 37°C.
5. Microconcentrators: with 10-kDa cutoff membranes, such as Centricon and Centriprep C10 devices (Amicon, Beverly, MA).
6. X-ray film and cassettes, such as those from Genetic Research Instrumentation, Essex, UK.
7. Standard microfuge and benchtop centrifuges.
8. Ultraviolet (UV)/visible spectrophotometer.

3. Methods

3.1. Design of Synthetic Peptide Immunogens

Both recombinant fusion proteins and synthetic peptides derived from a GPCR's amino acid sequence can make suitable immunogens. A major advantage of the use of synthetic peptides for this purpose is that they are directed against a defined sequence in the primary structure of these proteins. This permits the production of antisera selective for a particular isoform in a GPCR family by comparison of the sequences of these receptors using multiple alignment software (e.g., http://www.ebi.ac.uk/clustalw/ *[3]*) and choice of regions poorly conserved between family members. Furthermore, site-directed anti-synthetic peptide PAbs can be useful in structural analysis of proteins, such as in the investigation of transmembrane topology or of ligand binding sites. The features of an amino acid sequence that determine its antigenicity have been elucidated mainly by empirical methods. Synthetic peptide characteristics that promote immune responses include hydrophilicity, flexibility, charge, and surface exposure of the sequence in the native receptor. Potential antigenic sequences in GPCRs include the N- and C-termini, as well as the intracellular and extracellular loops. Transmembrane-spanning regions tend to make poor antigens. Typically, synthetic peptides for immunization are in the range of 12–30 residues in length, as smaller peptides may not be immunogenic, whereas larger ones are more difficult and expensive to synthesize. The selection of candidate synthetic peptide antigens can be assisted using predictive algorithms, such as the antigenic index of Jameson and Wolf *(4)*. Such tools are often included in sequence analysis software packages (*see* **Note 1**). However, antigenic index algorithms are not infallible, so ideally more than one synthetic peptide should be generated from the target GPCR amino acid sequence.

Once a synthetic peptide of interest has been selected, it should be compared to those in protein databases, using software such as Fasta3 (http://www.ebi.ac.uk/fasta33/), to avoid unwanted crossreaction of anti-GPCR synthetic peptide PAbs with proteins bearing similar sequences. Peptide synthesis is usually performed by commercial or in-house services, often combined with immunization and test bleeding of host animals. The cost of commercial services varies widely, so it is worth shopping around. Peptides are usually synthesized by fluorenylmethoxycarbonyl-(Fmoc)-polyamide solid support chemistry *(5)*, with a scale of 0.25 mmol being sufficient for immunization and screening of several animals. The purity of such peptides should be analyzed by mass spectroscopy or high-pressure liquid chromatography and should be supplied in a lyophilized, fully reduced form. Normally, a cysteine residue is present on the N-terminus of the peptide to facilitate conjugation to carrier molecules. If such a residue is not present in the selected peptide, it can be added on. Addition of two or three glycine residues between the sequence of choice and the terminal cysteine can increase antigenicity of the conjugated immunogen, acting as a spacer between the antigen and carrier. Covalently modified residues can be incorporated into synthetic peptides, to generate PAbs recognizing specific in vivo GPCR modifications, such as phosphorylation of tyrosine, serine, or threonine. Finally, several (usually four or eight) copies of a synthetic peptide antigen can be synthesized on a lysine core using the multiple antigen peptide system *(6)*. This generates a highly immunogenic macromolecule, which does not need to be conjugated to a carrier. Synthesis of multiple antigen peptides is available commercially.

3.2. Conjugation of Immunogens to Carriers

Coupling of antigens to carrier proteins, such as bovine serum albumin (BSA) or keyhole limpet hemocyanin (KLH), promotes the generation of high-titer antisera in host animals and is usually required for fusion proteins or synthetic peptides of less than 40 amino acid residues. It is useful to conjugate the target antigen to two distinct carriers, using one for immunization and the other for screening antisera for immunoreactivity, thereby avoiding problems associated with potential anti-carrier antibodies. A variety of bifunctional crosslinking reagents such as sulfo-succinimidyl 4-(*N*-maleimidomethyl) cyclohexane-1-carboxylate (sulfo-SMCC) can be employed to conjugate the antigen to several different carrier proteins. In the case of sulfo-SMCC, the peptide must be fully reduced and possess a cysteine residue at its coupling site, usually at the C-terminus.

1. All steps should be performed in glass vessels because peptides and conjugates tend to adhere to plasticware. Dissolve 8 mg of BSA or KLH in 0.5 mL of 10 m*M* sodium phosphate buffer, pH 7.2.
2. Add 2 mg of sulfo-SMCC and incubate for 1 h at 30°C with constant stirring.
3. Pass the mixture through an Exocellulose GF-5 column preequilibrated with column buffer A and elute the activated carrier with 20 × 1 mL steps of the same buffer. Measure absorbance at 280 nm and pool peak fractions.
4. Incubate 4 mL of the activated carrier with 8 mg of synthetic peptide, or the molar equivalent of protein, overnight at 4°C with continuous stirring.
5. Store the antigen–carrier conjugate in glass vials at –20°C.

3.3. Immunization of Host Animals

Rabbits or guinea pigs are commonly used for production of PAbs for research purposes, although donkeys, goats, or sheep can be used where very large volumes of antiserum are required. In the case of rabbits and guinea pigs, at least two animals should be immunized per antigen, because immune response can vary greatly between individuals even in highly inbred strains. In all cases, animals should be immunized and cared for by trained operatives, employing approved techniques. An animal license is often a legal requirement for performance of such experimental procedures, which must adhere to a defined schedule (*see* **Note 2**). In many research establishments, immunization and test bleeding of animals is an in-house service, or can be obtained commercially.

Rabbits or guinea pigs are typically immunized by subcutaneous injection, as this route provides for slow release of the antigen and reduces the risk of pathology to the animal. The antigen is often mixed with an adjuvant, a substance that nonspecifically promotes immune response. Freund's complete adjuvant (FCA) is used in primary immunizations and consists of heat-killed *Mycobacterium tuberculosis* in mineral oil. Freund's incomplete adjuvant (FIA) is used in booster immunizations and lacks heat-killed bacteria. These substances deposit the antigen in an emulsion that takes the host a long time to clear and contain material that stimulates the immune system (bacteria in the case of FCA). In a typical immunization schedule for rabbits, animals are injected subcutaneously at five different sites using a total of 100 μg of antigen in 0.5 mL of PBS mixed with the same volume of FCA. Adjuvant and antigen are mixed by passing them between two syringes joined with a Luer fitting. At 28-d intervals, animals are given booster immunizations with the same antigen preparation mixed with FIA instead of FCA, for up to three boosts. Test bleeds (~5 mL) are taken 14 d after each immunization, to assay antibody titer. Once the titer of the antiserum is sufficiently high, or after the final boost, animals are normally terminally bled by cardiac puncture, to obtain a large volume (20–60 mL) of antiserum.

3.4. Preparation of Antisera

Crude antisera can be readily prepared from whole blood by clotting. This removes cellular components and clotting factors from the blood.

1. Incubate fresh bleeds for 1 h at 37°C in 1-mL aliquots in 1.5-mL Eppendorf tubes.
2. Use a cocktail stick to move clotted material from the sides to the bottoms of the tubes.
3. Allow the clots to contract overnight at 4°C.
4. Pellet the clots by centrifugation at $1000g_{max}$ for 10 min using a microfuge.
5. Carefully collect the antiserum (supernatant) and aliquot in 0.5-mL fractions. The majority can be stored at –20°C for up to several years, but for short-term use, a few aliquots can be kept at 4°C in the presence of 0.02% (w/v) sodium azide. Avoid repeated freeze–thawing of frozen aliquots of antiserum.

3.5. Selection of Bacteriophage-Displayed scFvs

Selection of monoclonal scFv antibodies from a library expressed on coat proteins of bacteriophage ('phage-display) provides a powerful approach for the generation of monospecific probes against a variety of antigens, some of which would be unsuitable for immunization into host animals. Such targets against which such scFv antibodies have been selected include an epidermal growth factor receptor mutant *(7)*, steroids *(8)*, heparan sulfates *(9)*, and the toxin paraquat *(10)*. Furthermore, use of 'phage-display to select monospecific scFvs that recognize GPCRs is a proven approach, as this technique has been used to generate anti-neurotensin receptor antibodies *(11)*. Because a population of 'phage-displayed antibodies can readily be subcloned, this technique also permits selection of MAbs against a protein of interest present in a mixture of other proteins, such as cells expressing a receptor, or a membrane subfraction enriched in this target. Bacteriophage display expression systems can be obtained either commercially (Pharmacia Bioscience, Uppsala, Sweden), or from academic sources. These scFv libraries are either preconstructed, or can be polymerase chain reaction (PCR) cloned into the expression system from a suitable source, such as cDNA from splenocytes from an immunized mouse, or peripheral human B-lymphocytes. Antibody fragments are usually expressed in frame with a bacteriophage coat protein, such that they are presented on the surface of the viral particle. Antibodies interacting with the target antigen of choice can then be selected from such 'phage-scFv libraries, using a variety of approaches ***(2)***. Such rounds of selection can be repeated until a highly enriched population of 'phage-scFv recognizing the target is obtained. Monoclonal 'phage-scFv can then be subcloned from this population. Although the protocols for using such systems are extensive, simple bacteriological techniques are required; scFv PAbs can be selected in <2 wk and MAbs can be subcloned within days of this. The following is an example of a selection protocol used to obtain an scFv PAb against a target protein using a premade scFv library expressed in the pHEN2 phagemid vector ***(12)***. This phagemid vector contains an scFv cloned in frame with the M13 bacteriophage GIII coat protein, an *amber* stop codon for expression of soluble scFv fragments, and hexahistidine/c-*myc* tags for purification/detection of these recombinant antibodies. A diagrammatic representation of this selection protocol is given in **Fig. 2**.

1. Inoculate 2 mL of 2X TY broth with a single colony of *E. coli* TG1 cells and grow overnight at 37°C, shaking at approx 200 rpm. Transfer 20 μL of this overnight culture to 20 mL of 2X TY medium and grow at 37°C, shaking at approx 200 rpm.
2. Biotinylate the target molecule using a commercially available kit, such as EZ-link biotin-NHS from Pierce Biotechnology, Rockford, IL, USA, or equivalent.
3. Dilute 4×10^{13} transforming units (t.u.) of Griffin 'phage-scFv library in PBS to a total volume of 2.5 mL. Add 2.5 mL of 4% mPBS, 50 μL of Tween-20, and 50 nmol of antigen in 100 μL of PBS. Incubate for 1 h at room temperature, rotating at approx 60 rpm.
4. Pellet 1.5 mL of Dynabeads M-280 by incubation in an MPC-1 magnetic concentrator for 2 min. Remove the supernatant using an aspirator and resuspend the pellet in 1 mL of 2% mPBS. Collect the beads using the MPC-1 device, resuspend in 1 mL of 2% mPBS, and incubate at 37°C for 2 h.
5. Collect the Dynabeads, resuspend in 1.5 mL of 2% mPBS, and incubate with the antigen–'phage-scFv mixture for 15 min at room temperature, rotating at approx 60 rpm.
6. Collect 'phage-scFv-antigen bound beads using the MPC-1 device. Wash the beads 15 times with 1 mL of PBS, or 1 mL of 2% mPBS with every third wash, using PBS for the final two washes.
7. Incubate the Dynabeads M-280 with 1 mL of column buffer C for 10 min at room temperature. Neutralize by addition of 0.5 mL of column buffer B. Remove beads using the magnetic concentrator and collect the supernatant.

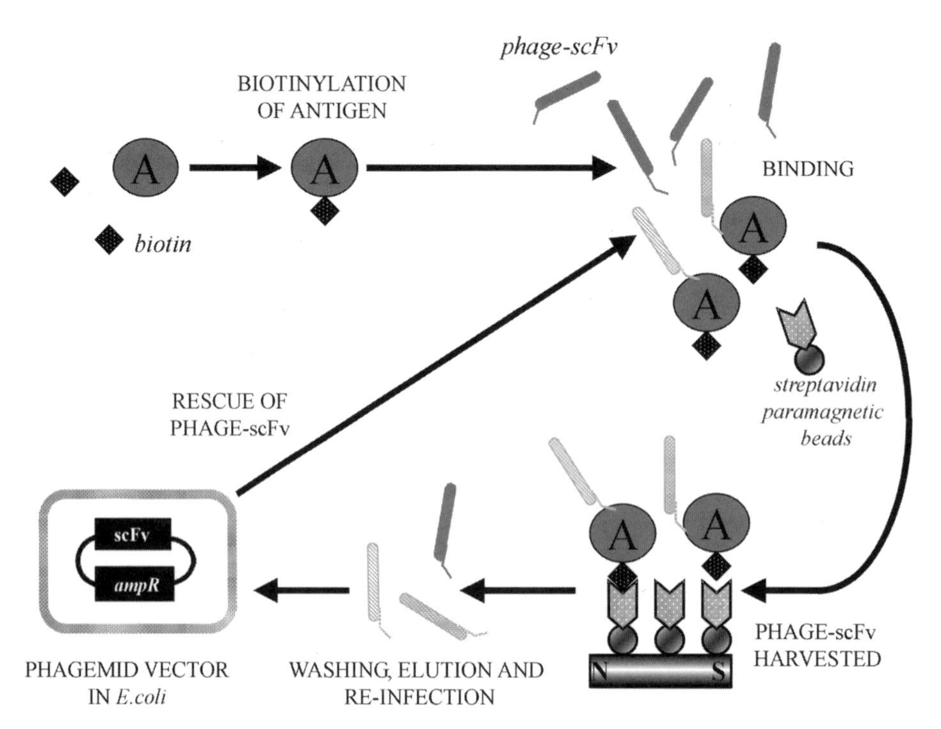

Fig. 2. Selection of phage-scFv recognizing a target antigen in solution. This is a diagrammatic representation of the protocol detailed in **Subheading 3.5.** The target antigen (A) is biotinylated and then incubated with a library of scFv expressed on the surface of bacteriophage. 'Phage-scFv binding to the target are harvested using streptavidin-coated paramagnetic beads, by virtue of the high-avidity interaction between streptavidin and biotin groups on the antigen. Nonspecifically bound 'phage-scFv are removed by extensive washing. Specifically bound 'phage scFv are eluted and used to infect untransformed *E. coli* host cells. Growth of these infected *E. coli* effectively amplifies the phagemid from antigen-binding 'phage-scFv. Packaged 'phage-scFv are then rescued from the host by coinfection with helper 'phage, then used in the next round of selection against the antigen. After several such rounds of binding, harvesting, washing, infection, and rescue, a high-titer anti-target polyclonal 'phage-scFv should be obtained.

8. Mix 0.75 mL of the eluted phage-scFv with 9.25 mL of *E. coli* TG1 cells grown to an optical density of approx 0.5 at a wavelength of 600 nm (OD_{600nm}). Incubate for 30 min at 30°C without shaking.
9. Remove 100 µL of phage-infected culture and plate 10-fold dilutions on 100-mm TYE plates containing 1% glucose and 100 µg/mL ampicillin. Grow overnight at 37°C.

30° to horizontal. This prevents bubbles being trapped underneath the comb, which would otherwise result in formation of uneven wells, as oxygen inhibits the polymerization of acrylamide. The stacking gel will take < 30 min to set.

4. Add 5 μL of 5X SDS-PAGE sample buffer to 20-μL samples of cell membranes, lysates, or fusion proteins in Eppendorf tubes. Make a small hole in the lid of each tube using a needle. Denature the samples for 5 min at 95°C, then centrifuge at $10,000g_{max}$ for 20 s.
5. Remove sample combs from the gels and assemble the apparatus according to the manufacturer's instructions. Fill the apparatus with 1X SDS-PAGE running buffer.
6. Load the samples into appropriate lanes, underneath the running buffer, preferably using long, narrow pipet tips that can reach the bottoms of the wells. Connect the apparatus to a power supply and electrophorese at 100 V. Once the bromophenol blue dye front has passed the stacking–separating gel interface (10–15 min), increase the voltage to 200 V. Under these conditions, the dye front will be within 1 cm of the bottom of the gel in approx 60 min.

3.7.2. Transfer for Western Blotting

The following protocol is for transfer of GPCRs from SDS-PAGE gels to nitrocellulose using a miniwet blotter. Protocols using other transfer systems, such as semidry blotters, vary considerably. Detailed instructions are often supplied by the manufacturers of such equipment. Semidry Western blotting tends to be more rapid than wet blotting, owing to the shorter distance between electrodes. However, wet blotting tends to give a more uniform transfer. Furthermore, semidry systems are unsuitable for extended transfer times, as they tend to overheat.

1. Cut four pieces of Whatman 3MM filter paper and one piece of nitrocellulose to the same area as each SDS-PAGE separating gel. Soak in transfer buffer.
2. Place two sheets of filter paper on top of each other, followed by the nitrocellulose, on the anode sponge of the transfer cassette. Apply the soaked sheets in a ∪-shape to the middle of the transfer sandwich, slowly lowering the edges. This pushes bubbles out, which could otherwise inhibit transfer.
3. Remove the SDS-PAGE minigel on which the samples have been resolved, using gloves with fingertips wetted in the transfer buffer. Place the gel on the nitrocellulose, making sure that it is flat and that there are no bubbles trapped underneath.
4. Place another two layers of soaked filter paper on top of the gel, close the cassette, and place it inside the transfer unit. Assemble the apparatus according to the manufacturer's instructions. Transfer at 60 V (200 mA maximum current) constant voltage for 1 h is suitable for most GPCRs, although transfer conditions should be optimized (*see* **Note 4**). For longer transfers, or higher voltages, cool the apparatus on ice during the run.

3.7.3. Immunostaining

1. Remove the nitrocellulose from the transfer cassette and place in 10 mL of 1X Ponceau S stain (diluted in distilled, deionized water) in a clean plastic tray. Wash the nitrocellulose once with distilled, deionized water and mark the positions of molecular weight standards, lane numbers, and proteins of interest with a soft pencil. Destain completely by extensive washing with PBS.
2. Incubate the nitrocellulose transfer in 5% mPBS for 1 h at room temperature with constant shaking.
3. Replace the mPBS with the antibody to be tested, diluted in 5% mPBS. A volume of 6 mL of antibody dilution is sufficient for a minigel blot in a plastic tray, although this can be reduced to 2 mL if the blot is placed in a close-fitting heat-sealed plastic bag. Bubbles should be squeezed out of such bags, which should then be completely sealed and placed in a tray containing approx 100 mL of distilled, deionized water. Floating the sealed bag in water permits free movement of any bubbles remaining inside, which could otherwise prevent interaction of antibodies with the blot. Incubate overnight at 4°C, or alternatively for 1 h at 37°C, with constant shaking.
4. Wash the blot for 3 × 5 min with PBS at room temperature, with constant shaking. If necessary, the diluted antibody can be reused, following storage for up to 1 mo at 4°C in the presence of 0.05% (w/v) sodium azide.
5. Incubate the blot in 6 mL of a 1:2000 dilution of the appropriate secondary antibody–HRP conjugate in 5% mPBS, for 1 h at room temperature with constant shaking.
6. Wash for 3 × 5 min with PBS at room temperature, with constant shaking. Develop the staining using an enhanced chemiluminescence (ECL) kit, according to the manufacturer's instructions.

3.8. Purification of Antibodies from Antiserum

The titer of PAbs can be improved by purification of antibody molecules. Two classes of technique are used for purification of immunoglobulins from antisera: (1) antigen affinity methods, which separate antibodies from a mixture of proteins on the basis of their interaction with target molecules and (2) methods that isolate antibodies, or specific classes of antibody, on the basis of biochemical properties other than their interaction with the antigen. Because MAbs are of single affinity and specificity, their titer in immunoassays cannot be improved using antigen affinity methods. However, purification of MAbs by other biochemical means can be of use for increasing signals in immunoassays, by virtue of increasing the concentration of immunoglobulin molecules.

3.8.1. Protein A–Agarose Chromatography

Staphylococcus aureus expresses a polypeptide on its cell wall called protein A, which binds to IgG antibodies from various mammalian species with high affinity. When covalently linked onto a solid support, this protein is extremely useful for isolation of the IgG class of immunoglobulins from crude antisera and hybridoma cell supernatants. Protein A affinity chromatography both concentrates and purifies the IgG class of immunoglobulins, which is useful because this antibody type tends to have the highest affinity for their target antigens. It should be noted that protein A affinity chromatography is very useful for purifying IgG from rabbit and guinea pig, although is not as suitable for use with sera from other hosts, such as rats and mice. However, other immunoglobulin-binding proteins, such as protein G, protein L, and protein LA, can be employed for the detection and purification of other classes of immunoglobulin, or of IgG from other hosts, including scFvs in the case of protein LA ***(13)***. The following describes a simple procedure for protein A chromatographic purification of IgG from rabbit or guinea pig antisera, which can be scaled up to allow isolation of large quantities of anti-GPCR IgG:

1. Wash 2 mL of unpacked protein A agarose with 10 mL of column buffer A, in a disposable plastic column.
2. Close off the column outlet. Add 4 mL of crude antiserum containing 400 µL of column buffer B. Close off the other end of the column.
3. Incubate the column overnight at 4°C with constant agitation on a rotator or shaker.
4. Bring the column to room temperature. Open up both ends of the column and collect the flowthrough.
5. Wash the column with 10 mL of column buffer A, then 10 mL of column buffer B.
6. Elute the IgG with ten 500-µL steps of column buffer C into tubes each containing 50 µL of column buffer B.
7. Wash the column with 10 mL of column buffer A, then store in 10 mL of column buffer D at 4°C for future reuse (starting from **step 2**).
8. Measure the optical density of the fractions at a wavelength of 280 nm in a spectrophotometer using a quartz cuvet.
9. Concentrate peak fractions approx 10-fold using Centricon C10 devices according to the manufacturer's instructions.
10. Add bovine serum albumin (BSA) and sodium azide to concentrations of 1 mg/mL and 0.05% (w/v), respectively. Assay by immunoblot and/or ELISA. Store at 4°C.

3.8.2. Affinity Purification of Antibodies on Protein Transfers

Anti-GPCR antibodies may be purified on the basis of their interaction with their target receptor, or receptor fusion protein, separated on an SDS-PAGE gel and transferred onto nitrocellulose ***(14)***. This technique is not suitable for large-scale isolation of antibodies, but is useful for rapid and easy preparation of anti-GPCR immunoglobulins.

1. Prepare SDS-PAGE minigels as described in **Subheading 3.7.1.**, except use a preparative comb in the stacking gel, or load multiple wells of a toothed comb with the same sample.
2. Load the gel with a denatured sample enriched in the protein of interest, such as a fusion protein (100 μg of protein/gel) or GPCR (1–10 μg of protein/gel of purified GPCR, or up to 500 μg/gel of lysate from a cell line overexpressing the GPCR). Run the gel and transfer onto nitrocellulose, as described in **Subheading 3.7.2.**
3. Stain the blot using 1x Ponceau S and excise a strip in the position of the GPCR/fusion protein using a scalpel blade. Destain the strip with H_2O and block using mPBS for 1 h at room temperature.
4. Incubate the strip with 4 mL of a 1:50 dilution of the crude antiserum in PBS for 1–2 h. The diluted antiserum can be incubated with one or two more strips if sufficient GPCR/fusion protein is available.
5. Incubate the strip in 2 mL of column buffer C for 10 min at room temperature. Neutralize with 200 μL of column buffer B, then add 1 mg/mL of BSA and 0.05% (w/v) sodium azide. Collect the eluted, neutralized PAb and assay its titer using ELISA and/or immunoblot, then store at 4°C.

4. Notes

1. Various software packages for analyses of biological sequences are commercially available. Although these packages tend to be expensive, they are useful for laboratories investigating the structure and function of GPCRs, assisting in the prediction of antigenic regions, sites of posttranslational modification, and secondary structure of these proteins. Examples of such software include DNAStar (DNASTAR, Inc., Madison, WI, USA) and MacVector (Accelrys, San Diego, CA, USA).
2. In most countries, a license is essential for performing procedures using mammalian laboratory animals. Application forms for such licenses are usually obtained from appropriate government departments and require a detailed description of the planned experimental schedule, as well as full training of the personnel involved.
3. Interaction between an antibody and its antigen obeys the same principles as receptor–ligand binding: it should be specific and saturable. It also follows that there will be some nonspecific antibody binding in any system, so it is essential that appropriate controls are incorporated into any immunoassay:
   a. Negative controls: (1) antigen block: preincubate anti-peptide or anti-fusion protein antibodies at the final assay concentration for 1 h in the presence of 1–10 μg/mL of peptide or fusion protein. (2) Secondary antibody control: perform the assay in the absence of the primary antibody, but with all other components of the detection system. (3) Knock-out control: perform the assay on cells or tissues that normally express the GPCR of interest, but in which the target has been depleted, such as in knock-out transgenic mice, or RNAi-treated cells. (4) Isoform control: screen potentially isoform-specific antibodies against related GPCR subtypes, such as those expressed in transfected mammalian cells.

b. Positive controls: (1) The fusion protein, or synthetic peptide against which the antibody was generated, (2) lysate or membranes prepared from a cell line overexpressing the receptor of interest, (3) a subcellular fraction or purified receptor preparation in which the GPCR is enriched, and (4) lysate or membranes from a cell line endogenously expressing the GPCR of interest. This positive control is particularly useful if the GPCR undergoes extensive post-translational modification in cells in which it is normally expressed.

4. Although the protocols for immunoblotting or ELISA described in this chapter are suitable for analysis of most GPCRs, these immunoassays can be modified to give optimal detection. Optimization can be performed at all steps:
   a. SDS-PAGE: for proteins <40 kDa, resolution will be improved by using a 15% gel; for proteins >200 kDa, 5% gels will give better separation.
   b. Blotting: very large proteins (>200 kDa) will tend to be retained in SDS-PAGE gels under the conditions given in **Subheading 3.7.2**. To improve transfer of large proteins, transfer time and voltage should be increased to 2–4 h and 100 V. Under these conditions, the transfer unit should be cooled on ice, or by using a circulating water bath (12°C) and the power of transfer should not exceed 75% of the maximum stated by the manufacturer. Blotting of hydrophobic proteins can be improved by incorporating 10% (v/v) methanol into the 1X transfer buffer. Small proteins (<20 kDa) can pass through nitrocellulose, so in this case transfer voltage should be reduced (to 40 V initially). In all cases, transfer efficiency can be monitored by Coomassie R250 staining of the SDS-PAGE gel after transfer and comparison to a gel run in parallel, although not blotted. Coomassie stain on a shaker for between 15 min and 2 h (depending on the relative abundance of the protein of interest), then wash with several changes of destain over an hour.
   c. Blocking: mPBS is inexpensive and suitable for blocking free sites in most ELISA and immunoblot assays, but can give a high general background signal with certain antibodies. Alternatives include mPBS-T, 3% (w/v) BSA (immunoglobulin-free) in PBS, or 1% (v/v) Tween-20.
   d. Antibody incubations (for ELISA and immunoblots): the dilutions of primary and secondary (HRP-conjugated) antibodies, incubation time with these antibodies, duration of washing steps, and abundance of target GPCR in the sample analyzed all contribute to the signal:noise ratio in an immunoassay. Consequently, all of these processes can be optimized. Initially, screen antibody dilutions against a fixed concentration of target antigen, as described in **Subheading 3.6., step 5**. For immunoblots, individual lanes can be excised from Ponceau S-stained nitrocellulose using a scalpel. Once the optimal dilution of antibody has been chosen, test this concentration against a range of secondary antibody concentrations (1:500, 1:1000, 1:2000, 1:5000).
   e. For very weak signals, increase the primary antibody concentration and incubation time; incubate overnight at 4°C, or decrease the duration of washing steps. Alternatively, try to increase the abundance of target GPCRs in the

sample of interest, for example, by immunoprecipitation; or purify the antibodies as described in **Subheading 3.8.**

f. For very strong signals with high backgrounds, decrease primary antibody concentration and incubation time, increase the duration of washing steps, or decrease the amount of target protein in the assay. If background signals persist, include Tween-20 in wash buffers, to a maximum of 0.5% (v/v).

References

1. Goding, J. W. (1987) *Monoclonal Antibodies: Principles and Practice*, 2nd edit., Academic, London.
2. Winter, G., Griffiths, A. D., Hawkins, R. E., and Hoogenboom, H. R. (1994) Making antibodies by phage display technology. *Annu. Rev. Immunol.* **12,** 433–455.
3. Higgins, D., Thompson, J., Gibson, T., Thompson, J. D., Higgins, D. G., and Gibson, T. J. (1994) CLUSTAL W: improving the sensitivity of progressive multiple sequence alignment through sequence weighting, position-specific gap penalties and weight matrix choice. *Nucleic Acids Res.* **22,** 4673–4680.
4. Jameson, B. A. and Wolf, H. (1988) The antigenic index: a novel algorithm for predicting antigenic determinants. *CABIOS* **4,** 181–186.
5. van Regenmortel, M. H. V., Briand, J. P., Muller, S., and Plaué, S. (1988) *Synthetic Peptides as Antigens*, Elsevier, New York.
6. Tam, J. P. (1988) Synthetic peptide vaccine design: synthesis and properties of a high-density multiple antigenic peptide system. *Proc. Natl. Acad. Sci. USA* **85,** 5409–5413.
7. Lorimer, I. A., Keppler-Hafkemeyer, A., Beers, R. A., Pegram, C. N., Bigner, D. D., and Pastan, I. (1996) Recombinant immunotoxins specific for a mutant epidermal growth factor receptor: targeting with a single chain antibody variable domain isolated by phage display. *Proc. Natl. Acad. Sci. USA* **93,** 14815–14820.
8. Dorsam, H., Rohrbach, P., Kurschner, T., et al. (1997) Antibodies to steroids from a small human naive IgM library. *FEBS. Lett.* **414,** 7–13.
9. van Kuppevelt, T. H., Dennissen, M. A., van Venrooij, W. J., Hoet, R. M., and Veerkamp, J. H. (1998) Generation and application of type-specific anti-heparan sulfate antibodies using phage display technology. Further evidence for heparan sulfate heterogeneity in the kidney. *J. Biol. Chem.* **273,** 12960–12966.
10. Devlin, C. M., Bowles, M. R., Gordon, R. B., and Pond, S. M. (1995) Production of a paraquat-specific murine single chain Fv fragment. *J. Biochem. (Tokyo)* **118,** 480–487.
11. Wick, B. and Groner, B. (1997) Evaluation of cell surface antigens as potential targets for recombinant tumor toxins. *Cancer Lett.* **118,** 161–172.
12. Nissim, A., Hoogenboom, H. R., Tomlinson, I. M., et al. (1994) Antibody fragments from a 'single pot' phage display library as immunochemical reagents. *EMBO J.* **13,** 692–698.
13. Svensson, H. G., Hoogenboom, H. R., and Sjobring, U. (1998) Protein LA, a novel hybrid protein with unique single-chain Fv antibody- and Fab-binding properties. *Eur. J. Biochem.* **258,** 890–896.

14. Smith, D. E. and Fisher, P. A. (1984) Identification, developmental regulation, and response to heat shock of two antigenically related forms of a major nuclear envelope protein in *Drosophila* embryos: application of an improved method for affinity purification of antibodies using polypeptides immobilized on nitrocellulose blots. *J. Cell Biol.* **99,** 20–28.

4

Immunocytochemical Identification of G-Protein-Coupled Receptor Expression and Localization

Laura C. Mongan and Blair D. Grubb

Summary

Immunocytochemistry exploits the incomparable specificity of the antibody–antigen interaction to form the basis of a flexible approach to the study of expression and localization of proteins both in model systems and their physiological context. This chapter details the theory and practice of the technique as well as lists the materials required. A general protocol is proposed, which can be adapted to suit the needs of individual investigators using the suggestions outlined in the **Notes**. The use of frozen tissue sections and cultured cells is described. Finally, the most common causes for failure of the technique are presented, along with likely solutions.

Key Words

Antibody, antigen, fluorescence microscopy, immunofluorescence, morphology.

1. Introduction

Immunocytochemical techniques make use of the unparalleled specificity of the antibody–antigen interaction. Recent advances in antibody technology have made antibody methods standard practice in both research and diagnostic settings.

The basic theory for immunocytochemical analysis was described more than 50 yr ago *(1)*. Immunocytochemistry requires recognition by an antibody of the target protein (antigen) of interest, then subsequent detection of the antibody–antigen complex. Detection can be achieved microscopically using either fluorescent markers or enzyme labels that produce an insoluble product deposited at the site of antibody–antigen interaction.

From: *Methods in Molecular Biology, vol. 259, Receptor Signal Transduction Protocols, 2nd ed.*
Edited by: G. B. Willars and R. A. J. Challiss © Humana Press Inc., Totowa, NJ

Methods have evolved in which either the primary antibody is labeled (direct method), or a labeled secondary antibody is used (indirect method). The direct method is relatively simple and involves fewer steps, whereas the indirect method is more involved and time consuming. The indirect method is favored by most researchers because it exhibits several clear advantages over the direct method. First, the effect of using a two-step protocol is to amplify the signal, resulting in greater sensitivity. Second, this method offers greater flexibility because a single labeled secondary antibody can be used to detect any number of primary antibodies of the species for which it is specific. Third, labeling of antibodies with fluorescent dyes or enzyme markers is costly and wasteful of primary antibodies that are available in finite supply.

Immunocytochemistry can be used to describe the expression and distribution of antigens under several conditions. It is possible to label antigens in cultured cells, cell smears, or cytocentrifuge preparations, as well as in tissue sections made from wax-embedded and frozen tissues. Effective processing of the starting material underlies successful immunocytochemistry. The material must be prepared such that the protein of interest is preserved and, importantly, retains its antigenicity, within material that retains structural integrity. Inadequate initial processing results in poor morphology and can contribute to high background staining. A set of conditions is developed that is appropriate for both the antigen and the tissue. For example, wax embedding of tissues results in excellent morphological preservation, yet utilizes high temperatures that are not especially favorable to the preservation of all antigens. Frozen sections provide a useful alternative. Tissue is rapidly frozen in a supporting compound and frozen sections are cut using a cryostat. Morphology is preserved and antigens survive the freezing process well.

Fixation can be achieved using a number of methods. The most commonly used for tissue sections is a low concentration of formaldehyde prepared fresh from paraformaldehyde polymer. Freshly prepared formaldehyde is also suitable for cultured cells, as are alcohols such as methanol or ethanol. The next stage in tissue preparation is permeabilization of the tissue, a process that allows the antibody intracellular access. This step can be omitted if it is desirable to study plasma-membrane localization of a G-protein-coupled receptor (GPCR) (*see* **Note 1**). The conditions outlined in **Subheading 3.4.** can be optimized to suit individual antibody–antigen combinations.

For most purposes, fluorescently labeled secondary antibodies are the reagent of choice. These produce labeling that is sensitive and offers excellent resolution of fine structure as compared to methods using enzyme-coupled secondary antibodies. In addition, it is possible to dual-label tissues and, using optical filters, to view the distribution of each antigen separately with minimal interference from the other.

Immunocytochemical protocols offer a flexible and adaptable strategy for many research applications. However, a note of caution is essential at this point regarding the manufacture or purchase and appropriate validation of antisera. The success of this technique is dependent solely on the availability of good quality, reliable, and credible antisera that have been systematically tested to prove antibody specificity. Strategies for the validation of antibodies are explained comprehensively in the preceding chapter, and a number of caveats are detailed (*see* **Note 2**).

Epifluorescence microscopy provides a serviceable, inexpensive, and accessible detection method. Although it is not without disadvantages, it is the method of choice for routine immunocytochemical investigations. Data can be acquired using conventional photography or digitally using either monochrome or color charge-coupled device (CCD) imaging. Immunofluorescence predated the introduction of confocal microscopy for biological applications by many decades. Currently, confocal microscopy is seen as a cure-all for problems associated with reduction of interference from out-of-focus fluorescence and (relatively) thick biological specimens. It also has the advantage of using a laser as a monochromatic light source, enhancing resolution, producing often spectacular images. Another advantage of confocal microscopy is that this configuration allows for collection of information from the *z*-plane, allowing 3D reconstruction.

Immunocytochemical localization of members of the GPCR superfamily presents inherent challenges. First, some investigations may depend on the discrimination between highly homologous receptors from within a subfamily. This is practically achievable, but strict antibody validation and inclusion of good controls is particularly important. In addition, the low expression levels of some GPCRs may be problematic, and although methods for signal amplification exist (*see* **Subheading 3.4.**), an alternative experimental approach may prove more productive.

It is difficult to produce a "definitive protocol," all the more so when the success of the application depends heavily on the variable properties of the antibody used. However, with the help of suggestions listed in the **Notes**, it is possible to mold a general protocol (*see* **Fig. 1**) into a method of use to a large number of investigators to produce high-quality, reliable data.

2. Materials

2.1. Specimen Preparation

1. Poly-L-lysine hydrobromide (Sigma-Aldrich, Poole, UK), dissolved to 0.1 mg/mL in dH_2O, then filter sterilized and aliquots stored at –20°C.
2. Molds (Polysciences Europe, e.g., "Peel Away" Square T12).
3. Tissue embedding compound: TissueTek® OCT™ Compound (Sakura Finetek).

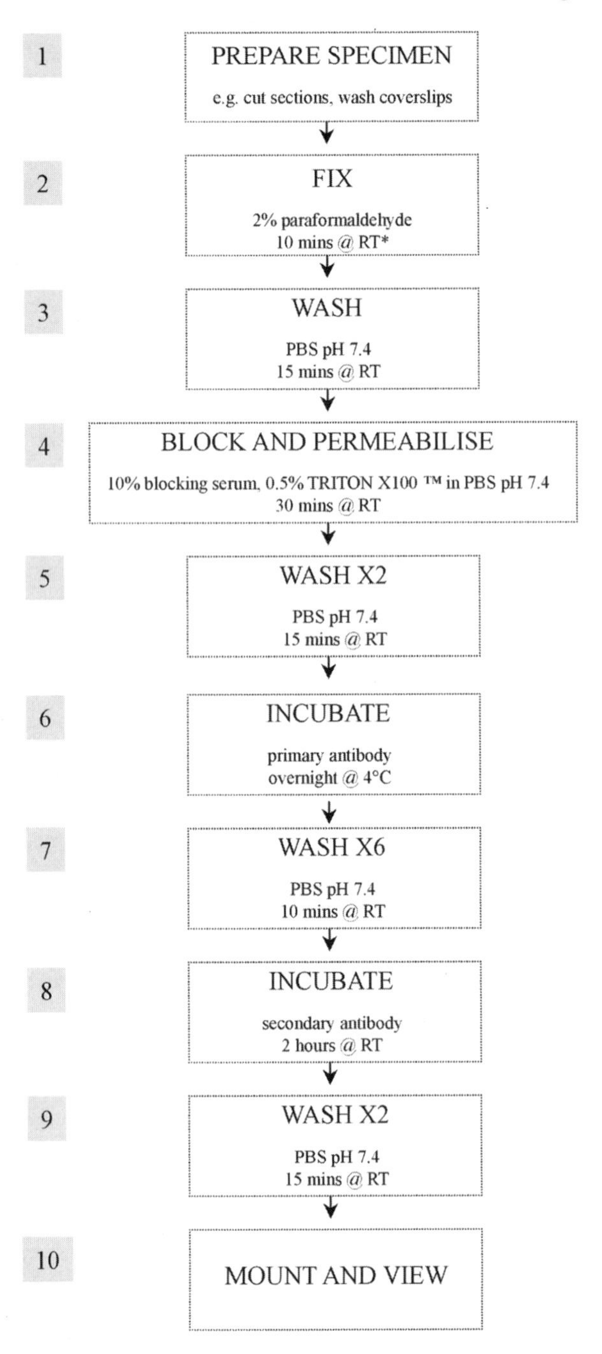

Fig. 1. Flow diagram describing the sequence of steps in a standard immunocytochemistry protocol. This general protocol can be used with or without modification to suit the distinct requirements of individual investigators. * RT, Room temperature.

4. Hexane (Fisher, UK) and dry ice (BOC, Guildford, UK).
5. Dewar flask.
6. ImmEdge™ pen (Vector Laboratories, Inc., Burlinghame, USA).
7. Cryostat (e.g., Bright Instruments, Huntington, UK, Model OTF).

2.2. Fixation

1. Slide racks and troughs (RA Lamb, Eastbourne, UK, or Fisher, UK).
2. Paraformaldehyde (Sigma-Aldrich, Poole, UK).
3. 0.2 *M* Na_2HPO_4.
4. 0.2 *M* NaH_2PO_4.
5. Methanol.

2.3. Permeabilization and Blocking

1. 10X Phosphate-buffered saline (PBS): 80.0g of NaCl, 11.5 g of Na_2HPO_4, 2 g of KH_2PO_4, 2 g of KCl per liter of dH_2O.
2. Blocking serum (*see* **Note 3**). Animal sera are available from several suppliers, for example, Sigma-Aldrich, or Jackson ImmunoResearch.
3. Triton X-100 (Sigma-Aldrich, Poole, UK).
4. Incubation box (RA Lamb, Eastbourne, UK).

2.4. Antibodies

Abcam (www.abcam.com), in addition to supplying antisera, provide a search engine that covers a wide range of antibody suppliers, and is a useful place to check the commercial availability of an antibody. Good quality conjugated secondary antibodies are available from a number of suppliers, for example, Jackson ImmunoResearch, Serotec, and Abcam.

2.5. Microscopy

1. Slides (e.g., RA Lamb, Eastbourne, UK).
2. Acetone (Fisher, UK).
3. Detergent (e.g., Decon 90®).
4. 3-Aminopropyltriethoxysilane (Sigma-Aldrich, Poole, UK).
5. Cover slips (RA Lamb, Eastbourne, UK).
6. Citiflour™ AF1 mountant (Citiflour Ltd, UK).
7. TSA™ Fluorescein System (PerkinElmer Life Sciences, e.g., NEL701).
8. Microscope fitted with an epifluorescence attachment.

3. Methods

3.1. Specimen Preparation

3.1.1. Cultured Mammalian Cells

The most accessible method for immunocytochemistry using cultured cells is to grow the cells on cover slips. Cover slips can be sterilized by flaming and can be treated with substrates such as poly-L-lysine to promote cell adherence if necessary. Apply enough poly-L-lysine solution to cover the surface to be treated, incubate at room temperature for 5 min, then aspirate, wash with sterile dH_2O and allow to dry before plating the cells. Treatment of cover slips in this way ensures adherence of cells throughout the extensive washing steps of the immunocytochemistry protocol (*see* **Note 4**).

3.1.2. Tissue Sections

The tissue of interest is quickly dissected, submerged in TissueTek OCT compound in a mold of appropriate size (*see* **Note 5**), and rapidly frozen over a dry ice and hexane slush (–73°C) in a Dewar flask.

Knowledge of the basic operation of a cryostat is assumed. Sections of 10–12 µm thickness produce optimum morphology and minimum background staining. Once the required number of sections has been melted onto the subbed slide (*see* **Subheading 3.5.1.** for preparation of subbed slides), the sections are circumscribed using an ImmEdge Pen, giving a hydrophobic perimeter that prevents runoff of reagents, and minimises the volumes required.

Careful specimen preparation (*see* **Note 6**) is crucial to successful immunocytochemical investigations, and efforts invested at this stage typically will be rewarded with a favorable outcome.

3.2. Fixation

3.2.1. Preparation of Fixative and PBS

To make 2% formaldehyde solution from paraformaldehyde polymer, add 8 g of paraformaldehyde powder to 160 mL of 0.2 *M* Na_2HPO_4 and dissolve by heating to approx 70°C until the solution becomes clear (≈20 min). Allow to cool to room temperature, then add 40 mL of 0.2 *M* NaH_2PO_4. Finally, add 200 mL of dH_2O and mix thoroughly.

3.2.2. Cultured Mammalian Cells

Cover slips are removed from the culture medium, and washed briefly and gently with PBS before incubation in fixative (*see* **Note 7**) for 10 min. The fixative is gently aspirated and the coverslips washed with PBS for 15 min.

3.2.3. Tissue Sections

Slides are placed in metal slide racks and submerged in a glass trough containing fixative for 10 min. The slides are then washed in PBS with gentle agitation (rocking platform) for 15 min. PBS should be filtered to remove any contaminating particulate material, and the pH is adjusted to 7.4 before use.

3.3. Permeabilization and Blocking

Excess PBS is carefully wiped away from the specimen (slide or cover slip). It is then placed in a humidified incubation box (*see* **Note 6**) and 0.5% Triton X-100 in 10% blocking serum (*see* **Note 3**) is applied for 30 min at room temperature. Following the incubation period, the specimen is washed twice in PBS for 15 min each, again with gentle agitation.

3.4. Antibodies

3.4.1. Titration

The dilution suitable for each tissue and antigen must be determined empirically by the user. Most commercially available antisera will be supplied with a datasheet giving information concerning its suitability for immunocytochemistry and a suggested dilution for use; note that both suitability and specificity must be verified by the user. It is advisable to perform a titration of the antibody in immunocytochemistry on the tissue of interest, using concentrations ranging around those suggested by the manufacturer. For antibodies that have been custom synthesized, the titer is determined by enzyme-linked immunosorbent assay (ELISA). A full description of the method for determining antibody titers is beyond the scope of this chapter, and details are available elsewhere *(2)*. In brief, an excess of the peptide used to immunize the animal is used to coat 96-well ELISA plates, then a range of dilutions of the antisera (usually 10^{-1}–10^{-7}) is applied.

For optimum results, all antisera (both primary and secondary) should be removed from appropriate storage (*see* **Note 8**) as near as possible to use, and diluted in 10% blocking serum in PBS. Incubation with primary antisera should be carried out at 4°C (*see* **Note 9**) overnight in an incubation box.

3.4.2. Dual Labeling

Dual labeling is simply achieved by incubating the sample concomitantly with two primary antibodies in the first instance, and then with two secondary antibodies. The two primary antibodies should be raised in different (*see* **Note 10**) species, combined with two secondary antibodies raised in the same species (*see* **Fig. 2**).

3.4.3. Washing and Detection

Following incubation with the primary antibody, thorough washing is imperative to remove nonspecifically bound antibody. A wash protocol of six changes of PBS for 10 min each after the primary antibody incubation gives good results. The preparation is now ready for incubation with the secondary (detection) antibody. As with primary antisera, the optimum concentration of all sec-

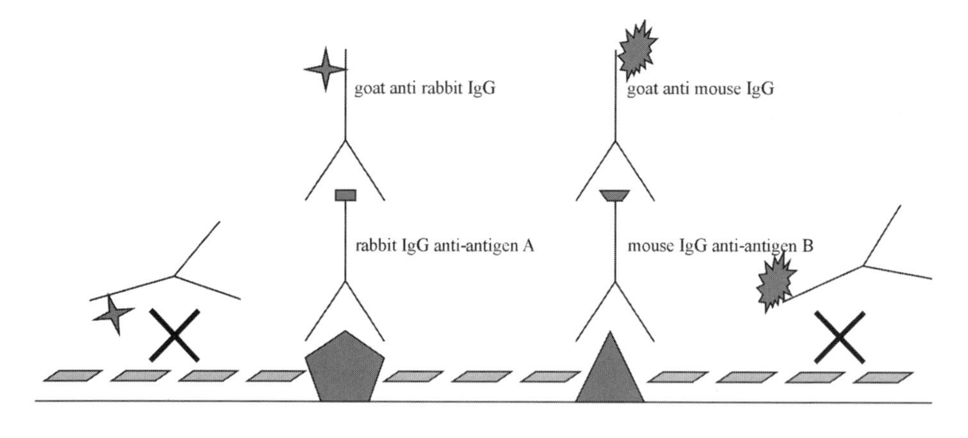

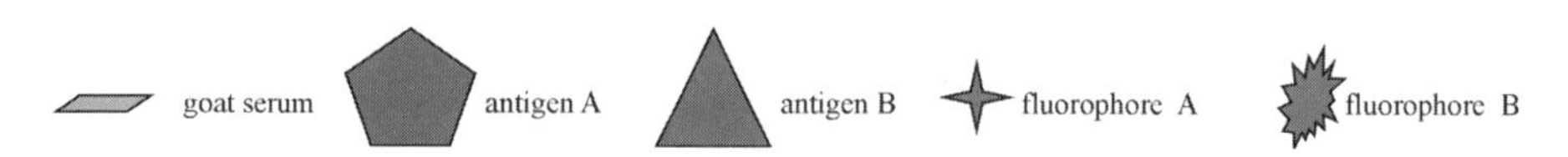

Fig. 2. Dual-labeling immunocytochemistry. To label two antigens within the same sample, two antisera raised in distinct species are required. Detection is achieved by using two secondary antisera, one specific for each of the two primary antisera. Both secondary antisera must be raised in the same species for the principles of the use of blocking serum to apply. For example, one of the primary antisera is a rabbit IgG raised against antigen A, and is detected by anti-rabbit IgG raised in goat that is labeled with fluorophore A. The other primary antiserum is a mouse IgG raised against antigen B, and is detected by anti-mouse IgG also raised in goat and labeled with fluorophore B. When nonspecific sites in the specimen are blocked by goat serum, then neither of the secondary antisera can crossreact with these sites.

ondary antisera must be determined by titration. Incubation with secondary antisera is carried out for 2 h at room temperature in an incubation box. The tissue preparation is then washed through two changes of PBS for 15 min each, and is then ready for mounting and viewing.

3.4.4. Signal Amplification Methods

There are a number of methods that are useful for the amplification of signals that are known or are suspected to be of low intensity (e.g., in the case of receptors expressed at low copy number such as some GPCRs). One such method is known as the Tyramide Signal Amplification (TSA)™ system. The protocol relies on the use of a horseradish peroxidase-labeled detection step, followed by incubation with a proprietary fluorophore tyramide amplification reagent. These reagents come in kit form, with full instructions for individual optimization.

Tyramide signal amplification has been used with success *(3)* to achieve dual labeling using a combination of primary antibodies raised in the same species. One antibody is used at a titer well below the optimum and this signal is amplified using the TSA method. The second antibody is then applied sequentially at its optimum titer, and detected conventionally using a secondary antibody labeled with a complementary fluorophore. The second detection step cannot detect the first antibody because the quantity present is insufficient to elicit a signal.

3.5. Microscopy

3.5.1. Preparation of Subbed Slides

To ensure that tissue sections remain firmly attached to the slide during the numerous washing stages of the immunocytochemistry protocol, it is necessary to prepare microscope slides by a process known as subbing. The racked slides are first soaked overnight in a 5% solution of Decon 90®, then washed under running hot water for 30 min, and dried for approx 30 min in a drying cabinet. The slides are then soaked for 1 min in a 2% solution of 3-aminopropyltriethoxysilane in acetone, followed by two 1-min washes in acetone, and finally two 1-min washes in distilled water, then thoroughly dried before use.

3.5.2. Mounting and Viewing

Glycerol-based mountants offer the best optical properties for cells or tissues prepared in aqueous solutions as in immunocytochemistry. These are not permanent mountants, although the coverslips can be sealed onto the slides using clear nail varnish applied carefully at the edges. Citiflour™ AF1 mountant contains antibleach and antifade reagents in glycerol–PBS solution, and is the reagent of choice in a number of laboratories where immunocytochemistry is carried out routinely. Excess PBS is blotted away from the preparation and a small quantity of the mountant is applied. A cover slip is then placed on the slide if sections have been processed, or if cells grown on cover slips have been labeled, the cover slip is inverted onto a slide. The preparations can then be viewed and documented (*see* **Notes 11** and **12** for methods used to manage high background staining and poor signal quality, respectively).

3.5.3. Image Analysis

A number of specialized software programs are available for image analysis, including Scion Image™ (*see* www.scioncorp.com), which is a modified version of the NIH Image software (*see* http://rsb.info.nih.gov/nih-image/download.html where PC and MacOS versions are available without charge). These can be used to measure parameters such as cell size, differential staining of certain cell types, or variations in fluorescence intensity across a tissue preparation.

4. Notes

1. Permeabilization is necessary to detect intracellular antigens. It is also worthwhile to note that permeabilization may be necessary even for molecules expressed at the cell surface, because the epitope used to generate the antibody may be intracellular. In other cases, permeabilization can be avoided if it is desirable to determine if a GPCR is expressed at the cell surface, providing the antibody has been raised against an extracellular epitope.
   Permeabilization also allows visualization of proteins at various stages of processing. Consequently, labeling of cell surface proteins in permeabilized specimens often results in the observation of a cytosolic labeling pattern consistent with detection of the antigen within, for example, the Golgi apparatus.
2. The inclusion of comprehensive controls underlies the validity of all immunocytochemical analysis. The techniques used in the validation of antisera have been covered in Chap. 3. Thorough validation is the obligatory first stage of immunocytochemistry (*see* **Fig. 3**), and a number of caveats are worth consideration. A given antiserum may recognize an antigen when overexpressed in a heterologous expression system, but may yield disappointing results when used in native systems. An antiserum that works only at a high titer in Western blotting (in which the antigen of interest is concentrated into a discrete band) may prove difficult to work with in tissues or cells in which the protein is expressed at normal levels. Once it has been established (by the individual investigator) that the antibody in question is specific for the antigen against which is was raised, immunocytochemistry can proceed.

   A number of controls for the immunocytochemical process itself must then be carried out. A preabsorption control is carried out by incubating the primary antibody with an excess (normally 10-fold by protein concentration) of the peptide against which was raised for approx 16 h at 4°C. This is then applied to the tissue preparation in parallel to antibody alone. Specific staining will be obliterated by preabsorption. In addition, a mismatched peptide should be used for "sham" preabsorption.

   A control for the secondary antibody must be carried out in every experiment. The primary antibody is omitted from the first incubation step, and all other processing is identical.

   Isotype controls can be performed by using a nonspecific immunoglobulin of the same isotype (or subclass in the case of monoclonal antibodies) used in parallel to the antibody of interest.

   Positive controls should be carried out wherever possible. A tissue known to express the antigen of interest can be used to verify the specificity of the antibody. Immunolocalization of an antigen known to be expressed in the tissue of interest will confirm the success of the method.

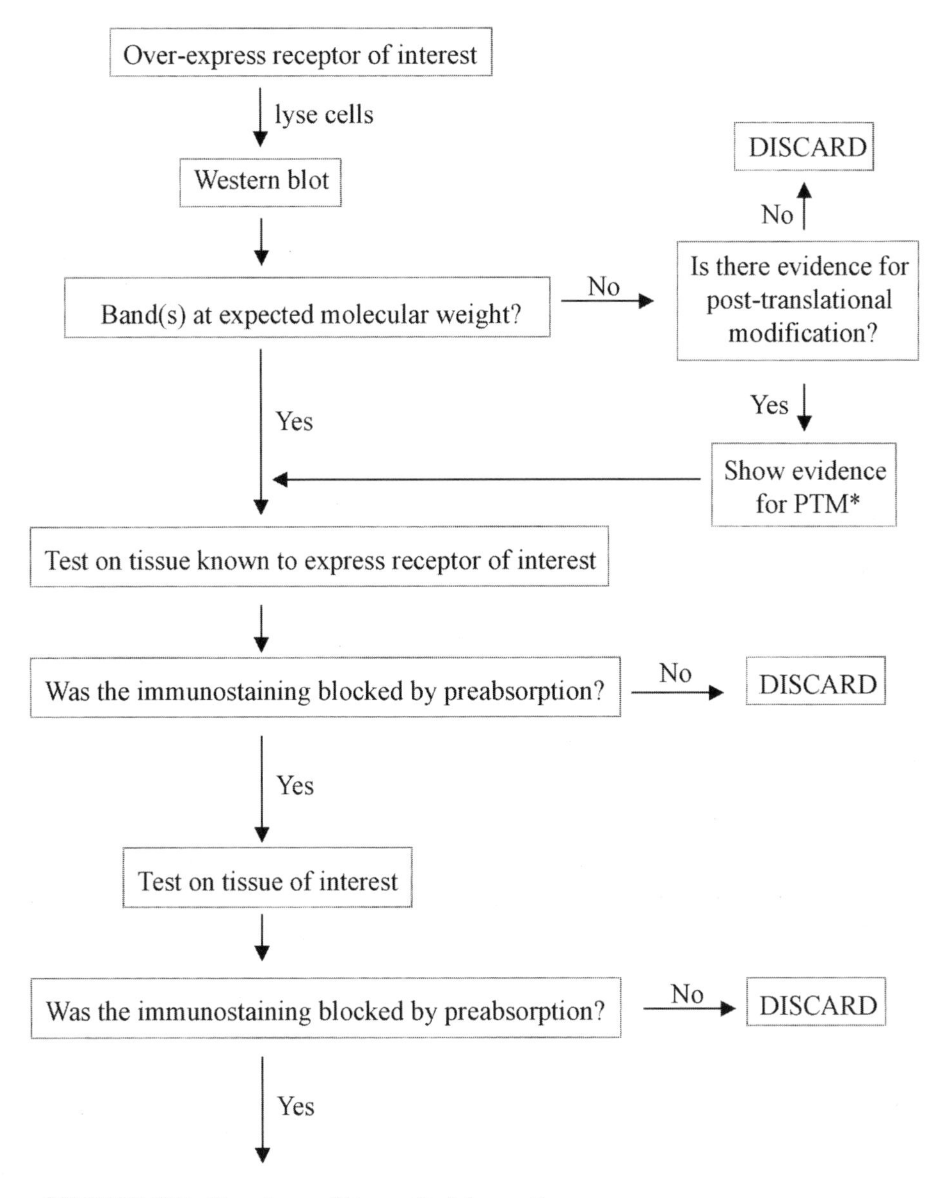

Fig. 3. Immunocytochemistry—tests and controls. This flow diagram depicts the necessary controls and evaluations required before the execution of an immunocytochemical investigation. The methods used to carry out these tests are described in the text and in the preceding chapter. Posttranslational modification (PTM) of GPCRs by glycosylation can be responsible for significant shifts in apparent molecular mass. Cells can be grown in the presence of tunicamycin to inhibit receptor glycosylation *(4,5)*. The reader is also referred to the **Notes** section, Chap. 3 for a discussion of the positive controls required in Western blotting assays, with reference in particular to PTM of GPCRs.

3. The purpose of the inclusion of blocking serum is to use an excess of nonspecific protein to block nonspecific binding sites that exist within all tissue preparations. Using a key principle of immunology, that is, self cannot recognize self, the choice of blocking serum is by necessity nonimmune serum collected from the species in which the detection (secondary) antibody was raised. For example, a secondary antibody raised in goat against rabbit IgG will bind to all rabbit IgG in the tissue, but not the goat serum proteins occupying sites that might otherwise have nonspecifically bound either the primary or secondary antisera.
4. Other variations such as the use of chamber slides (available from Nunc, Lab-Tek™ Chamber Slide™ System) exist that are useful for the simultaneous processing of a number of samples. For most applications however, it is sufficient to use sterilized cover slips that are placed in multiwell plates (one per well) or Petri dishes (several per dish). Raising the cover slip by placing it on a platform such as the lid of an Eppendorf tube facilitates the gentle application and removal of reagents and also prevents runoff of reagents.
5. The bulbs of plastic Pasteur pipets are very useful as freezing molds for small tissue samples such as peripheral ganglia for which standard molds may be too large. Frozen tissue can be stored indefinitely at –20°C or –80°C, although in practice for certain antigens, it may prove necessary to use recently frozen tissues for optimum results.
6. Ensure that the tissue is dissected as quickly as possible. This is especially important for neural tissues that deteriorate very rapidly postmortem.

   Ensure that the knife used for sectioning is sharp and without nicks. It is good practice to designate one end of the blade for trimming and coarse sectioning and the center of the blade for cutting sections for processing. Good upkeep of the knife is essential and it should be examined and sharpened regularly.

   Tissue preparations must NEVER be allowed to dry out, hence the need for incubation boxes to maintain a humid environment. Use of an ImmEdge pen ensures that reagents are not drawn off the slide by capillary action, and is particularly useful during the permeabilization step (lack of surface tension due to the presence of detergent).

   If tissues are to be taken from experimental animals, it is possible to fix the tissues by perfuse-fixation using 2% paraformaldehyde at normal blood pressure. This eliminates any potential postmortem deterioration and makes the tissue more rigid; consequently, the quality of the sections can be improved.
7. If it seems possible that either the preparation or the antibody–antigen combination under investigation is not suited to fixation using formaldehyde, a number of alternatives exist. For example, cells grown on cover slips can be fixed by incubation for 5 min in either methanol, or a 50% methanol–50% acetone mixture. Some investigators advocate the use of such solvent fixatives at low temperatures (4°C).
8. For long-term storage, antibodies should be stored in working aliquots (5–10 μL) at –80°C. Repeated free–thaw cycles should be avoided, therefore the use of frost-free freezers is inadvisable.

9. By lowering the temperature at which antibody–antigen binding occurs, the rate at which equilibrium is achieved is decreased. This results in an increase in specific binding, and a decrease in nonspecific binding. Conversely, by increasing the temperature (up to 37°C) the faster this equilibrium is reached, but with a concomitant increase in nonspecific binding. It is possible to carry out immunocytochemistry in one working day by carrying out both antibody incubations for 1 h at 37°C. In practice, the times and temperatures of antibody incubation can be varied to optimize individual requirements. Low temperatures are preferred, in particular for primary antibody incubations. The short incubation time variant works well with high-affinity antibodies.
10. When carrying out dual labeling, avoid combinations of primary antibodies raised in closely related species (e.g., rat and mouse). Secondary antibodies may not discriminate particularly well between them at the detection step.
11. To deal with high background:
    a. Washing: the importance of washing steps cannot be overemphasized. Strict adherence to washing protocols is essential for reproducibility as well as reducing background staining. For some tissues and/or antibodies, it may be necessary to increase the washing by extending the duration of the washes, and increasing the number of washes. Overwashing is unlikely to be problematic.
    b. Adequate fixation: deterioration in the quality of cells or tissues can significantly increase nonspecific binding.
    c. Titer of primary or secondary antibody may need to be adjusted. Check the secondary antibody control to determine which is contributing to the increase in background.
    d. Tissues that naturally express receptors for the Fc portion of immunoglobulin present a unique problem. This phenomenon is easily overcome by using $F(ab)_2$ fragments (i.e., IgG with Fc portion enzymatically removed) of the antibody of interest.
    e. A characteristic speckled pattern of "stars in night" is indicative of free fluorophore aggregates. These can sometimes be removed by centrifuging the secondary antibody at high speed in a benchtop centrifuge for 5 min before use.
    f. Autofluorescence derived from certain fixatives (e.g., some aldehyde fixatives such as glutaraldehyde) can occasionally give rise to high background fluorescence. Alternatively, some tissues are inherently autofluorescent (e.g., sympathetic ganglia).
12. To deal with no or low signal:
    a. Recheck the antibody titer. Titres suggested in manufacturers' information are intended as a guide only, and may vary significantly from that suitable for end-user conditions.
    b. If it is suspected that the signal is low due to low expression of the antigen within the tissue, it is possible to amplify the signal. A simple method is to introduce another step into the standard two-step immunocytochemistry protocol. Instead of using detectable label on the secondary antibody, this antibody can

be labeled with biotin. Biotin is then detected using enzyme- or fluorophore-labeled avidin.

c. Ensure that the appropriate secondary antibody was used.
d. Is the microscope filter set in use appropriate for the fluorophore?
e. Ensure that the preparations do not dry out.

References

1. Coons, A. H. and Kaplan, M. H. (1950) Localisation of antigen in tissue cells. II. Improvements in a method for detection of antigen by means of a fluorescent antibody. *J. Exp. Med.* **91,** 1–13.
2. Harlow, E. and Lane, D. (1988) Immunoassay, in *Antibodies, A Laboratory Manual*, Cold Spring Harbor Laboratory Press, Cold Spring Harbor, NY, pp. 553–612.
3. Amaya, F., Decosterd, I., Samad, T. A., et al. (2000) Diversity of expression of the sensory neuron-specific TTX-resistant voltage-gated sodium ion Channels SNS and SNS2. *Mol. Cell. Neurosci.* **15,** 331–342.
4. Weiss, H. M., Haase, W., Michel, H., and Reilander, H. (1998) Comparative biochemical and pharmacological characterization of the mouse $5HT_{5A}$ 5-hydroxytryptamine receptor and the human β_2-adrenergic receptor produced in the methylotrophic yeast *Pichia pastoris*. *Biochem. J.* **330,** 1137–1147.
5. Boer, U., Neuschafer-Rube, F., Moller, U., and Puschel, G. P. (2000) Requirement of *N*-glycosylation of the prostaglandin E_2 receptor EP3β for correct sorting to the plasma membrane but not for correct folding. *Biochem. J.* **350,** 839–847.

5

Generation and Use of Epitope-Tagged Receptors

R. A. Jeffrey McIlhinney

Summary

Epitope tagging of a receptor involves introducing a defined amino acid sequence, to which an antibody has already been produced, into the primary amino acid sequence of the receptor. The new sequence can be as short as 10–15 amino acids and the method allows the receptor to be monitored without having to raise an antibody specific to it, and so permits its biochemical characterization and immunolocalization within cells. Other related techniques involve the introduction of functional domains of proteins such as green fluorescence protein or one of its derivatives into receptors to allow their direct visualization. There are a wide range of amino acid tag sequences, and fluorescent proteins, available for use as tags, and choice of tag will depend on several factors including the availability of antisera and cost. The position for the introduction of the tag into the native receptor will depend on precisely what the requirements of the experiments are but it must always be such that the receptor is normally processed, trafficked, and remains functional after tagging. These considerations are discussed fully in this chapter, which also describes examples of strategies for introducing tags, and some general methods for use in the characterization of the tagged proteins

Key Words

Antibodies, dimerization, epitope, glutamate, glycosylation, immunofluorescence, membranes, metabotropic, polymerase chain reaction, primers, receptors, sodium dodecyl sulfate-polyacrylamide gel electrophoresis, tag, trafficking, transfection.

1. Introduction

What is an epitope tag? An epitope is a region of a molecule, such as a carbohydrate side chain, a fold in the structure, or a specific amino acid sequence, that can stimulate the production of antibodies. Epitope tags are short stretches of peptide that remain immunoreactive even when they are introduced into other proteins. These sequences are mostly derived from antigenic regions of particular proteins (e.g., the hemagglutinin protein or the *myc* oncogene),

From: *Methods in Molecular Biology, vol. 259, Receptor Signal Transduction Protocols, 2nd ed.*
Edited by: G. B. Willars and R. A. J. Challiss © Humana Press Inc., Totowa, NJ

Table 1
Commonly Used Epitope Tags

Tag	Sequence	Origin
FLAG tag	DYKDDDDK	Novel
HA tag	YPYDVPDYA	Flu hemagglutinin protein
Myc tag	EQKLISEEDL	c-myc protein
VSV G tag	YTDIEMNRLGK	VSV protein
Glu-Glu tag	EEEEYMPME	Polyoma middleT antigen
His tag	HHHHHH	Novel

The tag sequences are given using the single amino acid abbreviations. Monoclonal and commercial antibodies are available to all of these tags. A cell line secreting monoclonal antibodies to the Myc tag (antibody 9E10) is available from cell culture collections. The FLAG tag, however, has three monoclonal antibodies to it: M1, M2, and M5. These recognize the epitope in different positions in the protein. Thus M1 recognizes the tag sequence only when it is at the N-terminus of the protein and needs Ca^{2+} to bind; M2 recognizes the tag when it is within the protein sequence or at the C-terminus and is not Ca^{2+} dependent, and M5 recognizes the epitope when it is N-terminal but still has the initiator methionine in place. The histidine tag is most often used for purification of protein using metal chelate matrices.

although some are designed (e.g., FLAG). A wide range of epitope tags are currently available, of which some of the most commonly used sequences are shown in **Table 1**. Epitope tagging involves the introduction, by the application of appropriate molecular biology techniques, of one or more of these sequences into another protein. Commercially available monoclonal or polyclonal antibodies are available for use with most of these sequences.

The choice of a tag is governed principally by the nature of the project to be undertaken and will be dictated by past experience, availability and cost of reagents, and the systems being used to express the receptor of interest. Unless other constraints apply this means that one of the three most commonly used systems will be chosen, namely the HA, Myc, or FLAG tags. Because studies involving receptors generally use epitope tagging to monitor receptor assembly, trafficking, and interaction with other proteins, the tag chosen should not give spurious reactions with proteins present in the cell systems being used for these studies. Preliminary immunoblotting and immunofluorescence studies on the target cells, using the antibodies to different tags at different concentrations, can be a useful first screen in helping to determine which tag may be optimal for a particular study. Clearly the tag with the antibody giving either no, or the lowest, reactivity with the cells, or cell extracts, would be the best choice.

Although epitope tagging generally involves the use of short stretches of sequence to label the protein of interest, more recently fusion of larger molecules to receptor subunit domains have been used to produce chimeric proteins. In these the N-terminal domains of a transmembrane acceptor protein is fused with a receptor domain, usually C-terminal, to examine its targeting sequences. The acceptor proteins used have generally been single transmembrane domain proteins chosen for the availability of high-affinity and highly specific antibodies to their extracellular regions (e.g., telencephalin *[1]*, Tac *[2]*, and CD2 *[3]*). This approach has the advantage that it permits the function of the introduced domain to be examined in the absence of any effects of the rest of the receptor. It has proved especially valuable in monitoring receptor trafficking of a number of receptors as shown for mGluR7, the *N*-methyl-D-aspartate (NMDA) receptor subunit NR1, and mGluR1b *(1–3)*.

The concept of using chimeric proteins to study receptors has reached its most elegant development with the use of the green fluorescent protein (GFP) from the jellyfish *Aequorea victoria* and its derivatives to study the trafficking and cell localization of receptors. The protein contains 238 amino acids, as compared to 9–15 for a conventional epitope tag, and emits green, or in the case of its derivatives, other wavelengths of light, when excited by ultraviolet (UV) light. The fact that the tagged protein containing GFP can be monitored in real time allows the behavior of protein to be studied in living cells. This can only be done using this system, and the availability of different colored variants of the protein means that double-labeling studies can be performed, as can fluorescence resonance energy transfer (FRET) experiments to monitor protein–protein interactions (reviewed in **ref. *4*** and *see* Chapter 21). A range of commercially available vectors containing the genes for GFP and antibodies to the protein are now available. Indeed by combining the fluorescence emitted by GFP and its derivatives with the ability of the enzyme luciferase to emit light when provided with a substrate, novel systems to study the interactions of proteins in real time using bioluminescence resonance energy transfer (BRET) have been developed (reviewed in **ref. *5*** and *see* Chapter 20). Because both FRET and BRET involve the insertion of protein domains rather than short peptide sequences into proteins, they may not be regarded strictly as epitope tagging. They do, however, provide the only means available to monitor the kinetic changes between protein interactions in living cells, something that has led to their rapid adoption in the biosciences. In this chapter, however, I concentrate on the introduction of short epitope tags into proteins, noting that the same considerations of why tag, where should one tag, and what characterization of the tagged protein should one do are the same for these and their larger cousins. The perils of larger tags are illustrated with an example of the GFP-tagging of mGluR1a.

1.1. Why Tag Proteins?

It might at first sight seem strange to go to all the trouble of introducing an additional amino acid sequence into a protein, especially if one has just spent some time identifying and cloning the gene of interest. There are circumstances, however, in which it can be an advantage. First, not all proteins are immunogenic, and producing a specific antibody can be difficult and takes time. The ability to produce, relatively quickly, a tagged version of a protein, which will be recognized by commercially available antibodies, means that preliminary experiments can be done to characterize both the protein's behavior and its place in the cell. In addition, tagged proteins can be expressed in cells that normally contain the protein, and the behavior of the native protein inferred by monitoring that of the tagged form. Furthermore, the homo- and hetero-oligomerization of proteins can be studied by utilizing differently tagged versions of the proteins, and this has had considerable utility in the study of G-protein-coupled receptor (GPCR) dimerization (e.g., *[6]* and reviews *[7,8]*).

The ability to put a tag in different positions within the sequence of a membrane protein can also be useful because it allows the topology of the protein to be studied as illustrated in **Fig. 1**. By following the surface expression of the tag by immunofluorescence in intact cells, or monitoring its protease sensitivity, the accessibility of the tag at the cell surface can be determined. A similar approach can also be used using in vitro translation of the tagged protein in the presence of microsomes. Because extracellular epitopes will be in the lumen of the microsomes they will be protected from proteases added to the in vitro translation reaction. Therefore analysis of the protease-treated translated tagged products can be used to deduce which regions of the protein are extracellular. These methodologies have been useful in elucidating the transmembrane topology of the AMPA receptors *(9,10)*, other ion channels *(11)*, and some GPCRs *(12)*.

1.2. Where to Place the Tag

The first decision that one has to make is where to put the tag. In soluble proteins this is less of a problem because both the N- and C-termini are available for tagging. N-terminal tags offer the advantage that for most receptor subunits this region of the molecule is extracellular, and therefore imunofluorescence for the tag allows surface expression of the receptor to be monitored. Many receptor subunits, however, have signal peptides, subsequently removed, that are essential for their correct membrane insertion. This means that N-terminal tags need to be situated after the signal peptide, or within the N-terminal domain of the protein. As the precise end of the signal peptide can be estimated only from the protein sequence the optimum position for introducing the tag may not be easy to determine. It is worth noting that the

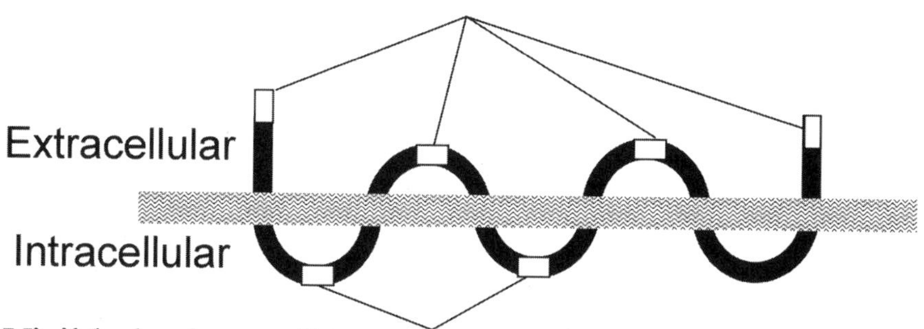

Fig. 1. Use of epitope tags to determine receptor topology. By placing the tag in the positions indicated and performing immunofluorescence and/or whole cell proteolysis on the expressed proteins the accessibility of the extracellular domains can be determined. Because access to the tags can be restricted in whole cell experiments, this methodology can be combined with in vitro translation of the tagged receptor in the presence of microsomes and proteolytic digestion of the products used to confirm the orientation of the protein.

metabotropic glutamate receptor mGluR1 has an unusually long signal peptide ***(13,14)***. These factors complicate the process of N-terminal tagging and it may be necessary to try a number of positions before finding a tolerated site for the insertion of an N-terminal tag. Additional factors to be considered and investigated are the position of the ligand binding site, the predicted structure of the N-terminus that might reveal any regions of relatively unfolded structure into which the tag sequence might be inserted, or potential domain structures between which the tag could be placed. Thus the siting of the tag at the N-terminus necessitates a bioinformatics-based analysis of the primary structure of the receptor subunit under study, together with the use of the available existing information as to the structure of related proteins, to decide where to put the tag.

C-terminal tagging is relatively simple as this site is readily accessible in most proteins, and for many receptors, removal or replacement of part of the C-terminus has little impact on function. Consequently this would be my first choice for the insertion of a tag into a novel receptor subunit. It should, however, be noted that while C-terminal tags have been widely used to monitor receptor behavior, this region of the molecule can contain phosphorylation sites that are important for the regulation of receptor function and elements necessary for receptor desensitization. In addition, the distal C-terminus of some receptors is responsible for their interaction with intracellular proteins ***(15–17)***. These interactions have been implicated in the trafficking and targeting ***(18,19)*** of these receptors and therefore adding tags to this region of the protein may have unforeseen consequences. All of these considerations apply especially forcefully to the positioning of the large GFP-related proteins, widely used to monitor receptor trafficking in living cells.

1.3. Methods for Tag Insertion

The introduction of an epitope tag requires the application of general molecular biological methods, for example, polymerase chain reaction (PCR) primer design, cloning, DNA purification, ligation, and so forth, that will not be dealt with in any detail here as they are covered in detail in the many cloning manuals currently available. Equally it is not possible to give specific protocols for the introduction of a tag into a generalized receptor, as the design of the primers, the possible restriction enzyme sites available for use in the ligation of the tag, and so forth, will be specific for each protein and expression vector, and therefore the basic strategies only will be outlined. In the methods considered it is assumed that the protein of interest has been cloned into a vector of some kind, if not the final expression vector. There are three methods available to tag receptors: use existing expression vectors containing tags, introduce the tag by PCR, or use an adaptor-duplex to put the tag in place.

1.3.1. Using Commercial Vectors

The commercial mammalian expression vectors usually contain a multiple cloning site followed by the tag sequence followed by a stop codon. These vectors are usually available for use in all three reading frames so that often it is necessary only to subclone the receptor into the appropriate vector to derive the tagged product. The ability to do this depends on the presence of a suitable restriction enzyme site lying within the C-terminal coding region to permit the subcloning. Note that if a suitable restriction site within the coding sequence is available, its use necessarily means that some of the C-terminus will be truncated (*see* caveats in **Subheading 1.2.**). If a suitable C-terminal restriction site for cloning is not present in the gene sequence, then this could be introduced by PCR or site-directed mutagenesis, although a different PCR-based strategy might be better employed under these conditions (*see* **Subheading 3.**). Similar vectors are also available to introduce N-terminal tags, but these have limited utility for receptor subunits as these have signal peptides as discussed in **Subheading 1.2.** Some of the commercial vectors are also multiply tagged, containing, for example, a His-tag for purification purposes together with another tag sequence for protein identification. They may also contain selective proteolytic cleavage sites to allow removal of one or both tags. Vectors containing tags are also available for use in yeast, bacteria, and insect cells.

1.3.2. Inserting Tags by PCR

This is without doubt the most adaptable method for tag insertion, as it can be used to introduce tags into many different regions of a protein. In this and later discussions the primer for the 5′-end of the gene will be termed the forward primer, while the primer used for the 3′-end will be called the reverse primer. Using PCR to introduce a tag into a protein involves making a pair of forward and reverse primers, one of which contains a portion of the relevant coding sequence of the receptor subunit (*see* **Note 1**) together with the nucleotides containing the tag sequence. It should also contain a sequence coding for a unique restriction enzyme site as illustrated for the introduction of a C-terminal HA-tag (**Fig. 2**). In the case of the C-terminal tag the reverse primer contains the additional oligonucleotides coding for the tag sequence, whereas in the case of an N-terminal tag these would be contained in the forward primer. The other primer does not contain the tag sequence and its position is dictated by the position of the other unique restriction site to be used for cloning. In designing the primers care should be taken to ensure that the insert containing the tag is in the correct reading frame. Suitable cloning sites may be contained in the vector as illustrated in **Fig. 2**. The introduction of the tag is achieved by using the PCR reaction to amplify the region of interest, isolating the amplified DNA, digesting it with appropriate restriction enzymes, and then inserting the amplified product, by ligation, into the compatible digested vector. This method can be used to insert both N- and C-terminal tags, or to insert a tag within the coding sequence of the gene of interest; however, in the latter cases the approach may need to be modified as described in the example in **Subheading 3.** With smaller receptors (up to 1–1.5 kb cDNA) PCR could be used to amplify the entire protein to introduce the tag. However, because of the possibility of introducing errors in the sequence during PCR, for larger receptors the approach may have to be modified and only a region of the protein amplified during the introduction of the tag as illustrated in **Subheading 3.** Errors in PCR can be minimized by using one of the proofreading polymerases available commercially (e.g., Pfu or Pfx). In the event of introducing a C-terminal tag the new sequence must contain a stop codon after the tag sequence to replace that removed during the process (as illustrated in **Fig. 2**).

1.3.3. Using Adaptor-Duplexes to Introduce a Tag

The introduction of a tag using an adaptor-duplex requires that the coding sequence of the protein contain a suitably positioned restriction site or pair of

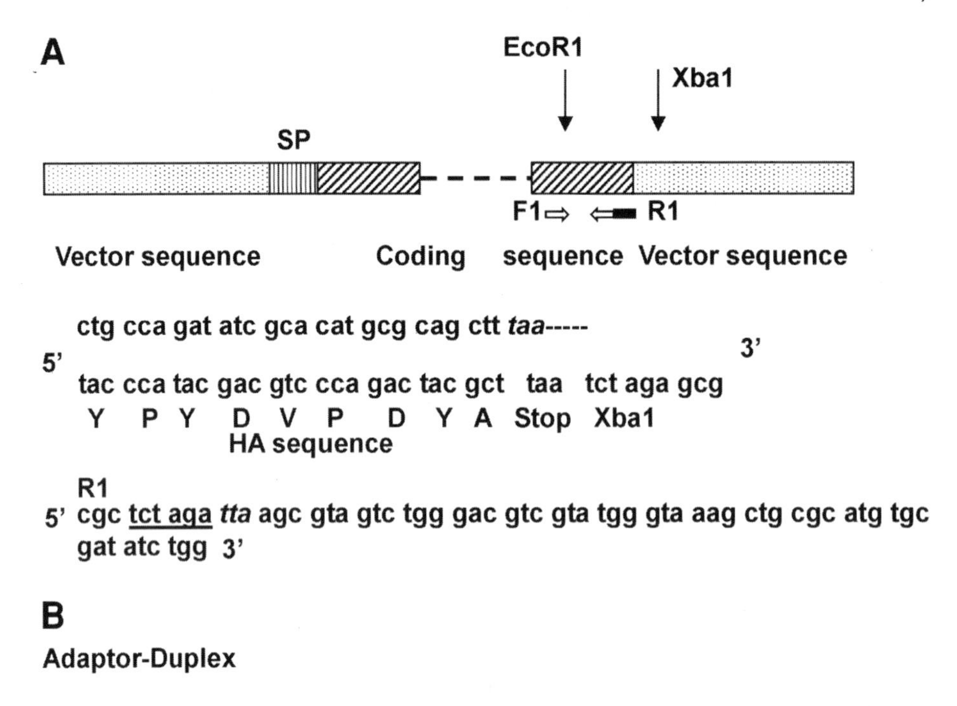

Fig. 2. Schematic illustration of C-terminal tagging. In the example shown, the receptor sequence has been cloned into an expression vector. The sequence has an *Eco*R1 site close to the C-terminus and the vector an *Xba*1 site that can be used for the cloning. The vector sequence is indicated by *stipling*, the coding sequence by the *crosshatching*, and the signal peptide (SP) indicated by the *vertical hatching* (**A**). The sequence close to the C-terminus including the stop codon is shown in *italics*. Below this is the HA-tag primer sequence with the new stop codon indicated together with the *Xba*1 site to allow cloning (*see* **Note 1**). The additional bases at the 3′-end of this sequence are present to allow efficient digestion of the amplified product (*see* **Note 2**).

Replacing the termination codon in the receptor sequence with the first triplet of the HA sequence and the additional bases illustrated gives the necessary primer (R1) to achieve the introduction of the tag, indicated by the *black bar* on the *arrow* labeled R1. As this is a 3′ or reverse primer the sequence to be synthesized is shown as R1 sequence, with the termination codon and *Xba*1 site *underlined*. The forward primer for the PCR (F1) can be any sequence 5′ to this region, which includes the *Eco*RI site and fulfills the necessary criteria for successful PCR (*see* **Note 1**). If the *Xba*1 and *Eco*RI site had been close to the termination codon as illustrated in **B** (SEQ . . .) where the restriction sites are sites close to the site for insertion of the tag (*see* **Fig. 2B**). These could be introduced by mutagenesis if needed and in the case of a C-terminal tag it might be possible to make use of 3′-restriction sites in the cloning vector as a source of one of the sites for this purpose. In this case a stop codon should be incorporated into the adaptor sequence. Then, after the adaptor sequence has been designed, both oligonucleotides are synthesized, annealed, and the duplex ligated into the appropriately digested vector. Although it is possible to use a single unique restriction enzyme site to incorporate the tag sequence, this can result in the incorporation of several tag sequences, or in the inserted sequence having the wrong orientation. This means that several clones may have to be sequenced and checked to derive a properly orientated and single tag insertion.

1.4. Introducing an N-Terminal FLAG Tag into mGluR1a

To illustrate how these basic protocols can be adjusted to introduce a tag into a receptor, a specific example is described. To be able to monitor the behavior of rat mGluR1a in neuronal cultures, which might contain endogenous receptors, a tagged version of the receptor was needed. Because it would be necessary to monitor surface fluorescence the tag had to be at the N-terminus, and as the FLAG epitope was already in use in the laboratory this was chosen as the tag. The cDNA coding for mGluR1a was already contained in the mammalian expression vector pcDNA3™ (Invitrogen) and this was used for the subsequent manipulations (*see* **Fig. 3A**). The *mGluR1a* gene had been cloned into the vector just after a *Kpn*1 site. Because mGluR1 is coded for by more than 2700 bp it would not be sensible to try and amplify the whole cDNA sequence, but the presence of a *Kpn*1 site close to the 5′-end of the cDNA within the coding sequence meant that the tag could be introduced by amplifying only a short stretch (about 500 bp) of the sequence containing the N-terminus and the final product cloned into the vector again. Given that mGluR1 has a long signal peptide the tag would have to be introduced after residue 32, and it was decided to place the tag between resides 57–58 of the sequence, both prolines, in the hope that this would be a relatively unfolded region of the molecule and not interfere with folding of the important N-terminal ligand binding domain. A modified PCR strategy was used as shown in **Fig. 3**. As illustrated, two pairs of primers were used, one pair (F1

Fig. 2. *(continued)* indicated in *italics* and the stop codon is *underlined*, then two oligonucleotides with the sequences illustrated below this could be synthesized and annealed to provide an adaptor-duplex that would introduce the tag sequence. These have been synthesized with the cut ends of the restriction sites in place so that the annealed fragment can be directly inserted into the vector following digestion with *Eco*R1 and *Xba*1.

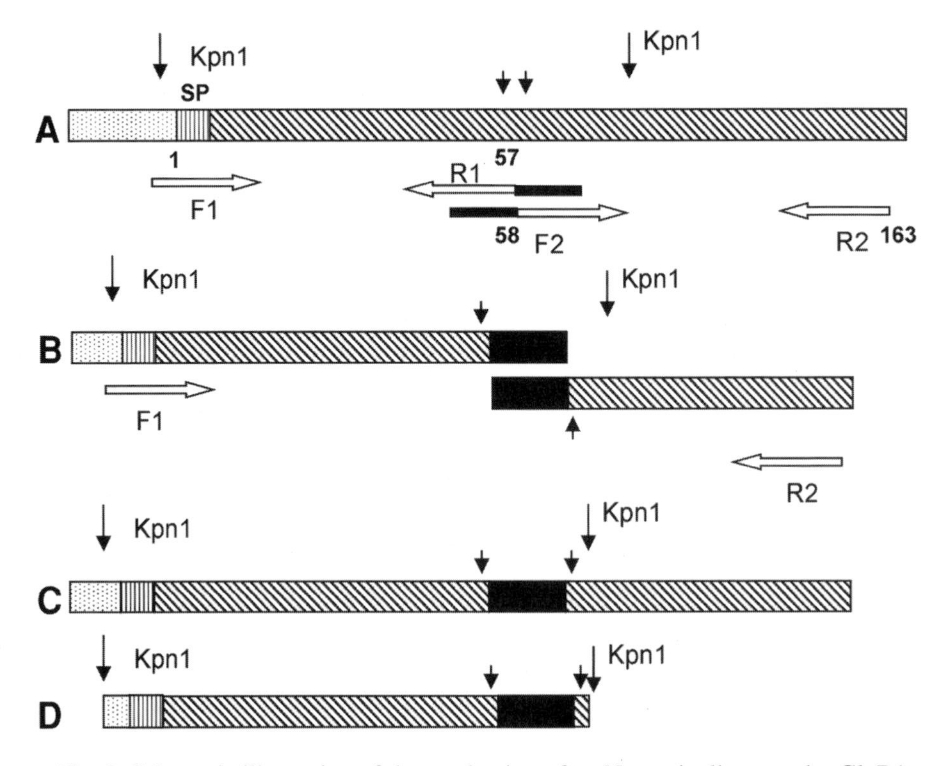

Fig. 3. Schematic illustration of the production of an N-terminally tagged mGluR1a. A portion of the N-terminus of mGluR1a cloned into pcDNA3 is shown with the *Kpn*1 sites used in the cloning indicated by the *large arrows*. The signal peptide is indicated by the vertical shading, vector sequence by *stippling*, gene sequence by the *cross-hatching*, and tag sequence by the *solid black bars*. The residues between which the tag was to be introduced are indicated by the *small arrows* **(A)**. For further details see text.

and R1) amplifying the vector sequence containing the *Kpn*1 site and the N-terminus of the mGluR sequence up to amino acid residue 57. The reverse primer of this pair (R1) was extended at the 3′-end with additional sequence coding for the FLAG epitope. The second pair of primers (F2 and R2) amplified from residue 58–163 with the forward primer (F2) extended at the 5′-end with residues encoding the FLAG epitope. Both pairs of primers were used to amplify their respective sequences in separate reactions and the amplicons purified. These were then mixed and used as the substrate for a last round of PCR with the forward primer from the first primer pair and the backward primer from the second pair (F1 and R2). The complementary nature of the FLAG sequence in both amplicons meant that they anneal as shown in **Fig. 3B**, and amplification produces the long PCR product containing the FLAG epitope inserted between residues 57 and 58 (**Fig. 3C**). This is then purified, digested with *Kpn*1 to yield the insert shown in **Fig. 3D**, and this ligated into the *Kpn*1-digested vector containing mGluR1. Because of the use of a single enzyme site, as noted above, this could lead to multiple inserts or inverted inserts so the clones of the final product were sequenced to find one with a single and correctly orientated epitope tag. It was fortuitous that the same two *Kpn*1 sites were present in the expression vector containing the splice variant of mGluR1a, mGluR1b, and therefore producing a FLAG-tagged version of this was a simple matter of subcloning the *Kpn*1 fragment into this vector also.

1.5. Characterization of the Tagged Receptor

Having produced the clone containing the tagged receptor it is essential to check the sequence to ensure that no mutations to the wild-type sequence have been introduced by the process. If short sequences are used in the PCR for the tagging and these are cloned into the original sequence then only these need be checked. If the sequence is satisfactory then check whether the protein is expressed by transient expression and immunoblotting for the receptor. This should be done using the antitag antibody but it may be useful to also check using an antireceptor antibody, if one is available, in case the tag cannot be detected. It is possible for adjacent residues to modify the detection of a specific tag. Because in the course of these experiments membranes from transfected cells will have been made (*see* **Subheading 3.1.**) it is convenient at the same time to determine if the receptor subunit is fully glycosylated (*see* **Subheading 3.2.**). Fully processed receptor subunits will, in general, be EndoH resistant, although some exceptions are known (e.g., the NMDA receptor subunit NR1 *[20,21]*). I would strongly advise checking for proper glycosylation of the receptor, as it is possible for incompletely glycosylated receptor to make it to the cell surface and be functional as shown for mGluR1b *(3)*. It is of course essential also to check for normal receptor function in terms of the concentration–response for the second-messenger responses and antagonist effects, to see if the tag has altered these. If it is possible to carry out binding studies these too should be performed to ensure that the tagged receptor is functioning normally. Finally, because it is becoming apparent that many GPCRs may be dimeric and possibly heterodimeric, forming complexes with other receptors that can alter receptor pharmacology and function, it may be necessary to examine the behavior of the tagged receptor in this regard using the appropriate biochemical techniques (for review *see [8]*). In the case of the FLAG-tagged mGluR1a discussed in **Subheading 1.2.**, the native receptor was known to be at the cell surface as a disulfide bonded dimer *(6)*. To check that the tag had not interfered with this, membranes prepared from HEK-293 cells transiently transfected with the receptor were analyzed on a nonreducing gel,

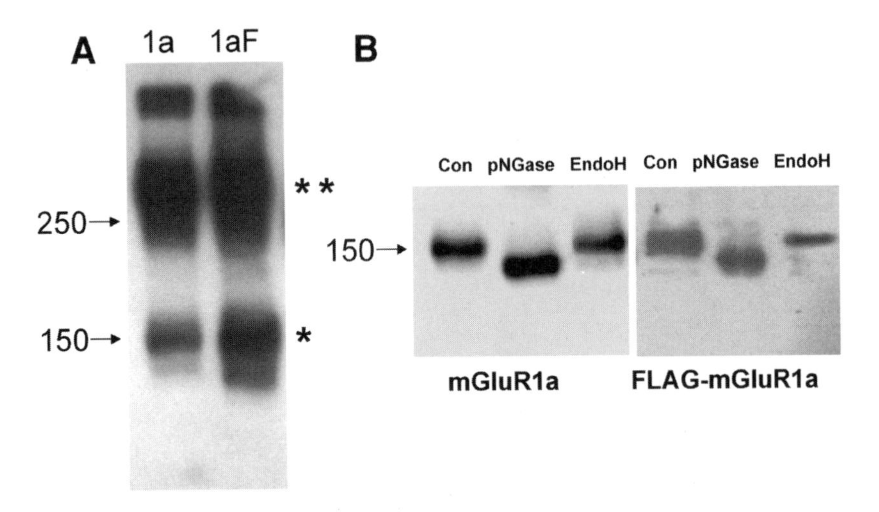

Fig. 4. Biochemical characterization of the N-terminally tagged FLAG-mGluR1a. For dimerization analysis, membranes were prepared from HEK-293 cells transfected with either mGluR1a (*lane 1a*) or FLAG-tagged mGluR1a (*lane 1aF*) and run unreduced (*see* **Subheading 3.1.**) on 5% SDS-polyacrylamide gels (**A**). The proteins were transferred to PVDF membranes and immunoblotted with a C-terminal specific mGluR1a antibody. Monomer and dimer of the protein are indicated by * or **, respectively. Separately, membranes were treated with endoglycosidases as described in **Subheading 3.2.** and the digests analyzed fully reduced on 6% SDS-polyacrylamide gels (**B**). The immunoblots were developed with an anti-mGluR1a antibody except for the glycosidase treated membranes from FLAG-mGluR1a expressing cells, which was developed with anti-FLAG antibody. Only the monomer of the mGluR1a is shown on this gel. The position of the 150 and 250 molecular weight markers in kDa are shown by *arrows*.

and the receptor detected with a C-terminal anti-mGluR1a antibody (*see* **Note 3**). The result showed clearly that the tagged receptor was indeed largely present in the cells as a dimer, as is the native receptor (**Fig. 4A**).

As shown in **Fig. 4B**, the tagged mGluR1a is also EndoH resistant, as is the wild-type receptor. Given that the tagged mGluR1a is a dimer and fully processed it is no surprise that it gave normal responses, measured as inositol phosphate production, to agonist when compared to the wild-type receptor *(22,23)*. Both the FLAG-tagged mGluR1a and mGluR1b were subsequently used successfully to monitor mGluR1 trafficking in cultured neurones and transfected cells *(3,24)*. The success of the N-terminal FLAG-tagging of mGluR1 contrasts with the effect of introducing GFP into a similar position in the mGluR1a N-terminus. This was undertaken to allow the trafficking of the receptor to be studied in live cells, and the N-terminus was chosen as the inser-

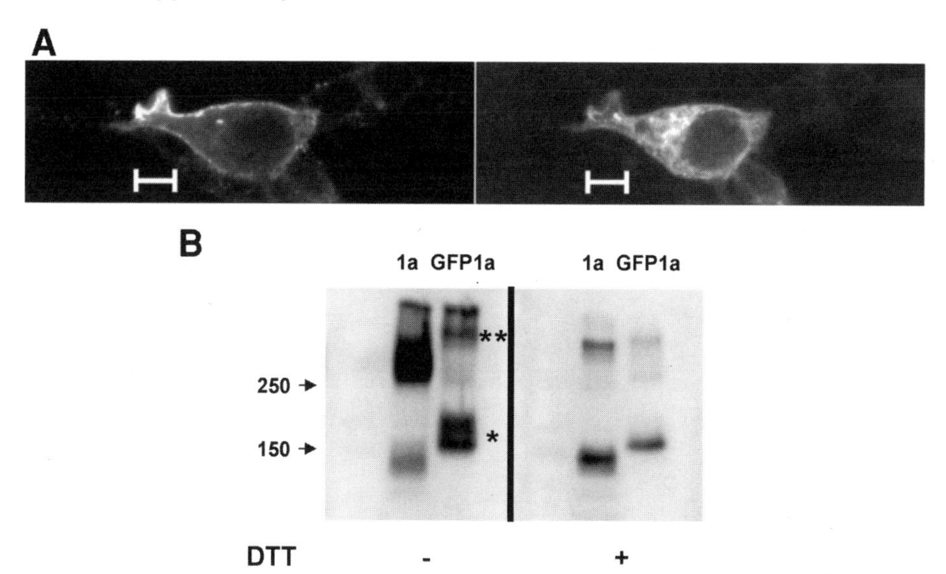

Fig. 5. Characterization of an N-terminally tagged GFP mGluR1a. HEK-293 cells were transiently transfected with GFP-tagged mGluR1a (*see* **Subheading 1.6.**). The cells were then fixed with paraformaldehyde and reacted with an anti-GFP antibody followed by a rhodamine coupled secondary anti-rabbit antibody to detect the surface bound anti-GFP (**A**). The cells were examined on a Leitz LSM 510 confocal microscope. The intrinsic GFP immunofluorescence (**B**) is seen to be present strongly in the cytoplasm of the cells and at the edges, whereas the surface immunofluorescence simply outlines the cells. The scale bar is 10 μm. Membranes from HEK-293 cells transfected with either mGluR1a (1a) or GFP-tagged mGluR1a (GFP1a) were analyzed reduced (+DTT) and unreduced (–DTT) on a 5% SDS-polyacrylamide gel (**B**). The proteins were transferred to a polyvinylidene fluoride (PVDF) membrane and immunoblotted with a C-terminal mGluR1a specific antibody. The increase in size of the GFP-tagged mGluR1a of 20 kDa owing to the introduced GFP can be seen in both the monomer (*) and dimer (**) of the protein.

tion site to minimize any interference with the binding of Homer proteins ***(16)*** to the C-terminus, since these had been found to contribute to receptor trafficking ***(24,25)***. The resulting N-terminal GFP-tagged mGluR1a was produced when transiently expressed in HEK-293 cells and appeared to be at the cell surface (**Fig. 5A**). It also showed some evidence of dimerization, although this was reduced compared to the native receptor (**Fig. 5B**). It was also very noticeable that the amount of GFP–mGluR1a monomer was much greater than that of the native receptor in the transfected cells, suggesting that the GFP-tagged receptor was folding inefficiently. Given these data it was no surprise that functional checks showed a significantly reduced maximal response for inositol phosphate production in the GFP–mGluR1a-transfected cells, with a response only 20% that of the untagged protein. Therefore this position in the N-terminus of mGluR1a does not appear to support the introduction of the larger GFP sequence into it. This highlights the fact that different tags can give different effects even at the same position, and therefore more than one tag position should be tried.

2. Materials

1. 10 m*M* $NaHCO_3$.
2. Phosphate-buffered saline (PBS).
3. 100 m*M* Sodium iodoacetamide in water.
4. Protease inhibitor mixture (Boehringer; made up as per the manufacturer's instructions).
5. 100 m*M* Phosphate buffer, pH 7.2.
6. 100 m*M* Sodium acetate buffer, pH 5.5.
7. Protein N-glycosidase F (pNGase F; 100 U/100 µL; Boehringer).
8. Endoglycosidase H (EndoH; 100 U/100 µL; Boehringer).
9. 10% (w/v) Sodium dodecyl sulfate (SDS) in water.
10. 10% (v/v) Triton X-100 in water.
11. 27-gage Hypodermic syringe needle and a 2-mL syringe.
12. 2X SDS sample buffer: 125 m*M* Tris-HCl, pH 6.8, 4% (w/v) SDS, 20% (v/v) glycerol, and 0.01% (w/v) bromophenol blue. For reduced samples add dithiothreitol (DTT) to give a final concentration of 20 m*M*.
13. Eppendorf tubes and a microfuge capable of 14,000 rpm.

3. Methods

3.1. Preparation of Membranes

1. Cells are grown in 25-cm^2 flasks. The method is designed for processing two of these, and can be scaled up as needed.
2. Wash the cells three times with 5 mL of PBS and carefully remove as much of the last wash as possible.
3. Add 1 mL of 10 m*M* $NaHCO_3$ to each flask and leave at room temperature for 5 min.
4. Pipet the cells from the flask using trituration with a 1-mL Oxford pipet tip or equivalent, and pool the cells from both flasks. Add iodoacetamide to give a final concentration of 10 m*M* and the protease inhibitors to the recommended concentration (*see* **Note 3**).
5. Carefully, avoiding foaming, pass the cell suspension through the 27-gage needle into the syringe and eject the suspension forcefully again. Repeat this nine more times.
6. Centrifuge the cell suspension at 2000*g* for 5 min at 4°C to remove unbroken cells.

7. Transfer the supernatant from this centrifugation to two Eppendorf tubes and centrifuge at 14,000*g* for 15 min at 4°C.
8. Resuspend the pellet in 100 μL of 10 m*M* $NaHCO_3$ containing 1 m*M* iodoacetamide and snap-freeze in liquid nitrogen. Store at –20°C.

3.2. Deglycosylation of Receptors

1. Rapidly thaw and resuspend the membranes prepared in **Subheading 3.1.**
2. Add 10 μL of 10% SDS and heat to 100°C for 5 min (*see* **Note 5**).
3. Cool the samples and place 30 μL in three clean Eppendorf tubes. To each tube add 15 μL of 10% Triton X-100, 2 μL protease inhibitor cocktail, and 53 μL of distilled water.
4. To two of the tubes add 50 μL of the 100 m*M* phosphate buffer, pH 7.2, and label one control and the other pNGaseF. To the third tube add 50 μL of the sodium acetate buffer, pH 5.5, and label this tube EndoH (*see* **Note 6**).
5. To the tube labeled pNGaseF, add 1 U of pNGaseF and likewise 1 U of EndoH to the tube labeled EndoH. Mix carefully all the tube contents.
6. Place all three tubes at 37°C overnight (20–25 h).
7. Add 25 μL of each sample to an equal volume of 2X SDS sample buffer, add DTT to give a final concentration of 20 m*M*, and heat at 60°C for 15 min.
8. Analyze by SDS-polyacrylamide gel electrophoresis (SDS-PAGE) and immunoblotting as normal.

4. Notes

1. When designing primers for these steps it is essential to take into account the codon usage of the species in which the protein is to be expressed. As a rule of thumb the primers will contain 18–28 of the bases coding for the receptor sequence with these designed to give melting temperatures between 55 and 80°C. Assume each GC pair contributes 4°C to the annealing temperature whereas each AT pair gives 2°C. Try and avoid primers with complementary 3′-ends as these can lead to primer dimer formation. Ensure that the sequence, including the epitope tag sequence being amplified, does not contain the restriction sites to be used in the cloning step. Primer design is most easily optimized using the primer design protocols in molecular biology software.
2. A few extra nucleotides, usually G and C, should be added to the ends of both primers to ensure efficient restriction digestion and prevent opening of the PCR product end.
3. The addition of iodoacetamide is to prevent the disulfide bonding of free SH groups in the cell proteins. As the cells are lysed these groups are exposed to an oxidising environment and any resulting disulfide bond formation could lead to artefactual oligomer formation. The iodoacetamide blocks any free SH groups by carboxymethylating them.
4. SDS-PAGE of membrane proteins is difficult as they have a tendency to aggregate, if heated at 100°C, during sample preparation. This property is especially

problematic for unreduced samples. The critical element in sample preparation is the heating of the sample. We have found that heating at 60°C for 15 min gives good solubilization of samples without aggregation of receptors. The preparation of membranes in the presence of iodoacetamide, and the use of 20 m*M* DTT rather than mercaptoethanol to reduce the samples, also helps prevent aggregation.

5. This step does several things. It solubilizes and denatures the membrane proteins and it denatures any proteases present in the sample. Because the endoglycosidases do not work on native proteins this step is essential.
6. Protein endoglycosidase F removes all the *N*-linked sugars from proteins. Thus treatment with this will generally yield the core amino acid sequence of the protein devoid of carbohydrate. Endoglycosidase H removes the unbranched oligomannose sugars that are added in the endoplasmic reticulum, from proteins. It will not remove the branched and complex sugars that are added in the Golgi apparatus. Thus if a protein is retained in the endoplasmic reticulum, it will be EndoH sensitive, whereas if it translocated through the Golgi apparatus it will not be. By comparing the size of a protein after pNGase F and EndoH treatment it is possible therefore to determine how much of the protein has passed through the endoplasmic reticulum.

Acknowledgments

The author would like to thank Drs. Mikhail Soloviev, Wai-Yee Chan, and F. Ciruela for their invaluable help in the preparation and characterization of the mGluR1a constructs described in this chapter, as well as Adrian Gray for his technical assistance.

References

1. Stowell, J. N. and Craig, A. M., (1999) Axon/dendrite targeting of metabotropic glutamate receptors by their cytoplasmic carboxy-terminal domains. *Neuron* **22,** 525–536.
2. Standley, S., Roche, K. W., McCallum, J., Sans, N., and Wenthold, R. J. (2000) PDZ domain suppression of an ER retention signal in NMDA receptor NR1 splice variants. *Neuron* **28,** 887–898.
3. Chan, W. Y., Soloviev, M. M., Ciruela, F., and McIlhinney, R. A. J. (2001) Molecular determinants of metabotropic glutamate receptor 1B trafficking. *Mol. Cell Neurosci.* **17,** 577–588.
4. Szollosi, J., Nagy, P., Sebestyen, Z., et al. (2002) Applications of fluorescence resonance energy transfer for mapping biological membranes. *J. Biotechnol.* **82,** 251–266.
5. Xu, Y., Johnson, C. H., and Piston, D. (2002) Bioluminescence resonance energy transfer assays for protein–protein interactions in living cells. *Methods Mol. Biol.* **183,** 121–133.
6. Romano, C., Yang, W. L., and Omalley, K. L. (1996) Metabotropic glutamate receptor 5 is a disulfide-linked dimer. *J. Biol. Chem.* **271,** 28612–28616.

7. Rios, C. D., Jordan, B. A., Gomes, I., and Devi, L. A. (2001) G-protein coupled receptor dimerization: modulation of receptor function. *Pharmacol. Ther.* **92,** 71–78.
8. Gomes, I., Jordan, B. A., Gupta, A., et al. (2001) G protein coupled receptor dimerization: implications in modulating receptor function. *J. Mol. Med.* **79,** 226–242.
9. Bennett, J. A. and Dingledine, R. (1995) Topology profile for a glutamate receptor: three transmembrane domains and a channel-lining reentrant membrane loop. *Neuron* **14,** 373–384.
10. Anand, R. (2000) Probing the topology of the glutamate receptor GluR1 subunit using epitope-tag insertions. *Biochem. Biophys. Res. Commun.* **276,** 157–161.
11. Murkerji, J., Haghighi, A., and Seguela, P. (1996) Immunological characterization and transmembrane topology of 5-hydroxytryptamine 3 receptors by functional epitope tagging. *J. Neurochem.* **66,** 1027–1032.
12. Xie, L. Y. and Abou-Samra, A. B. (1998) Epitope tag mapping of the extracellular and cytoplasmic domains of the rat parathyroid hormone PTH-related peptide receptor. *Endocrinology* **139,** 4563–4567.
13. Han, G. M. and Hampson, D. R. (1999) Ligand binding to the amino-terminal domain of the mGluR4 subtype of metabotropic glutamate receptor. *J. Biol. Chem.* **274,** 10008–10013.
14. Selkirk, J. V., Challis, R. A. J., and McIlhinney, R. A. J. (2002) Characterization of an N-terminal secreted domain of the type-1 human metabotropic glutamate receptor produced by a mammalian cell line. *J. Neurochem.* **80,** 346–353.
15. Niethammer, M., Kim, E., and Sheng, M. (1996) Interaction between the C terminus of the NMDA receptor subunits and multiple members of the PSD-95 family of membrane associated guanylate kinases. *J. Neurosci.* **16,** 2157–2163.
16. Brakeman, P. R., Lanahan, A. A., O'Brien, R., et al. (1997) Homer: a protein that selectively binds metabotropic glutamate receptors. *Nature* **386,** 284–288.
17. Kitano, J., Kimura, K., Yamazaki, Y., et al. (2002) Tamalin, a PDZ domain-containing protein, links a protein complex formation of group 1 metabotropic glutamate receptors and the guanine nucleotide exchange factor cytohesins. *J. Neurosci.* **22,** 1280–1289.
18. Sheng, M. and Wyszynski, M. (1997) Ion channel targeting in neurons. *Bioessays* **19,** 847–853.
19. Sheng, M. and Sala, C. (2001) PDZ domains and the organization of supramolecular complexes. *Annu. Rev. Neurosci.* **24,** 1–29.
20. Brose, N., Gasic, G., Vetter, D., Sullivan, J. M., and Heinemann, S. F. (1993) Protein chemical characterization and immunocytochemical localization of the NMDA receptor subunit NMDA R1. *J. Biol. Chem.* **268,** 22663–22671.
21. McIlhinney, R. A. J., Molnar, E., Atack, J. R., and Whiting, P. J. (1996) Cell surface expression of the human *N*-methyl-D-aspartate receptor subunit 1a (NMDAR1a) requires the co-expression of other NR2 subunits in transfected cells. *Neuroscience* **70,** 989–997.
22. Ciruela, F., Soloviev, M. M., and McIlhinney, R. A. J. (1999) Co-expression of metabotropic glutamate receptor type 1 alpha with Homer-1a/Vesl-1S increases the cell surface expression of the receptor. *Biochem. J.* **341,** 795–803.

23. Chan, W. Y. (2002) The targeting of metabotropic glutamate receptors, D. Phil Thesis Physiological Sciences. University of Oxford, Oxford.
24. Ciruela, F., Soloviev, M. M., Chan, W. Y., and McIlhinney, R. A. J. (2000) Homer-1c/Vesl-1L modulates the cell surface targeting of metabotropic glutamate receptor type 1 alpha: evidence for an anchoring function. *Mol. Cell Neurosci.* **15,** 36–50.
25. Ango, F., Pin, J. P., Tu, J. C., et al. (2000) Dendritic and axonal targeting of type 5 metabotropic glutamate receptor is regulated by Homer 1 proteins and neuronal excitation. *J. Neurosci.* **20,** 8710–8716.

6

Identification of G-Protein-Coupled Receptor mRNA Expression by Northern Blotting and *In Situ* Hybridization

Lucy F. Donaldson

Summary

G-protein-coupled receptor mRNAs are expressed at low levels and therefore present a challenge for the study of their sites and levels of expression. *In situ* hybridization (ISH) and Northern blotting are powerful methods for the localization of mRNAs and the study of regulation of mRNA expression. ISH combines the power of precise cellular localization with the ability to perform semiquantitative analysis of the mRNA level, whereas Northern blotting has the ability to identify genetic splice variants, or to study multiple RNA molecules sequentially in the same tissue samples.

These protocols give step-by-step instructions for the performance of these techniques, and the analysis of the data that can be obtained using them.

Key Words

Autoradiography, cRNA probe, *in situ* hybridization, in vitro transcription, microautoradiography, northern blotting, poly(A)$^+$ RNA selection, riboprobe, RNase-free conditions.

1. Introduction

In situ hybridization (ISH) and Northern blotting are techniques for the study of mRNA expression that both rely on the hybridization of a labeled probe to a target mRNA species, by Watson–Crick basepairing. In both methods, the label is then detected to localize the RNA. In this chapter I describe the use of single-stranded radiolabeled probes in the detection of G-protein-coupled receptor (GPCR) mRNAs using Northern blotting or ISH.

From: *Methods in Molecular Biology, vol. 259, Receptor Signal Transduction Protocols, 2nd ed.*
Edited by: G. B. Willars and R. A. J. Challiss © Humana Press Inc., Totowa, NJ

1.1. *In Situ Hybridization*

There are many reviews and published protocols of the application of ISH to the localization of mRNA species, for example, *(1)*. The hybridization of probes to RNA immobilized in tissue sections is not well understood and presents many problems in optimization as a result. In this protocol for ISH, I describe the use of radiolabeled RNA probes with unfixed frozen tissue for the *in situ* localization of receptor mRNA. ISH is a powerful technique for the localization of mRNA expression. The power of this technique lies in its ability to identify mRNA species in individual cells in tissue sections, allowing precise sites of mRNA expression to be localized using the light or electron microscope. In addition, as the level of isotope in any location is proportionally related to the amount of mRNA present (in the presence of excess probe), this technique can be used to obtain data on relative mRNA expression levels between different areas of tissues, different tissues, or between experimental and control samples.

In considering ISH and the pitfalls that await the unwary practitioner of this technique the following areas must be considered if good, interpretable data are to be generated: access to the target RNA (tissue preparation), possible degradation of either the target or probe RNA (sources of potential contamination by ribonucleases), specificity of signal (hybridization and washing conditions), and controls. The first two areas are discussed in detail in the **Notes** (**Subheading 4.**). Here I briefly consider the latter issues—specificity of signal and adequate controls for ISH using radiolabeled riboprobes.

The specificity of the signal obtained in ISH is determined by several factors including the following:

1. The homology between the target and the probe. This may be 100%, that is, an exact match, or less than this if, for example, a probe is used that represents a sequence from a different species, for example, mouse probe used to detect a rat target. Conditions must be altered if probes and targets are not 100% identical in sequence, as the hybridization is less stable under these circumstances and stringent washes may remove specifically hybridized probe from the target.
2. The stringency of the hybridization and washing conditions. Stringency is used to refer to the conditions under which hybridization and washing occur, and essentially means how conducive the experimental conditions are to nonspecific binding. High-stringency conditions allow less nonspecific binding to occur owing to conditions that destabilize the hydrogen bonds between the bases in the probe and target. Stringency is *increased* by raised temperature, lowered sodium concentration, or increased formamide concentration, all conditions that destabilize hydrogen bonds.
3. Other factors that may cause nonspecific binding. Some tissues seem to be "sticky" and retain RNA probes more readily than others do; these may require

additional treatments such as acetylation. Oxidation of the radiolabel in ^{35}S-labeled probes or accidental exposure of nuclear emulsion to ionizing radiation may also result in increased backgrounds.

Controls for ISH are a hotly debated topic. I usually recommend performing at least two of the controls listed here and in following paragraph to be sure of the specificity of the signal generated. The most commonly used and requested (by reviewers) negative control is the use of a "sense" probe, that is, a probe that is the same sequence as the target RNA, as opposed to the complementary sequence. Unfortunately it is becoming evident that many genes may be transcribed from both strands in different species and conditions ***(2–5)*** and a sense control may therefore detect a totally unrelated mRNA that is a specific signal in itself, but that can be erroneously interpreted as nonspecific binding. Other controls that may be included are predigestion of target RNA with ribonuclease A to determine that the signal is derived from binding to RNA ***(6)***, probes targeted to different parts of the same target molecule (which should give the same distribution of signal), probes targeted to different targets or members of a family of mRNAs that show different distributions in the same tissue ***(7,8)***, or competition for hybridisation using an excess of unlabeled sense RNA.

1.2. Northern Blotting

Our understanding of the hybridization of RNA immobilized on solid supports is now fairly good ***(9)***, and many protocols and reviews describing Northern blotting and the theory and application of the technique have been published (for one example *see* ***[10]***). Northern blotting involves the isolation of RNA and its size separation by electrophoresis on denaturing agarose gels, to separate different RNA species. RNA is then transferred onto membranes that enable probing of the blot multiple times with different probes. Thus, Northern blots can be used to examine the expression of several different mRNAs in a sample using sequential probing of the blot. In addition, Northern blotting can identify different transcripts or splice variants of the same gene, an area that is of increasing interest. The major problem with Northern blotting is that the RNA under study is extracted from sample lysate or homogenate. This creates two potential problems: first, tissues rarely consist of a single cell type and it is therefore impossible to know whether any change in mRNA level is specifically in one cell type; second, mRNA expressed in a small number of cells in the tissue sample may not be detectable owing to dilution in the pool of RNA from the whole sample. Like ISH, choice of probe affects the sensitivity of Northern blots, with single-stranded probes being more sensitive than double-stranded probes because of the lack of a competitor for hybridization in the sense strand of the probe. In addition, the sensitivity of Northern blots can be improved by the selection of mRNAs with poly(A)$^+$ tails using oligo(dT)$^-$ cellulose columns.

GPCR mRNA, along with many mRNAs, is usually expressed at low levels, in the region of 10–100 molecules per cell *(11)*. The method of mRNA localization must therefore be as sensitive as possible. For this reason I describe in this chapter the use of the most sensitive techniques for Northern blotting—poly(A)$^+$ RNA selection in combination with single-stranded DNA probes—and ISH—single-stranded RNA probes (also known as riboprobes or cRNA probes).

The procedures described for ISH and Northern blotting are multifaceted, taking several days to perform, and will be dealt with as the following separate procedures: (1) ISH: preparation of tissue and equipment; probe synthesis and labeling, hybridization, and washing; film and microautoradiography; optional/additional modifications of the techniques; and computer-aided image analysis of RNA expression. (2) Northern blotting: RNA isolation; poly(A)$^+$ RNA selection; blotting and RNA immobilization, determination of good RNA transfer; probe synthesis, hybridization, and signal visualization; and analysis of mRNA expression levels.

2. Materials

2.1. ISH

Required reagents, tools, solutions, and equipment for ISH using complementary RNA probes:

1. Tissue, either unfixed frozen (*see* **Subheading 3.1.1.1.**) or fixed, sectioned (*see* **Subheading 3.1.1.3.**), and mounted on subbed slides.
2. Subbed slides for mounting of sections (*see* **Subheading 3.1.2.**).
3. cDNA template cloned into a suitable plasmid vector and linearized with the appropriate restriction enzyme (*see* **Subheading 3.1.3.** and **Fig. 1**).
4. Slide troughs/containers large enough to hold the number of slides to be hybridized.
5. Large beakers (2 L).
6. Containers (plastic boxes with tight-fitting lids) for hybridization and RNase treatment.
7. Oven for incubation at 40–60°C. It is useful to also have an oven that can bake at 200°C.
8. Slide mailers or specialized dipping container for emulsion coating of slides.
9. Light-tight boxes with slide racks for exposure of slides.
10. Sterile and "clean" pipet tips and microfuge tubes (*see* **Note 1**).

2.1.1. Basic Precautions When Working With RNA

Many reagents in these protocols are extremely toxic, and protective clothing such as laboratory coats, goggles, and masks should be worn when neces-

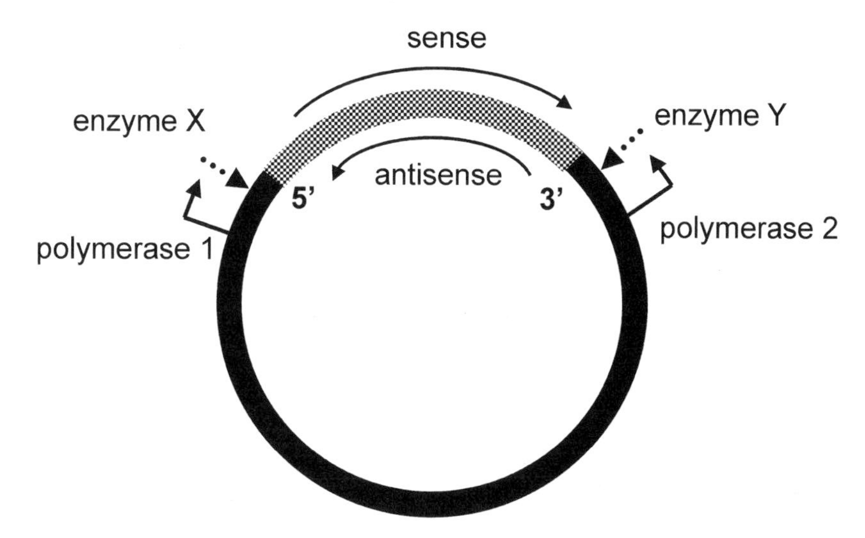

Fig. 1. Schematic outline of cloned cDNA and cRNA probe manufacture. The cloned cDNA (probe template) is depicted by the *hatched area* of the plasmid. For antisense probe generation the plasmid is linearized with restriction enzyme X, a unique enzyme that cuts the DNA only at the 5′-end of the cloned cDNA. The probe is then transcribed using RNA polymerase 2, which could be T7, T3, or SP6 depending on the plasmid used. Most plasmids now have polymerase start sites at either end of the multiple cloning site, although older plasmids may not. Sense control probes can be made by linearisation with enzyme Y and transcription with polymerase 1.

sary. Gloves should also be worn at all times when working with RNA to avoid ribonuclease (RNase) contamination. Gloves should therefore be worn when handling plasticware for sterilization.

All glassware should be baked at 200°C for a minimum of 3 h prior to use. Plasticware should be sterile and disposable where possible or kept strictly for use in ribonuclease-free conditions, as it is extremely difficult to decontaminate. Plastics can be decontaminated by rinsing in 0.1 *M* sodium hydroxide solution followed by thorough rinsing with diethylpyrocarbonate (DEPC)-treated water. I recommend baking glassware over NaOH and rinsing with DEPC-treated water as any traces of alkaline solution that come into contact with target or probe can cause RNA degradation by alkaline hydrolysis. There are also reagents on the market that will remove contaminating RNase or nucleic acids; these can be useful if preparation time is limited. Things to be aware of as potential sources of contamination of solutions are pH meter probes and spatulas used for weighing chemicals.

Reagents used for RNA detection do not necessarily need to be molecular grade and certified RNase-free, but should be dedicated for RNA work and handled only while wearing gloves, and weighed with RNase-free spatulas into sterile containers. Reagents that cannot be treated to render them RNase-free (*see* **Note 2**), such as Trisma, or reagents that are heat labile and cannot be autoclaved should be molecular grade and certified RNase-free.

2.1.2. Solutions and Reagents—RNase-Free

For most mixtures, the final concentrations of each reagent are given (*see* **Note 3**). For some mixtures, such as 2X hybridization buffer, recipes are given. Clean stock solutions should be used and diluted to final concentrations with DEPC-treated water.

All solutions (except those with primary amine groups such as Tris; *see* **Note 2**) should be treated with DEPC to render them RNase free. There are many protocols on DEPC treatment; we find that using 0.1% DEPC is adequate to treat solutions. As DEPC degrades to ethanol and carbon dioxide on contact with water, the lowest concentration possible should be used to avoid any effect of the contaminating ethanol becoming a problem. Add DEPC to the solution, shake vigorously to ensure mixing, and let stand for a minimum of 1 h. Autoclave solution to sterilize and hydrolyze any DEPC left in the solution.

1. General salt solutions—DEPC-treated: 5 *M* NaCl solution, 1 *M* Tris-HCl, pH 7.5 (no DEPC, *see* **Note 2**), 250 m*M* EDTA, pH 8.0.
2. 2X Hybridization buffer (2X HB): 6.2 mL of DEPC-treated water, 2.4 mL of 5*M* NaCl, 200 µL of 1 *M* Tris-HCl, pH 7.5, 66 µL of 6% Ficoll (*see* **Note 4**), 66 µL of 6% polyvinylpyrolidone (PVP), 334 µL of 6% bovine serum albumin (BSA), 80 µL of 250 m*M* EDTA, pH 8.0, 200 µL of salmon testes DNA (10 mg/mL), 2 g of dextran sulfate, and 40 µL of yeast tRNA (50 mg/mL).
   Mix well and aliquot in 500-µL amounts. Store at –20°C and avoid multiple freeze–thaw cycles. The solution is extremely viscous and the dextran sulfate settles, so vortex mix prior to use.
3. RNase A buffer: 0.5 *M* NaCl, 10 m*M* Tris-HCl, pH 7.5, 1 m*M* EDTA, pH 8.
4. RNase-free 20X saline sodium citrate (SSC): 3 *M* NaCl, 0.3 *M* sodium citrate.
5. Box buffer: 2 mL 20X SSC (RNase-free), 5 mL of formamide (deionized if necessary, *see* **Note 5**) and make up to 10 mL with DEPC-treated water.
6. 4% Paraformaldehyde in phosphate buffer (*see* **Note 5**): make 900 mL DEPC-treated water and autoclave. Make 0.1 *M* phosphate buffer in the DEPC-treated water by adding NaH_2PO_4 and Na_2HPO_4 (final concentrations 20 m*M* and 80 m*M*, respectively). Add 40 g of paraformaldehyde from a dedicated stock, and heat to a maximum of 80°C to dissolve the paraformaldehyde. Cool and store at 4°C for no more than 1 mo.
7. Deionized formamide (*see* **Note 6**): add 10 g of dedicated RNase-free mixed bed resin per 100 mL of formamide and mix for 1 h. Filter twice through Whatman No. 1 filter paper and store at room temperature. Protect from light.
8. Dithiothreitol (DTT).

3. Methods

3.1. ISH

3.1.1. Tissue Preparation

3.1.1.1. Unfixed Frozen Tissue (*See* **Note 8**)

Animals are usually killed by decapitation, as this removes trunk blood, which is rich in ribonucleases and may therefore accelerate RNA degradation. Decapitation also allows for rapid removal of the brain, but obviously has no advantage for the removal of other tissues.

Tissues should be dissected out as rapidly as possible to preserve the RNA. If possible dissect on a cool surface. Tissue should then be frozen rapidly either by direct placement on powdered dry ice or using an isopentane bath cooled by immersion in ethanol and dry ice (*see* **Note 9**). Tissue can then be stored in the –80°C freezer until sectioning.

3.1.1.2. Fixed Tissue

Animals should be fixed by transcardial perfusion under deep anesthesia with 4% paraformaldehyde (*see* **Note 10**).

3.1.1.3. Tissue Sectioning

Sections should be cut on a cryostat, preferably at 8–10 μm thickness at as cold a temperature as possible (chamber temperature –20°C) and then thaw mounted onto subbed slides. Sections can be stored in boxes in the –80°C freezer until use.

3.1.2. Subbing Slides

Use glass slides with frosted ends to allow labeling of slides (*see* **Note 11**).

1. D 1. Rinse slides sequentially in 0.2 *M* HCl in DEPC-treated water, DEPC-treated water, and acetone and dry in an oven at 60°C.
2. D 2. Dip slides in 0.15% gelatin, 0.03% sodium azide in DEPC-treated water for 5 min and dry overnight at 60°C.
3. D 3. Dip slides in poly-L-lysine solution (100 mg/500 mL in DEPC-treated water) for 2x 10 s, rinse in DEPC-treated water for 10 min, and dry overnight at 60°C.
4. Wrap slides in aluminum foil to keep free of dust and to prevent contamination.

3.1.3. Template Preparation

The cDNA of choice should be cloned into a suitable plasmid vector and then linearized with an appropriate restriction enzyme (*see* **Note 12** and **Fig. 1**). The final concentration of template DNA in the digest should be 1 μg/μL (*see* **Note 13**).

3.1.4. In Vitro Transcription for Riboprobe Synthesis

This riboprobe synthesis method is based on that routinely used in the laboratory of Professor J. R. Seckl, University of Edinburgh ***(12)***, which I have extensively modified.

All tips, plastic tubes, and solutions should be sterile and RNase free. As a general rule, the "hottest" probe is usually necessary for localization of GPCR mRNA. More abundant mRNA species can often be detected using probes of lower specific activity.

Reagent	Standard	Hot	Hottest
DEPC-treated water	2.7 μL	1.7 μL	1 μL
10X Transcription buffer	1 μL	1 μL	1 μL
Nucleotide mix (*see* **Note 14**)	1.5 μL	1.5 μL	1.2 μL
200 m*M* DTT (*see* **Note 15**)	0.5 μL	0.5 μL	0.5 μL
RNase inhibitor	0.3 μL	0.3 μL	0.3 μL
UTP	1 μL (175 μ*M*)	1 μL (100 μ*M*)	— (*see* **Note 16**)
DNA template (1 μg/μL)	1 μL	1 μL	1 μL
[35S]UTP	1 μL	2 μL	4 μL
RNA polymerase	1 μL	1 μL	1 μL

1. Mix ingredients in the above order in a clean microfuge tube. Incubate at 37°C (T3 and T7) or 40°C (SP6) for 1 h. (*See* **Fig. 1** for comment on RNA polymerases.)
2. Add: 8 μL of DEPC-treated water, 2 μL of yeast tRNA, and 1 μL of DNase.
3. Mix and incubate at 37°C for 10 min to remove the DNA template.
4. Add 20 μL of phenol–chloroform–isoamyl alcohol (25:24:1), vortex-mix, and remove aqueous (upper) layer into a fresh tube (*see* **Note 17**).
5. Add 27.4 μL 10 *M* ammonium acetate; 10 μL of 3 *M* sodium acetate, pH 5.2; 63 μL of DEPC-treated water; and 550 μL of 100% cold ethanol.
6. Vortex-mix and precipitate the RNA either at –20°C for a minimum of 1 h or at –80°C for 10 min (*see* **Note 18**).
7. Centrifuge at maximum speed in a microfuge for 15 min. Remove the supernatant and dry the pellet thoroughly to ensure all ethanol is removed, either under vacuum or by air-drying. Do not overdry or the pellet will not resuspend.
8. Check radioisotope incorporation roughly using a hand-held monitor to assess the relative amounts of isotope in the pellet and supernatant (*see* **Note 19**).
9. Resuspend the pellet in 50–100 μL of 10 m*M* DTT in DEPC-treated water. Count 1 μL of the probe in a scintillation counter to determine the total counts per minute (cpm) of the probe (*see* **Note 20**).
10. Store the probe in 10 m*M* DTT at –20°C for no more than 1 wk (*see* **Note 21**).

3.1.5. Limited Alkaline Hydrolysis (Optional)

The minimum length recommended for riboprobes is 100–150 bases. I generally find that probes of 300–500 bases work very well with high specificity for the target, but longer probes can also be used. Probes of 1 kb or over may be too long to penetrate the fixed tissue. Shorter probes can be made by subcloning a fragment of cDNA of an appropriate length, or by reducing the length of the probe itself after synthesis, by alkaline hydrolysis. We use a method based on that described by Angerer and Angerer ***(13)*** that works extremely well.

1. Start with 50 µL of probe, add 30 µL of 0.2 *M* Na_2CO_3 and 20 µL of $NaHCO_3$, and incubate at 60°C for time *t* as calculated from the equation below:

$$t = \frac{\text{(starting length in kb)} - \text{(desired length in kb)}}{0.11\text{(starting length in kb)(desired length in kb)}}$$

2. Stop hydrolysis by adding 3 µL of 3 *M* sodium acetate, pH 6, and 5 µL of 10% glacial acetic acid.
3. Add 1 µL of yeast tRNA and 2.5 volumes of 100% ethanol, incubate, and precipitate as described in **Subheading 3.1.4., steps 7** and **8**.
4. Resuspend the RNA pellet in 10 m*M* DTT and use probe immediately or store as described (*see* **Note 22**).

3.1.6. ISH Using Radiolabeled cRNA Probes

D 1:

1. Set up four baked slide troughs: 4% paraformaldehyde in 0.1 *M* phosphate buffer, and three washes of 2X SSC. Add 20 µL of DEPC to each trough (*see* **Note 23**).
2. Remove slides from the freezer. Do not allow to thaw but immerse directly in the paraformaldehyde and fix for 10 min.
3. Rinse slides three times in 2X SSC for 5 min each (*see* **Note 24**).
4. Make up enough hybridization mix (HM) for the number of slides to be hybridized using (for 1 mL of HM) 500 µL of 2X HB, 500 µL of formamide, and 10^7 counts/mL of probe (*see* **Notes 25** and **26**).
5. Denature the probe in HM by heating at 65°C for 5 min and quench on ice. Add 10 µL of 1 *M* DTT to each milliliter of HM.
6. Line hybridization boxes with 3MM paper (Whatman) and add box buffer so that the paper is wet but not "swimming."
7. Drain the slides and dry around the sections with lens tissue to remove excess 2X SSC. Add hybridization mix to the sections ensuring coverage of the tissue with HM (*see* **Note 27**) and place slides on the 3MM paper in the hybridization boxes (*see* **Note 28**).
8. Seal the box lids tightly with adhesive tape (*see* **Note 29**).
9. Hybridize overnight at T_m – 20°C according to the formula *in* **Note 30**.

D 2:

1. If slides were cover slipped, soak off the cover slips by immersion of the slides (in a slide rack) in 2X SSC.
2. Once cover slips have been removed, rinse the sections three times in 2X SSC to remove excess HM.
3. Drain slides and dry round sections with lens tissue. Pipet 100 μL of RNase A solution (30 μg/mL in RNase buffer) over the sections and incubate in a box humidified with RNase buffer for 1 h at 37°C.
4. Drain the slides and place in a slide rack. Suspend the slide rack using tape in a 2-L beaker containing 2X SSC/1 ml/L of β-mercaptoethanol to cover the slides completely. Stir the solution using a magnetic stirrer and wash the slides at room temperature for 30–60 min (*see* **Note 31**).
5. Wash slides stringently (0.5–0.1X SSC plus β-mercaptoethanol at the same concentration as in nonstringent washes) at 45–60°C for 30–60 min each wash, in the same manner as described in **step 4** (*see* **Note 32**).
6. Drain slides and dehydrate in 50%, 70%, and 90% ethanol in 0.3 *M* ammonium acetate.
7. Air-dry slides overnight (in a fume cabinet if using β-mercaptoethanol in the washes).
8. If using large sections such as brain, expose sections to film in an X-ray cassette (*see* **Note 33**).
9. For microautoradiography for direct cellular localization, dip slides in liquid nuclear emulsion (*see* **Subheading 3.1.7.**).

3.1.7. Microautoradiography for Tissue Localization of Receptor mRNA

In the dark room:

1. Warm approx 10 mL of distilled water to 42°C and melt a similar volume of emulsion at the same temperature in a water bath.
2. Mix the emulsion and water and pour into a slide mailer or dipping vessel.
3. Dip each slide into the emulsion slowly, twice. Leave to dry horizontally in the dark.
4. Store the slides in light-tight containers in the refrigerator for 2–6 wk (*see* **Note 34**).
5. Bring slides to room temperature. In a dark room, develop the slides in freshly made developer for 3 min.
6. Rinse thoroughly in distilled water.
7. Fix for 3 min in ordinary NOT rapid fixer.
8. Wash well in distilled water for 20 min with frequent changes (*see* **Note 35**).
9. Counterstain tissue if preferred (*see* **Note 36**).
10. Dehydrate in graded ethanols to 100% ethanol, clear, and cover slip.
11. Visualise signal under light or dark field illumination.

*3.1.8. Additional Procedures (*See **Note 37***)*

Prefixed tissue may need permeabilization (*see* **Note 38**), prehybridization and/or acetylation to reduce nonspecific background.

3.1.8.1. Proteinase K Permeabilization (Prefixed Material)

1. If wax embedded, dewax and rehydrate as usual.
2. Rinse slides with mounted sections in 2X SSC (DEPC-treated) for 5 min.
3. Incubate slides in 1 μg/mL of proteinase K in proteinase K buffer at 37°C for up to 30 min (*see* **Note 39**).
4. Rinse slides in 2X SSC.
5. Follow ISH protocol beginning at **step 4**.

3.1.8.2. Acetylation (*See* **Note 40**)

1. Fix the tissue as per **step 1** of the ISH protocol and rinse twice in 2X SSC.
2. Add acetic anhydride (liquid) to slide trough so that final concentration in triethanolamine buffer will be 0.25% when the slides are in the trough and covered.
3. Add freshly made triethanolamine buffer together with the slides and use slides to mix the solution in the trough.
4. Stand for 10 min at room temperature.
5. Rinse in 2X SSC and follow the ISH protocol from **step 4**.

3.1.8.3. Prehybridization

Some probes require a prehybridization step to reduce nonspecific binding.

1. Follow **steps 1–3** of the ISH protocol.
2. Add 1X hybridization buffer (dilute 2X HB with DEPC-treated water) and incubate at the hybridization temperature without probe for 1–3 h.
3. Drain slides and follow ISH protocol from **step 4** onwards.

3.1.8.4. Dehydration (*See* **Note 41**)

1. Fix and rinse tissue as in **step 1** of the ISH protocol.
2. Dehydrate tissue in graded ethanols (50%, 70%, 90%, 100%) made in DEPC-treated water.
3. Air-dry slides and store in a dry-dust free container until required for ISH.
4. Begin hybridization at **step 4** of the ISH protocol. There is no need to wet the tissue sections before adding the hybridization buffer.

3.1.9. Analysis of ISH Signal

ISH can be analyzed by performing silver grain counting of the precipitated grains in the emulsion overlying the tissue sections. This enables quantitation of mRNA levels per cell, compared to a control sample. This can be done by eye but this is extremely time consuming and fatiguing. Many commercial computerized image analysis systems are available and will perform this function very well. We find, however, that a good microscope with video camera attached to a computer capable of direct video capture works very well. We capture images under dark field, to prevent any counterstain being detected as a signal, and use either the density slice feature in NIH Image (Macintosh platform) or the threshold feature in ImageJ (PC and Macintosh platforms) to count the number of silver grains either in a field or per cell. ImageJ and NIH Image are the public domain programs developed at the U.S. National Institutes of Health and are available for download on the Internet at http://rsb.info.nih.gov/nih-image/.

3.2. Northern Blotting

This method of Northern blotting, poly(A)$^+$ selection, and probe synthesis is based on the method devised by Scott Van Patten ***(14)***, modified by me. See also protocols in Sambrook et al. ***(15)***.

3.2.1. Tissue Preparation and RNA Extraction

Tissues are rapidly dissected on ice if possible and snap frozen either in liquid nitrogen or on dry ice.

RNA extraction using the method of Chomczynski and Sacchi ***(16)*** generally gives the best results. We find that using the commercial product TriReagent (also known as TriZol) gives the highest RNA yields of the best quality RNA (*see* **Note 42**). Tissues are homogenized from frozen specimens using glass-on-glass hand held homogenizers, following the manufacturer's protocol. RNA is stored at –80°C until needed for poly(A)$^+$ selection.

3.2.2. Poly(A)$^+$ Selection (See **Note 43**)

1. Plug a 1-mL syringe with a small pellet of clean glass wool.
2. Fill the syringe with the slurry of oligo(dT)$^-$ cellulose and allow the buffer to drain through until the syringe is loaded with 0.25 mL of oligo(dT)$^-$ cellulose.
3. Denature the RNA sample at 65°C for 5 min, add an equal volume of 2X buffer A at 65°C, and quench on ice.
4. Apply the RNA sample to the column and allow to run through under gravity.
5. Wash the column with 2X 1 ml of 1X buffer A.
6. Wash the column with 2X 1 mL of buffer B.
7. Elute the RNA into a clean microfuge tube using 0.4 mL of buffer C. Squeeze the remaining buffer out of the column with the syringe plunger (*see* **Note 44**).
8. Split the RNA sample into two. To each tube add 20 µL of 3 *M* sodium acetate, pH 5.2, 1 µL of yeast tRNA (10 mg/mL), and 600 µL of 100% ethanol.
9. Precipitate the RNA by incubation at –20°C for a minimum of 1 h or –80°C for 30 min.
10. Centrifuge the RNA at full speed for 30 min, remove the ethanol supernatant, and air-dry the pellet.

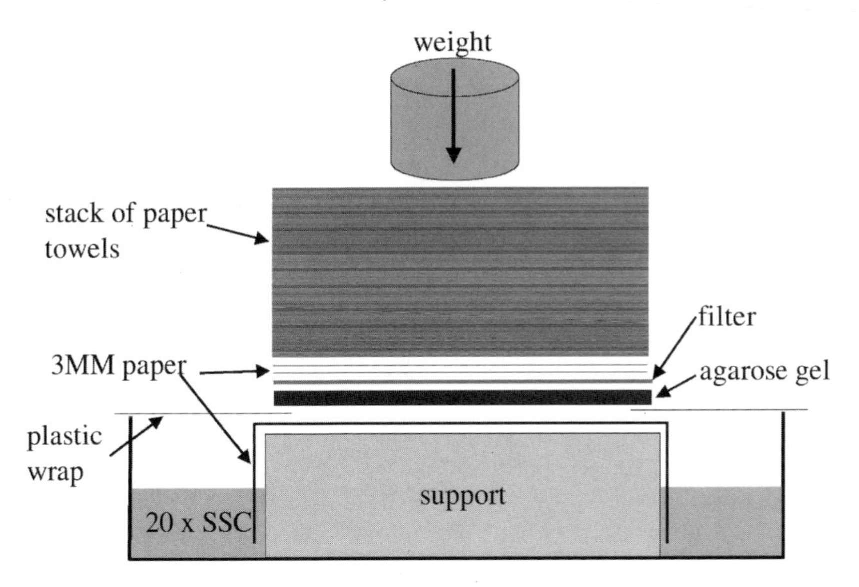

Fig. 2. Capillary blotting of RNA from agarose gels. Agarose gel is placed top down on a 3MM paper wick over a support in a reservoir of DEPC-treated 20X SSC. The gel is surrounded by plastic wrap to prevent wicking of the liquid around the gel. The filter is then placed onto the gel and a stack of 3MM paper followed by paper towels built on top. A weight is placed on the stack and left overnight for transfer to take place.

11. Wash pellets in 1 mL of 75% ethanol in DEPC-treated water. Centrifuge for 15 min.
12. Resuspend one pellet in 4 µL of DEPC-treated water and then use this RNA solution to resuspend the second RNA pellet from the same sample. Store at –80°C.

3.2.3. Northern Blotting

D 1:

1. Pour a denaturing 1% agarose gel in a fume cupboard.
2. To each RNA sample (4 µL; *see* **Note 43**) add 2 µL of 10X MOPS buffer, 4 µL of formaldehyde (40% solution), and 10 µL of formamide.
3. Mix and incubate at 65°C for 15 min. Cool on ice.
4. When the gel is set, age (allow to soak) the gel in the gel tank in 1X MOPS buffer while the RNA samples are denaturing.
5. Add 2 µL of loading buffer to each RNA sample and rapidly load each sample on the gel.
6. Run the gel at 6–10 V/cm until the dye fronts divide the gel into thirds.
7. Soak the gel in 20X SSC while preparing the blot (*see* **Note 45**).
8. Prepare the stack as described in **Fig. 2**. Place a solid support of at least the size of the gel in the reservoir and wrap a piece of 3MM paper around it to act as a

wick. Fill the reservoir with DEPC-treated 20X SSC and allow the 20X SSC to wet the paper fully.

9. Cover the support and reservoir in plastic film, and cut a hole in it over the support, just smaller than the gel.
10. Place the gel, upside down on the support with the plastic film just under the edge of the gel all the way around (*see* **Note 46**). Remove any air bubbles from between the gel and the 3MM paper.
11. Prewet a piece of nylon filter (e.g., Hybond N) the same size as the gel on the gel surface in 20X SSC (*see* **Note 47**).
12. Place a piece of 3MM paper on the nylon filter, then build a stack of folded paper towels until the stack is at least 10 cm high (**Fig. 2**).
13. Place a weight of 500 g (in a plastic container) on the top of the stack and leave overnight (*see* **Note 48**).

D 2:

1. Remove the stack from above the filter.
2. Crosslink the RNA either by exposure to UV light in a commercial crosslinker, or by baking in an oven at 60°C for 3 h.
3. Place the filter in a solution of methylene blue in sodium acetate for 10 min. Destain the filter in distilled water with several changes.
4. Photograph the wet stained filter on a light box as a record of the RNA transfer to the filter. If using an RNA ladder, cut off the lane and keep for future size determination.
5. Soak the filter in 0.1% SDS at room temperature to remove the methylene blue before probing (*see* **Note 49**).

3.2.3.1. Hybridization (*See* Note 50)

1. Prehybridize the filter with 10 mL of QuickHyb either in a rotating bottle in a hybridization oven or in a sealed bag on a shaking platform at 68°C (or temperature appropriate to the probe of use according to the manufacturer's instructions) for 1 h.
2. Mix the probe with 100 µL of salmon sperm DNA and 5 µL of 10 m*M* dCTP (*see* **Note 51**). Add the probe to the hybridization bottle and hybridize for 1–2 h at 68°C.
3. Discard the hybridization buffer and probe according to local rules for disposal of aqueous radioactive waste. Rinse the filter in 2X SSC/0.1% SDS at room temperature.
4. Wash the filter in 2X SSC/0.1% SDS at room temperature, then wash stringently in 0.2X SSC/0.1% SDS with agitation at 68°C for 15 min.
5. Monitor the filter after the wash and continue to wash until the surface counts on the filter are very low, typically less than 10 cpm.
6. Wrap the filter well in plastic film to prevent leakage of fluid (*see* **Note 52**) and expose to X-ray film in a cassette with an intensifying screen at –80°C (*see* **Note 53**).

3.2.4. Single-Stranded Antisense PCR Probe Synthesis

Begin probe synthesis on the day before you wish to probe the filter. I find it best to run the gel and make the probe together and then probe the filter the following day.

1. Generate a double-stranded PCR template using upstream (5′) and downstream (3′) primers for the sequence under study (*see* **Note 54**).
2. Gel purify the template and determine its approximate concentration using a DNA mass ladder.
3. Make up premix of the PCR reagents as below: 10 μL 10X PCR buffer, 5 μL 25 m*M* $MgCl_2$, 5 μL mix of 10 m*M* dATP, dGTP, dTTP (mixed in equal volumes), 5 μL downstream (3′) primer (20 μ*M*), 5 μL DNA template (6 ng/μL), and 10 μL sterile water. Aliquot 4 μL/tube and store at –20°C until needed (*see* **Note 55**).
4. To one aliquot add 5 μL of [α-^{32}P]dCTP (*see* **Note 56**) and 1 μL of *Taq* polymerase
5. Incubate in a thermocycler under the same cycling conditions as used for the DNA template but include an autoextension of the polymerase extension time of 20 s per cycle (*see* **Note 57**).

3.2.5. Reprobing of Northern Blots

Strip the filter by pouring boiling 0.1% SDS over the filter and agitating for 10 min. Check the probe is fully stripped by exposing the filter to X-ray film. If the probe is not completely removed, strip again (*see* **Note 52**).

3.2.6. Analysis of Northern Blot Data

Northern blots can be analyzed to generate data on the relative levels of RNA in each lane of the blot. If comparisons are to be made, all samples should be hybridized simultaneously under the same conditions using the same probe, and exposed for the same time. We use densitometry from scanned images of the films, and NIH Image to generate numerical data that can then be analyzed statistically. When comparing the relative amounts of RNA in a sample by Northern blot it is usual to probe the filter for a housekeeping gene such as actin and use this to control for any differences between lanes in the amount of RNA. We normally ratio the density of the signal from the gene of interest and the control gene, and use the ratios for comparison.

4. Notes

1. "Clean" is used throughout to mean RNase-free. Sterile items are not necessarily clean as autoclaving is not a reliable method of ribonuclease removal. Care should be taken to avoid contamination of plasticware and solutions prior to autoclaving by the use of gloves, clean glassware, and spatulas.

2. Chemicals with primary amine groups, such as Tris, react with DEPC and this method cannot therefore be used to treat these solutions. Use dedicated stocks of Tris for RNA work, and buy molecular grade certified RNase-free. Solutions such as Ficoll or DTT that are not thermostable should also be made from dedicated stock chemical certified nuclease-free, made up in DEPC-treated water and filter sterilized (0.2-µm syringe filter). Aliquots can often be stored in the freezer to eliminate the need to make solutions fresh each time.
3. Mixtures can be made from DEPC-treated stock solutions in sterile plastic or baked glass containers, made up to final volume with DEPC-treated water without the need for further DEPC treatment or autoclaving.
4. Six percent solutions of BSA, PVP, and Ficoll should be made up in sterile containers using DEPC-treated H_2O and then filter sterilized through a 0.2-µm syringe filter prior to use. Aliquots can be frozen and stored at –20°C.
5. Paraformaldehyde for use in ISH can be stored for up to 1 mo at 4°C, although it is better used as rapidly as possible. Although paraformaldehyde should be RNase-free, we find that if it is made in non-DEPC-treated water contamination problems occur more frequently and experiments fail.
6. New formamide usually does not require deionization but older stocks may need to be treated in this way. All glassware, spatulas, and magnetic fleas need to be RNase-free when making up/treating solutions that are not subsequently DEPC-treated.
7. The use of radioisotopes in these protocols carries inherent risks that must be evaluated in each laboratory according to local and national guidelines. Although ^{35}S β-particle emission is of relatively low energy, I recommend using shielding for the manufacture of the riboprobes as the specific activity of the probes is extremely high and therefore the exposure hazard is also relatively high. Standard 1-cm Perspex screens are adequate to reduce exposure levels to a minimum. Radioisotope waste must be disposed of according to local and national regulations.
8. We find that unfixed frozen tissue works best for ISH, although working with fixed tissue may be the only option. The compromise of working with fresh frozen tissue is that tissue morphology may not be as good as when using perfused fixed samples. We find that a ceramic tile on a bucket of ice makes a good cool dissecting surface.
9. Immersion in liquid nitrogen is not recommended; although some groups obtain very good results with this methodology we find that tissue can freeze-dry and therefore sectioning becomes more difficult. Direct placement on dry ice is a slightly slower method of freezing than immersion, but gives good tissue morphology and preserves the RNA well despite the slower freezing. Tissue that requires embedding can be frozen in embedding medium by either method.
10. Tissue fixed by perfusion probably needs permeabilization (*see* **Subheading 3.1.8.1.**) prior to hybridization. However, although fixation will result in better tissue morphology, the permeabilization steps may reverse this!
11. Slide troughs should be baked, slide racks should be baked if possible, or, if plastic, used solely for subbing and kept clean and dust free.

12. Always check that the cDNA template is completely linearized before using for in vitro transcription, by electrophoresis on a horizontal agarose gel. Run the uncut plasmid alongside to check for full linearization. Any contaminating circular plasmid will result in the generation of large transcripts containing plasmid sequences. This can lead to nonspecific binding and high background. Do not use restriction enzymes for linearization that result in a 3′ overhang following digestion, as this can result in continued transcription of the complementary strand of the template, thus generating a hybrid antisense/sense transcript.
13. Sometimes the template cannot be digested at a concentration of 1 µg/µL as either the DNA is too concentrated or the amount of enzyme required is too high and inhibition by glycerol cannot be avoided. In this situation the DNA should be digested in a more dilute solution and then precipitated and resuspended at 1 µg/µL. Restriction enzymes can be inactivated and removed after digest by heating and phenol–chloroform extraction prior to in vitro transcription, but it is not always necessary. Contaminating phenol can inhibit transcription if not completely removed.
14. Make a mix of the three nucleotides (ATP, CTP, GTP) in equal volumes of the 10 m*M* working solutions.
15. 200 m*M* DTT should be freshly made in DEPC-treated water.
16. For standard and hot probes use 1 µL of "cold" UTP of 175 µ*M* or 100 µ*M* solutions prepared by dilution of 10 m*M* working stocks in DEPC-treated water.
17. Discard the phenol layer according to local rules for the disposal of organic solvents.
18. Precipitation can be carried out either directly on dry ice, or in a –80°C freezer.
19. Incorporation of 90% of the isotope can be achieved in these reactions, but anything up to 50% incorporation is probably usable.
20. The first time a probe is synthesized it is worth checking the probe on a 4% polyacrylamide gel containing 7 *M* urea to check that there is no RNA degradation and that the radioisotope is incorporated into a single band. We routinely check our probes in this manner after synthesis.
21. Storage of the probe in 10 m*M* DTT ensures that the ^{35}S does not oxidize, as this can result in high background.
22. Probes can be stored in 10 m*M* DTT or in a 500-µL aliquot of 2x hybridization buffer. Probes degrade on storage due to radiolysis, and so cannot generally be stored for longer than 1–2 wk. Storage at –80°C can reduce probe degradation and therefore probes can last longer under these conditions. If probes are to be used in more than one hybridization reaction, aliquot the probe after synthesis to ensure that it is not frequently freeze–thawed. Often one freeze–thaw cycle (synthesis on d 1, storage overnight, and use the next day) may improve the signal achieved.
23. Adding fresh DEPC to the fix and wash solutions ensures that these solutions are clean, but also may inhibit background. It is thought that fresh DEPC may do this by inhibition of electrostatic binding to charged moieties in a manner similar to acetylation, and therefore obviates the necessity for the acetylation step. This may be why I have never found acetylation to be necessary.

24. Slides can be left in the last wash as long as necessary before draining and probe application.
25. Make enough HM for all the slides to be hybridized by multiplication of the number of slides by the amount of HM per slide (typically 100–200 μL), and make up HM in aliquots of 1 mL. Make more HM than calculated as the viscous solution is difficult to pipet accurately.
26. If probes are stored in aliquots of 2x HB, calculate the counts per microliter based on total probe counts diluted into 500 μL. Add an appropriate volume of probe to 500 μL of formamide and make up to 1 mL with 2x HB.
27. Cover slips can be used to ensure good probe overlay and prevent evaporation, but reduce the amount added to each slide to a maximum of 100 μL. If using cover slips make sure they are baked at 200°C to ensure they are clean.
28. Lining the lid of the box with 3MM paper and placing the slides on this surface is often easier than trying to place slides in deep boxes with radioactive solutions balanced on the tissue sections.
29. The lid of the box should be sealed with tape to prevent evaporation of the probe and/or box buffer during overnight incubation. Plastic boxes with lids that "snap" on can often pop open during high temperature incubations and are best avoided. Keep boxes horizontal during hybridization to make sure solutions do not run off the sections. Boxes used for hybridization should be kept separate from general glass and plasticware to prevent ribonuclease contamination.
30. Melting temperature (T_m) for 100% homologous cRNA probes

    $$= 79.8 + 58.4(F_{GC}) + 11.8(F_{GC})^2 + 18.5 \log[Na^+] - 820/bp - 0.35(\% \text{ formamide})$$

    Note: for nonhomologous probes subtract % mismatch

    F_{GC} = mole fraction G + C (typically 0.4–0.5). $[Na^+]$ in 1X SSC is 0.165 *M*.

    Although hybridization temperature can be calculated by the above formula, we optimize hybridization temperature by initial incubation at 45°C. This temperature is then increased (or decreased, although this is rare) as necessary to improve the signal to noise ratio. Hybridization temperatures above 60°C may cause coagulation of tissue proteins and affect tissue morphology. Lower hybridization temperatures will give higher nonspecific binding which may be difficult to remove. Higher temperatures reduce the specific binding as T_m is approached.
31. 5 m*M* DTT can be used as a reducing agent in all washes instead of β-mercaptoethanol if preferred. A reducing agent should be added to all washes of slides when using ^{35}S-labeled probes to prevent high backgrounds associated with probe oxidation.
32. The stringency of a wash is determined by temperature, salt concentration, and formamide concentration. High stringency is achieved at high temperatures, low salt concentrations, and high formamide concentrations as these conditions all destabilize hydrogen bonds involved in Watson–Crick basepairing. New probes are washed at increasing stringency (0.5X SSC, 0.2X SSC, 0.1X SSC at 50°C), removing slides at each stage to determine optimum salt concentration and strin-

gency. These are then used for future experiments. With established probes, maximally stringent washes are performed immediately after the nonstringent (2X SSC) wash. Wash volumes should be large; we use at least 1.5 L of SSC for each wash. Maximum wash temperature can be calculated as T_m – 5°C (remember to alter [Na^+] according to the SSC concentration of the wash). However, we rarely wash according to the T_m but begin at 50°C and never go above 60°C. Maximum wash temperature that we use is 60°C as protein coagulation above this temperature affects tissue structure and morphology. If higher stringency is needed than 0.1X SSC at 60°C we introduce formamide into the washes. However it is rare that this is required.

33. Exposure times vary between 1 and 3 wk depending on specific activity of the probe and mRNA abundance. For GPCRs we find that 2 wk exposure to film is often required for a discernible signal on film.
34. Exposure times are determined empirically for each probe depending on probe specific activity, age of isotope, and mRNA abundance. We tend to include additional slides that are developed after 2 wk to determine the optimal exposure time.
35. Use only distilled water for dilution of emulsion, fixer, and developer. Use of tap water in hard water areas can cause the emulsion to lift off the slides during development, thus ruining the entire experiment.
36. Tissue can be counterstained or left unstained. We tend to counterstain with haemotoxylin and eosin. Eosin will stain the emulsion layer on the back of the slide and this may need to be removed by scraping with a razor blade. Caution should be used with some counterstains, such as toludine blue, as contaminants may cause the silver grains to disappear. For more information on this and other potential problems with microautoradiography *see* http://www.ilford.com/html/us_english/prod_html/nuclear/faq1.html#Sect6
37. The protocol outlined in this chapter is the basic methodology with which we always begin when localizing a new target. Typically we try this method first, then add in these additional steps if we cannot achieve good signal and low background. Permeabilization, acetylation, or dehydration steps should be added before hybridization if used.
38. ISH is always a balance between good fixation of the tissue to make sure that mRNA remains intact and at the site of synthesis, and access of the probe that can be reduced by crosslinking fixative such as paraformaldehyde. Previously fixed tissue such as archival material will need permeabilization.
39. The time of digestion with proteinase K should be optimized for each fixation method, as overlong digestion will result in disruption of tissue morphology, whereas too little permeabilization will prevent access of the probe to the target RNA.
40. Acetylation reduces electrostatic binding, and therefore background by aceylating positively charged groups in the tissue. I have never found acetylation to be necessary.
41. Some tissues give better hybridization signals if dehydrated and dried prior to hybridization. These include tissues that have a high lipid content such as human

nervous tissues or fetal tissues. Tissues that have been dehydrated are stable at room temperature if kept dry and dust free. Slides can be hybridized directly from dry; indeed hybridization is often better as the probe is not diluted out by any 2X SSC left on the slide.

42. RNA yields are usually higher using TriZol than older methods.
43. Poly(A)$^+$ selection is not always necessary. Some GPCRs can be detected using single-stranded probes and total RNA. If using the latter approach use up to 20 μg of total RNA in 4 μL total volume per sample when running the Northern gel.
44. Poly(A)$^+$ RNA can be stored at –80°C at this point.
45. Commercial blotting apparatus can be used to transfer the RNA onto filters but we find that blotting by capillary action works extremely well and can be performed with the minimum of apparatus.
46. Turn the gel upside down to make sure the filter is on a smooth surface, that is, the undersurface of the gel. Make sure to mark one corner of the gel and filter to be able to orientate the filter after staining and hybridization. Make sure the plastic film is under the gel all the way around to prevent the stack of paper towels touching the 3MM paper on the support under the gel, and thereby allowing the 20X SSC to wick up the stack around the gel.
47. Nylon or nitrocellulose and charged or uncharged filters have all been used for Northern blotting. I find that Amersham Hybond N uncharged nylon filters work very well and are easily stripped and reprobed.
48. Use an unbreakable weight on top of the stack, as occasionally the stack tilts and falls over overnight.
49. The filter can be probed immediately—I find this often gives the best results—or it can be stored dry until needed. Air-dry the filter thoroughly and store at room temperature. Hydrate the filter in 2X SSC before prehybridization. After the filter has been probed, it can be stripped and reprobed multiple times provided it is kept wet between probings.
50. There are many protocols on Northern blotting giving details of different hybridization buffers (e.g., *[15]*). We find that the commercial rapid hybridization buffers such as QuickHyb (Stratagene) markedly reduce the time needed for an experiment, as probing and washing can be carried out in a single day.
51. The unincorporated radiolabeled dCTP is not removed from the probe after synthesis so we include excess unlabeled dCTP to prevent any background generated by the presence of [^{32}P]dCTP.
52. Do not allow the filter to dry once probed. If dried, the hybridized probe will not be removable.
53. Exposure times need to be determined empirically for each probe as they depend on the specific activity of the probe and the amount of target RNA. I usually expose Northern blots for GPCR for 1 wk, develop the film, and reexpose for a shorter or longer time as necessary. Note that due to the short half-life of ^{32}P, exposure times of much greater than 3 wk in total may not result in an improved signal.
54. I find that longer sequences are better for these probes (1–1.5 kb) as the specific activity of the probe is higher.

55. Keep premix for 4 wk only, as diluted nucleotides are not stable for long periods of time.
56. Do not use "Redivue" isotopes as they are not well incorporated by *Taq* polymerase.
57. An autoextension can be programmed into most thermocyclers. This is required as the radiolabeled nucleotide is limiting and the autoextension ensures that isotope is incorporated without premature termination of the probe toward the end of the cycle run. Run the reaction overnight as it can take 8 h. Label the thermocycler as containing radioisotope to warn other users of the potential hazard.

References

1. Wilkinson, D. G., (1992) *In Situ Hybridization: A Practical Approach*, Vol. 109. *The Practical Approach Series* (Hames, B. D., ed.), Oxford University Press, Oxford.
2. Masumoto, N., Esaki, T., and Sirotnak, F. M. (2001) Additional organizational features of the murine gamma-glutamyl hydrolase gene. Two remotely situated exons within the complement C3 gene locus encode an alternate 5′ end and proximal ORF under the control of a bidirectional promoter. *Gene* **268,** 183–194.
3. Michel, U., Stringaris, A. K., Nau, R., and Rieckmann, P. (2000) Differential expression of sense and antisense transcripts of the mitochondrial DNA region coding for ATPase 6 in fetal and adult porcine brain: identification of novel unusually assembled mitochondrial RNAs. *Biochem. Biophys. Res. Commun.* **271,** 170–180.
4. Van Den Eynde, B. J., Gaugler, B., Probst-Kepper, M., et al. (1999) A new antigen recognized by cytolytic T lymphocytes on a human kidney tumor results from reverse strand transcription. *J. Exp. Med.* **190,** 1793–1800.
5. Mise, N., Goto, Y., Nakajima, N., and Takagi, N. (1999) Molecular cloning of antisense transcripts of the mouse *Xist* gene. *Biochem. Biophys. Res. Commun.* **258,** 537–541.
6. Donaldson, L. F., McQueen, D. S., and Seckl, J. R. (1995) Neuropeptide gene expression and capsaicin-sensitive primary afferents: maintenance and spread of adjuvant arthritis in the rat. *J. Physiol.* **486,** 473–482.
7. Donaldson, L. F., Haskell, C. A., and Hanley, M. R. (2001) Messenger RNA localization and further characterisation of the putative tachykinin receptor NK4 (NK3B). *Recept. Channels* **7,** 259–272.
8. Donaldson, L. F., Harmar, A. J., McQueen, D. S., and Seckl, J. R. (1992) Increased expression of preprotachykinin, calcitonin gene-related peptide, but not vasoactive intestinal peptide messenger RNA in dorsal root ganglia during the development of adjuvant monoarthritis in the rat. *Brain Res. Mol. Brain Res.* **16,** 143–149.
9. Meinkoth, J. and Wahl, G. (1984) Hybridization of nucleic acids immobilized on solid supports. *Anal. Biochem.* **138,** 267–284.
10. Krumlauf, R. (1994) Analysis of gene expression by northern blot. *Mol. Biotechnol.* **2,** 227242.
11. Hastie, N. D. and Bishop, J. O. (1976) The expression of three abundance classes of messenger RNA in mouse tissues. *Cell* **9,** 761–774.

12. Moisan, M. P., Seckl, J. R., and Edwards, C. R. (1990) 11-Beta-hydroxysteroid dehydrogenase bioactivity and messenger RNA expression in rat forebrain: localization in hypothalamus, hippocampus, and cortex. *Endocrinology* **127,** 1450–1455.
13. Angerer, L. M. and Angerer, R. C. (1992). In situ hybridization to cellular RNA with radiolabelled RNA probes, in *In Situ Hybridization. A Practical Approach* (Wilkinson, D. G., ed.), Oxford University Press, Oxford, pp. 15–32.
14. Turgeon, J. L., Van Patten, S. M., Shyamala, G., and Waring, D. W. (1999) Steroid regulation of progesterone receptor expression in cultured rat gonadotropes. *Endocrinology* **140,** 2318–2325.
15. Sambrook, J., Fritsch, E. F., and Maniatis, T. (1989) *Molecular Cloning: A Laboratory Manual*, 2nd edit., Cold Spring Harbor Laboratory Press, Cold Spring Harbor, NY.
16. Chomczynski, P. and Sacchi, N. (1987) Single-step method of RNA isolation by acid guanidinium thiocyanate–phenol–chloroform extraction. *Anal. Biochem.* **162,** 156–159.

7

Electron Microscopic Studies of Receptor Localization

Rafael Luján

Summary

The localization and density of G-protein-coupled receptors (GPCRs) on the cell surface is a critical factor for specifying signaling within and between neurons. GPCRs and their associated signaling components are localized to specific neuronal compartments that give rise to a variety of functional implications. Therefore, information regarding the precise localization of GPCRs is a prerequisite for studies designed to elucidate their contribution to neuronal function, and can be achieved only by immunoelectron microscopy. Three main techniques for immunoelectron microscopy exist: the preembedding immunoperoxidase, the preembedding immunogold, and the postembedding immunogold techniques. The preembedding immunogold method is reliable for the localization of receptors at extrasynaptic and perisynaptic sites. The preembedding immunoperoxidase method provides valuable information on regional distribution of receptors. Finally, to localize any receptor at synaptic sites the only reliable method is the postembedding immunogold technique. Therefore, the three immunocytochemical methods provide complementary information about the cellular, subcellular, and subsynaptic location of any receptor.

Key Words

Electron microscopy, extrasynaptic receptor, freeze-substitution, $GABA_BR$, mGluR, high-resolution techniques, immunohistochemistry, metabotropic receptor, perisynaptic receptor, preembedding immunogold, preembedding immunoperoxidase, postembedding immunogold, protocols, quantification, synaptic receptor.

1. Introduction

This chapter describes the use of pre- and postembedding immunocytochemical techniques for the localization of G-protein-coupled receptors (GPCRs) at an electron microscopic level. The experimental procedures provided are derived from studies on metabotropic glutamate and γ-aminobutyric acid (GABA) receptors, the mGlu and $GABA_B$ receptors respectively ***(1,2)***. However, the general principles and methods can be applied to other GPCRs,

From: *Methods in Molecular Biology, vol. 259, Receptor Signal Transduction Protocols, 2nd ed.*
Edited by: G. B. Willars and R. A. J. Challiss © Humana Press Inc., Totowa, NJ

or to ionotropic receptors (*see*, e.g., **refs.** ***3,4***). Furthermore, the same methodology can also be successfully applied to elucidate the distribution of GPCRs during development ***(5–7)*** and in adulthood ***(1,2,8)***.

GPCRs are cell surface receptors with seven predicted transmembrane-spanning domains. The binding of a ligand to its associated seven transmembrane receptor produces a conformational change in the structure of the receptor, thus binding and activating specific heterotrimeric G-proteins (consisting of α, β, and γ subunits). G-proteins couple the activation of the surface receptor to changes in the activity of effector enzymes (e.g., adenylyl cyclases, phospholipases) and effector channels (e.g., ion channels, such as potassium or calcium channels). These signaling systems play a role in a variety of cell physiological functions ***(9)***.

Specialized neuronal functions are determined by the highly precise arrangements of axons and synapses. GPCRs and their associated signaling components are localized to specific post- or presynaptic sites. The location of GPCRs within different neuronal compartments has a variety of functional implications. Therefore, the precise localization and density of GPCRs on the cell surface seems to be a critical factor for specifying signaling within and between neurons, which can be determined only using high-resolution electron microscopic immunohistochemistry.

Immunohistochemical techniques can be broadly divided into two groups: preembedding techniques, whereby the immunolabeling steps take place before samples are embedded, and postembedding techniques, where the immunolabeling steps take place on embedded and sectioned samples. At an electron microscopic level, there are three main techniques: (1) preembedding immunoperoxidase, (2) preembedding immunogold, and (3) postembedding immunogold **(Fig. 1)**. Each of the three techniques has several advantages and disadvantages:

1. The preembedding immunoperoxidase method is the most sensitive procedure and provides valuable information on the regional distribution. However, owing to the diffusible nature of the peroxidase reaction end product **(Fig. 1A)**, this method is not suitable for localizing GPCRs, or any other receptors at synaptic sites, because we cannot exclude the possibility that the labeling of the synaptic junction originated from extrasynaptic sites ***(1)***.
2. The preembedding immunogold method produces a nondiffusible label, so the precise site of the reaction can be determined **(Fig. 1B)**. Furthermore, this method is reliable for the localization of receptors at extrasynaptic and perisynaptic sites in single and serial sections **(Fig. 2)**, and has been successfully used to quantify GPCRs along the plasma membrane ***(2,5,8)***. However, synaptic receptors at putative glutamatergic synapses are not generally detected using this method. Only

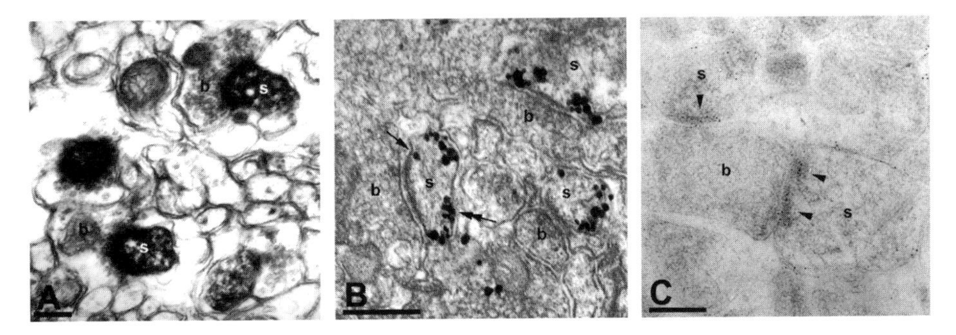

Fig. 1. Electron micrographs showing immunoreactivity for neurotransmitter receptors as demonstrated by three different methods at synapses between Purkinje cell dendritic spines (s) and parallel fiber terminals (b) in the molecular layer of the cerebellum. **(A)** Preembedding immunoperoxidase method. Peroxidase reaction endproduct for the metabotropic glutamate receptor type 1α (mGluR1α) fills dendritic spines of Purkinje cells and also covers the postsynaptic densities. **(B)** Preembedding immunogold method. Immunoparticles for mGluR1α are located along the plasma membrane of Purkinje cells *(double arrows)* spines and at the edge of postsynaptic densities *(arrow)*. **(C)** Postembedding immunogold method. Immunoparticles for the GluR2/3 subunits of the AMPA type ionotropic glutamate receptor are concentrated within postsynaptic densities *(arrowheads)*, and they also occur at extrasynaptic sites. Scale bars: 0.2 μm.

inhibitory synapses seem to be labeled for GPCRs using the preembedding immunogold method ***(10)***.

3. The postembedding immunogold method appears to be less sensitive than the preembedding methods. For instance, a much lower density of extrasynaptic receptors is revealed with the postembedding method than under preembedding conditions ***(8)***. However, this is the only reliable method to localize any receptor at synapses **(Fig. 1C)**, because the entire cut length of the plasma membrane is uniformly exposed to the antibodies as the sections are directly floated on to the antibody-containing solutions ***(11–13)***.

In conclusion, the three immunohistochemistry methods provide complementary information about the cellular, subcellular, and subsynaptic location of GPCRs and are best used in combination to achieve conclusive results.

2. Materials

2.1. Tissue and Fixation

1. Paraformaldehyde (supplied as powder). Keep at room temperature in a dry place. This is very toxic by contact with the skin, inhalation, and ingestion. Always wear protective clothing and gloves, and handle in a fume hood.

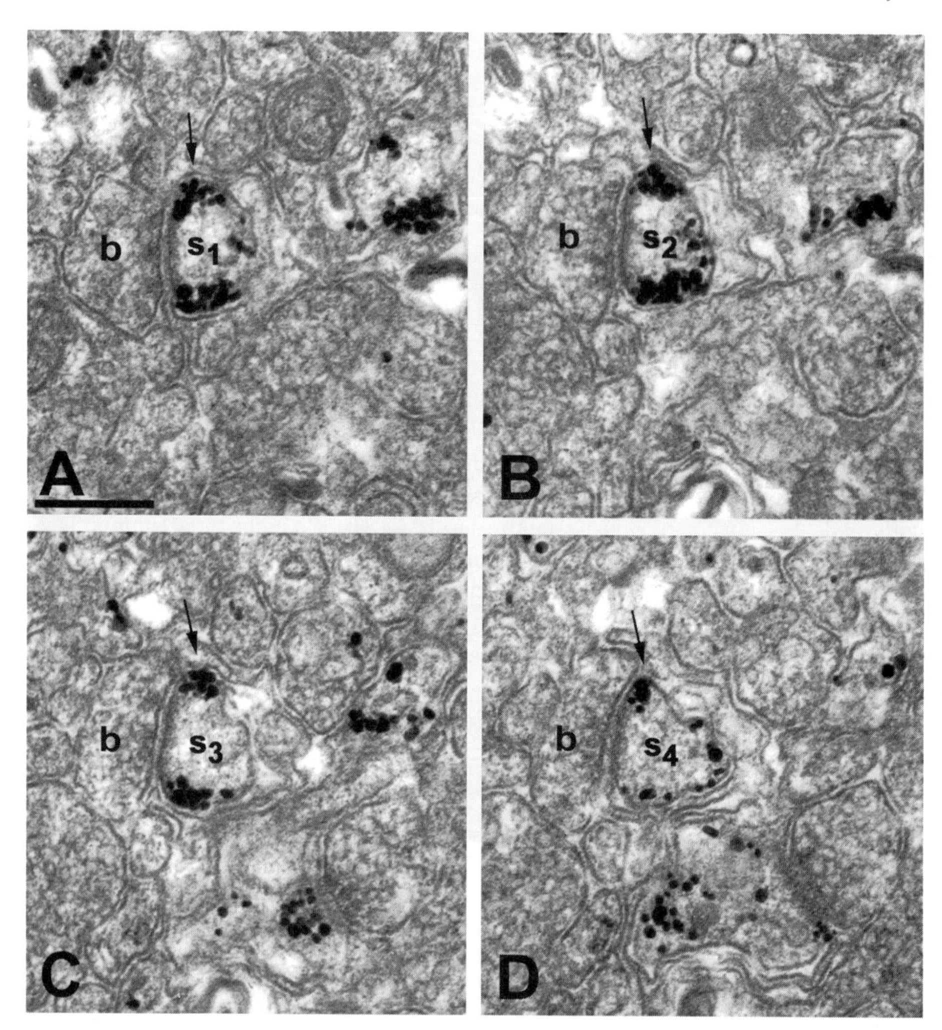

Fig. 2. Electron micrographs showing immunoreactivity for the metabotropic glutamate receptor type 1α (mGluR1α) in Purkinje cells dendritic spines as demonstrated by a preembedding immunogold in the molecular layer of the cerebellum. **(A–D)** Immunoparticles for mGluR1α are consistently located along the plasma membrane in serial sections of the same dendritic spine. Scale bars: 0.2 μm. Label definitions, same as in **Fig. 1**.

2. Glutaraldehyde (EM grade, supplied as a 25% aqueous solution). Can be stored at 4°C. Follow safety notes described for paraformaldehyde.
3. Buffer: 0.2 *M* sodium phosphate buffer (PB), pH 7.4. A mixture of 35.6 g/L of $Na_2HPO_4 \cdot 2H_2O$ and 31.2 g/L of $NaH_2PO_4 \cdot 2H_2O$, each at 0.2 *M*, in the ratio 4:1 (v/v), has a pH of approx 7.4. Can be stored at 4°C for several weeks.
4. Sodium hydroxide: 1 *M* in distilled water. Keep at 4°C for several weeks.

2.2. Membrane Permeabilization

2.2.1. Triton X-100

1. Triton X-100.
2. Buffer: 0.05 *M* Tris-buffered 0.9% saline, pH 7.4 (TBS) (*see* **Note 1**). Can be stored at 4°C for several weeks.

2.2.2. Freeze–Thaw

1. Sucrose: cryoprotectant solutions of 0.5 *M* and 1 *M* in 0.1 *M* PB.
2. Buffer: 0.1 *M* PB, pH 7.4. Prepared by diluting equal volumes of the 0.2 *M* PB stock solution with distilled water.
3. Liquid nitrogen. Can be stored in a Dewar flask.

2.3. Freeze–Substitution

1. Dry absolute methanol.
2. Sucrose: cryoprotectant solutions of 1 *M* and 2 *M* in 0.1 *M* PB.
3. Uranyl acetate: 0.5% in dry absolute methanol. Prepare 30 min before use, as it takes time to dissolve. Light sensitive. Filter before use. Very toxic, as it is a heavy metal and radioactive. Wear protective clothing and gloves, and handle in a fume hood.
4. Lowicryl resin (*see* **Note 2**). Avoid inhaling the vapors from the Lowicryl resin, as they are toxic. Wear protective clothing and gloves, and use a well-ventilated fume hood for mixing the components of the resin.

2.4. Preembedding Immunoperoxidase

1. Buffers: (a) 0.1 *M* PB, pH 7.4; (b) 0.05 *M* Tris buffer, pH 7.4 (TB); (c) 0.05 *M* TBS.
2. Monoclonal or polyclonal primary antibodies.
3. Normal goat serum. Can be aliquoted and stored at –20°C.
4. Biotinylated secondary antibodies: working dilution of 1 : 100 in TBS.
5. Avidin–biotin–peroxidase (ABC) complex: Vectastain kit (Vector Laboratories).
6. 3,3′-Diaminobenzidine tetrahydrochloride (DAB): 0.05% in TB. Light sensitive. Can be stored at –20°C.
7. Hydrogen peroxide: working solution of 1% in distilled water to give a final concentration of 0.01%.
8. Osmium tetroxide (supplied in the form of crystals in sealed ampoules): 1% in 0.1 *M* PB, pH 7.4 (*see* **Note 3**). Very toxic and volatile, so always wear protective clothing and gloves, and handle in a fume hood. Can be stored at 4°C for several weeks.
9. Uranyl acetate: 1% in double-distilled water. Follow safety notes described in **Subheading 2.3.**
10. Propylene oxide. It is extremely toxic by contact with the skin, inhalation, and ingestion, and also volatile and flammable. This will dissolve protective gloves. Handle with care in a fume hood using plastic pipets.

11. Graded series of ethanol in water (50%, 70%, 90%, 95%, and 100%, v/v).
12. Dry absolute ethanol: absolute ethanol with anhydrous cupric sulfate.
13. Epoxy resin (*see* **Note 4**). Some components of the epoxy resin are toxic, so always wear protective clothing and gloves, and handle in a fume hood.

2.5. Preembedding Immunogold

1. Buffers: (a) 0.1 *M* PB, pH 7.4; (b) 0.1 *M* phosphate-buffered 0.9% saline, pH 7.4 (PBS), prepared by mixing 0.9% NaCl with PB; (c) 0.05 *M* TBS, pH 7.4.
2. Primary antibodies and normal goat serum.
3. Secondary antibodies conjugated to 1.4-nm gold particles (Nanoprobes Inc.).
4. Glutaraldehyde: 1% in PBS. Follow safety notes described in **Subheading 2.1.**
5. HQ Silver intensification kit (Nanoprobes Inc.).
6. Material listed in **Subheading 2.4.** from **steps 8–13**.

2.6. Postembedding Immunogold

1. Buffers: (a) 0.05 *M* TBS, pH 7.4; (b) TBS containing 0.1% Triton X-100 (TBST).
2. Sodium ethanolate (*see* **Note 5**). Filter before use. Light sensitive.
3. Human serum albumin (supplied as powder, Sigma). Can be stored at 4°C.
4. Monoclonal or polyclonal primary antibodies.
5. Secondary antibodies conjugated to 10-nm colloidal gold particles.
6. Uranyl acetate: 40–60 μL of a saturated solution in distilled water. Filter before use. Follow safety notes described in **Subheading 2.3.**

2.7. Staining of Ultrathin Sections

1. Lead citrate (*see* **Note 6**). Store at 4°C. Wear protective gloves.
2. Sodium hydroxide pellets.
3. Laboratory film (e.g., Parafilm).

3. Methods

3.1. Tissue and Fixation

1. Prepare a fixative solution composed of 4% paraformaldehyde and 0.05% glutaraldehyde in 0.1 *M* PB, pH 7.4 (*see* **Note 7**).
2. Perfuse–fix the animal with the fixative solution (*see* **Note 8**).
3. Remove brain from the skull (or remove the tissue of interest) and wash in 0.1 *M* PB, pH 7.4.
4. Cut areas of interest into blocks. Wash blocks four times for 15 min each with 0.1 *M* PB, pH 7.4.
5. Cut blocks using a vibrating microtome obtaining sections that are 50–80 μm thick.
6. Store sections in 0.1 *M* PB at 4°C.

3.2. Membrane Permeabilization

To enhance the penetration of immunoreagents, it may be necessary after fixation to disrupt the plasma membranes to some extent. This can be done either by incubating the sections with the detergent Triton X-100, which dissolves the lipid layers of membranes, or by freeze–thawing, which mechanically disrupts the membranes by producing small ice crystals.

3.2.1. Triton X-100

1. Prepare a blocking solution of 10% normal serum in TBS containing 0.05% of Triton X-100.
2. Incubate sections and shake gently for 1 h at room temperature.
3. Wash the sections four times for 15 min each with TBS.

3.2.2. Freeze–Thaw

1. Prepare cryoprotectant solutions: (1) 0.5 *M* sucrose in 0.1 *M* PB; (2) 0.1 *M* sucrose in 0.1 *M* PB (*see* **Note 9**).
2. Place the sections in 0.5 *M* sucrose in 0.1 *M* PB and shake gently until they sink.
3. Place the sections in 1 *M* sucrose in 0.1 *M* PB and shake gently until they sink.
4. Drain off the excess sucrose and flatten the sections in the vial with a paintbrush.
5. Drop the vial quickly into liquid nitrogen and allow the tissue sections to freeze completely. Then let the sections thaw in 0.1 *M* PB at room temperature.
6. Wash the sections twice for 15 min each with 0.1 *M* PB.

3.3. Freeze–Substitution

1. Wash freshly fixed tissue four times for 15 min each in 0.1 *M* PB, pH 7.4.
2. Place sections (400 μm thickness, cut with a Vibratome) in cryoprotectant solutions of 1 *M* sucrose, until they sink, and then in 2 *M* sucrose, overnight at 4°C (*see* **Note 10**) ***(11–13)***.
3. Carry out the slam-freezing of the sections (Leica EM CPC). Follow the manufacturer's instructions.
4. Place sections in the chamber of the freeze-substitution machine (Leica AFS) set at –80°C. In order to program and to operate the machine follow the manufacturer's instructions. All of the following steps are carried out in this machine.
5. Place sections in 0.5% uranyl acetate in dry absolute methanol at –80°C overnight.
6. Wash four times for 2 h each with dry absolute methanol at –80°C. Subsequently, warm up to –50°C at 10°C/h.
7. Resin infiltration: infiltrate with a mixture of dry methanol–Lowicryl (1:1, v/v) for 2 h; change to a mixture of dry methanol–Lowicryl (1:2, v:/v) for 2 h; and then infiltrate in pure Lowicryl for 2 h. Leave in pure Lowicryl overnight at –50°C.
8. Embedding process: transfer the sections into freshly prepared Lowicryl resin in embedding molds at –50°C.
9. Start UV light (Leica AFS) polymerization for 48 h at –50°C. Subsequently, warm up to 18°C at 10°C/h under continuous UV irradiation. Further UV polymerization at 18°C for 24 h.

3.4. Preembedding Immunoperoxidase

1. Wash sections three times for 15 min each with 0.1 *M* PB to remove any remaining fixative. Then rinse sections with TBS for 30 min.
2. Block nonspecific binding by incubating sections in blocking solution, TBS containing 10% normal animal serum, for 1 h.
3. Incubate sections in primary antibody at 4°C for 12–24 h diluted in TBS containing 1% normal goat serum (*see* **Note 11**).
4. Rinse the sections four times for 15 min each with TBS.
5. Incubate sections in biotinylated secondary antibody in TBS containing 1% normal serum for 2–4 h at room temperature.
6. Rinse the sections three times for 15 min each with TBS.
7. Incubate sections in ABC in TBS for 2 h at room temperature (*see* **Note 12**).
8. Rinse the sections twice for 15 min each with TBS.
9. Rinse the sections twice for 15 min each with TB.
10. Incubate sections for 15 min in a solution of 0.05% DAB in TB at room temperature.
11. Add hydrogen peroxide to a final concentration of 0.01% to the DAB solution and incubate for a further 2–3 min (*see* **Note 13**).
12. Stop reaction in TB and then rinse the sections four times for 10 min each with TB.
13. Rinse the sections three times for 15 min each with 0.1 *M* PB.
14. Postfix sections with 1% osmium tetroxide solution in 0.1 *M* PB for 30 min (*see* **Note 14**).
15. Rinse the sections five times for 8–10 min each with 0.1 *M* PB to remove all traces of osmium, and then rinse them once for 8–10 min with distilled water.
16. Stain with 1% uranyl acetate solution in distilled water for 30 min.
17. Dehydrate sections with 50%, 70%, 90%, 95%, and 100% ethanol for 10–15 min each, and then 100% dry ethanol for 10–15 min (*see* **Note 15**).
18. Add propylene oxide twice for 10–15 min each.
19. Infiltration in epoxy resin (e.g., Durcupan) for 4 h, and flat-embedding (*see* **Note 16**).
20. Polymerize the resin by heating in an oven at 60°C for 48 h.

3.5. Preembedding Immunogold

1. Perform **steps 1–4** of **Subheading 3.4.**
2. Incubate sections in 1.4-nm gold compound-conjugated secondary antibody in TBS containing 1% normal serum for 2–4 h (*see* **Note 17**).
3. Rinse the sections four times for 15 min each with TBS.
4. Rinse the sections twice with PBS for 15 min each.
5. Postfix sections in PBS containing 1% glutaraldehyde for 10 min.
6. Rinse the sections three times with PBS for 10 min each.
7. Rinse the sections twice with distilled water for 10 min each.
8. Silver intensification according to the manufacturer's instructions (*see* **Note 18**).

9. Stop reaction in distilled water and then rinse the sections four times for 10 min each with distilled water.
10. Perform **steps 14–20** of **Subheading 3.4.**

3.6. Postembedding Immunogold

1. Place grids in a saturated solution of sodium ethanolate for 1 s (*see* **Note 5**).
2. Rinse grids quickly three times for several seconds each in double-distilled water.
3. Rinse grids twice for 5 min each in TBS (*see* **Note 19**).
4. Incubate grids in blocking solution: 2% human serum albumin (HSA) in TBST for 30 min at room temperature.
5. Incubate grids in primary antibody in TBST with 2% HSA at 27°C overnight (*see* **Note 11**).
6. Wash grids three times for 15 min each with TBST.
7. Incubate grids in 2% HSA in TBST for 10 min at room temperature.
8. Incubate sections in 10-nm gold compound-conjugated secondary antibody (*see* **Note 20**) in TBST with 2% HSA for 2 h at room temperature.
9. Wash grids four times for 10 min each in TBS.
10. Let grids dry at room temperature.
11. Place grids in a saturated solution of uranyl acetate in distilled water for 35 min at room temperature.
12. Rinse grids three times for several seconds each in double-distilled water and then let grids dry at room temperature.

3.7. Staining of Ultrathin Sections

1. Prepare the lead citrate solution (*see* **Note 6**).
2. Take a plastic or glass Petri dish with a piece of laboratory film and place a few sodium hydroxide pellets in the dish.
3. Take up the lead citrate solution with a Pasteur pipet, placing individual drops on the laboratory film.
4. Place grids with ultrathin section down on the drops of the lead citrate. Leave for 1.5–3 min.
5. Rinse grids three times for several seconds each in double-distilled water and then let grids completely dry with warm air using a hair dryer. Place grids in a grid box.

4. Notes

1. Tris buffer can be prepared in several ways. For precise accuracy and reproducibility, it is recommended to obtain a stock solution of 0.05 *M* TB, pH 7.4 at 25°C, by dissolving 6.61 g of Trizma HCl (Sigma) and 0.97 g Trizma Base (Sigma) in 1000 mL of distilled water. Desiccate the reagents before weighing. Once this stock solution is ready, TBS is prepared by dissolving 9 g/L of NaCl in TB.
2. Lowicryl resins consist of four variants (K4M, HM20, K11M, and HM23), which differ in their hydrophobic and hydrophilic properties and the temperature at

which they freeze. These acrylic resins provide improved antigenicity preservation over epoxy resins for the high-resolution localization of neurotransmitter receptors using postembedding immunohistochemistry. The most commonly used acrylic resin for GPCR localization is Lowicryl HM20 (Chemische Werke Lowi GMBH). It has three components: Crosslinker D, Monomer E, and Initiator C. The resin is prepared by weighing 2.98 g of Crosslinker D and 17.02 g of Monomer E, and mixing gently by bubbling a continuous stream of dry nitrogen gas into the mixture with a disposable glass pipet. Add 0.1 g of Initiator C to the final solution and mix again gently in the same way. Humidity will interfere with the polymerization of the resin. Another acrylic resin giving good results for postembedding immunocytochemistry is Unicryl (BioCell International) ***(11)***.

3. Usually, a stock solution of 4% osmium tetroxide in double-distilled water is made first. To prepare the stock solution, open a 1-g ampoule of osmium tetroxide and dissolve in 25 mL of double distilled water. The osmium tetroxide needs about 24 h to dissolve properly, so make sure to prepare the stock solution in advance. Osmication of the sections labeled with immunoperoxidase and immunogold methods is usually with a solution of 1% osmium tetroxide in 0.1 *M* PB. That percentage can be increased or reduced depending on the thickness of the sections, as the penetration of osmium tetroxide into brain tissue is very slow (0.5 mm/h).
4. One of the most commonly used epoxy resins is Durcupan ACM (Fluka). This resin is composed of four components. Mix the components A–B–D–C in that order and in the ratios 10:10:0.3:0.3 by wt, in a disposable plastic beaker. Mix thoroughly with a disposable plastic pipet. Durcupan resin easily facilitates flat-embedding (*see* **Note 15**) of the samples. To use other epoxy resins, follow the manufacturer's instructions.
5. Sodium ethanolate is used to remove Lowicryl resin from the ultrathin sections, allowing full accessibility of all immunoreagents to the antigen. We use a saturated solution of sodium ethanolate, which is prepared by mixing 50 g of NaOH with 300 mL of absolute ethanol. Leave the solution in the dark for several weeks until it turns brown before use. The degree of Lowicryl resin removed by this solution depends on the exposure time. It is recommended to perform some trials, but to avoid poor ultrastructure preservation and ultimately destruction of the tissue, do not leave grids with the ultrathin sections for longer than 4 s.
6. Although tissue sections can be examined in the electron microscope after immunohistochemical reactions, it is usually necessary to counterstain them using heavy metal solutions to improve contrast. The most commonly used stain is Reynold's lead citrate. To prepare the lead citrate solution: weigh out 0.133 g of lead nitrate and 0.176 g of trisodium citrate. Add 4.8 mL of double-distilled water to the lead nitrate. Shake gently to dissolve. Add the trisodium citrate to the lead nitrate solution. Shake gently until a milky suspension is formed and then rapidly add 0.2 mL of 4 *M* NaOH. Note that sodium carbonate (an insoluble electron-dense precipitate) is formed when the lead citrate solution is exposed to air.

7. Brain tissue must be fixed before pre- or postembedding immunocytochemistry. An appropriate fixation is very important for preserving antigenicity and ultrastructure. However, optimal conditions for good ultrastructural preservation contrast with the optimal conditions required for good preservation of antigenicity. Therefore, the degree of fixation involves a compromise between these two factors. As a fixative solution, a combination of paraformaldehyde and glutaraldehyde is most widely used. To prepare 1000 mL of 4% paraformaldehyde–0.05% glutaraldehyde in 0.1 *M* PB, pH 7.4: dissolve 40 g of paraformaldehyde in 300 mL of distilled water, make up to 500 mL with distilled water, filter, add 500 mL of 0.2 *M* PB, and add 2 mL of 25% glutaraldehyde. For safety reasons, all these steps should be carried out in a fume hood.
8. Fixation of the tissue can be carried out by immersion or by vascular perfusion. Immersion–fixation is generally used for postmortem or biopsy tissue, but for most purposes it is recommended to perform perfusion–fixation. The perfusion–fixation method allows the removal of blood from the animal, giving rise to a very rapid and uniform fixation. It consists of administering the fixative to deeply anesthetized animals through the heart into the aorta, either using a peristaltic pump or a gravity-powered mechanism, for 15–20 min. For postnatal animals, postfixation in the same fixative for 2–5 h should be carried out. A saline solution administered for about 1 min prior to the fixative solution may be used to rinse the vasculature, hence reducing the amount of blood. The flow rate of saline and fixative solutions should be approximately the same as the heart rate of the animal.
9. Instead of using the two sucrose solutions consecutively, it is possible to use a solution of 25% sucrose and 10% glycerol in 0.1 *M* PB, pH 7.4, until the sections sink. This alternative may slightly reduce the time for cryoprotection while giving rise to the same qualitative results.
10. Cryoprotection can also be carried out by placing sections in 10%, 20%, or 30% glycerol in 0.1 *M* Tris-maleate buffer, pH 7.4, overnight ***(11)***. The choice of whether to use glycerol instead of sucrose in the cryoprotection solution depends on the nature of the antigen. It is recommended to do trials, first using cryoprotection with 2 *M* sucrose. If the preservation of antigenicity is not appropriate, then cryoprotection with glycerol can be used instead.
11. The optimal dilution of any antibody should be determined in a test trial using a wide range of concentrations. In general, the final concentration of antibodies may be approx 0.5–4 µg/mL for preembedding conditions and approx 8–12 µg/mL for postembedding conditions. The binding of any antibody to an antigen is affected by the concentration of the antibody, the number of sections incubating per volume of antibody solution, and the duration and temperature of the incubation. It is therefore recommended to optimize all these parameters in each case, taking into account that short incubations should be carried out at room temperature and long incubations should be carried out at 4°C.
12. In the case of kits from Vector Laboratories, the ABC complex is made by mixing solution A (avidin) diluted 1:100, with solution B (biotinylated peroxidase)

diluted 1:100 in TBS. This complex should be made 30 min before use. There are several kits available from Vector Laboratories; the Elite kit has more sensitivity, but it is more expensive and gives a higher background reaction if the antibody binds nonspecifically to the tissue.

13. Monitor the peroxidase reaction by watching the development of color within the sections. Immunolabeled areas should appear dark brown, whereas areas lacking the specific antigen, as well as control sections (lacking primary or secondary antibodies), should appear white. The intensity of reaction can be checked with the light microscope, but because the DAB reaction is light sensitive, exposure to intense light should be very short. The speed of the peroxidase reaction depends on the antigen and concentration of immunoreagents. Hence it requires monitoring, then stop the reaction quickly with TB.
14. When osmium tetroxide is added to the vials, the sections will turn black rapidly and become rigid and fragile, so vigorous agitation should be avoided. One main problem associated with the osmium tetroxide step is that a strong osmication may reverse the silver intensification, and thus immunogold labeling disappears during processing. There are three possible solutions: (1) after osmication, do not leave the sections washing in 0.1 *M* PB overnight, but continue the processing and leave sections infiltrating in resin; (2) osmication may be reduced to 0.5% for 10–15 min; and (3) prepare a new stock solution of 4% osmium tetroxide in distilled water.
15. If the sections are intended to be flat-embedded on slides, they must be completely flat before polymerization. In this case, dehydration with 50% and 70% ethanol is carried out in vials, and then the sections transferred to a slide using a paintbrush. Cover the slide with a cover slip and add 90%, 95%, and 100% ethanol for 10–15 min each (ethanol will reach the section by capillarity). The sections are then transferred again to vials, to carry out the second 100% ethanol and next steps.
16. To perform the infiltration of the resin, transfer the sections to aluminum boats using a paintbrush. Add 1–2 mL of freshly prepared resin to each aluminum boat and ensure that the sections do not float on the resin. To perform the flat-embedding step, use a toothpick to very carefully remove the sections from the aluminum boats and place them on a slide. Add a few drops of resin on top of the sections and then place a cover slip and remove with filter paper the excess of resin.
17. Two different types of gold reagents are commercially available that can be used for preembedding conditions. One uses colloidal gold particles (ranging in size from 1 to 40 nm), which are based on hydrophobic interactions with the immunoglobulins. However, this association is very weak, and dissociated immunoglobulins lacking gold particles can be formed. The second type consists of 1.4-nm gold particles bound covalently to the immunoglobulin (Nanoprobes Inc.), avoiding the formation of IgG lacking gold particles. Because the main problem with the preembedding immunogold method is the limited penetration of immunoreagents, it is recommended to use the latter reagent.

18. There are several protocols for silver intensification. The HQ Silver intensification kit (Nanoprobes, Inc.) is one of the most widely used. It produces small silver particles that are approx 20–40 nm in diameter. It is easy to use, as it is prepared by thoroughly mixing equal volumes of three components (initiator, moderator, and activator). However, this kit is light sensitive, so should be used in a dark room. The reaction is quite fast, so to achieve a good size of the silver particles it normally takes 8–10 min, but any longer yields large immunoparticles. Alternatively, it is possible to use the silver intensification kit from AURION ***(14)***, which is light insensitive, has low viscosity, the size of silver particles can be easily controlled under the light microscope, and has high reproducibility. It also has three components (enhancer, initiator, and activator) that are easy to use following the manufacturer's instructions.
19. Glutaraldehyde may result in a poor penetration of immunoreagents and a high background staining. To prevent these problems, after the sodium ethanolate step and before immunolabeling steps, grids may be exposed to 0.1% sodium borohydride and 50 m*M* glycine in TBS containing 0.1% Triton X-100 (TBST) for 10 min ***(11,12)***. Sodium borohydride reduces the excess of aldehyde groups and thus may improve the sensitivity of the immunolabeling, but also may increase the background staining, so it is advisable to carry out test trials.
20. Gold particles of 10 nm size are easy to visualize on the ultrathin sections at the electron microscope level. Gold particles of a bigger size (e.g., 15 or 20 nm, up to 40 nm) are also available, that are easier to visualize at electron microscope level, but appear to provide less sensitivity to the postembedding method. The highest sensitivity is achieved using secondary antibodies conjugated to 1.4-nm gold particles. However, these are too small to visualize directly at the electron microscope level. Therefore, a silver intensification step, similar to that described for preembedding immunogold conditions (*see* **Subheading 3.4.**), should be carried out. For double-labeling experiments it is recommended to use a combination of 10-nm and 15-nm size gold particles. Owing to the low sensitivity of the postembedding immunogold method, we may add polyethyleneglycol (5 mg/mL), which dissolves lipids, to the final solution to facilitate the penetration of the 10-nm colloidal gold secondary antibodies and therefore to improve immunolabeling in our ultrathin sections.

References

1. Luján, R., Nusser, Z., Roberts, J. D. B., Shigemoto, R., and Somogyi, P. (1996) Perisynaptic location of metabotropic glutamate receptors mGluR1 and mGluR5 on dendrites and dendritic spines in the rat hippocampus. *Eur. J. Neurosci.* **8,** 1488–1500.
2. Kulik, A., Nakadate, K., Nyiri, G., et al. (2002) Distinct localization of $GABA_B$ receptors relative to synaptic sites in the rat cerebellum and ventrobasal thalamus. *Eur. J. Neurosci.* **15,** 291–307.
3. Baude, A. and Shigemoto, R. (1998) Cellular and subcellular distribution of substance P receptor immunoreactivity in the dorsal vagal complex of the rat and cat: a light and electron microscopic study. *J. Comp. Neurol.* **402,** 181–196.

4. Bernard, V., Somogyi, P., and Bolam, J. P. (1997) Cellular, subcellular, and subsynaptic distribution of AMPA-type glutamate receptor subunits in the neostriatum of the rat. *J. Neurosci.* **17,** 819–833.
5. López-Bendito, G., Shigemoto, R., Luján, R., and Juiz, J. M. (2001) Developmental changes in the localization of the mGluR1α subtype of metabotropic glutamate receptors in Purkinje cells. *Neuroscience* **105,** 413–429.
6. López-Bendito, G., Shigemoto, R., Fairén, A., and Luján, R. (2002) Differential distribution of group I metabotropic glutamate receptors during rat cortical development. *Cereb. Cortex* **12,** 625–638.
7. López-Bendito, G., Shigemoto, R., Kulik, A., Paulsen, O., Fairén, A., and Luján, R. (2002) Expression and distribution of metabotropic GABA receptor subtypes $GABA_BR1$ and $GABA_BR2$ during rat cortical development. *Eur. J. Neurosci.* **15,** 1766–1778.
8. Luján, R., Roberts, J. D. B., Shigemoto, R., and Somogyi, P. (1997) Differential plasma membrane distribution of metabotropic glutamate receptors mGluR1α, mGluR2 and mGluR5, relative to neurotransmitter release sites. *J. Chem. Neurochem.* **13,** 219–241.
9. Stryer, L. and Bourne, H. R. (1986) G-proteins: a family of signal transducers. *Annu. Rev. Cell Biol.* **2,** 391–419.
10. Hubert, G. W., Paquet, M., and Smith, Y. (2001) Differential subcellular localization of mGluR1a and mGluR5 in the rat and monkey substantia nigra. *J. Neurosci.* **21,** 1838–1847.
11. Nusser, Z., Luján, R., Laube, G., Roberts, J. D. B., Molnar, E., and Somogyi, P. (1998) Cell type and pathway dependence of synaptic AMPA receptor number and variability in the hippocampus. *Neuron* **21,** 545–559.
12. Landsent, A. S., Amiri-Moghaddam, M., Matsubara, A., et al. (1997) Differential localization of δ glutamate receptors in the rat cerebellum: coexpression with AMPA receptors in parallel fiber-spine synapses and absence from climbing fiber-spine synapses. *J. Neurosci.* **17,** 834–842.
13. Matsubara, A., Laake, J. H., Davanger, S., Usami, S., and Ottersen, O. P. (1996) Organization of AMPA receptor subunits at a glutamate synapse: a quantitative immunogold analysis of hair cell synapses in the rat organ of Corti. *J. Neurosci.* **16,** 4457–4467.
14. Wisden, W., Cope, D., Klausberger, T., et al. (2002) Ectopic expression of the $GABA_A$ receptor α6 subunit in hippocampal pyramidal neurons produces extrasynaptic receptors and an increased tonic inhibition. *Neuropharmacology* **43,** 530–549.

8

Generation of Model Cell Lines Expressing Recombinant G-Protein-Coupled Receptors

Emmanuel Hermans

Summary

The molecular cloning of the cDNA sequences encoding most G-protein-coupled receptors, including those from humans, allows their study in a variety of recombinant systems. In this respect, transfected mammalian cell lines constitute the most frequently used model for investigating the pharmacological and biochemical properties of these receptors. Several protocols have been described (based on the use of calcium phosphate precipitation, DEAE dextran, cationic lipids, and electroporation), allowing their transient or stable expression in diverse cell lines. This chapter gives a brief overview of the different techniques and provides methodology for the generation of transiently transfected cells and for selection, isolation and maintenance of stable transfected cell lines.

Key Words

Butyrate, calcium phosphate, cationic phospholipids, CHO cells, COS cells, DEAE dextran, electroporation, G-418, hygromycin B, mammalian cells, selection, stable, transfection, transient.

1. Introduction

Since the initial cloning of the prototypical G-protein-coupled receptors (GPCRs) (adrenergic and muscarinic receptors) ***(1–3)***, transfected cells expressing these cell surface receptors have probably became the most widely used model system in biochemical and pharmacological studies (for the first reports, *see* **refs. *4*** and ***5***). Cell transfection refers to the technique allowing the transfer of foreign nucleic acid (DNA or RNA) into living cells by nonviral approaches. By introducing the appropriate cDNA sequence cloned under the control of a viral promoter inside the cytosol, robust expression of the corresponding receptor has been obtained in a large variety of mammalian cell lines. In many cases, experiments are conducted on transiently transfected cells,

From: *Methods in Molecular Biology, vol. 259, Receptor Signal Transduction Protocols, 2nd ed.*
Edited by: G. B. Willars and R. A. J. Challiss © Humana Press Inc., Totowa, NJ

taking advantage of the high expression levels obtained after a couple of days, thanks to the efficient transcription of the cytosolic circular plasmid DNA harboring the sequence of interest. Indeed, many recent commercially available reagents facilitate the efficient transfer of foreign DNA in unexpectedly high proportions of growing cells. Stable transfection relies on the accidental and random integration of the foreign DNA into the chromosomes of the host cells. Although this event occurs at an extremely low frequency, the integrated DNA sequence will obviously be inherited by all cells deriving from further cell divisions. After the proliferation and selection of such cells, experiments can be performed on isolated clones of stably transfected cells. Although the generation, selection, and propagation of stable transfectants inevitably requires a prolonged period of time, these cells constitute valuable models for laboratory research. This chapter provides a brief overview of the different techniques generally used for the generation of transiently transfected cells as well as the selection, isolation, and maintenance of stable transfected cell lines for the study of GPCRs.

2. Materials

2.1. Nucleic Acid

2.1.1. Vector

The minimal requirements for a vector to be used in mammalian cell transfection are a promoter sequence, a multiple cloning site (MCS), and a polyadenylation site (pAD) (*see* **Note 1**). To ensure expression of the cloned sequence in any cell type, vectors for mammalian expression generally contain viral promoters such as the cytomegalovirus immediate early promoter (CMV) and the simian virus 40 early enhancer/promoter (SV40) DNA sequences (*see also* **Note 2**). CMV usually generates higher expression levels than SV40. The cDNA sequence encoding the receptor is generally amplified by the polymerase chain reaction (PCR) and only the coding sequence is inserted into the vector. To enhance the efficiency of transcription and translation, it is of critical importance that a TATA box and a Kozak sequence are present between the promoter and the translation initiation site ***(6)***.

Since the molecular cloning of the cDNA containing the coding sequence of the GPCR to be expressed is performed in *Escherichia coli*, the vector must also contain an origin of replication for prokaryotes (ORI) and an antibiotic resistance gene (e.g., ampicillin resistance, AmpR or tetracycline resistance, TetR) that can be used as selection markers after transformation of the bacterial strain. A prototypic example of such a plasmid is pSVK3 (available from Pharmacia in the early 1990s). Using such vectors means that the isolation of stable transfected cells requires the simultaneous transfection with a selection marker (most frequently a gene encoding resistance to an antibiotic that is toxic for mammalian cells). Efficient transfection have been obtained by cotransfecting cells with two distinct vectors that respectively drive the expression of the receptor and the selection marker (e.g., pSV2neo) (*see* **Note 3**). However, many vectors designed for cloning and mammalian cell transfection now contain the gene conferring antibiotic resistance, avoiding the need for cotransfection. The prototypical example of such a vector is pCDNA1 (from Invitrogen) which has now been replaced by improved versions while many similar vectors are available from several companies. A multitude of vectors can be found that differ in the nature of the promoter sequence, the nature of the selection marker or in the possibility to facilitate the insertion of PCR-amplified DNA sequences.

Unless the cloning procedure or specific questions to be addressed require the use of specialized vectors, the pCDNA3.1 vector (or similar) constitutes an appropriate tool for either transient or stable expression of GPCRs in mammalian cells. In many recent studies, the receptors are expressed as epitope-tagged proteins, thanks to the presence of additional DNA sequence at the vicinity of the cloning site within the vector or that can be added to the receptor sequence before insertion into the transfection plasmid. Epitope tagging facilitates the specific immunodetection or immunoprecipitation of the receptor expressed after transfection. Although appropriate controls need to be conducted, small epitopes (c-*myc*, polyhistidine, hemagglutinin) fused at the N- (extracellular) or C (intracellular)-terminus of the GPCR often do not affect cell surface expression, ligand binding, or functional coupling with intracellular G-proteins (*see* Chapter 5). In contrast, the fusion with larger peptide sequences may alter the properties of the receptor or the fused protein (*see* **Note 4**). Thus, although fusion of GPCRs with fluorescent proteins is widely used to characterize receptor trafficking in living cells *(7)*, only C-terminal fusion have been reported so far.

2.1.2. Quality of DNA

The purity of the plasmid DNA preparation is a critical determinant of transfection efficiency. Depending on the transfection protocol used, residual RNA, proteins, or a high salt concentration can affect the efficiency of mammalian cell transfection. Organic solvents (phenol, chloroform) used in some DNA extraction protocols are known to interfere with transfections mediated by the lipid-based reagents. Although transfection quality plasmid DNA can be obtained from large scale bacterial cultures followed by alkaline lysis and precipitation with polyethylene glycol and/or purification through a cesium chloride equilibrium gradient *(8)*, these techniques are laborious and time consuming. Many suppliers of molecular biology reagents produce kits for the purification of plasmid DNA based on anion-exchange chromatography that allow rapid preparation of high-purity DNA. Plasmid DNA is best conserved at a high concentration in aqueous solution (sterile nuclease-free water or Tris-HCl/EDTA buffers) at –20°C or –80°C (up to at least 1 yr) (*see* **Note 5**).

2.2. Cell Line

Transient or stable expression of many GPCRs by transfection has been reported in a variety of cell lines. The choice of a cell line devoid of any endogenously expressed related GPCR allows the study of a particular receptor type that is well characterized at the molecular level. Obviously, some biochemical properties of the receptor will directly depend on the cell line used. Fibroblast cell lines (e.g., CHO cells, HEK-293 cells, COS cells, A9 cells, BHK-21 cells, NIH/3T3 cells, L-929 cells, Rat-1 cells; *see* **Note 6** for details) are the most common hosts for pharmacological and biochemical studies of GPCRs in transfected cells. These cell lines generally grow fast and are thus easy to maintain. They show high efficiency of transfection and express few endogenous GPCRs. However, expression of GPCRs by transfection in cell lines from selected tissues is also frequently reported, especially those deriving from the nervous system (e.g., N1E-115, NG108-15 cells, C6 cells, PC12 cells, SH-SY5Y cells; *see* **Note 6** for details) ***(9–12)***. The choice of a cell line may also be dictated by specific properties (e.g., structural/functional polarization) ***(13–15)***. Several commercially available reagents designed for transfection of mammalian cells have been reported to enable transfection of many cell types in primary culture (including astrocytes and neurons). However, the efficiency is generally limited and only a few examples relate to the expression of GPCRs ***(16–18)***. Obviously, only transient transfection applies to cells in primary culture and virus-mediated gene transfer appears to provide better results ***(19)*** (*see* Chapter 9).

Among the four different methods described in the following subheadings, differences in transfection efficiency can be observed between cell types. Transfection with cationic lipids can be used for any cell type and is the method of choice for cells in primary culture. Calcium phosphate precipitation is frequently used for CHO or HEK-293 cells and we have used it successfully with N1E-115, C6, and PC12 cells. With some exceptions ***(20)***, the use of DEAE-dextran is generally restricted to the transient transfection of COS cells. Electroporation may be helpful for large scale cell transfection or when other techniques are inefficient. Because of the large amount of cells required and as cells have to be used in suspension (after monolayer harvesting), it is a technique generally not applicable to cells in primary culture.

2.3. Transfection Reagents and Equipment

2.3.1. Transfection by Calcium Phosphate Precipitation

1. Culture medium without antibiotics.
2. Sterile water.

3. 2X HBS: 50 m*M* 4-(2-hydroxyethylpiperazine)-1-ethanesulfonic acid (HEPES), 280 m*M* NaCl, 1.5 m*M* Na_2HPO_4 with pH adjusted to 7.1, sterile filtered (0.22-µm filter).
4. 2 *M* $CaCl_2$ in water, sterile filtered (0.22-µm filter).
5. Phosphate-buffered saline (PBS), sterile (autoclaved or filtered): 137 m*M* NaCl, 2.7 m*M* KCl, 4.3 m*M* Na_2HPO_4, 1.47 m*M* KH_2PO_4, pH 7.4.

The pH of the 2X HBS solution is critical and should be adjusted at room temperature. Large volumes of this solution are prepared and dispensed into small aliquots for storage at –20°C for long periods of time (more than a year). Each new batch should be tested using a thawed aliquot. For ease of use, we also prepare frozen aliquots of sterile filtered water and 2 *M* $CaCl_2$. However, for efficient transfection, thawed solutions need to be used at room temperature.

2.3.2. Transfection Using DEAE-Dextran

1. Culture medium without antibiotics and without serum.
2. Nu-serum (Becton Dickinson, Franklin Lanes, NJ, USA), which is a growth medium supplement as a low-protein alternative to newborn calf serum.
3. Diethylaminoethyl-dextran (DEAE-dextran), aqueous solution prepared at 40 mg/mL (sterile filtered) (Sigma, St. Louis, MO, USA).
4. Chloroquine diphosphate (Sigma, St. Louis, MO, USA), aqueous solution prepared at 10 m*M*.
5. Dimethyl sulfoxide (DMSO).
6. PBS, sterile (autoclaved or filtered): 137 m*M* NaCl, 2.7 m*M* KCl, 4.3 m*M* Na_2HPO_4, 1.47 m*M* KH_2PO_4, pH 7.4.

2.3.3. Transfection by Electroporation

1. Electroporator, available from many manufacturers (e.g., Bio-Rad Gene Pulser).
2. Bench centrifuge running 50-mL conical tubes (Falcon tubes) at 700*g*.
3. Sterile electroporation cuvets (gap width: 0.4 cm).

2.3.4. Transfection Using Cationic Lipids

This technique requires the use of a solution of cationic lipids that is available from several suppliers as "ready for use." Protocols may differ slightly between suppliers, but in most case, all reagents are provided. In some cases, it is suggested that the transfection reagent is diluted in a defined medium (e.g., Opti-MemI for Lipofectamine reagents—Invitrogen, Paisley, UK).

2.4. Selection Antibiotics

1. G-418 disulfate (Geneticin® from Invitrogen, Paisley, UK) (*see* also **Note 7**).
2. Hygromycin B (Cabiochem-Novabiochem, La Jolla, CA, USA) (*see* also **Note 7**).

A variety of other antibiotics have been developed recently for the selection of stable transfectants harboring the corresponding resistance gene (e.g., puromycin for selection of cells expressing puromycin-*N*-acetyltransferase, bleomycin derivatives for selection of cells expressing the bleomycin-resistance gene *[Sh ble]*, and blasticidin S for selection of cells expressing blasticidin deaminase).

3. Methods

3.1. Transfection

The protocols indicated below are optimized for the development of stable cell lines and are generally performed in 35-mm wells (i.e., 6-well plates). However, volumes and DNA quantity can be adapted for transfection in 10-cm dishes which may be more convenient for transient transfections. Protocols should be adapted on the basis of the area of the culture support (*see* **Note 8**). In transient transfections, GPCRs are sometimes studied in 24- or 96-well plates and transfection can be directly performed in these cultures plates or in larger dishes from which cells can be trypsinized and replated 24–36 h after transfection.

3.1.1. Calcium Phosphate Precipitation

It is convenient to transfect cells at the end of the afternoon so as to leave cells overnight in contact with the precipitate. This technique was originally developed by Graham and van der Eb ***(21)*** and has been widely used with some minor modifications. The following protocol is simple and gives good results.

1. Seed the cells the day before transfection (d 1) in order to obtain 50% confluence on the day of transfection. For some cells growing or adhering slowly (experienced with some batches of HEK-293), cells can be plated 2 d before transfection (*see* **Note 9**). Cells are plated in 6-well plates with 2 mL of medium/well.
2. At 4 h before transfection, change the medium (2 mL/well). Also take the aliquots of 2X HBS and $CaCl_2$ out of –20°C, and allow them to equilibrate at room temperature.
3. Approximately 1 h before transfection, prepare the precipitate using small containers (7-mL sterilins are quite convenient). The entire procedure has to be performed in sterile conditions, in a laminar flow hood.

   The following volumes are adapted for the transfection of 3 wells of a 6-well plate.

   Sterilin A: 0.5 mL of 2X HBS.

   Sterilin B: prepare in the following order: water (so that the final volume will be 0.5 mL), 10 μg of supercoiled DNA, 65 μL of $CaCl_2$. Mix well. While holding (firmly) sterilin A on a vortex, add progressively (dropwise) the content of sterilin B (a 1-mL micropipet is convenient). Instead of vortexing, continuous mixing may be achieved by bubbling air with a 1-mL sterile pipet using an electric pipettor. In these conditions, the final concentration of DNA is 3 μg/well, but can be increased to 4–10 μg/well.

4. Leave the tube in the laminar flow hood for 35–45 min. A small precipitate will form, which is often difficult to see. In some cases, the solution will just appear opalescent.
5. Before adding the precipitate to the cells, remix the tube. Add the precipitate (300 μL/well) dropwise to the well, while gently swirling the plate.
6. Place the plate back into the incubator.
7. After 8 h or an overnight incubation, rinse the plates twice with PBS or culture medium and add fresh culture medium.

3.1.2. DEAE-Dextran

Although this technique (initially developed by Vaheri and Pagano *[22]*) has been used for the transient or stable transfection of other cell types (including HeLa cells and NIH 3T3 cells), it is almost exclusively used for the transient transfection of COS cells. As indicated above, transient transfection is usually performed in 10-cm plates or in 24-well plates.

1. Seed the cells on the day before transfection to obtain 50% confluence on the day of transfection (*see* **Note 9**).
2. Before transfection, prepare the following mix: 9 mL of culture medium (without serum, antibiotics, and amino acids), 1 mL of Nu serum, 10–30 μg of supercoiled DNA, 100 μL of chloroquine phosphate (10 m*M* aqueous solution), and 100 μL of DEAE-dextran (40 mg/mL aqueous solution) (*see* **Note 9**).
3. Aspirate the culture medium and rinse twice with medium without serum.
4. Add the transfection mixture at 10 mL/10 cm plate and place back into the incubator for 4 h. (For other culture supports *see* **Note 8** for adapted volumes.)
5. Aspirate the medium and add 5 mL of PBS supplemented with 10% DMSO (*see* **Note 8**).
6. Leave for 2 min at room temperature (in the culture laminar flow hood).
7. Aspirate and replace with 10 mL of culture medium (with serum, antibiotics, and other supplements as required) (*see* **Note 8**).
8. Leave cells in the incubator until use. Allow a minimum of 12 h before passaging the cells.

3.1.3. Electroporation ***(23)***

1. Each transfection requires 10^7 cells. Better results are obtained with dividing cells and cells should therefore be obtained from culture flasks before reaching maximal confluence.
2. Harvest cells from a sufficient number of flasks by centrifugation (700*g*, 5 min at room temperature) in a 50-mL tube (Falcon). Resuspend and pellet the cells in a small volume of culture medium (with serum, antibiotics, and other supplements as required) (*see* **Note 9**).
3. After cell counting, dilute the cells to obtain 10^7 cells/450 μL.
4. Place 500 μL of cell suspension in the electroporation cuvet and add 20–50 μg of DNA diluted in 50 μL of water.

5. Electroporate cells. Although electroporation conditions should be optimized for each cell line, most protocols suggest the following guidelines: voltage 250 V; capacitance 960 μF; one single pulse.
6. Leave the cells in the cuvet for 10 min at room temperature.
7. Transfer cells from the cuvet to two 10-cm culture dishes containing 10 mL of prewarmed culture medium.
8. Allow cells to recover overnight in the incubator and renew medium the next morning.

3.1.4. Cationic Lipids (or Cationic Polymers)

Tranfection of mammalian cell monolayers with cationic lipids is generally rapid and straightforward. Detailed protocols are always supplied by the manufacturer. Basically, supercoiled DNA is usually mixed with an appropriate volume of a cationic lipid solution (*see* **Note 5**). After incubation at room temperature for 30–60 min, this mix is added to growing cells, often after elimination of the serum (*see* **Note 9**). After a few hours, culture medium and serum are added to increase the volume and reach the optimal serum concentration for cell growth (the transfection mix is not removed). Finally, the culture medium is renewed after 24 h. Initially, the cell density/DNA/lipid ratio as well as the incubation time has to be optimized for each cell type. Recent formulations appear to be more flexible and supplier guidelines are applicable for most cell types.

3.2. Selection

As indicated above (*see* **Subheading 2.1.1.**), isolation of stable transfectants requires the specific elimination of nontransfected (or transiently transfected) cells which is usually based on the simultaneous transfer of a gene encoding the resistance to an antibiotic. The most widely used selection marker is the resistance to neomycin since the *NeoR* gene is present in the majority of mammalian cell transfection vectors. Alternatively, vectors encoding the resistance to hygromycin B are available and are useful for transfecting cells that already carry the *NeoR* gene resistance (i.e., for two consecutive transfections). Selection is generally started 2 or 3 d after transfection by supplementing the culture medium with G-418 or hygromycin B. The concentration of the antibiotic should not be increased progressively, but instead, the appropriate concentration should be used immediately. Generally, toxicity and cell death are noticeable only after 3–4 d. At that time, renewal of the culture medium is necessary every day or every other day to remove dead cells and debris. It can be useful to set up sister wells where DNA was omitted during the transfection protocol, allowing the efficiency of the selection procedure to be evaluated. Interestingly, it has been suggested that the activity of G-418 is decreased when the classical antibiotics penicillin and streptomycin are present in the culture medium.

Not all cell lines are equally sensitive to G-418 or hygromycin B and large variations are sometimes observed between different batches of the same cell line. The optimal selection concentration (killing nontransfected cells) has to be determined before transfection. Several cultures dishes are seeded at low density in culture medium supplemented with increasing concentrations of the antibiotic. The optimal concentrations of G-418 and hygromycin B for the selection of mammalian cells carrying the corresponding resistance gene range from 400 to 1000 μg/mL and 100 to 400 μg/mL, respectively (*see* **Note 10**).

3.3. Isolation and Propagation of Clones

Because of the rather low efficiency of stable transfection, the population of cells remaining after selection is highly heterogeneous regarding the level of expression of the transfected gene. Isolation of clonal cell lines is required to obtain constant expression levels and therefore reproducible functions. The following techniques describe how to isolate and transfer clonal cells in 24-well plates. After isolation and proliferation, the cells will be successively transferred to 6-well plates and medium sized flasks before providing enough material to determine the expression of the recombinant GPCR. Growth is rather slow at the beginning and depending on the cell type, these proliferation steps will take at least 3–6 wk.

3.3.1. Picking Colonies

1. Antibiotic-resistant cells generally grow as colonies that are easily detected after elimination of all nontransfected dead cells and cell debris. On visualization of the colonies to be picked (with the microscope or sometimes with the naked eye), mark their positions by drawing circles on the bottom of the dish.
2. Aspirate the medium and using forceps, place small, sterilized glass cylinders (cloning disks) on the circled colonies (it is sometimes helpful to glue the disks with a small amount of silicon grease).
3. Add 50 μL of trypsin to the inside of the cylinders and leave in the laminar flow hood for 5 min.
4. Using fresh tips for each colony, collect the contents from each disk and transfer to a single well of a 24-well plate containing 2 mL of culture medium.

To obtain pure clonal cell lines, it is critical to pick colonies as early as possible once the nontransfected cells have been eliminated. Otherwise, colonies will become too large and may fuse, or cells from the middle of colonies may detach and reseed at a distance, contaminating neighboring colonies.

An alternative to the use of cloning disks is to directly pick the colonies using a pipettor set to 10 μL and sterile pipet tips. There is no need to remove the culture medium from the plate. Instead, just incline the entire plate to scrape the marked colonies which are then directly transferred to the 24-well plate containing culture medium. This technique offers the advantage of being fast and avoiding drying out the rest of the cells. If needed, culture medium can be renewed after colony picking and the entire population can then be screened again later by limited dilution.

3.3.2. Limited Dilution

1. After elimination of nontransfected dead cells and cell debris, allow the colonies to grow for 2–3 d to increase the total number of cells.
2. Trypsinize the entire population of transfected cells and collect them in culture medium. After counting, prepare 2 mL of cell suspension adjusted to 1000 cells/mL. The remaining cells can be frozen or seeded in order to obtain a "total population."
3. Transfer 100 μL of the cell suspension into the first row of a 96-well plate (flat-bottom) and put 100 μL fresh culture medium in the seven other rows.
4. Using a 12-multichannel pipettor, transfer 30 μL from row 1 to row 2. After gentle mixing by repetitive pipetting, transfer 30 μL from row 2 to row 3 and proceed for the other rows until row 8. It is not necessary to change tips between rows.
5. Leave in the incubator for approx 1 wk.
6. Using a microscope at low magnification, screen every well and spot those containing a single round shaped colony (likely to be found in the middle rows). *Note:* the first row rapidly reaches confluence and this might be helpful for setting the focus.
7. Trypsinize the cells from the well containing single colonies and transfer to single wells of a 24-well plate containing culture medium.

3.4. Determining GPCR Expression (see also *Note 11*)

A single transfection may give rise to dozens of clones to be tested, requiring laborious and time-consuming cell culture of multiple cell lines. Variation in the expression level between clones is frequently observed and this could be related to differences in the site of insertion within the chromosomes or from the number of inserted copies of the transgene. Also, despite their resistance to the selection antibiotic, some clones may show low or no expression of the transgene. The goal is to detect the positive clones as early as possible. The most quantitative evaluation of the expression of a GPCR is to measure the specific binding of an appropriate radioligand. However, this requires relatively large amounts of cells and therefore frequently delays the elimination of negative clones. In contrast, functional assays or immunodetection can be performed on a very small number of cells (*see also* **Note 4**). However, these approaches do not provide quantitative information that allows the selection of high or low expressing clones among the positive ones. For some GPCRs, radioligand binding can be measured on intact cells seeded in 24-well plates. This requires fewer cells than the preparation of a crude homogenate and is a good compromise between delay of selection and quantitative screening.

3.4.1. Radioligand Binding on Homogenates (see also *Chapter 1*)

At an early stage of the screening of clones, the selectivity of this radioligand is of minor importance since the target receptor is cloned and pharmacologically characterized. In addition, nontransfected cells will constitute a negative control. For most receptors, a simple radioligand-binding assay (determination of the specific and nonspecific binding in triplicate) can be performed on the crude homogenate from a single 10-cm dish prepared as indicated below. A single concentration of radioligand should be tested (close to the K_D value) and a well characterized unlabeled competitor should be used for determining the nonspecific binding (blank).

Preparation of a crude homogenate from a single 10-cm culture dish:

1. Aspirate the culture medium and rinse the cells twice with ice-cold PBS.
2. Scrape the cells in 1 mL of PBS and collect the suspension in a 2-mL microtube. Rinse the plate with another 1 mL of PBS, collect, and transfer to the same tube. Centrifuge at maximal speed in a microfuge for 3 min at 0–4°C and discard the supernatant.
3. Resuspend the pellet in 1 mL of 50 m*M* Tris-HCl, pH 7.4, using a 1 mL syringe and a 26-gauge needle. Centrifuge at maximal speed for 15 min at 0–4°C and discard the supernatant. Repeat step 3 twice.
4. The final pellet is resuspended in 200 µL 50 m*M* Tris-HCl, pH 7.4. This crude homogenate can be used for protein and radioligand-binding assays after dilution in the appropriate binding buffer.

3.4.2. Radioligand Binding on Intact Cells (see also *Chapter 1*)

In this case, cells are seeded in 24-well plates (6 wells per clone: 3 for total binding and 3 for nonspecific binding). At confluence, cells are rinsed with PBS or isotonic binding buffer and incubated with the radioligand in optimized conditions (buffer, temperature, incubation time). Wells are rapidly rinsed twice with ice-cold PBS and cells are finally dissolved in 0.1 *N* NaOH or detergents. Radioactivity associated with the cells is counted, providing a gross estimation of receptor expression. A detailed overview of the technique of radioligand binding on intact cells can be found elsewhere (*see* **refs. *24–26***).

3.4.3. Functional Assay

The presence of a GPCR in most cells can be tested by evaluating the functional response to a specific agonist. A variety of techniques have been developed to easily evaluate the usual responses associated with GPCR stimulation. The expression of GPCRs coupled to the activation of phospholipase C can easily be detected by screening clones for agonist-mediated increases in the intracellular calcium concentration (*see* **Note 10**). The use of several intracellular calcium probes (e.g., Fura-2, Fluo-3, Indo) and dynamic fluorescence microscopy allows the detection of functional responses from very few cells. This approach can therefore be used shortly after isolation of clones. High-throughput screening of clones can be performed using specialized equipment allowing simultaneous fluorescence recording from 96-well plates *(27)*.

3.4.4. Immunodetection of GPCR (see also *Chapter 4)*

When specific antibodies raised again the receptor are available (*see* Chapter 3) or when an epitope tagged receptor is expressed (*see* Chapter 5), both immunoblotting and immunocytochemistry can help to screen positive clones. Although immunoblotting may provide semiquantitative data concerning the expression level of the GPCR, the preparation of the sample and the quantitation of protein concentration may require a relatively large amount of cells. Unless no radioligand for the receptor is available, immunoblotting does not offer major advantages in comparison to binding assays. In contrast, immunocytochemistry can be performed with a minimal amount of cells. However, this technique provides limited information since it is nonquantitative and may frequently give false-negative results.

4. Notes

1. The transcriptional activity of the viral promoter ensures particularly high expression levels of the cloned receptor. In transient transfection, even higher expression levels can be observed in cell lines that ensure the replication of the plasmid (permissive cells—well documented in the case of COS cells). Such extrachromosomal replication requires the presence of critical DNA sequences in the transfection vector. In such cases, the transfected cells can be used only for a couple of days, as the accumulation of plasmid DNA usually results in cell death.
2. The use of clonal cell lines ensures homogeneous expression of the GPCR in the experimental model. In theory, the constant transcriptional control by the viral promoter ensures the stable expression of the receptor, independently of the cell cycle or other environmental conditions. Recently, inducible systems for expression of proteins in mammalian cells have been developed. In these systems, the sequence of the viral promoter controlling the expression of the cloned gene contains additional regulatory elements allowing the exogenous manipulation of the transcriptional activity *(28)*. Using such systems, the expression level of the cloned gene can be tightly controlled through the addition or elimination of a given reagent in the culture medium. Different systems are now commercially available and there are now many examples of their use in the generation of stable transfected cell lines in which the expression of a given GPCR can be induced *(29–33)*.
3. As indicated under **Subheading 2.1.1.**, the generation of stable transfectants using vectors lacking any selection marker requires the cotransfection with a second plasmid such as pSV2neo (Clontech, Palo Alto, CA, USA). The success of this

approach results from the relatively high probability that when stable chromosomal integration occurs, this happens to both plasmids. However, to avoid the selection of cells being resistant to G-418 without expressing the gene of interest, it is suggested to work with a plasmid ratio (pSV2neo/vector containing the transgene) of 1:9.

4. An advantage of using transfected cells for the study of GPCRs is that the cDNA sequence encoding the receptor can be manipulated to modify the structure of the recombinant protein. This approach has been widely used to assess the molecular determinants of ligand binding and G-protein coupling of a multitude of receptors (e.g., Chapter 19). As mentioned above, it also allows the generation of epitope-tagged receptors or chimeric proteins (*see* Chapter 5). However, these structure or sequence modifications may affect radioligand binding or functional properties and thereby interfere with the experimental detection of positive clones.
5. Supercoiled plasmid DNA is generally used for transfection of mammalian cells. Although some data from the literature reports on the linearization of the plasmid before stable transfection (perhaps to avoid intracellular linearization within the sequence of interest before chromosomal integration), this was not found to improve the efficiency of transfection. Plasmids should not be linearized for transient transfection, in particular when its replication in mammalian cells is expected.
6. Most studies on GPCRs expressed after transfection of mammalian cells were conducted in fibroblasts cell lines (including fibroblast-like cells) and in cell lines derived from the central nervous system. **Table 1** gives a brief description of the most widely used cell lines.
7. The antibiotic G-418 is not available as 100% pure powder and the accurate concentration has to be calculated on the basis of purity. Some manufacturers supply "ready to use" standardized solutions of G-418. Hygromycin B purity is also variable and since its activity is frequently expressed in units, the conversion factor (unit per milligram) must be known. Hygromycin B is highly irritating to eyes and skin and should be used with extreme caution.
8. The quantity of plasmid DNA to be used in most transfection protocols depends on the number of cells and, in the case of adherent cells growing as monolayers, is proportional to the area of the culture plates. For cationic lipids and DEAE-dextran mediated transfections, the final concentration of the reagents in the culture medium also needs to be considered. The following table indicates the area of plasticware commonly used in mammalian cell transfection and shows the appropriate volume of culture medium to be used during transfection.

Culture surface	Area (cm^2)	Volume of medium (mL)
10-cm dish	58.95	10
6-cm dish	19.5	3
6-well plate	9.6	2
12-well plate	3.8	1
24-well plate	2	0.5
96-well plate	0.32	0.1

Table 1
Mammalian Fibroblast Cell Lines Used to Study Recombinantly Expressed GPCRs

Cell line	ATCC reference	Description
A9	CCL-1.4	Murine subcutaneous connective tissue
BHK-21	CCL-10	Fibroblast from hamster kidney
C6	CCL-107	Rat glial tumor
CHO-K1	CCL-61	Fibroblasts from hamster ovary
COS-1	CRL-1650	Fibroblast-like cell line established from CV-1 cells (which derive from African green monkey kidney)
HEK-293	CRL-1573	Human embryonic kidney fibroblasts
L-929	CCL-1	Murine subcutaneous connective tissue
N1E-115	CRL-2263	Mouse neuroblastoma
NG108-15	HB-12317	Fusion of mouse neuroblastoma cells with rat glioma cells
NIH/3T3	CRL-1658	Fibroblast-like cells established from mouse embryo cultures
PC-12	CRL-1721	Rat pheochromocytoma
Rat-1	CRL-2210	Fibroblast from rat connective tissue
SH-SY5Y	CRL-2266	Human neuroblastoma

9. Dividing cells show the highest transfection efficiency. Therefore, transfection is generally performed 24 h after seeding at a density of approx 50–70%. Although some protocols allow the presence of serum during transfection, the critical steps are frequently performed in serum-free conditions. Cell transfection techniques introduce large DNA fragments into the cells and are potential source of stress for the cells. Therefore, it is advisable to supplement the culture medium with serum as soon as possible after transfection.
10. Once the transfected cells have been selected and clones isolated, the selection antibiotic is added to the culture medium for maintenance and propagation although the final concentration can be decreased by 75%. However, when cells are seeded for GPCR studies, it is generally better to eliminate antibiotics from the culture medium. Indeed, as other aminoglycosides, G-418 theoretically could inhibit phospholipase activity ***(34)***.
11. Sodium butyrate is known to arrest cell growth and enhances the activity of the cytomegalovirus promoter ***(35,36)***. It has been sometimes added to culture media of transfected cells (5 m*M* for 6–48 h before the experiment) to increase the expression of recombinant GPCRs in HEK cells ***(37)*** and CHO cells ***(38,39)***. Sodium butyrate (*n*-butyric acid sodium salt) can be prepared as aqueous solution and is available from Sigma (St. Louis, MO, USA).

References

1. Frielle, T., Collins, S., Daniel, K. W., Caron, M. G., Lefkowitz, R. J., and Kobilka, B. K. (1987) Cloning of the cDNA for the human beta 1-adrenergic receptor. *Proc. Natl. Acad. Sci. USA* **84,** 7920–7924.
2. Kubo, T., Fukuda, K., Mikami, A., et al. (1986) Cloning, sequencing and expression of complementary DNA encoding the muscarinic acetylcholine receptor. *Nature* **323,** 411–416.
3. Chung, F. Z., Lentes, K. U., Gocayne, J., et al. (1987) Cloning and sequence analysis of the human brain beta-adrenergic receptor. Evolutionary relationship to rodent and avian beta-receptors and porcine muscarinic receptors. *FEBS Lett.* **211,** 200–206.
4. Brann, M. R., Buckley, N. J., Jones, S. V., and Bonner, T. I. (1987) Expression of a cloned muscarinic receptor in A9 L cells. *Mol. Pharmacol.* **32,** 450–455.
5. Fraser, C. M., Chung, F. Z., and Venter, J. C. (1987) Continuous high density expression of human β2-adrenergic receptors in a mouse cell line previously lacking beta-receptors. *J. Biol. Chem.* **262,** 14,843–14,846.
6. Kozak, M. (1987) At least six nucleotides preceding the AUG initiator codon enhance translation in mammalian cells. *J. Mol. Biol.* **196,** 947–950.
7. Kallal, L. and Benovic, J. L. (2000) Using green fluorescent proteins to stud G-protein-coupled receptor localization and trafficking. *Trends Pharmacol. Sci.* **21,** 175–180.
8. Sambrook, J., Fritsch, E. F., and Maniatis, T. (1989) *Molecular Cloning: A Laboratory Manual*, 2nd ed. Cold Spring Harbor Laboratory Press, Cold Spring Harbor, NY.
9. Ann, D. K., Hasegawa, J., Ko, J. L., Chen, S. T., Lee, N. M., and Loh, H. H. (1992) Specific reduction of δ-opioid receptor binding in transfected NG108-15 cells. *J. Biol. Chem.* **267,** 7921–7926.
10. DeBernardi, M. A., Seki, T., and Brooker, G. (1991) Inhibition of cAMP accumulation by intracellular calcium mobilization in C6-2B cells stably transfected with substance K receptor cDNA. *Proc. Natl. Acad. Sci. USA* **88,** 9257–9261.
11. McDonald, R. L., Balmforth, A. J., Palmer, A. C., Ball, S. G., Peers, C., and Vaughan, P. F. (1995) The effect of the angiotensin II (AT1A) receptor stably transfected into human neuroblastoma SH-SY5Y cells on noradrenaline release and changes in intracellular calcium. *Neurosci. Lett.* **199,** 115–118.
12. Sadot, E., Gurwitz, D., Barg, J., Behar, L., Ginzburg, I., and Fisher, A. (1996) Activation of m1 muscarinic acetylcholine receptor regulates τ phosphorylation in transfected PC12 cells. *J. Neurochem.* **66,** 877–880.
13. Okusa, M. D., Huang, L., Momose-Hotokezaka, A., Huynh, L. P., and Mangrum, A. J. (1997) Regulation of adenylyl cyclase in polarized renal epithelial cells by G protein-coupled receptors. *Am. J. Physiol.* **273,** F883–F891.
14. Becker, B. N., Cheng, H. F., Burns, K. D., and Harris, R. C. (1995) Polarized rabbit type 1 angiotensin II receptors manifest differential rates of endocytosis and recycling. *Am. J. Physiol.* **269,** C1048–C1056.

15. Schulein, R., Lorenz, D., Oksche, A., et al. (1998) Polarized cell surface expression of the green fluorescent protein-tagged vasopressin V2 receptor in Madin Darby canine kidney cells. *FEBS Lett.* **441,** 170–176.
16. Jolimay, N., Franck, L., Langlois, X., Hamon, M., and Darmon, M. (2000) Dominant role of the cytosolic C-terminal domain of the rat 5-HT1B receptor in axonal-apical targeting. *J. Neurosci.* **20,** 9111–9118.
17. Ango, F., Albani-Torregrossa, S., Joly, C., et al. (1999) A simple method to transfer plasmid DNA into neuronal primary cultures: functional expression of the mGlu5 receptor in cerebellar granule cells. *Neuropharmacology* **38,** 793–803.
18. Nicot, A. and DiCicco-Bloom, E. (2001) Regulation of neuroblast mitosis is determined by PACAP receptor isoform expression. *Proc. Natl. Acad. Sci. USA* **98,** 4758–4763.
19. Slack, R. S. and Miller, F. D. (1996) Viral vectors for modulating gene expression in neurons. *Curr. Opin. Neurobiol.* **6,** 576–583.
20. Holter, W., Fordis, C. M., and Howard, B. H. (1989) Efficient gene transfer by sequential treatment of mammalian cells with DEAE-dextran and deoxyribonucleic acid. *Exp. Cell Res.* **184,** 546–551.
21. Graham, F. L. and van der Eb, A. J. (1973) A new technique for the assay of infectivity of human adenovirus 5 DNA. *Virology* **52,** 456–467.
22. Vaheri, A. and Pagano, J. S. (1965) Infectious poliovirus RNA: a sensitive method of assay. *Virology* **27,** 434–436.
23. Andreason, G. L. and Evans, G. A. (1988) Introduction and expression of DNA molecules in eukaryotic cells by electroporation. *BioTechniques* **6,** 650–660.
24. Koenig, J. A. (1999) Radioligand binding in intact cells. *Methods Mol. Biol.* **106,** 89–98.
25. Keen, M. (1997) Radioligand-binding methods for membrane preparations and intact cells. *Methods Mol. Biol.* **83,** 1–24.
26. Bylund, D. B. and Toews, M. L. (1993) Radioligand binding methods: practical guide and tips. *Am. J. Physiol.* **265,** L421–L429.
27. Sullivan, E., Tucker, E. M., and Dale, I. L. (1999) Measurement of [Ca^{2+}] using the Fluorometric Imaging Plate Reader (FLIPR). *Methods Mol. Biol.* **114,** 125–133.
28. Gossen, M., Bonin, A. L., and Bujard, H. (1993) Control of gene activity in higher eukaryotic cells by prokaryotic regulatory elements. *Trends Biochem. Sci.* **18,** 471–475.
29. Howe, J. R., Skryabin, B. V., Belcher, S. M., Zerillo, C. A., and Schmauss, C. (1995) The responsiveness of a tetracycline-sensitive expression system differs in different cell lines. *J. Biol. Chem.* **270,** 14,168–14,174.
30. Van Craenenbroeck, K., Vanhoenacker, P., Leysen, J. E., and Haegeman, G. (2001) Evaluation of the tetracycline- and ecdysone-inducible systems for expression of neurotransmitter receptors in mammalian cells. *Eur. J. Neurosci.* **14,** 968–976.
31. Choi, D. S., Wang, D. X., Tolbert, L., and Sadee, W. (2000) Basal signaling activity of human dopamine D2L receptor demonstrated with an ecdysone-inducible mammalian expression system. *J. Neurosci. Methods* **94,** 217–225.

32. Minneman, K. P., Lee, D., Zhong, H., Berts, A., Abbott, K. L., and Murphy, T. J. (2000) Transcriptional responses to growth factor and G protein-coupled receptors in PC12 cells: comparison of α(1)-adrenergic receptor subtypes. *J. Neurochem.* **74,** 2392–2400.
33. Hermans, E., Young, K. W., Challiss, R. A. J., and Nahorski, S. R. (1998) Effects of human type 1alpha metabotropic glutamate receptor expression level on phosphoinositide and Ca^{2+} signalling in an inducible cell expression system. *J. Neurochem.* **70,** 1772–1775.
34. McDonald, L. J. and Mamrack, M. D. (1995) Phosphoinositide hydrolysis by phospholipase C modulated by multivalent cations La(3+), Al(3+), neomycin, polyamines, and melittin. *J. Lipid. Mediat. Cell. Signal* **11,** 81–91.
35. Archer, S., Meng, S., Wu, J., Johnson, J., Tang, R., and Hodin, R. (1998) Butyrate inhibits colon carcinoma cell growth through two distinct pathways. *Surgery* **124,** 248–253.
36. Cockett, M. I., Bebbington, C. R., and Yarranton, G. T. (1990) High level expression of tissue inhibitor of metalloproteinases in Chinese hamster ovary cells using glutamine synthetase gene amplification. *Biotechnology* **8,** 662–667.
37. Gazi, L., Bobirnac, I., Danzeisen, M., et al. (1999) Receptor density as a factor governing the efficacy of the dopamine D-4 receptor ligands, L-745,870 and U-101958 at human recombinant D-4.4 receptors expressed in CHO cells. *Br. J. Pharmacol.* **128,** 613–620.
38. Nash, M. S., Selkirk, J. V., Gaymer, C. E., Challiss, R. A. J., and Nahorski, S. R. (2001) Enhanced inducible mGlu1 alpha receptor expression in Chinese hamster ovary cells. *J. Neurochem.* **77,** 1664–1667.
39. Pindon, A., van-Hecke, G., van-Gompel, P., Lesage, A. S., Leysen, J. E., and Jurzak, M. (2002) Differences in signal transduction of two 5-HT4 receptor splice variants: compound specificity and dual coupling with Gαs- and Gαi/o-proteins. *Mol. Pharmacol.* **61,** 85–96.

9

Viral Infection Protocols

Antonio Porcellini and Antonio De Blasi

Summary

This chapter describes the protocol for preparation of recombinant adenoviruses and infection of target cells to express transiently G-protein-coupled receptors or other proteins of interest. Adenoviruses are nonenveloped viruses containing a linear double-stranded DNA genome. Their life cycle does not normally involve integration into the host genome, rather they replicate as episomal elements in the nucleus of the host cell and consequently there is no risk of insertional mutagenesis. The wild-type adenovirus genome is approx 35 kb, of which up to 30 kb can be replaced by foreign DNA. Adenoviral vectors are very efficient at transducing the gene of interest in target cells in vitro and in vivo and can be produced at high titers ($>10^{11}$/mL). The viral infection has a number of useful features: (1) the efficiency of gene transduction is very high (up to 100% in sensitive cells). (2) the infection is easy and does not alter physically the cell membrane for gene transduction. (3) it is possible to infect cells that are resistant to transfection with plasmids (including nondividing cells).

Key Words

Adenovirus, infection, TSH receptor.

1. Introduction

This chapter describes the protocol for preparation of recombinant adenoviruses and infection of target cells to transiently express G-protein-coupled receptors (GPCRs) or other proteins of interest.

Viruses are obligate intracellular parasites, designed through the course of evolution to infect cells, often with great specificity for a particular cell type. They tend to be very efficient at transfecting their own DNA into the host cell, and the DNA is expressed to produce new viral particles. By replacing genes that are needed for the replication phase of their life cycle (the nonessential genes) with foreign genes of interest, the recombinant viral vectors can trans-

From: *Methods in Molecular Biology, vol. 259, Receptor Signal Transduction Protocols, 2nd ed.*
Edited by: G. B. Willars and R. A. J. Challiss © Humana Press Inc., Totowa, NJ

duce the gene of interest into the cell type it would normally infect. To produce such recombinant viral vectors the nonessential genes are provided in *trans*, either integrated into the genome of the packaging cell line or on a plasmid. A number of viruses have been developed, but four types are mostly used: retroviruses (including lentiviruses), adeno-associated viruses, herpes simplex virus type 1, and adenoviruses.

Retroviruses are a class of enveloped viruses containing a single-stranded RNA molecule as genome. Following infection, the viral genome is reverse transcribed into double-stranded DNA, which integrates into the host genome and is expressed as proteins. A requirement for retroviral integration and expression of viral genes is that the target cells should be dividing. Lentiviruses are a subclass of retroviruses that are able to infect both proliferating and nonproliferating cells. They are considerably more complicated than retroviruses.

Adeno-associated viruses are nonpathogenic human parvoviruses, dependent on a helper virus, usually adenovirus, to proliferate. They are capable of infecting both dividing and nondividing cells, and in the absence of a helper virus integrate into a specific point of the host genome (19q 13-qter) at a high frequency *(1)*.

Herpes simplex virus type 1 (HSV-1) is a human neurotropic virus: consequently interest has largely focused on using HSV-1 as a vector for gene transfer to the nervous system. The wild-type HSV-1 virus is able to infect neurons and either proceed into a lytic life cycle or persist as an intranuclear episome in a latent state. Latently infected neurones function normally and are not rejected by the immune system. Although the latent virus is transcriptionally almost silent, it does possess neuron-specific promoters that are capable of functioning during latency.

This chapter focuses on adenoviruses. Adenoviruses are nonenveloped viruses containing a linear double-stranded DNA genome (**Fig. 1**). Although there are more than 40 serotype strains of adenovirus, most of which cause benign respiratory tract infections in humans, subgroup C serotypes 2 or 5 are predominantly used as vectors. They are capable of infecting both dividing and nondividing cells. The life cycle does not normally involve integration into the host genome; rather they replicate as episomal elements in the nucleus of the host cell and consequently there is no risk of insertional mutagenesis. The wild-type adenovirus genome is approx 35 kb, of which up to 30 kb can be replaced by foreign DNA *(2)*. There are four early transcriptional units (E1, E2, E3, and E4), which have regulatory functions, and a late transcript, which codes for structural proteins. In progenitor vectors either the E1 or E3 gene is inactivated, with the missing gene being supplied in *trans* either by a helper virus, by a plasmid, or by being integrated into a helper cell genome (HEK-293 cells,

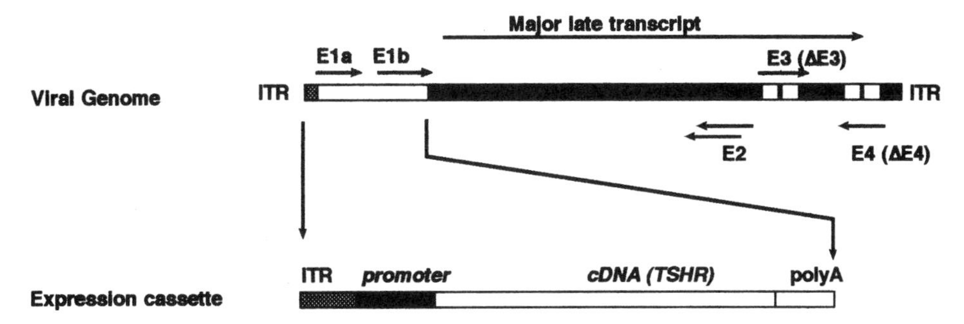

Fig. 1. Schematic representation of the adenoviral genome and of the expression cassette. The sites encoding for the four early transcriptional units (E1, E2, E3, and E4) are indicated. To obtain the recombinant adenovirus the regions containing the E1a and E1b sites are deleted and the expression cassette is inserted. The expression cassette contains the cDNA of interest the (TSH-R in this case). The deletion of the E1a and E1b cassette prevents the transcription of the major late transcript. This renders the viruses defective for replication and incapable of producing infectious viral particles in target cells. The possible deletion of E3 and/or E4 present in some commercially available viruses is indicated (ΔE3, ΔE4). ITR, inverted terminal repeats.

[3]). Second-generation vectors in addition use an E2a temperature-sensitive mutant or an E4 deletion. Viruses in which E4 is deleted have to be replicated in 911E4 cell lines that can complement E4. The most recent "gutless" vectors contain only the inverted terminal repeats (ITRs) and a packaging sequence around the transgene, all the necessary viral genes being provided in *trans* by a helper virus *(4)*. In recent protocols, recombination occurs in prokaryotic cells to improve the yield of recombinant viruses and to facilitate their screening *(5,6)*. Adenoviral vectors are very efficient at transducing the gene of interest in target cells in vitro and in vivo and can be produced at high titers ($>10^{11}$/mL). With the exception of one report *(7)*, which showed prolonged transgene expression in rat brains using an E1 deletion vector, transgene expression in vivo from progenitor vectors tends to be transient *(2)*.

The essential steps for generating the viral vector are shown in **Fig. 2**. The viral infection has a number of useful features:

1. The efficiency of gene transduction is very high (up to 100% in sensitive cells).
2. The infection is easy and does not alter physically the cell membrane for gene transduction.
3. It is possible to infect cells that are resistant to transfection with plasmids (including nondividing cells).
4. The viral vectors can be used for infection in vivo (including gene therapy) and potentially can be targeted with cell specificity.

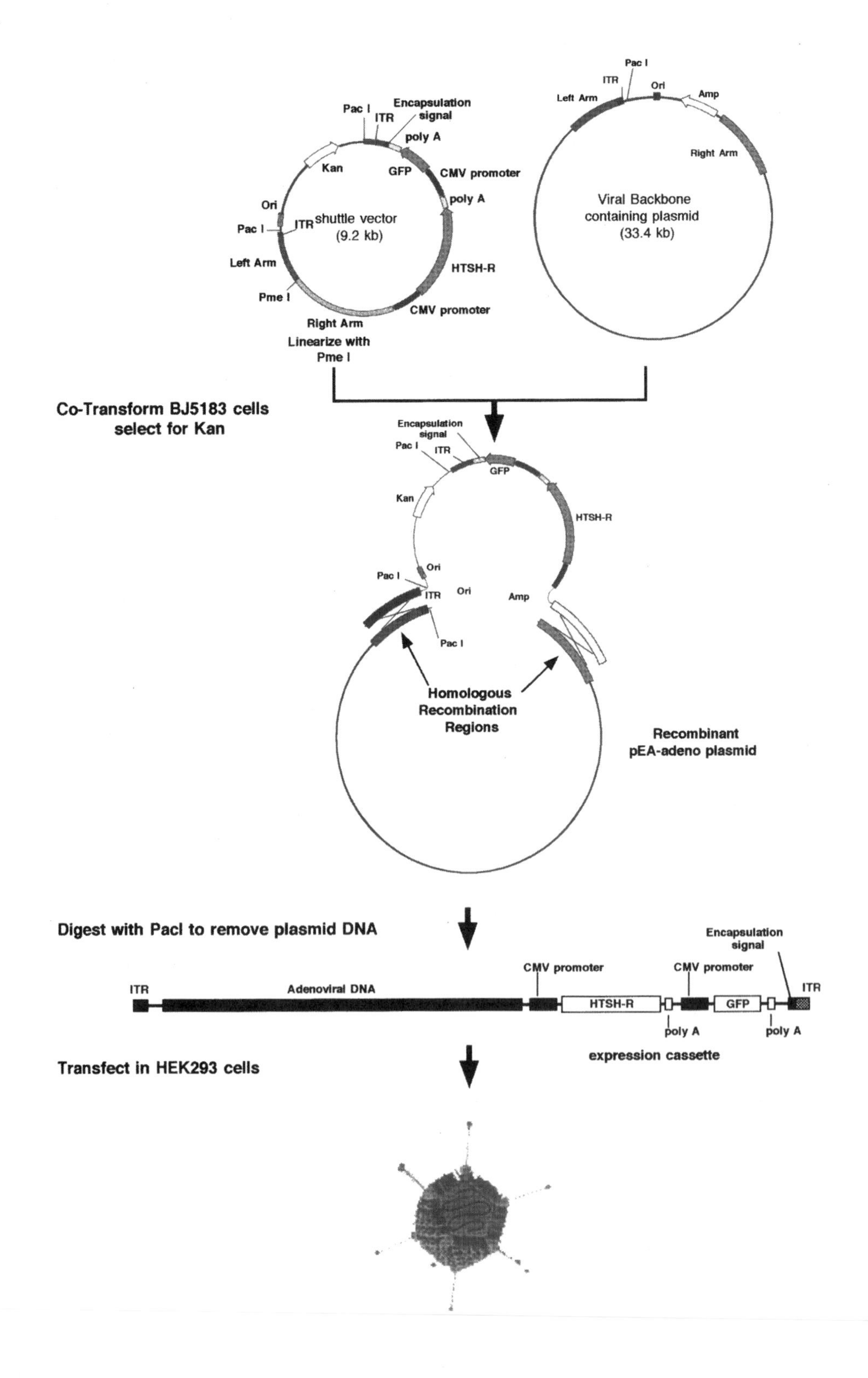
Pac I
ITR
Encapsulation signal
poly A
Kan
GFP
CMV promoter
Ori
poly A
Pac I
ITR
shuttle vector
(9.2 kb)
Left Arm
HTSH-R
Pme I
CMV promoter
Right Arm
Linearize with Pme I
Pac I
ITR
Ori
Left Arm
Amp
Right Arm
Viral Backbone containing plasmid
(33.4 kb)
Co-Transform BJ5183 cells select for Kan
Encapsulation signal
Pac I
ITR
GFP
Kan
HTSH-R
Ori
Pac I
ITR
Ori
Amp
Pac I
Homologous Recombination Regions
Recombinant pEA-adeno plasmid
Digest with PacI to remove plasmid DNA
Encapsulation signal
CMV promoter
CMV promoter
ITR
Adenoviral DNA
ITR
HTSH-R
GFP
poly A
poly A
expression cassette
Transfect in HEK293 cells

The protocol used is a modification of the method described by He and colleagues *(5)*. The example presented here refers to the thyrotropin receptor (TSH-R) expressed using the pAdTrack-CMV vector.

2. Materials

1. Shuttle vectors: pShuttle-CMV vendors: Stratagene cat. no. 240006 or Q-Biogene cat. no. AES-1021; pAd-TRK-CMV *(6)*.
2. pAdEasy-1 vendors: Stratagene cat. no. 240005 or Q-Biogene cat. no. AES-1010.
3. BJ5183 electrocompetent cells (*see* **Note 6** or purchase from Stratagene cat. no. 200154 or Q-Biogene cat. no. AES-1005).
4. HEK-293 cells ATCC cat. no. CRL-1573.
5. *Pac*I and *Pme*I restriction endonucleases available from New England Biolabs.
6. L–broth (LB): 10 g/L of Bacto-tryptone, 5 g/L of Bacto-yeast, 5 g/L of NaCl; for LB-agar add 15 g/L of Bacto-agar (Difco); autoclave.
7. SOC medium: 20 g/L of Bacto-tryptone, 5.5 g/L of Bacto-yeast, 10 m*M* of NaCl, 10 m*M* of KCl; autoclave then add glucose to 20 m*M*, $MgCl_2$, and $MgSO_4$ to 10 m*M* each.
8. Electroporation cuvets, 0.2-cm gap.
9. Lipofectin (available from Invitrogen, cat. no. 18292-011).

3. Methods

3.1. Preparation of Shuttle Plasmid, Adenoviral Backbone Vector, and Competent Cells

1. Subclone the cDNA encoding for the gene of interest (in our case the TSH-R) in the shuttle vector. In this protocol we refer to the pAdTrack-CMV, which contains green fluorescent protein (GFP) as a tracer and the cytomegalovirus (CMV) promoter *(5)* (*see* **Note 1**). Prepare the recombinant shuttle vector at high purity (transfection grade) (*see* **Note 2**) for the next step. It can be stored at 4°C for up to 6 mo.

Fig. 2. *(see facing page)* Schematic representation of the essential steps for generating the recombinant adenovirus. The cDNA of interest (the human TSH-R) is cloned into the shuttle plasmid. The viral backbone containing plasmid is derived from the pBR322 and includes the sequence of the adenovirus (Ad5) without the region E1 and E3 (*see* **Fig. 1**) plus one sequence derived from the pBR322 containing the *E. coli* Ori and the Amp resistance. The BJ5183 cells are cotransformed with the linearized shuttle plasmid and the viral backbone containing plasmid for homologous recombination. After recombinant selection, the DNA is digested with *Pac*I to remove plasmid DNA. The resulting recombinant contains the viral backbone plus the expression cassette and is ready for HEK-293 transfection. HEK-293 will provide the transcription factors E1a and E1b for transcription of capsid protein RNAs and viral replication.

2. Linearize the recombinant shuttle vector by incubating 1 μg of DNA at 37°C for 1 h with the enzyme *Pme*I (4 U) in 100 μL of reaction buffer (provided by the manufacturer) (*see* **Note 3**). After extraction and purification (*see* **Note 4**), resuspend in ultrapure water (15 μL). It can be stored at –20°C. Run 2 μL on an agarose gel to determine the extent of digestion (*see* **Note 5**).
3. The adenoviral backbone vector can be obtained ready-to-use from the manufacturer. This reagent can also be amplified by transforming competent DH5α (as in **Subheading 3.2.** but using ampicillin and not kanamycin for LB agar) and purified for further experiments (*see* **Note 2**).
4. Prepare competent cells (BJ5183) *(6)* (*see* **Note 6**) and aliquot in 20 μL/tube. At least four aliquots are needed for one recombination. Competent cells can be stored at –80°C (not in liquid nitrogen). Competent cells can be obtained from the manufacturer.

3.2. Generation of Recombinant Adenoviral Plasmids (Fig. 2)

1. To 20 μL of competent cells add 3 μL of linearized shuttle vector and 1 μL (containing 100 ng) of adenoviral backbone vector and mix with the pipet. Transfer into a cuvet (*see* **Note 7**) for electroporation. All the reagents, the mix and the cuvette must be on ice.
2. Electroporate (Gene Pulser, Bio-Rad) at 2,500 V, 200 Ω, 25 μF for one pulse.
3. Add to the cuvet 500 μL of prewarmed (37°C) SOC medium, mix, transfer to a fresh tube (15 mL), and grow at 37°C for 20 min.
4. Centrifuge (800*g* for 10 min at room T), resuspend in 200 μL of LB (SOC is fine), and plate in two 10-cm Petri dishes with LB agar containing 25 μg/mL of kanamycin (*see* **Note 8**). Incubate overnight at 37°C.
5. On the next day colonies should be visible. Pick up 5–10 colonies (*see* **Note 9**) and grow in 3 mL of LB plus antibiotic for 15–18 h. Pellet and prepare miniprep (*see* **Note 10**). Digest 10 μL of miniprep with *Pac*I (5 U) for 1 h and run on an 0.7% agarose gel. This shows that recombination has occurred **(Fig. 3)**.
6. Take DNA from colonies positive for recombination (as assessed by *Pac*I digestion) and digest for the presence of the insert. For the TSH-R we took 10 μL of recombinant-positive colony and digested it with *Bst*EII (5 U) for 1 h. Run on an 0.8% gel and examine for the positive band **(Fig. 3)** (*see* **Note 11**).
7. At this stage you have checked that the viral backbone has recombined correctly with the plasmid (i.e., it contains the cDNA of interest [the TSH-R] plus the antibiotic resistance) but you do not know whether this construct is able to generate infectant viral particles and to express the protein of interest (the TSH-R) in eukaryotic cells. This is analyzed in **Subheading 3.3.** For this purpose transfer (at least) four different insert-positive clones (the 10-μL aliquot remaining after the two digestions) into electrocompetent (*see* **Note 6**) RecA(–) *E. coli* strains (such as DH5α) (*see* **Note 12**). Prepare 100–500 μg of transfection-grade purified plasmid (*see* **Note 2**) and save samples from corresponding colonies.

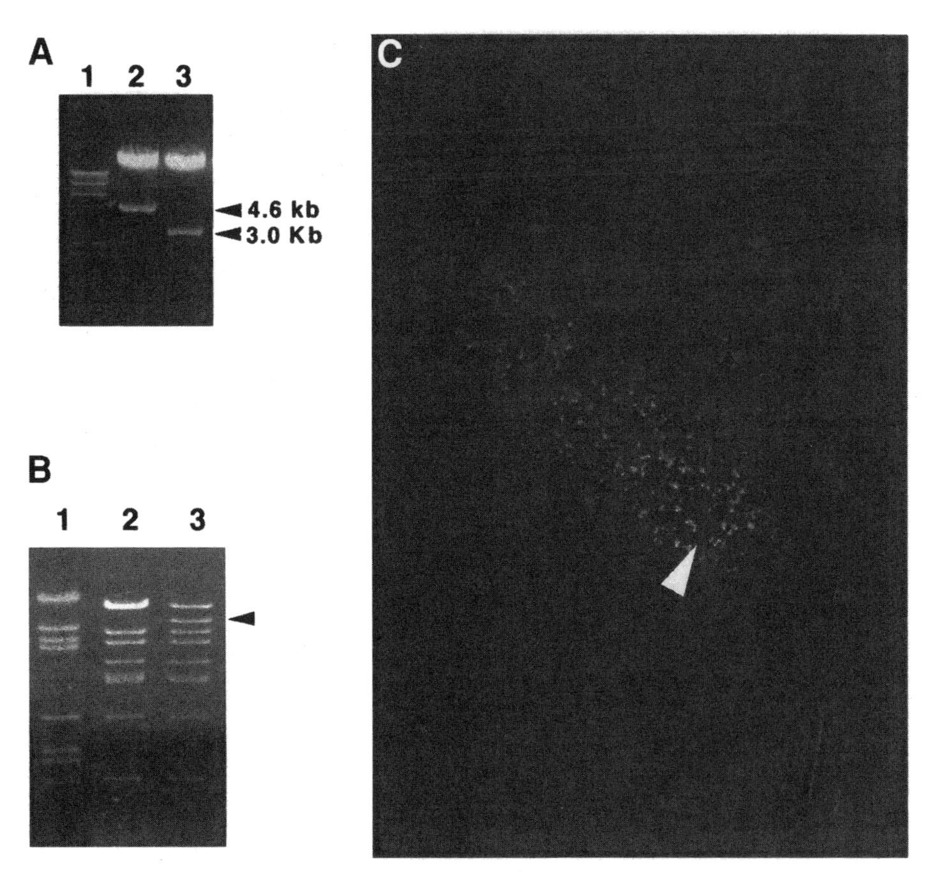

Fig. 3. **(A)** *Pac*I digestion of candidate recombinant clones. This digestion generates two bands: the 33-kb recombinant viral DNA, and the 3-kb fragment which contains the kanamycin resistance. In some clones we found instead a 4.6-kb fragment (likely generated by asymmetric recombination) which turned out to generate infective virus. The presence of a 3–4.6 band indicates positive recombination. **(B)** The same clones digested with *Bst*EII to analyze the presence of the insert (TSH-R cDNA). The digestions shown are with the viral backbone containing plasmid *(lane 1)*, the recombination without insert *(lane 2)*, and the positive clone *(lane 3)*. The 9.78-kb band containing the TSH-R cDNA is indicated by the *arrow*. **(C)** Adenovirus-generated foci in HEK-293 cells at 7–10 d after transfection. The *arrow* indicates the area of cell lysis and the surrounding cells are infected and express GFP. Note that at this stage you cannot see the focus unless you have a tracer (such as GFP).

3.3. Virus Production in Eukaryotic Cells (see Note 13)

1. At 24 h before transfection plate HEK-293 cells (5×10^6/T75 flask). Cells should be 50–70% confluent at the transfection.
2. Digest 50 µg of transfection-grade purified plasmid with *Pac*I (100 U) in a 250-µL final volume. Run 5 µL on an agarose gel to verify the digestion. Extract and precipitate DNA (*see* **Note 4**). Ethanol must be removed under sterile conditions. Resuspend in 100 µL of sterile ultrapure water.
3. Transfect HEK-293 cells (*see* **Note 14**) with digested DNA using Lipofectin. For each flask prepare 20 µL of *Pac*I-digested plasmid in 1.25 mL of OptiMem and 25 µL of Lipofectin in 1.25 mL of OptiMem and leave for 15–40 min at room temperature (RT). Mix gently these two solutions and leave for at least 10 min (stable for 30–40 min). Wash the HEK-293 cells with serum-free medium at least four times, add 3.5 mL of OptiMem, and equilibrate for 15 min at 37°C in a CO_2 incubator. Add the mix to the flask containing OptiMem and leave in the incubator for 4 h.
4. Remove the transfection medium and add 10 mL of complete medium with serum. Change the medium every other day.
5. At 7–10 d after transfection, scrape the cells into the medium and pellet the cells for virus extraction. At this time, unless you have a tracer protein (GFP) usually you will not see clear lysis plaques. The absence of clear plaques at this time does not indicate the absence of recombinant virus (*see* **Note 15**) (*see* **Fig. 3**).
6. Wash the pellet twice in PBS, resuspend in 2 mL of PBS, and transfer to 1.5- or 2-mL tubes. Freeze and thaw three times in dry ice–ethanol (*see* **Note 16**). For thawing, place the frozen tube at 37°C until it starts to thaw and then vortex-mix immediately. Avoid complete thawing at 37°C.
7. Centrifuge at 12,000*g* for 30 min and collect the supernatant.
8. Use 1 mL of the supernatant (primary lysate) for the amplification. Save the other 1-mL aliquots at –80°C for the next amplification (stable for up to 6 mo).

3.4. Preparation of High-Titer Viral Stocks

1. Grow HEK-293 cells in flasks to more than complete confluence (better to let them grow 3–4 d after confluence).
2. Wash the cells gently with prewarmed serum-free medium.
3. Add 1 mL of primary lysate to the cells and gently rock for 2 h at 37°C. Make sure that all the flask surface is in contact with the primary lysate.
4. Remove the primary lysate and add 10 mL of medium with serum. Leave cells at 37°C for 72 h.
5. Wash the pellet twice in PBS, resuspend in 2 mL of PBS, and transfer to 1.5- or 2-mL tubes. Freeze and thaw three times in dry ice–ethanol (*see* **Note 16**). For thawing, place the frozen tube at 37°C until it starts to thaw and then vortex-mix immediately. Avoid complete thawing at 37°C.
6. Centrifuge at 12,000*g* for 30 min and collect the supernatant.
7. Titrate the virus in the lysate (*see* **Note 17**). The expected titer is >107 plaque-forming units (PFUs)/mL.

8. If the viral titer is as expected (*see* **Note 18**) infect five (or more) T75 flasks of HEK-293 cells using this material. Use 2 mL of PBS containing 5–10 PFUs/cell and proceed as described in **steps 3–6**.
9. After centrifugation you should have 8–10 mL of lysate (about 2 mL used for freeze and thaw from five flasks). The expected titer is >10^9 PFUs/mL. This preparation can be used to infect cells. If you need to purify the virus suspension further or if you need a higher titer (e.g., to infect in vivo experimental animals) proceed to cesium purification (next step).
10. Add 4.5 g of cesium cloride to 8.5 mL of lysate and mix. Transfer this solution to an ultrafuge tube and centrifuge at 150,000*g* (SW 41 rotor) for 18–20 h at 10–12°C.
11. Use a needle to collect the viral fraction (within the cesium gradient). Dilute the collected material 1:2 with storage buffer (expected total volume is 1–2 mL), prepare 200-μL aliquots, and store at –20°C (stable for years).
12. Titrate either by plaque assay, by GFP fluorescence determination, or by OD determination of viral DNA (*see* **Note 19**).

3.5. Infection of Target Cells

1. The efficiency of infection and protein expression depends on the target cell type. You should perform preliminary experiments to determine the optimal MOI (multiple of infection = ratio between PFU and number of cells) and infection procedure.
2. For NIH3T3 cells, subconfluent cells must be covered with the minimal volume (i.e., 1.8–2 mL for a 100-mm dish) of serum-free medium (*see* **Note 20**) containing the virus at the MOI of 100 PFUs/cell (*see* **Note 21**). This MOI is referred to the minimal volume of virus-containing medium on subconfluent cells.
3. After 2 h of incubation at 37°C (possibly with rocking) remove the virus-containing medium and add the medium with serum. If needed, the cells can be harvested and plated 24 h after infection.
4. The protein (TSH-R) is functionally expressed on the cell surface 36–48 h after infection and cells can be used for experiments.

4. Notes

1. Different shuttle vectors are available commercially. For example, they may contain the GFP or β-Gal reporter gene. They may also have different promoters or be devoid of promoters to allow the cloning of one promoter of interest to direct the expression of the protein.
2. Use commercially available column purification or CsCl banding.
3. The linearization allows the recombination with the viral backbone and avoids the background of kanamycin-resistant colonies generated by the circular plasmid.
4. After digestion, dilute to 500 μL with TE and extract with one volume of phenol–chloroform–isoamidic 25:24:1 followed by ethanol precipitation. For ethanol precipitation add 1/10 vol of 4 *M* LiCl, mix, and add 2.5 volumes of ice-cold ethanol. Li is preferred, as it does not interfere with ligase or with electroporation transfection efficiency.

5. If the digestion is incomplete (i.e., <95%) you can purify the linearized DNA from the gel. In this case extreme care must be taken to avoid any agarose (and perchlorate if used) residual in the final DNA solution; otherwise this would impair the electroporation transfection efficiency.
6. Use a fresh colony or frozen stock of DH5α cells to inoculate 10 mL of LB medium in a 50 mL tube (for BJ5183 cells inoculate 10 mL LB containing 30 μg/mL of streptomycin). Grow cells in a shaker overnight at 37°C. On the next day, dilute 1 mL of cells into 500 mL of LB medium (for BJ5183, use streptomycin-containing LB medium) in a 2-L flask. Grow for 2–4 h with vigorous aeration at 37°C, until A_{550} is approx 0.5 for DH5α and A_{550} is approx 0.7 for BJ5183. Stop cell growth by incubation on ice for 10 min–1 h (the longer the cells are incubated the higher the competency will be). Collect the cells in two 250-mL conical centrifuge tubes. Pellet cells by centrifuging at 2600*g* at 4°C for 10 min. Wash the pellet by resuspending in 500 mL of sterile ice-cold WB (WB = 10% ultrapure glycerol, 90% distilled water [v/v]). Centrifuge the cell suspension at 2500*g* for 30 min. Wash the pellet by resuspending in 250 mL of sterile ice-cold WB. Centrifuge the cell suspension at 2500*g* for 15 min. Pour off the supernatant gently, leaving about 30 mL. Resuspend and transfer the cell suspension to a 50 mL tube. Centrifuge at 2500*g* for 10 min, and pipet out all but 5 mL of the supernatant (for BJ5183 cells, the final total volume should be limited to 2–3 mL). Resuspend the cell pellet in the WB remaining in the tube. Aliquot 20–40 μL/tube (the tubes should be prechilled at –80°C) and store the aliquots at –80°C.
7. Use a 2-mm cuvet and not a 5-mm cuvet.
8. We suggest that you do not use 50 μg/mL of kanamycin, as at this concentration the growth of the recombinant adenoviral plasmid is inhibited.
9. Pick the small colonies, as the recombinant-containing colonies are usually smaller than those of the shuttle plasmid.
10. Pellet 2 mL overnight culture in 2-mL Eppendorf tubes, and centrifuge for 1 min. Discard the supernatants. Add 100 μL of buffer A (50 m*M* glucose; 25 m*M* Tris-HCl, pH8.0; 10 m*M* EDTA, pH 8.0), and vortex-mix briefly. Add 200 μL of lysis buffer (0.2 *N* NaOH and 1% sodium dodecyl sulfate [SDS]), gently mix by inverting the tubes several times. Add 150 μL of precipitation solution (for 100-mL mix: 60 mL of 5 *M* potassium acetate, 11.5 mL of glacial acetic acid, and 28.5 mL of water); mix well by inverting the tubes several times. Centrifuge the tubes at top speed in a benchtop microfuge for 3 min. Pour the supernatant into a fresh 1.5-mL tube. Add 400 μL of 2-phenol–chloroform (1 : 1). Centrifuge at top speed for 5 min (at RT). Transfer the upper phase to a fresh 1.5-mL tube. Add 1 mL of ethanol, and leave at RT for 10 min. Centrifuge at top speed for 5 min (at RT). Discard the supernatants. Add 500 μL of 70% ethanol, vortex-mix and centrifuge for 5 min. Discard the supernatants. Briefly centrifuge and aspirate the residual liquid in the tubes. Add 30 μL of TE/RNase A (5 μg/mL) to resuspend the DNA.
11. Given the size of the recombinant adenoviral plasmid (~40 kb) and the limited number of diagnostic restriction sites, the digestion may not be sufficient to assess

the presence (and the orientation) of the insert. This point should then be addressed by Southern blotting the digest.

12. Do not use BJ5183 strain, as these cells allow further recombination of the recombinant adenoviral plasmid.
13. Safety rules for the use of adenoviruses can be found in laboratory manuals or in websites (e.g., www-ehs.ucsd.edu/bio/biobk/bioap10.htm, or www.cdc.gov/od/ohs/biosfty/bmbl4/bmbl4s3.htm).
14. Use care when transfecting HEK-293, as these cells tend to detach from the flask, particularly when they are superconfluent. Do not expose cells to Lipofectin for >4 h.
15. At this stage a small number of cells that were infected by the virus have been lysed. After the lysis of these cells the virus is released and infects the surrounding cells. This is the optimal time to collect the cells to extract the virus (i.e., after infection of surrounding cells and before their lysis, usually 7–10 d after transfection). If you have GFP tracer you will see green fluorescent transfected cells at d 2 and green infected surrounding cells at d 7–10. If there is no tracer you can use control plates to see the lysed plaques at d 15–20 (as in **Note 19**). This control plate is no longer useful for virus collection (as many infected cells are lysed), but indicates that the clone is able to generate lytic virus.
16. You must use polypropylene tubes.
17. For titration you should determine the PFU by standard methods. Alternatively, if the virus contains GFP tracer, simply infect superconfluent HEK-293 cells with various viral dilutions and count green cells after 24 h. This gives the number of infecting particles, which in our hands corresponds to 5–10 times the PFU values. Determination of PFU is less simple but more quantitative.
18. If the titer is lower, check the transfection efficiency (should be >20%) or start from a different clone.
19. For virus titration, remove all but 2 mL of medium per well from 6-well plates containing 80–90% confluent HEK-293 cells. Infect with appropriately diluted virus (1 mL) for 2 h. Infect cells with six different dilution titers (e.g., 10^{-3}–10^{-8}). Prepare the overlay agar as follows: autoclave 100 mL of 2.8% Bacto-agar (Difco) and keep warm in a 45°C water bath. To 36 mL of 2.8% Bacto-agar add 50 mL of prewarmed 2X BME (Gibco), 10 mL of fetal bovine serum (FBS), 1.25 mL of 1 *M* $MgCl_2$, and 2 mL of 1 *M* 4-(2-hydroxyethyl)piperazine-1-ethanesulfonic acid (HEPES). Mix well and swirl in a 37°C water bath; add 4 mL/well for a 6-well plate. Leave plates at RT for 30 min–1 h. Return the plates to the 37°C CO_2 incubator. On d 5–7, overlay 2–3 mL of agar containing neutral red (from 100X stock, available from Gibco-BRL) to each well. Plaques should be visible 16–30 h after the neutral red overlay.
20. The serum must be carefully removed by washing two or three times. For cells that grow in multilayers (such as PC12) it can be difficult to wash out the medium. We suggest infecting these cells in suspension (in polypropylene tubes) as follows: harvest the cells (using trypsin if needed), wash two or three times, and resuspend in a minimal volume (10^7 cells/mL) of virus-containing medium.

Incubate for 2 h at 37°C, remove the virus, and resuspend in medium with serum and plate.

21. For different cell types the MOI can vary quite substantially. In our hands, to obtain 80–100% of cells expressing the GFP, we have calculated the following MOI: for U251, 50 PFUs/cell; for PC12, 250 PFUs/cell; for U87MG, 5–10 PFUs/cell; for COS7, 50 PFUs/cell; for T98G, 150 PFUs/cell; for HEK-293, 5 PFUs/cell.

References

1. Kotin, R. M., Siniscalco, M., Samulski, R. J., et al. (1990) Site-specific integration by adeno-associated virus. *Proc. Natl. Acad. Sci. USA* **87,** 2211–2215.
2. Verma, I. M. and Somia, N. (1997) Gene therapy—promises, problems and prospects. *Nature* **389,** 239–242.
3. Graham, F. L., Smiley, J., Russell, W. L., and Nairn, R. (1997) Characterization of a human cell line transformation by DNA from adenovirus 5. *Gen. Virol.* **36,** 59–72.
4. Chen, H., Mack, L. M., Kelly, R., Ontell, M., Kochanek, S., and Clemens, P. R. (1997) Persistence in muscle of an adenoviral vector that lacks all viral genes. *Proc. Natl. Acad. Sci. USA* **94,** 1645–1650.
5. He, T. C., Zhou, S., da Costa, L. T., Yu, J., Kinzler, K. W., and Vogelstein, B. (1998) A simplified system for generating recombinant adenoviruses. *Proc. Natl. Acad. Sci. USA* **95,** 2509–2514.
6. Hanahan, D. (1983) Studies on transformation of *Escherichia coli* with plasmids. *J. Mol. Biol.* **166,** 557–580.
7. Geddes, B. J., Harding, T. C., Lightman, S. L., and Uney, J. B. (1997) Long term gene therapy in the CNS: reversal of hypothalamic diabetes insipidus in the Brattleboro rat by using an adenovirus expressing arginine vasopressin. *Nat. Med.* **3,** 1402–1404.

10

Expression of G-Protein Signaling Components in Adult Mammalian Neurons by Microinjection

Stephen R. Ikeda

Summary

Although methods for expressing foreign proteins in clonal cell lines are well established, mature neurons remain a difficult preparation for the introduction of foreign genes. Microinjection is a reliable method for producing robust targeted expression in neurons that has advantages over conventional transfection/infection methodologies. Here, I describe procedures for expressing signaling proteins in adult rat sympathetic neurons by direct microinjection of cRNA and cDNA into the cytoplasm and nucleus, respectively. The methods are applicable to a wide variety of peripheral and central neuron preparations, as well as clonal cell lines.

Key Words

G-proteins, heterologous expression, ion channels, microinjection, sympathetic neurons, tissue culture.

1. Introduction

Heterologous expression of cloned proteins, both wild-type and mutant, is a major technique used to explore G-protein signaling pathways. The near completion of genome sequencing for several organisms facilitates the use of this methodology and reveals the presence of novel signaling components with unknown functions. Although methods for expressing foreign proteins in clonal cell lines are well established, mature neurons remain a difficult preparation for the introduction of foreign genes. In recent years, the number of options for heterologous expression in neurons has increased with the introduction of viral vectors, improved transfection reagents, and biolistic approaches. Microinjection of cRNA and cDNA, however, remains a reliable method for producing robust expression in neurons.

From: *Methods in Molecular Biology, vol. 259, Receptor Signal Transduction Protocols, 2nd ed.*
Edited by: G. B. Willars and R. A. J. Challiss © Humana Press Inc., Totowa, NJ

What are the advantages of microinjection in regard to gene expression in neurons? First, the technique, once mastered, is straightforward and widely adaptable. Microinjection can be successfully performed on most adherent (freshly plated or after several weeks in tissue culture) peripheral or central neurons using common mammalian expression vectors or in vitro transcribed cRNA. This obviates the requirements for cloning the desired insert into complex viral vectors, production of virus, and determining viral titers. Moreover, microinjection can also be used to introduce signaling proteins (purified or in vitro translated), antibodies, and peptides into neurons. In general, successfully injected neurons are healthy (once having survived the injection procedure and assuming no toxicity from the expressed protein) as evidenced by normal electrophysiological properties and the ability to extend processes during extended culture. Second, expression levels can be titrated by simply adjusting the concentration of cDNA or cRNA to be injected. Controlling expression level is a major concern with viral vectors—especially those that result in massive amplification of subgenomic viral RNA *(1)*. Under these circumstances, decreasing viral titer results in fewer infected cells rather than decreased single-cell expression. Third, multiple constructs can be injected with the expectation that each construct will produce expression in the large majority of cells. We have successfully expressed up to seven individual proteins in neurons *(2)*. A caveat with expressing such complex mixtures is the indeterminate nature of how each protein may influence the expression or targeting of other components of the mixture. Finally, the cell type to be injected is identified visually, thereby restricting expression to the desired target. For example, expression of protein in glial cells is avoided with microinjection whereas most viruses and transfection reagents tend to favor such background elements. In addition, the ability to target individual neurons that form synaptic connections in culture suggests the possibility of selectively expressing proteins in pre- vs postsynaptic neurons. At this point, microinjection remains one of the few techniques available that targets individual identified neurons. However, single-cell electroporation *(3,4)* and laser-mediated transfection *(5)* technologies offer enticing glimpses of alternative procedures that, once mature, may possess similar capabilities.

As with all techniques, microinjection techniques introduce some disadvantages and limitations. First, and foremost, the number of successfully injected cells is quite small. This is not a severe limitation for single-neuron assays such as electrophysiology, immunohistochemistry, or optical assays. However, biochemical assays requiring large amounts of protein are better suited to mass transfection or infection techniques. Second, the initial cost of setting up a microinjection workstation is high, requiring a quality inverted microscope, automated injection apparatus, and micropipet puller. Finally, the technique is somewhat tedious and time consuming. Although investigators new to the technique can produce successful injections on the first trial, introduction of microinjection techniques into laboratories that do not routinely employ micropipets and microscopy for other purposes (e.g., electrophysiology) can be problematic. Because microinjection is very much a "tactile" skill, the fastest way to gain familiarity is to visit a laboratory that is routinely using it.

Our laboratory has utilized microinjection-based heterologous expression in neuronal "surrogates" to probe G-protein-coupled receptor (GPCR) signaling pathways that modulate voltage-gated ion channels. Receptors, signaling components (e.g., G-protein subunits and regulators of G-protein signaling [RGS] proteins), and effectors (e.g., ion channels) have all been successfully expressed in neurons. Examples include: (1) expression of GPCRs, such as metabotropic glutamate ***(6)*** and cannabinoid receptors ***(7)***, to help define how molecularly defined receptor subtypes modulate voltage-gated Ca^{2+} and K^+ channels; (2) expression of defined G-protein subunits to further our understanding of N-type Ca^{2+} channel modulation ***(8–10)***; and (3) expression of GIRK-type K^+ channels to define signaling pathways linking neuronal GPCRs to K^+ channels ***(11,12)***. Moreover, mutant Gα subunits have been used to explore the coupling specificity of defined G-protein heterotrimer subunits ***(13)*** and how RGS proteins influence the kinetics and magnitude of GPCR-mediated ion channel modulation ***(2,14)***. Finally, signaling probes such as the Pertussis toxin S1 (i.e., catalytic) subunit have been expressed in neurons ***(15)***. Looking forward, the expression of fluorescently tagged G-protein subunits to examine dynamic protein–protein interactions with technologies such as fluorescence resonance energy transfer (*see* Chapter 21) ***(16,17)*** and the introduction of small interfering RNA ***(18)*** into cells to ablate gene expression provide additional uses for microinjection techniques in neurons. Here I describe the preparation of dissociated adult rat superior cervical ganglion (SCG) neurons and methods for expressing heterologous proteins in these neurons by cytoplasmic and nuclear microinjection of cRNA and cDNA, respectively. The described methods are readily adaptable to other peripheral neurons (e.g., dorsal root ganglion neurons) with minor modifications. Microinjection of CNS-derived neurons (e.g., hippocampal) follow similar principles although isolation and culture techniques are different *(unpublished observation)*.

2. Materials

2.1. Neuron Dissociation and Short-Term Culture

1. Hanks' balanced salt solution (HBSS): prepare from 1 L of powder mix (Sigma, St. Louis, MO, USA) per the manufacturer's instructions. Store at 4°C in a 1-L bottle.
2. Modified Earle's balanced salt solution (mEBSS): prepare 100 mL by adding 10 mL of 10X concentrated liquid EBSS (cat. no. E7510, Sigma) to about 70 mL of deionized water. Add 1.0 mL of 1 *M* 4-(2-hydroxyethyl)piperazine-

1-ethanesulfonic acid (HEPES) (cat. no. H0887, Sigma), 0.36 g of glucose, and adjust the pH of the solution to 7.4 with 1 *N* NaOH. Add 0.220 g of $NaHCO_3$ and bring up the volume to 100 mL in a volumetric flask with deionized water. Sterilize the solution by passing through a 0.2-µm filter and store 10-mL aliquots in tightly sealed 15-mL sterile polypropylene centrifuge tubes at 4°C.

3. Enzyme solution: prepare the enzyme solution just prior to the incubation step (*see* **Subheading 3.1.2., item 1**). To 10 mL of mEBBS (brought to room temperature), add 6 mg of collagenase D (cat. no. 1088882, Roche Applied Science, Indianapolis, IN), 4 mg of trypsin (type TRL, Worthington Biochemical Corp., Lakewood, NJ), and 1 mg of DNase I, Type II (cat. no. D5025, Sigma). Invert the centrifuge tube gently (to avoid foaming) several times to insure that the enzymes are properly dissolved. The concentrations of collagenase and trypsin are critically dependent on specific lots of enzyme and thus must be determined empirically for each new lot (*see* **Note 1**).
4. Culture media: minimal essential medium (MEM) with Earle's salts supplemented with 1% glutamine (200 m*M* stock), 1% penicillin–streptomycin solution, and 10% fetal calf serum (all from Invitrogen Life Technologies, Carlsbad, CA).
5. Poly-L-lysine-coated 35-mm dishes. Dissolve 10 mg of poly-L-lysine (cat. no. P1274, Sigma) in 100 mL of 0.1 *M* sodium borate buffer, pH 8.4 (with HCl) and pass the solution through a 0.2-µm filter. Although the borate buffer is stable for long periods of time, we prepare the poly-L-lysine solution just prior to use. Add 1 mL of solution to each 35-mm dish and allow to stand (usually in the culture hood) for several hours (overnight is convenient). Aspirate off the poly-L-lysine solution and rinse the dishes twice with sterile distilled water. After the dishes are dry, they can be stored indefinitely. We place six to eight 35-mm dishes in a 150 × 25 mm tissue culture dish to facilitate handling.
6. Equipment: dissecting microscope (Model SMZ645, Nikon, Melville, NY) with fiber optic light source; Dumont no. 5 fine forceps (Fine Science Tools, Foster City, CA, USA); microspring scissors (cat. no. 15024-10, Fine Science Tools); a centrifuge capable of maintaining a low speed (~50*g*; Eppendorf 5804R, Brinkmann Instruments, Westbury, NY); tissue culture incubator; and a laminar flow culture hood (optional).

2.2. Microinjection

1. 0.1% Fluorescein dextran solution: weigh out approx 1 mg of 10,000 mol wt fluorescein dextran (cat. no. D1821, Molecular Probes, Eugene, OR, USA) in an autoclaved 1.5-mL microcentrifuge tube and then add molecular biology grade water to a final concentration of 1 mg/mL. Filter the solution through a 0.2-µm syringe filter and store in 100-µL aliquots in microcentrifuge tubes at 4°C. Because this solution is used with RNA, standard precautions should be taken to prevent RNase contamination. Thus, forceps used to handle the fluorescein dextran are wrapped in aluminum foil and autoclaved or baked (350°C for 2–3 h); all plasticware (pipet tips and centrifuge tubes) is autoclaved, and gloves are worn while preparing or handling the solution.

2. Microinjection equipment: our microinjection station consists of a Nikon Diaphot TMD inverted microscope equipped with ×20 and ×40 phase-contrast objective lenses, an epifluorescence unit consisting of a 100-W Hg lamp illumination source and a Nikon B2A filter cube (excitation filter 450–490 nm, dichroic mirror 510 nm, emission filter 520 nm), a remote head CCD camera (Model 6410, Cohu Inc., San Diego, CA, USA), a 9-in. B&W video monitor (Model PVM-122, Sony Corp., Japan), and an automated microinjection system consisting of an Eppendorf FemtoJet microinjection unit and 5171 micromanipulator system (Brinkmann Instruments). The microscope and associated equipment are mounted on a vibration isolation table (Technical Manufacturing Corporation, Peabody, MA, USA).
3. Microinjection pipets: two options are available. Commercially prepared injection micropipets can be purchased (Eppendorf Femtotips I & II, Brinkmann Instruments) that are convenient, sterile, and RNase free. However, the pipets are expensive and the tip geometry is fixed by the manufacturer. Alternatively, micropipets for injection can be manufactured in the laboratory (*see* **Note 7**) if a micropipet puller is available.
4. Microtubes made from hematocrit capillaries: plain (i.e., unheparinized) glass hematocrit capillaries are cleaned with consecutive washes in acetone, methanol, and deionized water and then dried in an oven. The hematocrit tubes are cut into 1-in. segments and then fire-polished shut on one end (be sure that the end is completely sealed) with a Bunsen burner and placed in a 60-mm tissue culture dish (after cooling). The tubes can be placed in a glass beaker and baked at 350°C for several hours to destroy RNase activity if necessary.
5. Microcentrifuge with swinging bucket rotor (Eppendorf 5417C, Brinkmann Instruments).

3. Methods

*3.1. Preparation of Dissociated Neurons (*see *also Notes 1–5)*

3.1.1. Dissection of the Rat Superior Cervical Ganglion

1. Adult male Wistar rats weighing between 250 and 350 g are decapitated with a laboratory guillotine. The animals are anesthetized with CO_2 before they were killed.
2. The head is first placed in a 100-mL plastic beaker containing cold HBSS to rinse blood from the dissection area. Subsequently, the head is placed ventral side up on a dissecting plate, a 150-mm plastic culture dish filled one half full with Sylgard 184 silicone elastomer (World Precision Instruments, Sarasota, FL), and a midline longitudinal incision made in the ventral neck skin with scissors. The skin is then retracted laterally and pinned to the bottom of the dissecting plate, thereby exposing the neck musculature.
3. Under a dissecting microscope, the common carotid artery is identified bilaterally and followed cranially until the carotid bifurcation is encountered. The superior cervical ganglion (SCG) lies on the medial side of the carotid bifurcation between the internal and external carotid arteries (*see* **Note 3**). Both carotid bifurcations,

with SCG attached, are carefully dissected from surrounding tissue and placed in a 60-mm culture dish containing cold (4°C) HBSS. During removal of neck musculature to expose the SCG, flush the dissection field often with cold HBSS from a Pasteur pipet to clear blood, to remove K^+ released from damaged muscle, and to cool the dissection area.

4. Further dissection is carried out with the carotid bifurcation (which is Y-shaped) secured, with the SCG facing upwards, to the bottom of a small dissecting plate (60-mm culture dish half filled with Sylgard) with 100-µm Minuten stainless steel insect pins (part no. WW-65-4366, Carolina Biological Supply, Burlington, NC, USA). During the dissection, the preparation is covered with cold HBSS, which can be periodically replaced to remove debris and keep the tissue cool.
5. Under a dissecting microscope, a longitudinal slit is made in the connective tissue capsule enclosing the ganglion with fine spring scissors. The ganglion is then "teased out" of the connective tissue capsule using blunt dissection with no. 5 Dumont forceps. Nerves entering and leaving the ganglion (usually three major ones) are cut close to the ganglion body with spring scissors. The desheathed ganglion is then transferred, using a large-bore fire-polished Pasteur pipet, to a 60-mm culture dish filled with cold HBSS and sitting on ice.
6. When both ganglia have been desheathed, multiple parallel slits about 1 mm apart are made perpendicular to the long axis of the ganglia using spring scissors to facilitate enzyme penetration. The free floating ganglia can usually be gently pushed into the scissors blades with the tips of the Dumont forceps without actually "grabbing" the tissue. We prefer to make slits approx 75% of the way through the ganglia, thus facilitating the transfer of the tissue as one piece. However, equivalent results are produced by cutting the ganglia into small pieces (~1 mm^3) using a small scalpel blade. The procedure is done using the dissecting microscope at fairly high magnification (×50).
7. At this point, the ganglia are ready for enzyme dissociation. Although the dissection should be carried out expeditiously, peripheral neurons seem quite hardy as long as they are bathed in cold (4°C) HBSS, and thus haste, at the expense of gentle handling of the tissue, does not seem warranted.

3.1.2. Enzymatic Dissociation and Culture

1. A 10-mL enzyme solution is prepared as described in **Subheading 2.1.3.** Six milliliters of the enzyme solution is pushed through a 0.2-µm syringe filter (Millex-GV, Millipore, Bedford, MA) into a 25-cm^2 tissue culture flask with a plug seal cap (e.g., Falcon no. 3013). The ganglia are then transferred to the flask with a fire-polished Pasteur pipet, taking care to transfer as little HBSS to the enzyme solution as possible.
2. The flask is flushed with a mixture of 95% O_2–5% CO_2 for approx 1 min and the cap tightly sealed.
3. Place the flask in a 35°C shaking water bath with the long axis of the flask aligned parallel to the stroke direction. The flask is shaken fairly rapidly (240 strokes/min) but not so rapidly as to cause solution to slosh up the sides of the flask.

4. After 1 h, remove the flask from the water bath. At this point the ganglia should be in small separate pieces of approx 1 mm^3. The pieces are then dispersed into single neurons by grasping the neck of the flask between the thumb and first finger and vigorously shaking the flask for 10 s with a motion similar to that used to shake the mercury down into the bulb of a thermometer.
5. Following the shake, the pieces should be completely dispersed with at most one or two barely visible pieces.
6. Five milliliters of culture media (preheated to 37°C) are added to the flask and the contents transferred to a 15-mL centrifuge tube.
7. The solution of dispersed neurons is centrifuged for 6 min at 50*g* in a swinging bucket centrifuge at room temperature, the supernatant removed, 10 mL of culture media added to resuspend the small pellet, and the centrifugation repeated.
8. Following removal of the supernatant, the pellet (which may be barely visible) is gently resuspended with 1.8 mL of culture media.
9. Neurons are plated into six 35-mm poly-L-lysine coated tissue culture plates (Corning no. 25000). To keep the neurons centered in the dish, an autoclaved glass 10 × 10 mm cloning ring (cat. no. 2090-01010, Bellco Glass Inc., Vineland, NJ) treated with Sigmacote (Sigma) is placed in the center of each 35-mm dish and 0.3 mL of the dispersed neuron solution placed carefully in the cloning ring. Two milliliters of culture media is then carefully added to the 35-mm culture dish, taking care not to dislodge the cloning ring.
10. After about 60 min incubation in a tissue culture incubator (37°C, humidified atmosphere of 5% CO_2 in air), the cloning rings are carefully removed from the dishes with sterile forceps.
11. Immediately after dissociation the neurons usually have several small stumps of amputated processes and are loosely attached to the substrate of the dish. However, after 2–3 h of incubation, the neurons resorb the processes, attain a spherical geometry, and attach firmly to the dish.
12. Neuron somata are usually 20–40 µm in diameter with a large nucleus containing two or three nucleoli.

3.2. Injection of cRNA Into the Cytoplasm of Neurons (see also Notes 7–10, 14–16)

1. In all procedures involving RNA, precautions are taken to avoid RNase contamination. Thus gloves are worn while handling the sample, all plasticware (microcentrifuge tubes and pipet tips) that comes in contact with the sample is autoclaved, glassware is baked (350°C for several hours), and solutions are made from molecular biology grade or diethyl pyrocarbonate (DEPC)-treated H_2O.
2. In vitro transcribed capped cRNA coding for the desired protein is stored at –80°C, usually at a concentration of 1–5 µg/µL in water (*see* **Note 11**).
3. Mix 1–3 µL of cRNA solution with 2–4 µL of 0.1% fluorescein dextran solution. We find that a final concentration of 1 µg/µL is a good starting point. The small volumes can be mixed on a piece of Parafilm while they are observed with a dissecting microscope. Apply a small square of Parafilm to a surface with a

drop of water (uncovered side down). Carefully remove the paper backing and use the resulting clean surface for mixing the drops of solution by pipetting up and down several times with a pipettor.

4. Transfer the 5 μL of solution to a microtube (*see* **Subheading 2.2.4.**) with an autoclaved Eppendorf microloader pipet tip (cat. no. 930 00 100-7, Brinkmann Instruments). Place the microtube in a standard 1.5-mL microcentrifuge tube.
5. Centrifuge the tube for 30–60 min at maximum speed (~10,600*g*) at room temperature in a microcentrifuge equipped with a swinging bucket rotor. We find that fixed-angle rotors often cause the microtubes to break.
6. Place the microtube on ice. We use a 1 × 1 × 6-in. brass bar with holes drilled in it to hold the microtubes. The bar is placed in a small styrofoam container filled with ice and serves as both a holder and heat sink.
7. Pipet 1 μL of cRNA solution into a microloader tip, being careful to remove the solution from the top of the fluid column. Avoid agitating the solution near the bottom of the tube as this invariably stirs up particulate matter, which then clogs the submicron opening of the microinjection pipet.
8. Backfill the injection micropipet with the 1 μL of cRNA solution by depositing the solution into the shank of the pipet. The glass capillary in the pipet will cause the remainder of the micropipet tip to fill by capillary action. Sometimes a small bubble forms that can be dislodged by gently "flicking" the micropipet with a finger.
9. Place a dish of dissociated neurons (3–6 h after isolation) on the microscope stage, attach the injection micropipet to the holder, and position the micropipet about 100 μm above the bottom of the dish using the 5171 micromanipulator.
10. We typically use a ×20 phase-contrast objective to view the neurons. The image is detected with a charge-coupled device (CCD) video camera attached to the microscope and displayed on a video monitor. Although video equipment is not required, the inherently high contrast of the B&W video system enhances the phase contrast.
11. When the injection micropipet is lowered into the culture media, the phase-contrast image can be greatly degraded as result of the meniscus formed at the glass–fluid interface. The image can be dramatically improved by slightly rotating the turret holding the phase rings one way or the other (just off the "detent" position) to realign the phase rings.
12. Cytoplasmic injections are carried out using the axial injection mode of the 5171 micromanipulator at the default injection velocity (300 μm/s). Injection pressure (P2) usually ranges from 100 to 200 hectoPascals (hPa; 1 hPa ≈ 0.015 psi) and injection duration from 0.3 to 0.5 s depending on pipet size. Holding pressure (P_3) is set to about 45 hPa (not critical).
13. The axial injection *z*-axis limit is set with the following procedure. First, the micropipet tip is positioned as close to the bottom of the dish as possible using the "step mode" of the manipulator. This is facilitated by focusing on one of the very flat fibroblast-like cells (Schwann cells) that accompany the neurons in most preparations. The coordinates of the micromanipulator are reset to zero, the micropipet moved upwards 5 μm (as observed on the LCD display of the micromanipulator),

and the *z*-axis limit set. The pipet is now positioned about 30–32 μm above the bottom of dish so that the tip will clear the top of the neurons (*see* **Note 9**).

14. Neurons are positioned under the tip of the injection micropipet using the *x–y* movement of the microscope stage. The focus of the microscope is adjusted back and forth between the neuron and the pipet tip to assess the relative position.
15. The injection sequence is initiated when the neuron is positioned with the pipet tip about 0.25 diameters from the edge of the neuron. A successful injection results in a distinct blanching (i.e., a change in contrast) of the cell interior and rapid increase in the diameter of the neuron (about 10%). Injection can be confirmed by checking the cell for fluorescence of the coinjection marker. Sometimes repositioning the neuron such that the pipet tip is closer to the center of the cell (where it is in a more upright position) results in a successful injection when the initial attempt was unsuccessful.
16. Subsequent neurons in the dish are injected by moving the stage to reposition the next neuron. Use a "scanning pattern" to cover systematically a given area of the dish. On a good day, approx 50 neurons in a dish can be injected in 10–15 min. Neurons are returned to the incubator following injection. Although injections take place under only "pseudo-sterile" conditions, microbial contamination has not been a problem in the short term (24–36 h) when antibiotics are included in the culture media.
17. Following overnight incubation, previously injected neurons are identified from the fluorescence of the coinjection marker. It is prudent to minimize the exposure of the preparation to the intense epifluorescent illumination, as products resulting from photobleaching might be toxic.
18. Patch-clamp recordings are performed on the injected neurons 8–24 h after isolation. At this time, the neurons are still relatively spherical and usually devoid of significant processes, thus facilitating whole-cell voltage-clamp recordings.

3.3. Injection of cDNA Into the Nucleus of Neurons (see also Notes 6–10, 13–16)

1. Injection of cDNA into the nucleus of sympathetic neurons utilizes methodology common to that described in **Subheading 3.2.** for cytoplasmic injections; thus only major differences in protocol are described in detail.
2. DNA coding for the desired protein, subcloned into an appropriate mammalian expression vector (*see* **Note 12**), is stored at –20°C at a concentration of 1 μg/μL in TE buffer (10 m*M* Tris, 1 m*M* EDTA, pH 8.0).
3. A vector encoding enhanced green fluorescent protein, EGFP-N1 (BD Biosciences Clontech, Palo Alto, CA), is coinjected as a reporter construct.
4. One microliter of plasmid DNA (supercoiled) encoding the target protein (1 μg/μL of stock) and EGFP-N1 (0.05 μg/μL stock) are mixed with 8 μL of TE buffer, pH 8.0. The solution is centrifuged as described in **Subheading 3.2.5.**
5. DNA solutions are kept at room temperature during the injection session.
6. Injection pressure (P_2) usually ranges from 100 to 200 hPa and injection duration from 0.2–0.3 s depending on pipet size.

7. Injections are carried out as for cytoplasmic injections with the following differences. (1) The neuron is positioned such that the pipet tip is centered over the middle of the nucleus. (2) Successful nuclear injections are difficult to confirm visually. The best indication seems to be a "bright flare" at the end of the pipet, which forms in the center of the nucleus. If the nucleoli move laterally during the injection or a very discrete "bright dot" is evident, this usually means the nuclear membrane was "dented" but not penetrated. (3) Pronounced swelling of the cell almost always indicates a cytoplasmic injection. Significant swelling of the nucleus is poorly tolerated and neurons experiencing such treatment often die. (4) Injection pipets tend to clog much more frequently when compared with cytoplasmic RNA injections. This (maddening) tendency seems to result from both the properties of plasmid DNA (less soluble) and the consistency of the nucleus (very viscous). We sometimes incubate the stock cDNA (after removal from the freezer) at 60°C in a dry heat block for a few minutes prior to mixing and centrifugation if clogging is a problem. Stock solutions can also be centrifuged through 0.1-μm filters (Amicon Ulftafree-MC, Millipore) to remove large particulates.
8. Following overnight incubation, neurons that received successful nuclear injections are identified from the fluorescence of the heterologously expressed EGFP-N1. The green fluorescence is usually easily detected under the conditions described in **Subheading 2.2., item 1** (i.e., Nikon B2A filter cube). If the fluorescence is dim, the culture medium should be removed and replaced with a solution that does not contain phenol red (which increases the background fluorescence).
9. In our hands, nuclear injection of adult rat sympathetic neurons is much more difficult than cytoplasmic injection. We estimate that perhaps 10–20% of attempted nuclear injections result in a satisfactory expression (vs 80–100% for cytoplasmic injections). The problem seems to arise from locating the nucleus in the *z*-axis. Thus the large size of the neurons, an asset for cytoplasmic injections, is actually a hindrance for nuclear injections. It should be noted, however, that several constructs that did not produce functional expression following cRNA injection generated a robust response following nuclear injection of cDNA.

4. Notes

1. The enzyme concentrations must be adjusted empirically for each lot of collagenase and trypsin. Thus, enzymes should be purchased in large quantities (several grams) once suitable lots are identified. For long-term storage (years), we aliquot the enzymes and store them at –80°C. The working aliquot of collagenase and trypsin is stored at –20°C. Collagenase and trypsin are used at a concentration of 0.5–1.0 and 0.3–0.5 mg/mL, respectively. The concentration of DNase is not critical.
2. We use a nearly identical procedure to isolate adult rat dorsal root and nodose ganglion sensory neurons (dissection procedure not described here). For sensory neurons, the trypsin concentration is reduced to approx 0.1 mg/mL. Dorsal root ganglia do not require desheathing, which is difficult because of their small size.

3. The anatomy of the rat autonomic nervous system, including superior cervical ganglion, is illustrated by Gabella ***(19)***.
4. The "vigorous shake" method of neuron dissociation was developed by the author and Dr. G. G. Schofield ***(20)***. We prefer this method to the more commonly used (and dignified) method involving trituration of the cell suspension with a fire-polished Pasteur pipet although this method will also work.
5. If large clumps of tissue remain after shaking the flask, a higher concentration of enzymes is required. The concentration of trypsin seems to have the greatest influence on the quality of the preparation and thus this should be increased first. Conversely, the trypsin concentration should be decreased if the neurons: (1) perish during the first few hours of incubation; (2) assume a "shrunken" appearance after dissociation; or (3) retain multiple long (several cell diameters) processes.
6. One of the keys to successful cDNA microinjection is clearly visualizing the nucleus. Prior to starting an injection session, make sure that the phase rings are properly aligned. If the objective is fitted with a correction collar, it should be adjusted to obtain the sharpest focus through the relatively thick bottom of the culture dish. Some modern phase objectives seem to produce much less contrast, perhaps to optimize light transmission for fluorescence imaging, than older models. Consult your microscope dealer to see if objective lenses optimized for phase contrast are available. Once the injection pipet enters the fluid in the dish, the meniscus will disrupt the alignment of the phase ring with the objective phase ring. This can be quickly improved by slightly rotating the phase ring (above the condenser) slightly out of the fixed position while monitoring the image on the monitor. Injection buffers with low ionic strength (e.g., TE buffer) seem to provide improved visualization of the injection process.
7. Initially, it is probably easiest to use commercially manufactured pipets (e.g., Femtotips) for injection. Manufacturing your own pipets, however, is considerably less expensive (if you have access to a pipet puller) and allows for some customization of pipet geometry. Start by observing the overall geometry of a Femtotip under a microscope at low (×60) and high power (×600). Then try to reproduce the sharply tapering geometry with a pipet puller. We currently pull injection micropipets from thin-walled (outer diameter 1.2 mm, inner diameter 0.9 mm, 100 mm length) borosilicate capillary-filled microelectrode blanks (cat. no. TW120F-4, World Precision Instruments, Sarasota, FL) using a Flaming/Brown type pipet puller (P-97; Sutter Instrument Co., Novato, CA). The pipet glass can be cleaned and baked as previously described for hematocrit capillaries (*see* **Subheading 2.2.4.**), although we often skip this step for cDNA injection. A two-stage pull is used with settings of (heat, pull, velocity, time): (1) 560, 115, 12, 250; (2) 580, 130, 65, 250. These settings vary considerably with puller (even of the same model), batch of glass, and filament type (we use a box-type platinum–iridium filament) but may represent a reasonable starting point. A recent techniques article ***(21)*** provides an excellent introduction to the effects of puller setting on pipet geometry. Because the opening of the micropipet is too small to

observe with a light microscope, the suitability of micropipets must be tested functionally (i.e., by injecting a few cells). Resist the temptation to use water or buffer (e.g., TE) for these tests as the viscosity of cDNA-containing solutions greatly influences flow through small openings.

8. Because culture dishes are not perfectly flat, the *z*-axis injection limit usually has to be adjusted if the dish is moved far from where the initial setting was determined. This can be a major irritation if the distance between individual neurons is great. The use of a cloning ring to confine the neurons to the center of the dish during plating minimizes this problem.
9. Readjusting the *z*-axis limit requires many key strokes on the 5171 micromanipulator. A deficiency of the 5171 manipulator controller is the inability to store a sequence of keystrokes as a "macro." Recently, we have taken advantage of the serial port connections on both the 5171 micromanipulator and Femtojet to integrate commonly used keystroke combinations with a programmable video-editing controller (ShuttlePro, Contour A/V Designs, Windham, NH). Buttons on the ShuttlePro (USB connection) are coded as keystroke commands that are intercepted by Igor Pro (version 4.05A, WaveMetrics, Lake Oswego, OR) running on an iBook (Apple Computer, Cupertino, CA). Each key push or rotation of the shuttle ring initiates a subroutine that sends a predetermined series of keystrokes (as ASCII characters via the VDT XOP included with Igor Pro) to either the 5171 manipulator or the FemtoJet via a USB four-port serial adapter (Keyspan, Richmond, CA, USA). Integrating several functions onto a single controller greatly improves the ergonomics of the injection process. Details of the programming are beyond the scope of this chapter (the author can be contacted for further details).
10. Clogging of the injection pipet is a frequent problem, especially with DNA solutions. Sometimes the offending particle can be ejected with a brief pulse of high pressure (P_1 button). A more serious maneuver is to touch gently the tip of the pipet to the bottom of the culture dish using the "step mode" of the micromanipulator. This maneuver usually enlarges the opening of the tip, thus requiring a reduction of injection pressure. The trick often works well for cytoplasmic injections but less often for nuclear injections.
11. We use mMESSAGE mMACHINE in vitro transcription kits (Ambion, Austin, TX) as per the manufacturer's instructions to prepare cRNA. As most vectors do not code for a poly(A) tail and the presence of such is known to prolong mRNA lifetime, the addition of a poly(A) tail, using the (Poly(A) tail kit (Ambion), may increase expression levels. However, we have no experience with this procedure.
12. Mammalian expression vectors containing a CMV promoter, such as pCI (Promega, Madison, WI) or pcDNA3.1 (Invitrogen), work well in SCG neurons. High-level expression seems to require injection of supercoiled plasmid DNA into the nucleus. Linearized plasmid DNA or injection of plasmid DNA into the cytoplasm produces little or no expression. We use QIAGEN (Chatsworth, CA) midiprep columns to purify plasmid DNA.
13. Cytoplasmic injection of cDNA results in very inefficient expression, possibly owing to nucleases present in the cytosol ***(22)***. It has been reported that mixtures of cDNA

with a polycation such as polyethylenimine *(23)* or nuclear targeting peptides *(24)* increase the expression efficiency of cytoplasmic cDNA. Such strategies may be useful for increasing the number of expressing cells following cDNA injection.

14. The decision of whether to use cytoplasmic injection of cRNA or nuclear injection of cDNA for expression in sympathetic neurons should be based on the following considerations. In general, cytoplasmic injections are much easier to perform and thus many more neurons are usually available for study. However, the manufacture of cRNA requires an additional step and the susceptibility of RNA to degradation requires more careful handling and storage. Conversely, nuclear injections are more difficult to perform and thus might provide a frustrating introduction to microinjection technique. Significantly, however, we have been able to obtain functional expression of several constructs (receptors and G-protein subunits) only from nuclear injection. Our impression is that expression levels resulting from nuclear injection of DNA are much higher than those obtained with cytoplasmic injection of RNA. Consequently, nuclear injection may be preferable when attempting to express dominant negative mutations. We speculate that the higher level of expression results from: (1) the stability of plasmid DNA in the nucleus, (2) the strong constitutive promoter used (CMV), and (3) the increased stability of nuclear-derived RNA due to posttranscriptional processing, for example, by addition of a poly(A) tail. At present, we use nuclear injection for nearly all expression experiments. However, the recent advent of siRNA technology, in which small ds-RNAs are introduced into the cytoplasm to ablate gene expression *(18)*, may be an appropriate use for cytoplasmic injections.
15. Expression from either RNA or DNA can sometimes be enhanced by removing most of the untranslated regions from the clone *(25)* and placing the start codon within a strong context *(26)*. We find that out-of-frame start codons in the 5′ untranslated region are particularly troublesome *(27)*.
16. An excellent source of signaling protein clones is the Guthrie cDNA Resource Center (www. cdna.org), which provides sequence-verified wild-type and mutant human cDNAs in the vector pcDNA3.1. The Guthrie Research Institute is a non-profit organization. The author is the former Director of the Guthrie cDNA Resource Center.

References

1. Lundstrom, K., Schweitzer, C., Richards, J. G., Ehrengruber, M. U., Jenck, F., and Mülhardt, C. (1999) Semliki Forest virus vectors for in vitro and in vivo applications. *Gene Ther. Mol. Biol.* **4,** 23–31.
2. Jeong, S. W. and Ikeda, S. R. (2001) Differential regulation of G protein-gated inwardly rectifying K^+ channel kinetics by distinct domains of RGS8. *J. Physiol. (Lond.)* **535,** 335–347.
3. Hass, K., Sin W. C., Javaherian, A., Li, Z., and Cline, H. T. (2001) Single-cell electroporation for gene transfer in vivo. *Neuron* **29,** 583–591.
4. Rae, J. L. and Levis, R. A. (2001) Single-cell electroporation. *Pflügers Arch. Eur. J. Physiol.* **443,** 664–670.

5. Tirlapur, U. K. and König, K. (2002) Targeted transfection by femtosecond laser. *Nature* **418,** 290–291.
6. Ikeda, S. R., Lovinger, D. M., McCool, B. A., and Lewis, D. L. (1995) Heterologous expression of metabotropic glutamate receptors in adult rat sympathetic neurons: subtype specific coupling to ion channels. *Neuron* **14,** 1029–1038.
7. Pan, X., Ikeda, S. R., and Lewis, D. L. (1996) The rat brain cannabinoid receptor modulates N-type Ca^{2+} channels in a neuronal expression system. *Mol. Pharmacol.* **49,** 707–714.
8. Ikeda, S. R. (1996) Voltage-dependent modulation of N-type calcium channels by G protein $\beta\gamma$-subunits. *Nature* **380,** 255–258.
9. Jeong, S. W. and Ikeda, S. R. (1999) Sequestration of $G\beta\gamma$ subunits by different G protein α subunits blocks voltage-dependent modulation of Ca^{2+} channels in rat sympathetic neurons. *J. Neurosci.* **19,** 4755–4761.
10. Ruiz-Velasco, V. and Ikeda, S. R. (2000) Various combinations of G-protein $\beta\gamma$ subunits produce voltage-dependent inhibition of N-type calcium channels in rat superior cervical ganglion neurons. *J. Neurosci.* **20,** 2183–2191.
11. Jeong, S. W. and Ikeda, S. R. (1998) G protein α-subunit $G_Z\alpha$ couples neurotransmitter receptors to ion channels in sympathetic neurons. *Neuron* **21,** 1201–1212.
12. Ruiz-Velasco, V. and Ikeda, S. R. (1998) Heterologous expression and coupling of G protein-gated inwardly rectifying K^+ (GIRK) channels in adult rat sympathetic neurons. *J. Physiol. (Lond.)* **513,** 761–773.
13. Jeong, S. W. and Ikeda, S. R. (2000) Effect of G-protein heterotrimer composition on coupling of neurotransmitter receptors to N-type Ca^{2+} channel modulation in sympathetic neurons. *Proc. Natl. Acad. Sci. USA* **97,** 907–912.
14. Kammermeier, P. J. and Ikeda, S. R. (1999) Expression of RGS2 alters the coupling of metabotropic glutamate receptor 1a to M-type K^+ and N-type Ca^{2+} channels. *Neuron* **22,** 819–829.
15. Ikeda, S. R., Jeong, S. W., Kammermeier, P. J., Ruiz-Velasco, V., and King, M. M. (1999) Heterologous expression of a green fluorescent protein–pertussis toxin S1 subunit fusion construct disrupts calcium channel modulation in rat superior cervical ganglion neurons. *Neurosci. Lett.* **271,** 163–166.
16. Ruiz-Velasco, V. and Ikeda, S. R. (2001) Functional expression of green fluorescent protein fused to G-protein β and γ subunits in adult rat sympathetic neurons. *J. Physiol. (Lond.)* **537,** 679–692.
17. Janetopoulous, C., Jin, T., and Devreotes, P. (2001) Receptor-mediated activation of heterotrimeric G-proteins in living cells. *Science* **291,** 2408–2411.
18. Elbashir, S. M., Harborth, J., Lendeckel, W., Yalcin, A., Weber, K., and Tuschl, T. (2001) Duplexes of 21-nucleotide RNAs mediate RNA interference in cultured mammalian cells. *Nature* **411,** 494–498.
19. Gabella, G. (1995) Autonomic nervous system, in *The Rat Nervous System*, 2nd ed. (Paxinos, G., ed.), Academic Press, New York, pp. 81–103.
20. Ikeda, S. R., Schofield, G. G., and Weight, F. F. (1986) Na^+ and Ca^{++} currents of acutely isolated adult rat nodose ganglion cells. *J. Neurophysiol.* **55,** 527–539.

21. Miller, D. F. B., Holtzman, S. L., and Kaufman, T. C. (2002) Customized microinjection glass capillary needles for P-element transformations in *Drosophila melanogaster*. *Biotechniques* **33,** 366–374.
22. Pollard, H., Toumaniantz, G., Amos, J. L., et al. (2001) Ca^{2+}-sensitive cytosolic nucleases prevent efficient delivery to the nucleus of injected plasmids. *J. Gene Med.* **3,** 153–164.
23. Ludtke, J. J., Sebestyen, M. G., and Wolff, J. A. (2002) The effect of cell division on the cellular dynamics of microinjected DNA and dextran. *Mol. Ther.* **5,** 579–588.
24. Subramanian, A., Ranganathan, R., and Diamond, S. L. (1999) Nuclear targeting peptide scaffolds for lipofection of nondividing mammalian cells. *Nat. Biotech.* **17,** 873–877.
25. Kaufman, R. J. (1990) Vectors used for the expression in mammalian cells. *Methods Enzymol.* **185,** 487–511.
26. Kozak, M. (1986) Point mutations define a sequence flanking the AUG initiator codon that modulates translation by eukaryotic ribosomes. *Cell* **44,** 283–292.
27. Pantopoulos, K., Johansson, H. E., and Hentze, M. W. (1994) The role of the 5′ untranslated region of eukaryotic messenger RNAs in translation and its investigation using antisense technologies. *Prog. Nucleic Acid Res. Mol. Biol.* **48,** 181–238.

11

Covalent Modification of G-Proteins by Affinity Labeling

Martin Hohenegger, Michael Freissmuth, and Christian Nanoff

Summary

The activation of heterotrimeric G-proteins is tightly regulated by the exchange of GTP for GDP in the α-subunit; mostly—but not exclusively—seven-transmembrane receptors function as the guanine nucleotide exchange factors (GEFs). A research goal may be to determine which G-protein α-subunit is activated by the receptor under investigation. In a membrane preparation obtained from cells or tissues this can be achieved in a seemingly straightforward manner by determining if the receptor increases the covalent incorporation of GTP analogs into G-protein α-subunits. Because the GTP analogs may be labeled to high specific radioactivity the α-subunit can then be identified with the use of specific antibodies (*see* Chapter 13). One of the compounds we present here (2′,3′-dialdehyde-GTP) can also be employed to block receptor-mediated G-protein activation and to disrupt the cognate signaling pathway.

Key Words

Azidoanilido-GTP; 2′,3′-dialdehyde GTP; 2′,3′-dialdehyde GTPγS; G-protein α-subunit; guanine nucleotide exchange reaction; receptor-dependent activation.

1. Introduction

The basic mechanism of G-protein-mediated transmembrane signaling was elucidated in the late 1970s and early 1980s. Subsequently, molecular cloning has identified a large array of closely related receptors (R), G-protein subunits (G), and effectors (E). In cellular signaling, an important research goal therefore is to understand the factors that, at each step of the cascade link R to G to E, and govern signal transfer and signal integration following activation of a specific receptor *(1)*. Receptors, in particular, appear to differ widely in their ability to interact with the different G-protein oligomers and examples for both (1) stringent coupling of a receptor to a single species of a heterotrimer composed of a defined α-, β-, and γ-subunit as well as (2) promiscuous interaction with a wide variety of different G-protein subunits exist *(2)*.

From: *Methods in Molecular Biology, vol. 259, Receptor Signal Transduction Protocols, 2nd ed.*
Edited by: G. B. Willars and R. A. J. Challiss © Humana Press Inc., Totowa, NJ

Several experimental strategies have been employed to identify the G-protein subunits that interact with a given receptor (*see* Chapter 12). Here, we focus on methods to covalently, label G-protein α-subunits, which can be used to study receptor G-protein coupling by exploiting receptor-promoted guanine nucleotide exchange. Two GTP analogs are well suited for affinity labeling of the guanine nucleotide binding pocket (*see* **Note 1**), namely (1) GTP-azidoanilide (azidoanilido-GTP) and (2) the 2′,3′-dialdehyde analogs of GTP and GTPγS (*o*-GTP, *o*-GTPγS). When bound to a G-protein, GTP-azidoanilide cannot be hydrolyzed and therefore remains tightly associated with the protein. Covalent incorporation is achieved by ultraviolet (UV) photolysis. *o*-GTP and *o*-GTPγS form a Schiff's base with the ε-amino group of lysine residues in the guanine nucleotide binding pocket of the protein; reduction with borohydride results in the formation of a covalent bond. Compared with other affinity probes (e.g., 8-azido-GTP) these ligands offer the advantage of binding in a quasi-irreversible manner. Hence, the free ligand can be removed; this greatly reduces background labeling and improves the signal-to-noise ratio.

2. Materials

The laboratory should be equipped for standard biochemical techniques involving radioactive reagents; i.e., possess a fanned hood reserved for handling radioactive substances, lead bricks, Plexiglas shields, and so forth, liquid scintillation counter, slab gel electrophoresis, high-performance liquid chromatograph (HPLC) equipped with a gradient-forming pump system, fraction collector, and filtration manifold.

2.1. Synthesis and Purification of [α-^{32}P]GTP-Azidoanilide

1. 1 mCi of [γ-^{32}P]GTP, specific activity 3000 Ci/mmol (NEN).
2. 4-Azidoaniline (Sigma-Aldrich).
3. 1,4-Dioxane (Merck).
4. *N*-(3-Dimethylaminopropyl)-*N*′-ethylcarbodiimide hydrochloride (Merck).
5. 2-(*N*-Morpholino)ethanesulfonic acid (Biomol).
6. 1 *M* Triethylammonium carbonate (TEAC, Sigma-Aldrich).
7. Ethanol.
8. Supelcosil LC 308 (RP C18—5 μm) HPLC cartridge (Supelco).
9. Neutral alumina (Merck).
10. Plexiglas shields, Plexiglas shades, Geiger–Müller counter; HPLC, and lyophilizer reserved and adapted for handling radioactivity in microCurie quantities.

2.2. Synthesis and Purification of 2′,3′-Dialdehyde Analogs of GTP and GTPγS

1. Radioactively labeled [^{35}S]GTPγS and [α-^{32}P]GTP (NEN) (*see* **Note 2**).
2. Unlabeled guanine nucleotides: GTP, GTPγS, GDP (Boehringer Mannheim or Sigma-Aldrich). Commercial preparations of GTPγS, and of Gpp(NH)p, another

hydrolysis-resistant GTP analog), are rarely 100% pure; most often the contamination is GDP; this can be removed by chromatography over an appropriately sized DEAE-Sephacel (Amersham Pharmacia Biotech) column as described in the following paragraphs.

In brief, pour a 20-mL column, wash with 1 *M* LiCI in 30 mL of water, equilibrate in 100 mL of water; dissolve 50 mg of GTPγS in 5 mL of water, apply onto column, wash with 15 mL of water, elute guanine nucleotides with a 80-mL gradient of 0–1 *M* LiCl; collect 2-mL fractions; follow elution either with on-line UV-detector or by diluting samples 1:100 in water and determining the absorbance at 252 nm in a UV/Vis-spectrophotometer (quartz cuvet)—typically two peaks will be detected, one at 0.4 *M* LiCI corresponding to GDP (~20%) and one at 0.7 *M* LiCl corresponding to GTPγS (~80%). Pool GTPγS-containing fractions and dispense into two to four centrifuge tubes; add 9 parts of ice-cold ethanol, store overnight at –20°C (lithium salts of nucleotides precipitate readily in ethanol); centrifuge, discard supernatant, lyophilize pellet to dryness, and dissolve in 5 mL of water (adjust the pH to 7–8 with NaOH); add dithiothreitol (DTT) to 2.5 m*M* final concentration; the concentration is best determined spectrophotometrically using serial dilutions (1:200, 1:400, and so on) and the molar extinction coefficient $\varepsilon = 13.7\ mM^{-1}\ cm^{-1}$ at 252 nm.

3. $NaBH_4$, $NaCNBH_3$, $NaIO_4$ (Merck). Prepare fresh solutions immediately prior to use; for concentration requirements *see* **Subheading 3.5.**
4. Glycerol (Biomol).
5. LiCl (Merck): prepare 2 *M* stock.
6. DTT (Sigma-Aldrich): prepare 1 *M* stock; most GTPγS binding reaction buffers contain 1 m*M* DTT; we omit DTT in the current applications for several reasons: DTT can interfere (1) with photoaffinity labeling by GTP-azidoanilide; (2) with oxidation/reduction by $NaIO_4$/$NaBH_4$; (3) DTT can activate some G-protein-coupled receptors (GPCRs) in the absence of agonists.
7. Sephadex G-10 (Amersham Pharmacia Biotech).
8. Polyethyleneimine (PEI) cellulose thin-layer chromatography (TLC) plates (Merck).

2.3. *Affinity Labeling of G-Protein α-Subunits*

1. Purified oligomeric G-proteins or G-protein α-subunits: alternatively, appropriate membrane preparations or partially purified G-proteins from native tissue are an adequate substrate. Membranes prepared from the cerebral cortex (gray matter) of mammals are a rich source of heterotrimeric G-proteins where G_o represents approx 1% of the total membrane proteins.
2. 4-(2-Hydroxyethyl)piperazine-1-ehtanesulfonic acid (HEPES)/NaOH, pH 7.6: make up as a 1 *M* stock solution.
3. Tris-HCl, pH 8.0: make up as a 1 *M* stock solution.
4. EDTA: make up as 0.1 *M* stock (adjust pH to 7.0 with NaOH; EDTA will dissolve as the pH rises).
5. $MgSO_4$ and $MgCl_2$: make up as 1 *M* stocks.

6. NaCl; make up as 4 *M* stock.
7. Lubrol PX (Sigma-Aldrich): make up as a 10% aqueous solution (w/v): add 5 g of mixed bed resin AG50 1-X8 (Bio-Rad)/1 L of Lubrol solution and stir overnight at room temperature; filter in the morning; dispense into aliquots and store frozen at –20°C.
8. Binding reaction buffer: make up 50 mL from stock solutions: 50 m*M* HEPES, pH 7.6; 1 m*M* EDTA; 10 m*M* $MgSO_4$; Lubrol 0.1%. Keep refrigerated.
9. Binding stop solution: make up 2 L from stocks: 20 m*M* Tris, pH 8.0; 20 m*M* $MgCl_2$; 150 m*M* NaCl. Keep refrigerated.
10. Filtration manifold (Bio-Rad).
11. BA85 nitrocellulose filters (Schleicher & Schuell) to trap purified G-protein α-subunits in binding assays, or glass fiber filters (Whatman GF/C, GF/B) for membrane-bound proteins.
12. UV hand lamp (254 nm; 100–200 W).
13. Antisera for the identification of G-protein α-subunits (*see* Chapter 13).

2.4. Affinity Labeling of Membrane-Bound G-Proteins by In Situ *Oxidation–Reduction*

1. Prepare reaction buffer: 20 m*M* HEPES–NaOH, pH 7.6, 1 m*M* EDTA, 2 m*M* $MgSO_4$. Alternatively, if Na^+ is to be avoided prepare: 20 m*M* Tris-HCl, pH 7.6, 1 m*M* EDTA, 2 m*M* $MgCl_2$. Make up binding buffer (50 μL/assay) by adding [^{35}S]GTPγS (1×10^6 cpm) and the desired concentration of unlabeled GTPγS (0–2 μ*M* final concentration); check total activity by scintillation counting and use to calculate specific activity; hold on ice. The choice of free Mg^{2+} concentration, final GTPγS concentration, incubation times, and supplementation with GDP is critical, if the receptor-catalyzed exchange is to be monitored (*see* **Note 3**).
2. Prepare 20 m*M* Mg buffer (reaction buffer containing 20 m*M* $MgCl_2$) and GTP stop buffer (0.1 m*M* GTP in 20 m*M* Mg buffer; 1 mL/assay); maintain both on ice.

3. Methods

3.1. Synthesis of [α-^{32}P]GTP-Azidoanilide

The synthesis of [α-^{32}P]GTP-azidoanilide was originally described by Pfeuffer *(3)*. The method outlined below is adapted from Offermanns et al. *(4)* with some modifications. An alternative chromatography procedure for isolation of [α-^{32}P]GTP-azidoanilide was proposed by Anis et al. *(5)*. Recently, [α-^{32}P]GTP-azidoanilide has been made available from Affinity Labeling Technologies (Alt. Inc, Lexington, KY).

1. [α-^{32}P]GTP is freeze-dried in a plain glass tube and dissolved in 50 μL of 0.1 *M* 2-(*N*-morpholino)ethanesulfonic acid, pH 5.6, containing 1.5 mg of *N*-(3-dimethyl-aminopropyl)-*N*′-ethylcarbodiimide hydrochloride. For most applications, the specific activity can be diluted to 300 Ci/mmol by the addition of 1.44 nmol of carrier GTP.

2. Five milliliters of 1,4-dioxane is made peroxide free by passing through approx 1 mL of neutral alumina. The alumina is filled into a polypropylene column and dioxane passed through by gravity feed.
3. Incubate [α-^{32}P]GTP with an excess of 4-azidoaniline, that is, 2.4 mg suspended (i.e., not dissolved) in 30 µL of dioxane, for 4 h at room temperature. Rotate the reaction tube slowly. During the entire procedure, the samples are best shielded from light.
4. [α-^{32}P]GTP-azidoanilide is purified by HPLC on a Supelcosil™ C_{18} reversed phase cartridge. The HPLC is operated at room temperature with a flow of 1 mL/min and the column is equilibrated with 2.8% ethanol in 0.1 *M* TEAC, pH 8.5. Instead of gassing triethylamine solutions with CO_2 until neutral, we use the commercially available (premade) TEAC as buffer. Although the pH is alkaline, this does not markedly reduce the efficiency of separation, but avoids installing bubble traps in the high-pressure liquid delivery system which would be required if "carbonated" buffer solutions were employed.
5. After application of the sample, the column is washed with equilibration buffer (for 15 min) to remove [α-^{32}P]GTP (and contaminating GDP) which appears typically at approx 5 min after injection. [α-^{32}P]GTP-azidoanilide is eluted by an ethanol gradient from 2.8–88% in 0.1 *M* TEAC and 1-mL fractions are collected. The gradient is developed over a period of 18 min followed by a 10-min wash with 88% ethanol in TEAC. The desired reaction product appears 16–17 min after the start of the gradient.

   Alternatively, GTP-azidoanilide can be isolated in a low-pressure chromatography system using a DEAE-cellulose column (DE52, Whatman). A TEAC gradient (50–450 m*M* in 10% ethanol) is applied for elution ***(5)***.
6. Chromatographic resolution is satisfactory if the peak of radioactivity (containing [α-^{32}P]GTP azidoanilide) is separated from the following (within the tail of the radioactivity peak) fractions with a yellowish tinge. These samples contain aniline derivatives that interfere with [α-^{32}P]GTP-azidoanilide labeling; only poor labeling of G-protein α-subunits is obtained with this material. The color of the pertinent samples can be inspected in dim light after counting the radioactivity in 1-µL aliquots.
7. About 75% of the radioactivity applied is incorporated into the GTP-azidoanilide. The clear, untinged radioactivity peak (typically over three 1-mL fractions) is combined and frozen in liquid nitrogen for lyophilization. Freeze-drying of 3 mL takes approx 2 h. The dried sample is resuspended in 0.1 mL of water yielding a concentration of approx 1 µ*M* at a specific activity of approx 3 µCi/µL.
8. [α-^{32}P]GTP-azidoanilide is kept in the dark and stored at –20°C. The stability of the compound is good and the "shelf-life" is limited by radioactive decay of the α-^{32}P.

3.2. Synthesis of o-GTP and of o-GTPγS

1. For the synthesis of the unlabeled compounds, a typical reaction volume is 0.5–1.0 mL; make 100 m*M* GTP (buffered to pH 7.0–8.0 with NaOH) and add $NaIO_4$ to 110 m*M*.

2. The reaction is carried out for 1 h on ice in the dark; stop by adding glycerol, 1/10 of the reaction volume (final concentration 10% v/v).
3. Since the reaction products (iodate and aldehydes generated from glycerol) may interfere in subsequent measurements (nonspecific effects in particular at high concentrations) a parallel mock synthesis is recommended; $NaIO_4$ is first reacted with glycerol, then GTP is added; obviously, this GTP remains GTP; if this solution behaves differently from authentic GTP in subsequent measurements, the observed effects are induced by reaction products.
4. Gel filtration: G_{10}-Sephadex (Amersham Pharmacia Biotech) is suspended in water and allowed to swell, and is then transferred into spin columns (bed volume 1.1–1.6 mL). The column is precentrifuged at approx 500*g* for 3 min in a swing-out rotor. Aliquots of 0.1–0.15 mL (=10% of bed volume after prespinning) of reaction mixture are added and the columns are centrifuged again. If larger amounts of *o*-GTP are to be purified, a column (column dimensions ~10 × 2 cm/ 1 mL of 100 m*M* *o*-GTP) is poured in a cold room and equilibrated in water; the reaction mixture is applied and fractions collected. Highly purified *o*-GTP is typically found in the ascending portion of the nucleotide peak. The elution should be monitored either with an on-line UV-monitor or by determining optical density at 252 nm in a UV-spectrophotometer in individual fractions (1:100 dilution of fractions; quartz cuvet). The concentration is determined as detailed in **Subheading 2.2.**
5. Yield: typically, >95% of GTP and of GTPγS are converted to the 2′,3′-dialdehype species; the compounds are dispensed into aliquots and stored frozen at –80°C. We avoid repeated freezing and thawing and do not use stocks that are older than 1 mo. While we have not systematically characterized the stability of the compounds, prolonged storage of oxidized guanine nucleotides results in loss of biological activity.
6. The synthesis of radioactively labeled oxidized GTP analogs is carried out in a manner analogous to that described in **steps 1–5** preceding for the unlabeled species; [α-^{32}P]GTP or [^{35}S]GTPγS are supplemented with unlabeled GTP or GTPγS to reach a final concentration of approx 1 m*M* and reacted with a 1.1-fold molar excess of $NaIO_4$. If the concentration of guanine nucleotides is reduced further, the rate of conversion of the parent compound to the oxidized species drops progressively. The unreacted compounds interfere with subsequent affinity labeling, we therefore do no recommend the sacrifice of purity for higher specific activity.

3.3. Chemical Identification of the Products

The most rapid and convenient way to analyze the reaction products is by TLC on PEI-cellulose-F plates. The unlabeled guanine nucleotides are visualized with a UV lamp, radioactively labeled nucleotides by autoradiography.

1. GTP-azidoanilide. Mobile phase: 0.2 *M* Tris-HCl, pH 7.2, 1 *M* LiCl. One microliter of each HPLC-purified [α-^{32}P]GTP-azidoanilide and of [α-^{32}P]GTP are spotted onto the starting line of the TLC plate; 1 μL of unlabeled GTP and GDP

(10 m*M*) each are also applied as a reference. GTP-azidoanilide (R_f ~0.5) migrates slightly ahead of GDP (R_f ~0.45), while the migration of the parent compound GTP is slower (R_f ~0.25).

2. *o*-GTP. Mobile phase: 1 *M* LiCl, 0.5 *M* Tris (unbuffered); 1 µL of each 10 m*M* *o*-GTP and GTP are applied to the starting line on the plate. Prepare a parallel sample, in which *o*-GTP is first reduced with $NaBH_4$ (add a 10-fold molar excess over *o*-GTP; 15 min on ice). Visualize spots under a UV lamp: *o*-GTP does not leave the origin. After reduction with $NaBH_4$, *o*-GTP (R_f ~0.8) migrates slightly ahead of GTP (R_f ~0.75). GTP does not change its migration after incubation with $NaBH_4$. R_f values for the pair GTPγS and *o*-GTPγS are approx 0.56 and approx 0.65, respectively.

3.4. Biochemical Assays for the Synthesized GTP Analogs

Several tests can be employed to verify that the products of the synthesis are indeed functionally active and thus useful for subsequent experiments. Depending on whether radioactively labeled or unlabeled compounds are to be tested, the following approaches can be used:

1. Competition of the unlabeled compounds for binding of [^{35}S]GTPγS to purified G-proteins or membrane-bound G-proteins.
2. Direct incorporation of the radioactively labeled compounds into purified G-proteins or membrane-bound G-proteins.

Several additional straightforward methods can be used: two examples are given in **Note 2**. We perform binding to purified recombinant $G_s\alpha$ as a standard procedure to assess the functional integrity of GTP-analogs; $G_s\alpha$ exchanges prebound GDP rapidly and its kinetics for guanine nucleotide binding and hydrolysis as well as its interaction with receptors are well characterized ***(6,7)***, but other well characterized α-subunits such as $G_o\alpha$ or appropriate membrane preparations can be used.

3.4.1. Competition of Unlabeled GTP Analogs With [^{35}S]GTPγS Binding

GTP-azidoanilide, oxidized GTP analogs, and the radioligand [^{35}S]GTPγS bind to G-protein α-subunits in a quasi-irreversible manner; in contrast GTP is hydrolyzed to GDP when bound to the protein and released again; the apparent affinity of GTP thus depends on the incubation time and decreases progressively with increasing incubation times. Hence, a second, parallel incubation can be set up for four times longer than the original time (e.g., 2 h). The potency of GTP will decrease while that of GTP-azidoanilide and oxidized GTP analogs is not altered ***(8)***.

1. Use 3-mL polypropylene tubes to minimize adsorptive losses of protein. Shield GTP-azidoanilide containing solutions from light.

2. Prepare an incubation medium (30 μL/reaction) containing binding buffer: 50 m*M* HEPES, pH 8.0, 1 m*M* EDTA, 10 m*M* $MgSO_4$, 0.1% Lubrol, and 1.67 μ*M* [^{35}S]GTPγS (specific activity 10–20 cpm/fmol). Check total activity by scintillation counting (0.5–1 × 10^6 cpm); maintain on ice.
3. Prepare a serial dilution of guanine nucleotides covering the range of 0.3–100 μ*M* as fivefold concentrated working solutions in binding buffer; use serial dilutions of GTP and/or GTPγS as a control; maintain on ice.
4. Dilute $G_s\alpha$ (or $G_o\alpha$) in binding buffer to obtain approx 2 pmol of α-subunit/ 10 μL; maintain on ice.
5. [^{35}S]GTPγS is assayed in duplicate in a 50-μL incubation. Combine 30 μL of [^{35}S]GTPγS-containing medium **(step 2)** with 10 μL of buffer (control binding) or increasing concentrations of guanine nucleotides (*see* **step 3**) and 10 μL of G-protein solution (*see* **step 4**). The filter blank is determined in the absence of G-protein.
6. Incubate the control binding reactions for 30 min at 20°C; for the remaining tubes, the binding reaction is allowed to proceed for 2 h. The reaction is stopped by the addition of 2 mL of stop solution (20 m*M* Tris-HCl, pH 8.0, 20 m*M* $MgCl_2$, 150 m*M* NaCl, 0.1 m*M* GTP).
7. Pour diluted reaction mixture over BA85 nitrocellulose filters and rinse filters with 20 mL of wash buffer (20 m*M* Tris-HCl, pH 8.0, 20 m*M* $MgCl_2$, 150 m*M* NaCl); allow filters to dry, place in scintillation vials, add scintillation cocktail, and determine the radioactivity.

Under these assay conditions, that is, with purified G-proteins, the nonspecific binding corresponds to the filter blank and is generally very low (<100 cpm/10^5 cpm added) if the filters are rinsed thoroughly. An analogous experiment can also be set up with membranes as the source of G-proteins; in this case, detergents are to be omitted from the reaction, glass-fiber filters and a semi-automatic filtration apparatus (Skatron or Brandel) can be used to trap the protein-bound radioactive ligand (*see* Chapter 12).

3.4.2. Binding and Affinity Labeling of Purified G-Proteins With [α-^{32}P]GTP-Azidoanilide

Binding of [α-^{32}P]GTP-azidoanilide to $G_s\alpha$ (or $G_o\alpha$) is assayed as outlined under **Subheading 3.4.1.** with the following modifications:

1. Prepare duplicate incubations in 25 μL containing 50 m*M* HEPES, pH 8.0, 1 m*M* EDTA, 10 m*M* $MgSO_4$, 0/1% Lubrol; [α-^{32}P]GTP-azidoanilide at a final concentration of 0.1–0.2 μ*M* and a molar excess of $G_s\alpha$ (e.g., 5 pmol/assay).
2. To determine nonspecific incorporation, a parallel incubation is set up that contains 100 μ*M* unlabeled GTPγS. The assay tubes are wrapped in aluminum foil.
3. After an incubation of 1 h at 20°C, 5 μL aliquots are withdrawn and spotted onto a metal plate covered with a double layer of parafilm (press the conical end of an Eppendorf tube firmly onto the Parafilm to create a well from which the sample

can be easily retrieved). Irradiate for 0.5 min with UV light from a distance of 3 cm. These samples are subjected to sodium dodecyl sulfate-polyacrylamide gel electrophoresis (SDS-PAGE), and the labeled bands identified by autoradiography of the dried gel, excised, and counted for radioactivity.

4. The remaining incubations are diluted by the addition of 2 mL of ice-cold stop solution (20 m*M* Tris-HCl, pH 8.0, 20 m*M* $MgCl_2$, 150 m*M* NaCl, 0.1 m*M* GTP) and filtered over BA85-nitrocellulose filters (*see* **step 7** of **Subheading 3.4.1.**).
5. The filters are rinsed with 20 mL of stop solution, dried, taken up in liquid scintillation cocktail, and the radioactivity is determined.

Because an excess of G-protein α-subunit over GTP-azidoanilide is present in the incubation, essentially all of the radioactivity ought to be bound to the protein as determined by filter binding. However, because the nitrene generated by UV irradiation can also react with water, the covalently incorporated amount of radioactivity is generally much lower. Using the comparison outlined above, we have found that approx 5% of the bound radioactivity is indeed incorporated into the protein.

3.4.3. Affinity Labeling of G-Proteins With [α-^{32}P]o-GTP and [^{35}S]o-GTPγS

1. $G_s\alpha$ (10 pmol) is incubated in 25 µL of buffer (50 m*M* HEPES, pH 7.6; 1 m*M* EDTA; 10 m*M* $MgSO_4$; Lubrol 0.1%) containing 10 µ*M* [α-^{32}P]*o*-GTP or [^{35}S]*o*-GTPγS (specific activity 20 cpm/fmol).
2. After an incubation period of 1 h at 20°C, add 5 µL of a freshly prepared aqueous solution of $NaBH_4$ to reach a final concentration 0.1–1 m*M* (minimum 10-fold excess over *o*-GTP/*o*-GTPγS); the reduction is carried out for 1 h on ice.
3. Add Laemmli sample buffer containing 40 m*M* DTT, denature by heating (10 min at 50°C), and apply to SDS-polyacrylamide gel; the labeled band is identified by autoradiography of the dried gel, excised, and counted for radioactivity. Expect a labeling stoichiometry of approx 0.5.

3.5. Affinity Labeling by In Situ *Oxidation–Reduction*

Membrane-bound GTP-binding proteins, GTP-binding proteins in cellular extracts, and purified G-proteins can be affinity labeled after they have been allowed to bind radioactively labeled GTP or GTPγS ***(8–10)***. In most G-proteins the dissociation of GTPγS approaches zero in the presence of millimolar free magnesium concentrations, while GTP is hydrolyzed and the GDP formed is released. Hence, [^{35}S]GTPγS is recommended for affinity labeling. For purified G-proteins, the labeling reaction is a logical extension of the GTPγS-binding assay. We will therefore only outline the method used to label membrane-bound G-proteins.

1. Weigh out the required amount of $NaIO_4$, $NaCNBH_3$, and $NaBH_4$ to obtain 10 × 4 m*M*, 2 × 80 m*M*, and 10 × 10 m*M* in water. Do not add water yet.

2. Dilute membranes into reaction buffer to yield 100 μg/50 μL (*see* **Note 3** for an estimate of the required amount); maintain on ice.
3. Combine the membrane dilution (50 μL) and [^{35}S]GTPγS-containing 50 μL of binding buffer in 1.5-mL microfuge tubes. If receptor agonists/antagonists are to be evaluated, combine appropriate dilutions first with membranes. For short incubations, prewarm both the membrane suspension and the binding buffer to desired temperature (20°–30°C).
4. Incubate for the desired period; add ice-cold 1 mL GTP stop buffer (*see* **step 3**) to stop the binding reaction; centrifuge in a refrigerated centrifuge at for 20 min at 50,000*g*; toward the end of the centrifuge run, dissolve preweighed $NaIO_4$, $NaCNBH_3$, and $NaBH_4$ in water.
5. Quickly withdraw supernatant and resuspend pellet in 0.1 mL of Mg-buffer (*see* **step 3**) by rapidly pipetting the solution up and down or by briefly immersing the tubes in a sonicating water bath filled with ice-cold water.
6. Add 11 μL of $NaIO_4$ (4 m*M* final); incubate for 1 min at 20°C.
7. Add 110 μL $NaCNBH_3$ (80 m*M* final); incubate for 1 min at 20°C.
8. Add 22 μL of $NaBH_4$ (10 m*M* final); incubate for 30 min at 0°C.
9. Centrifuge (microcentrifuge, maximum speed), withdraw supernatant, dissolve the pellet in Laemmli sample buffer containing 40 m*M* DTT and 1% SDS, heat denature, and apply to the SDS-polyacrylamide gel. Identify labeled bands by autoradiography. Alternatively, dissolve membrane in appropriate buffer for immunoprecipitation with specific antisera.

3.6. Affinity Labeling of Membrane-Bound G-Proteins With [α-^{32}P]GTP-Azidoanilide

The labeling reaction is carried out by following the appropriate steps outlined under **Subheading 3.5.**, that is, **steps 1** and **2** and **steps 3–7**. Thereafter the samples are spotted onto a metal plate covered with a double layer of parafilm (*see* **step 3** of **Subheading 3.4.2.**) and photolyzed with UV light. Labeled proteins are identified by SDS-PAGE or immunoprecipitation.

4. Notes

1. Oxidized GTP analogs vs GTP-azidoanilide: whether oxidized GTP analogs or GTP-azidoanilide are employed depends on the application. For the analysis of receptor-dependent activation of G-proteins, [α-^{32}P]GTP-azidoanilide has been widely used because it can be obtained at very high specific activity. However, not all G-proteins bind GTP-azidoanilide readily ***(11)***. The method of *in situ* oxidation–reduction with [^{35}S]GTPγS lends itself to the identification of unknown GTP-binding proteins. In addition, under appropriate conditions unlabeled *o*-GTP can be used to block endogenous GTP-binding proteins irreversibly; the disrupted signaling pathway may subsequently be reconstituted by the exogenous addition of purified G-proteins ***(12)***.

2. Synthesis: we recommend that the synthesis be carried out first with unlabeled GTP or GTPγS. This allows the investigator to become familiar with the procedure without the additional stress of handling microCurie quantities of radioactivity. Similar to GTPγS, and Gpp(NH)p, *o*-GTPγS and GTP-azidoanilide are not hydrolyzed and thus irreversibly activate G-protein α-subunits and their effectors; this feature therefore can also be exploited to verify that the synthesis and purification yields active compounds. For instance, membrane preparations can be used to compare the activation of adenylyl cyclase in the presence of *o*-GTPγS and GTP-azidoanilide with the stimulation induced by GTPγS (*see* **refs. *3*** and ***8***). Similarly, guanine nucleotides destabilize high-affinity binding of most agonist radioligands to GPCRs in membrane preparations (*see* Chapter 1 and **ref. *13***).
3. Receptor-dependent activation: the goal of most researchers is to determine which G-protein α-subunit is activated by the receptor under investigation in a membrane preparation obtained from cells or tissues. This can be achieved in a seemingly straightforward manner by determining if the receptor increases the covalent incorporation of GTP analogs into G-protein α-subunits. However, it is important to design the experiment appropriately and the following points ought to be considered:
   a. How abundant are the candidate G-proteins? Typically G_o and G_i are expressed at higher levels than G-proteins of the G_s, G_q, and $G_{12/13}$ classes. The expression level determines how much membrane protein is to be used in the incubation and how high the specific activity of the radioactively labeled GTP analogue must be to obtain detectable labeling. G-protein levels can be estimated by immunoblotting with appropriate antisera (*see* Chapter 13). It is obviously wise to use appropriate standards (purified proteins or, alternatively, membrane preparations, which have previously been characterized with respect to their expression levels).
   b. How fast do the G-proteins exchange guanine nucleotides spontaneously, that is, without activation by an agonist-activated receptor? The rates at which G-protein α-subunits release prebound GDP differ widely even among closely related proteins; for example, $G_o\alpha$ exchanges very rapidly, while the retinal G-protein transducin ($G_t\alpha$) has an extremely slow rate of exchange. The rate of exchange in the heterotrimeric G-protein is also influenced by the magnesium concentration (which accelerates GDP release) and the NaCl concentration (which decreases GDP release). The agonist-activated receptor induces a conformational change such that GTP and GTP analogs are bound preferentially even in the presence of GDP. Finally, once bound, hydrolysis-resistant GTP analogs are bound quasi-irreversibly as long as the Mg^{2+} concentration is kept in the millimolar range.
   c. When does the receptor-induced activation reach its maximum? During membrane preparation, receptors and G-proteins are physically separated. The ratio between receptor and G-protein, which are in the same vesicle, can be assumed to approximate a random distribution. G-proteins are in general present in

excess of the receptor density; some vesicles will therefore contain G-proteins and no receptor while the reverse will be a less likely event. An activated receptor in a vesicle will rapidly exhaust the supply of accessible G-proteins. Thus, long incubation periods favor the detection of low basal guanine nucleotide exchange.

d. If reasonable guesses are not available for the system under investigation, preliminary experiments are recommended in which agonist-dependent [^{35}S]GTPγS binding is determined in membranes; the optimal signal-to-noise ratio (i.e., a large agonist-induced increment of [^{35}SGTPγS-binding) can be assessed from the time course of activation as well as from the effect of GDP, Mg^{2+}, and NaCl (*see* Chapter 12); combined with the concentration of receptor derived from binding assays with an appropriate radioligand, the molar turnover number of the receptor can be estimated, that is, number of G-protein molecules activated per receptor (R) in a given time interval (mol of GTPγS bound/mol R/min).

References

1. Milligan, G. (1993) Mechanisms of multifunctional signalling by G protein-linked receptors. *Trends Pharmacol. Sci.* **14,** 239–244.
2. Offermanns, S. and Schultz, G. (1994) Complex information processing by the transmembrane signaling system involving G proteins. *Naunyn Schmiedebergs Arch. Pharmacol.* **350,** 329–338.
3. Pfeuffer, T. (1977) GTP-binding proteins in membranes and the control of adenylate cyclase activity. *J. Biol. Chem.* **252,** 7224–7234.
4. Offermanns, S., Schultz, G., and Rosenthal, W. (1991). Identification of receptor-activated G proteins with photoreactive GTP analog, [α-^{32}P]GTP-azidoanilide. *Methods Enzymol.* **195,** 286–302.
5. Anis, Y., Nürnberg, B., Visochek L., Reiss, N., Naor, Z., and Cohen-Armon, M. (1999) Activation of G_o-proteins by membrane depolarization traced by in situ photoaffinity labeling of $G\alpha_o$-proteins with [α^{32}P]GTP-azidoanilide. *J. Biol. Chem.* **274,** 7431–7440.
6. Graziano, M. P., Freissmuth, M., and Gilman, A. G. (1989) Expression of Gsα in *E. coli:* purification and characterization of two forms of the protein. *J. Biol. Chem.* **264,** 409–418.
7. Freissmuth, M., Selzer, E., Marullo, S., Schutz, W., and Strosberg, A. D. (1991) Expression of two human α-adrenergic receptors in *E. coli:* functional interaction with two forms of the stimulatory G protein. *Proc. Natl. Acad. Sci. USA* **88,** 8548–8552.
8. Hohenegger, M., Nanoff, C., Ahom, H., and Freissmuth, M. (1994) Structural and functional characterization of the interaction between 2′,3′-dialdehyde guanine nucleotide analogues and the stimulatory G protein α-subunit. *J. Biol. Chem.* **269,** 32,008–32,015.
9. Low, A., Faulhammer, H. G., and Sprinzl, M. (1992) Affinity labeling of GTP-binding proteins in cellular extracts. *FEBS Lett.* **303,** 64–68.

10. Hohenegger, M., Mitterauer, T., Voss, T., Nanoff, C., and Freissmuth, M. (1996) Thiophosphorylation of the G protein α-subunit in membranes: evidence against a direct phosphate transfer reaction to Gα-subunits. *Mol. Pharmacol.* **49,** 73–80.
11. Fields, T. A., Linder, M. E., and Casey, P. J. (1994) Subtype-specific binding of azidoanilido-GTP by purified G protein α-subunits. *Biochemistry* **33,** 6877–6883.
12. Nanoff, C., Boehm, S., Hohenegger, M., Schütz, W., and Freissmuth, M. (1994) 2′,3′-Dialdehyde-GTP as an irreversible G protein antagonist. *J. Biol. Chem.* **269,** 31,999–32,007.
13. Jockers R., Linder, M. E., Hohenegger, M., et al. (1994) Species difference in the G protein selectivity of the human and the bovine A_1-adenosine receptor. *J. Biol. Chem.* **269,** 32,077–32,084.

12

Measurement of Agonist-Stimulated [^{35}S]GTPγS Binding to Assess Total G-Protein and Gα-Subtype-Specific Activation by G-Protein-Coupled Receptors

Mark R. Dowling, Stefan R. Nahorski, and R. A. John Challiss

Summary

On activation, G-protein-coupled receptors (GPCRs) exert many of their cellular actions through promotion of guanine nucleotide exchange on the Gα-subunit of heterotrimeric G-proteins to release free Gα-GTP and βγ-subunits. In membrane preparations, GTP can be substituted by ^{35}S-labeled guanosine 5′-*O*-(3-thio)triphosphate ([^{35}S]GTPγS) and on agonist stimulation a stable [^{35}S]GTPγS–Gα complex will form and accumulate. Separation of ^{35}S-bound GTPγS–Gα complexes from free [^{35}S]GTPγS allows differences between basal and agonist-stimulated rates of [^{35}S]GTPγS–Gα complex formation to be used to obtain pharmacological information on receptor–G-protein information transfer. Further, by releasing Gα-subunits into solution following the [^{35}S]GTPγS binding step, Gα-subunit-specific antibodies can be used to investigate the Gα-protein subpopulations activated by receptors by immunoprecipitation of [^{35}S]GTPγS–Gα complexes and quantification by scintillation counting. Here we describe a total [^{35}S]GTPγS binding assay and a modification of this method that incorporates a Gα-specific immunoprecipitation step.

Key Words

[^{35}S]-GTPγS, GDP, G-protein-coupled receptor, guanine nucleotide, receptor-G-protein-coupling, G-protein-specific immunoprecipitation.

1. Introduction

The use of agonist-promoted binding of [^{35}S]GTPγS to evaluate the interactions between G-protein-coupled receptors (GPCRs) and G-proteins emerged from early experimental approaches designed simply to label guanine nucleotide binding proteins *(1)*. However, it quickly became apparent that the binding of [^{35}S]GTPγS could be used to assess receptor–G-protein interactions, initially in reconstitution studies *(2)*, but soon in membrane preparations *(3)*.

From: *Methods in Molecular Biology, vol. 259, Receptor Signal Transduction Protocols, 2nd ed.*
Edited by: G. B. Willars and R. A. J. Challiss © Humana Press Inc., Totowa, NJ

Thus, receptor activation stimulates GTP for GDP exchange at the Gα-subunit of G-protein heterotrimers. If GTPγS is present it will compete with GTP for the Gα–GTP binding site and will form a quasi-irreversible Gα–GTPγS complex, as the phosphorothioate group of GTPγS makes this analog highly resistant to Gα–GTPase activity. Thus, basal and agonist-stimulated rates of GDP/[^{35}S]GTPγS exchange can be compared to gain information about the initial post-ligand binding step in the GPCR signal transduction process.

The [^{35}S]GTPγS binding assay is extremely attractive as it is relatively simple to perform, can be adapted for high throughput, and provides an efficacy readout proximal to the receptor–ligand binding interaction. Thus, the original Gα-protein binding assay has been widely used to profile pharmacologically agonists, antagonists, and inverse agonists at a number of receptors and has been adapted for thin slice/autoradiography, as well as membrane preparations ***(4,5)***. The method is generally satisfactory for studies involving GPCRs that preferentially couple to $G_{i/o}$ proteins. The method has been less used, however, to study GPCRs that couple preferentially to the G_q, G_s, and G_{12} subfamilies of G-protein, as signals arising from these other G-proteins can be masked by high basal GDP/[^{35}S]GTPγS exchange rates at the predominant $G_{i/o}$ protein subpopulation ***(6)***.

More recently, elaborations on the original [^{35}S]GTPγS binding assay have been applied to gain semiquantitative information on the Gα-protein subpopulation(s) activated by the ligand–receptor interaction ***(7–9)***. One strategy (discussed in Chapter 11) is to use a photolabile GTP analog that can be covalently attached to its Gα-protein binding partner(s). Alternatively, an antibody-based method can be used that avoids the need to synthesize and utilize ^{32}P-labeled GTP analogs. This method relies on the availability of Gα-protein subtype-specific antibodies that can be used to immunoprecipitate Gα-subunits to quantify specific [^{35}S]GTPγS binding. In addition to providing important information on receptor-Gα protein coupling preferences, the method may also substantially improve signal-to-noise in the assay as it may markedly reduce basal Gα-[^{35}S]GTPγS binding, opening the way to studies of GPCR coupling to G_q, G_s, and G_{12} subfamilies ***(9–11)***. Further, the assay can be adapted to a homogeneous assay format not requiring a separation step ***(12)***.

2. Materials

1. [^{35}S]GTPγS (approx 1250 Ci/mmol) (*see* **Note 1**).
2. Cell preparation. Lifting buffer: 10 m*M* 4-(2-hydroxyethyl)piperazine-1-ethanesulfonic acid (HEPES); 0.9% NaCl; 0.2% EDTA, pH 7.4.
3. Homogenization buffers: Buffer A—10 m*M* HEPES; 10 m*M* EDTA, pH 7.4; buffer B—10 m*M* HEPES; 0.1 m*M* EDTA, pH 7.4 (*see* **Note 2**).
4. Assay buffer: 10 m*M* HEPES; 100 m*M* NaCl; 10 m*M* $MgCl_2$, pH 7.4, ± guanosine 5′-diphosphate (GDP) (*see* **Note 3**).
5. GDP, Na salt, GTPγS.

6. Filter-wash buffer: 10 m*M* HEPES; 100 m*M* NaCl; 10 m*M* $MgCl_2$, pH 7.4 at 4°C.
7. Solubilization buffer: 100 m*M* Tris-HCl, 200 m*M* NaCl, 1 m*M* EDTA, 1.25% Igepal CA 630 (*see* **Note 4**), pH 7.4; ± 0.2% (w/v) sodium dodecyl sulfate (SDS).
8. Normal rabbit serum.
9. Protein A–Sepharose: 1.5 g diluted in 50 mL of 10 m*M* Tris-HCl, 1 m*M* EDTA, pH 7.4. The suspension should be stored at 4°C and replaced every 2 wk.
10. Gα-protein-specific antibodies (*see* **Note 5**).
11. Scintillant: standard liquid scintillation counting cocktails (e.g., SafeFluor, Lumac-LSC, Groningen, The Netherlands) are adequate and should give 90%+ counting efficiency.
12. Equipment. (1) Total [^{35}S]GTPγS assays: manual filtration manifold/vacuum pump or 24/48/96 place cell harvester; scintillation β-counter. (2) [^{35}S]GTPγS/Gα-immunoprecipitiation assays: temperature controlled water bath; high-speed, refrigerated centrifuge; shaker/roller; scintillation β-counter.

3. Methods

3.1. Preparation of Membranes

The [^{35}S]GTPγS binding method has been applied most commonly to membranes derived from cultured cell lines expressing either endogenously or recombinantly the receptors and/or G-protein subpopulations being investigated.

1. Medium from confluent flask(s) of cells is removed and the cell monolayer briefly washed with ice-cold "lifting buffer" (2 × 10 mL). All subsequent steps are conducted at 0–4°C.
2. Ten milliliters of lifting buffer is then added for approx 5–10 min (or as long as it takes for the cell monolayer to detach from the plastic) and the cells poured into 30-mL tubes. The flask is rinsed with a further 5 mL of lifting buffer and this too is added to each tube.
3. The cell suspension is centrifuged (approx 250*g* for 5 min at 4°C) and the supernatant aspirated.
4. The cell pellet is dispersed in 5 mL of buffer A and homogenized (Polytron, speed ≤20,000 rpm, 5 × 10-s bursts). The homogenate is then transferred to centrifuge tubes diluted to approx 50 mL with buffer A and centrifuged (5000*g*, 10 min, 4°C).
5. The supernatant is carefully decanted into a fresh centrifuge tube and centrifuged (50,000*g*, 15 min, 4°C). The supernatant is discarded, and the pellet is resuspended in buffer B, rehomogenized, and recentrifuged as above (50,000*g*, 15 min, 4°C).
6. The final P_2 pellet is resuspended/rehomogenized in buffer B and its protein concentration determined. The solution is diluted to the required concentration (either 1 or 3 mg of protein/mL) and if not used immediately snap-frozen in liquid nitrogen and stored at –80°C until required (*see* **Note 6**).

3.2. Total [^{35}S]GTPγS Binding Assay

1. Fresh or frozen membranes should be diluted to an appropriate protein concentration (determined empirically; *see* **Note 7**) in ice-cold assay buffer. Typically

20–100 μg of protein per tube is used, depending on the level of receptor expression, the agonist utilized (e.g., partial vs full), and the predominant Gα-protein to which the receptor couples.
2. As the assay is to be terminated by rapid vacuum filtration, relatively high volume tubes (5 mL) are used. GDP is added to assay buffer to give the correct final concentration and [^{35}S]GTPγS is added to give a concentration of approx 0.1 n*M* in a final assay volume of 100 μL (i.e., approx 100,000 dpm per assay). To provide an indication of nonspecific binding, excess GTPγS (10 μ*M*) is added to some tubes.
3. Both assay mixture and membranes are brought to 30°C, and membranes added to each assay tube. Reactions are initiated by adding either agonist or vehicle (i.e., the diluent for the agonist) and tubes incubated at 30°C for a predetermined time (usually 15–60 min).
4. Toward the end of the incubation period glass fiber filters (GF/B, Whatman) are wetted with ice-cold filter-wash buffer. Incubations are terminated by addition of 3 mL of ice-cold filter-wash buffer and rapid transfer on to the filters. Tubes/filters are then washed with 3 × 3 mL filter-wash buffer.
5. Filters containing the membrane–Gα–[^{35}S]GTPγS complex are allowed to dry (either under vacuum or air-dry), transferred to vials, and scintillant added. Aliquots of the stock [^{35}S]GTPγS solution are also taken to determine precisely how much radioactivity was added to each assay tube.

An example of the data obtained using the above protocol is shown in **Fig. 1**, which illustrates the importance of optimizing the GDP concentration to maximize the signal-to-noise ratio in the assay.

3.3. Gα-Specific [^{35}S]GTPγS Binding Assay

The assay described in **Subheading 3.2.** gives a reliable indication of receptor activation that reflects an overall G-protein population activation. Using pertussis toxin (PTx) pretreatment it has been possible to show that many receptors can couple to both PTx-sensitive and -insensitive Gα subunits *(13)*. Furthermore, recent pharmacological observations have provided evidence that different agonists may be able to activate different Gα-subunit complements at the same receptor subtype *(14)* and has been termed "agonist-directed trafficking of receptor stimulus" *(14,15)*. Many of these studies have measured distal, functional readouts of receptor activation (e.g., Li^+-enhanced [^{3}H]inositol phosphate accumulations; reporter gene assays) where additional inputs such as pathway crosstalk and amplification may affect interpretation. The ability to measure an immediate consequence of receptor activation, Gα-protein GDP/GTP exchange, and further to dissect the Gα species involved can provide crucial and definitive pharmacological information.

1. Fresh or frozen membranes (prepared as described in **Subheading 3.1.**) should be diluted to an appropriate protein concentration (*see* **Note 8**); typically 50–100 μg per tube is used.

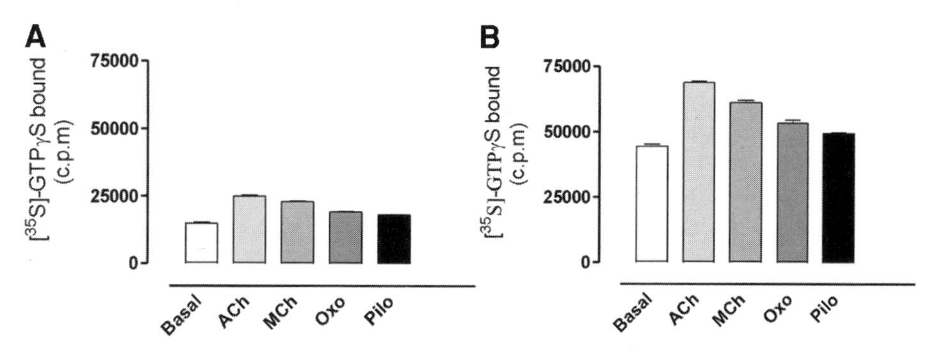

Fig. 1. Membranes prepared from Chinese hamster ovary cells, recombinantly expressing the human M_1 muscarinic acetylcholine receptor (CHO-m1; 25 µg; M_1 receptor density = 3.6 pmol/mg protein), were incubated in the presence of either 10 µ*M* GDP (**A**) or 1 µ*M* GDP (**B**), 300 p*M* [^{35}S]GTPγS, and various agonists (ACh, acetylcholine; MCh, methacholine; Oxo, oxotremorine; Pilo, pilocarpine) or vehicle (basal) at 30°C. Basal and stimulated binding was measured by rapid filtration after 40 min. The histograms show that although the percentage increase of [^{35}S]GTPγS binding is not affected by lowering the GDP concentration, the absolute amount of [^{35}S]GTPγS bound (expressed in cpm) is significantly elevated. This may allow for easier distinction of full and partial agonists. However, this is not the case for CHO-m2 (M_2 receptor density = 1.1 pmol/mg of protein) membrane preparations where under these conditions little or no [^{35}S]GTPγS binding above basal is seen at 1 µ*M* GDP and statistically significant [^{35}S]GTPγS binding levels are observed only at 10 µ*M* GDP. This figure illustrates the need for careful optimization of GDP concentrations to observe [^{35}S]GTPγS binding above basal, to distinguish full and partial agonist responses and to distinguish these from antagonists.

2. As the assay requires several rolling and centrifugation steps, the most convenient tubes to employ are 1.5-mL Eppendorf-type tubes. Tubes containing assay buffer, and appropriate concentrations of GDP, agonist, and [^{35}S]GTPγS (*see* **Note 9**) in a volume of 50 µL are maintained at 30°C. Membranes (50 µL) are then added to start the assay.
3. Incubations are continued at 30°C for a predetermined time (*see* **Note 10**).
4. Reactions are terminated by the addition of 1 mL of ice-cold assay buffer and rapid transfer to a refrigerated centrifuge (prechilled to 4°C). Samples are centrifuged (20,000*g*, 6 min, 4°C).
5. The supernatant is then aspirated to leave a membrane pellet and 50 µL of ice-cold solubilization buffer (containing 0.2% SDS) is added. The pellet is left (on ice) for 15–30 min before attempting to aid the solubilization process (*see* **Note 11**).
6. Following complete solubilization of the pellet, the SDS concentration is reduced by the addition of 50 µL of ice-cold solubilization buffer (minus SDS).
7. The solution is then precleared by addition of 13 µL of normal rabbit serum (prediluted 1:10 in assay buffer) and 30 µL of the 1.5% protein A–Sepharose suspension.

8. Samples are capped and rolled at 4°C for a minimum of 90 min.
9. Samples are transferred to a refrigerated centrifuge and the protein A–Sepharose conjugated material pelleted (20,000*g*, 2–3 min, 4°C).
10. Transfer 100 μL of each supernatant to a fresh tube containing an appropriate dilution of anti-Gα-subunit antibody (*see* **Note 12**).
11. Vortex-mix and roll overnight at 4°C.
12. Add 70 μL of protein A–Sepharose suspension (*see* **Note 13**) and roll as before for 90 min.
13. Centrifuge 20,000*g*, 2–3 min, 4°C.
14. Aspirate the supernatant and add 1 mL of ice-cold solubilization buffer (minus SDS).
15. Thoroughly vortex-mix and centrifuge as before.
16. Repeat **steps 13–15** so that the beads are washed four times in solubilization buffer (*see* **Note 14**).
17. After the final wash, the supernatant is removed and the protein A beads resuspended in 1.1 mL of FloScint IV (or similar scintillant) and vortex-mixed. Radioactivity is detected by liquid scintillation spectrometry.

3.4. Data Analysis

If the specific activity of the [^{35}S]GTPγS is calculated daily, to ensure a consistent final concentration of radioactivity is added, the data are usually comparable to allow a simple plotting of specific cpm/dpm (minus nonspecific binding) vs agonist concentration. Log concentration–response curves can then be analyzed by nonlinear regression using an appropriate commercially available program (e.g., Prism 3.0, GraphPad Software, San Diego, USA) to produce EC_{50}, E_{max}, and slope factor values. Alternatively, if the activity of the [^{35}S]GTPγS is routinely counted and the protein added per tube is accurately known, data may be converted into fmol of [^{35}S]GTPγS bound/mg of protein (or fmol of receptor). If there is daily variability between datasets this may be improved by referencing the results obtained to a maximum concentration of a known full agonist (obtained on the day) and converting the data into a percentage of this maximum. A concentration–response curve for the methacholine-stimulated $G_{q/11}\alpha$ coupling response in Chinese hamster ovary cell membranes recombinantly expressing the M_1 muscarinic acetylcholine receptor is shown in **Fig. 2**.

4. Notes

1. High specific activity [^{35}S]GTPγS (40–50 TBq/mmol) is supplied by NEN Life Science Products (NEG030H) and Amersham Biosciences (SJ1320/SJ1308). The NEN product is shipped in a 20-μL volume, which is diluted to 250 μL on arrival with 10 m*M* Tricine, 10 m*M* dithiothreitol (DTT), pH 7.6, to give an approx 800 n*M* stock solution, which is then dispensed as 20-μL aliquots and stored at –80°C.

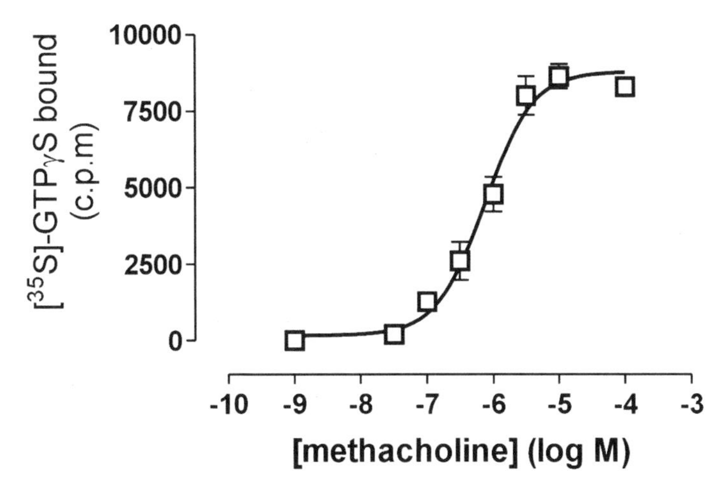

Fig. 2. Concentration dependency of methacholine (MCh)-stimulated [^{35}S]GTPγS binding to $G_{q/11}\alpha$-specific proteins in CHO-m1 membranes. CHO cell membranes (M_1 receptor density = 3.6 pmol/mg of protein) were incubated with MCh (10^{-9}–10^{-4} *M*) under optimal assay conditions. Data are shown as means ± SEM for four separate experiments carried out in duplicate. A basal value (approx 1000 dpm) has been subtracted. The EC_{50} value for MCh-stimulated [^{35}S]GTPγS-$G_{q/11}\alpha$ binding was 0.8 μ*M* (pEC_{50}, 6.12 ± 0.07).

The relatively short half-life of ^{35}S (87.4 d) needs to be accounted for so as to ensure that similar amounts of radioactivity are used in the [^{35}S]GTPγS assay. It is recommended that [^{35}S]GTPγS is used only for 1–2 mo after its activity date.

2. Protease inhibitor cocktails can be included in the homogenization buffer. In our experience, however, they are unnecessary in the cell/receptor systems we have studied.
3. Variations in assay concentrations of Na^+, Mg^{2+}, and GDP have been reported by a number of different groups. Preliminary experiments should be performed to assess the effects of varying Na^+ and Mg^{2+} concentrations. Most assays use a high salt (Na^+) level (routinely 100 m*M*), although low salt (Na^+ = 10 m*M*) has been proposed to favor some phenomena (e.g., observation of inverse agonist activity). Similarly, a relatively high [Mg^{2+}] (10 m*M*) is used in most studies. In common with other groups, we have found [GDP] to be a crucial determinant of whether an agonist-stimulated increase in [^{35}S]GTPγS binding is observed (*see* **Fig. 1**) and in setting the signal-to-noise ratio of the experimental system. A rule of thumb is that optimal conditions for observing receptor–$G_{i/o}$ protein interactions tend to occur at higher concentrations of GDP (10 μ*M*) than for receptor-$G_{q/11}$ protein interactions (0.1–1 μ*M*); however, preliminary experiments should always be performed to ensure that an optimal signal-to-noise ratio is achieved. Note also that it may be necessary to use more than one [GDP] to gain a complete experimen-

tal picture, for example, where PTx is used to study receptor coupling to PTx-sensitive ($G_{i/o}$) and PTx-insensitive G-protein subpopulations.

4. To obtain reproducible results from the [^{35}S]GTPγS–Gα/immunoprecipitation method it is essential that the membrane pellet is solubilized efficiently. The non-ionic detergent Igepal CA-630 ([octylphenoxy]polyethoxyethanol, Sigma-Aldrich), which has replaced nonidet P-40 (no longer available), together with SDS appear to give efficient solubilization without destroying the [^{35}S]GTPγS–Gα association.
5. There is an ever increasing array of commercially available Gα-protein-specific antibodies. We have achieved satisfactory results with the following Gα-protein antisera from Santa Cruz Biotechnology (Santa Cruz, USA): "pan" $G_{i1-3}\alpha$ antibody ($G_i\alpha$ common; sc-262); $G_{q/11}\alpha$ (sc-392); $G_o\alpha$ (sc-13532), and $G_s\alpha$ (sc-823). A number of high-quality Gα antibodies marketed by New England Nuclear (Brussels, Belgium) are no longer available. If an extensive program of work is to be based on the [^{35}S]GTPγS/immunoprecipitation strategy then it may be financially prudent to generate an antiserum "in-house." For example, we have generated our own immunoprecipitating $G_{q/11}\alpha$ antibody in rabbits against the C-terminal $G_{q/11}\alpha$-common sequence (C)LQLNLKEYNLV (where the cysteine has been added for conjugation purposes). This sequence is not only common to human $G_q\alpha$ and $G_{11}\alpha$, but is also conserved across mammalian species. When working with uncharacterized antibodies, experiments must be performed to assess the selectivity/specificity of the antibody for a specific Gα-subunit over the others in the cell Gα population.
6. This membrane preparation generally allows parallel radioligand-binding studies to be conducted to quantify receptor expression levels.
7. Preliminary experiments need to be performed to establish whether similar [^{35}S]GTPγS binding data are achieved using fresh or frozen membranes. In addition, there are a number of variables in the [^{35}S]GTPγS binding assay that need to be investigated to optimize signal-to-noise ratio (i.e., the amount of agonist-stimulated [^{35}S]GTPγS binding relative to the basal level). The most influential assay constituent is GDP. To optimize assay conditions, a sufficient level of GDP needs to be used to suppress basal [^{35}S]GTPγS binding to G-proteins and to reveal the maximal agonist-stimulated [^{35}S]GTPγS binding compared to basal [^{35}S]GTPγS binding activity. The concentration of GDP may also influence the time that the incubation needs to be continued to achieve best results. For a $G_{i/o}$-coupled GPCR a concentration of GDP of 10 μM is commonly used, while for $G_{q/11}$-coupled GPCRs lower concentrations (0.1–1 μM) are generally employed.
8. As this assay ultimately relies on immunoprecipitation (which may not be 100% efficient) and separation of a subfraction of the total Gα-protein population activated, it is often necessary to use a higher protein concentration than may be employed in the total [^{35}S]GTPγS assay. The optimization of protein concentration should take into account the balance between obtaining a sufficient capture of Gα–[^{35}S]GTPγS complexes with the need to solubilize the protein pellet thoroughly, which becomes increasingly problematic as amount of material used, and hence the size of the membrane pellet, increases.

9. For the reasons given under **Note 8** it is also necessary to increase the final concentration of [^{35}S]GTPγS. Typically a final concentration of 2–10 n*M* is employed (i.e., up to approx 2.5×10^6 dpm [^{35}S]GTPγS per assay tube).
10. Because a higher [^{35}S]GTPγS concentration is employed, the rate of [^{35}S]GTPγS-Gα association is more rapid. Careful optimization of the reaction time needs to be established as maximal [^{35}S]GTPγS binding may be achieved in as little as 2 min ***(9)***. There is also evidence, at least for some Gα subtypes, that GTP/GTPγS-liganded Gα may also be delipidated in vivo. This would allow membrane dissociation and nonrecovery in the initial membrane pellet. For this reason agonist incubation periods tend to be minimized.
11. Careful attention needs to given to the solubilization process, as any unsolubilized membrane fragments will partition with immunoprecipitated complexes and can therefore dramatically affect the quality and reproducibility of data. We have found that 60 min in a vortex-shaker, following by a brief vortex-mixing by hand and a subsequent 60 min in a vortex-shaker produces full solubilization without any detrimental effect on the maintenance of [^{35}S]GTPγS–Gα complexes.
12. The final concentration of antibody used needs to be optimized, not only for each antibody, but also for each antibody batch if the antibody is supplied commercially. The optimization procedure must balance the need to immunoprecipitate a significant fraction of the relevant Gα-protein against cost considerations. Typically, a good antibody should produce efficient immunoprecipitation of the Gα-protein at a final dilution of 1:100 or greater.
13. Protein A or protein G beads may produce higher affinity antibody capture, depending on the species of primary antibody. Suppliers (e.g., Amersham Biosciences) provide detailed information allowing optimal protein A/G-bead selection.
14. The number of washes may well influence the basal-to-stimulated signal; depending on the Gα-subunit and initial protein concentration, the beads may require two to five washes to produce consistent data. Not carrying out sufficient washes of the bead–antibody–Gα complex is a common cause of low signal-to-noise ratio and variability problems with this assay.

References

1. Northup, J. K., Smigel, M. D., and Gilman, A. G. (1982) The guanine nucleotide-activating site of the regulatory component of adenylate cyclase. *J. Biol. Chem.* **257,** 11,416–11,423.
2. Kurose, H., Katada, T., Haga, T., Haga, K., Ichiyama, A., and Ui, M. (1986) Functional interaction of purified muscarinic receptors with purified inhibitory guanine nucleotide regulatory proteins reconstituted in phospholipid vesicles. *J. Biol. Chem.* **261,** 6423–6428.
3. Hilf, G., Gierschik, P., and Jakobs, K. H. (1989) Muscarinic acetylcholine receptor-stimulated binding of guanosine 5′-*O*-(3-thiophosphate) to guanine-nucleotide-binding proteins in cardiac membranes. *Eur. J. Biochem.* **186,** 725–731.
4. Sim, L. J., Selley, D. E., and Childers, S. R. (1997) Autoradiographic visualization in brain of receptor-G protein coupling using [^{35}S]-GTPγS binding. In *Methods in*

Molecular Biology, Vol. 83: *Receptor Signal Transduction Protocols* (Challiss, R. A. J., ed.), Humana Press, Totowa, NJ, pp. 117–132.

5. Lazareno, S. (1997) Measurement of agonist-stimulated [^{35}S]-GTPγS binding to cell membranes, in *Methods in Molecular Biology*, Vol. 83: *Receptor Signal Transduction Protocols* (Challiss, R. A. J., ed.), Humana Press, Totowa, NJ, pp. 107–116.
6. Milligan, G. (2003) Principles: extending the utility of [^{35}S]-GTPγS binding assays. *Trends Pharmacol. Sci.* **24,** 87–90.
7. Friedman, E., Butkerait, P., and Wang, H. Y. (1993) Analysis of receptor-stimulated and basal guanine nucleotide binding to membrane G proteins by sodium dodecyl sulphate-polyacrylamide gel electrophoresis. *Anal. Biochem.* **241,** 171–178.
8. Burford, N. T., Tolbert, L. M., and Sadee, W. (1998) Specific G protein activation and μ-opioid receptor internalization caused by morphine, DAMGO and endomorphin I. *Eur. J. Pharmacol.* **342,** 123–126.
9. Akam, E. C., Challiss, R. A. J., and Nahorski, S. R. (2001) $G_{q/11}$ and $G_{i/o}$ activation in CHO cells expressing human muscarinic acetylcholine receptors: dependence on agonist as well as receptor-subtype. *Br. J. Pharmacol.* **132,** 950–958.
10. Carruthers, A. M., Warner, A. J., Michel, A. D., Feniuk, W., and Humphrey, P. A. (1999) Activation of adenylate cyclase by human recombinant sst_5 receptors expressed in CHO-K1 cells and involvement of $G_s\alpha$ proteins. *Br. J. Pharmacol.* **126,** 1221–1229.
11. Young, K. W., Bootman, M. D., Channing, D. R., et al. (2000) Lysophosphatidic acid-induced Ca^{2+} mobilization requires intracellular sphingosine 1-phosphate production: potential involvement of endogenous *Edg-4* receptors. *J. Biol. Chem.* **275,** 38,532–38,539.
12. DeLapp, N. W., McKinzie, J. H., Sawyer, B. D., et al. (1999) Determination of [^{35}S]guanosine-5′-*O*-(3-thio)triphosphate binding mediated by cholinergic muscarinic receptors in membranes from Chinese hamster ovary cells and rat striatum using an anti-G protein scintillation proximity assay. *J. Pharmacol. Exp. Ther.* **289,** 946–955.
13. Burford, N. T., Tobin, A. B., and Nahorski, S. R. (1995) Coupling of muscarinic m1, m2 and m3 acetylcholine receptors, expressed in Chinese hamster ovary cells, to pertussis-toxin sensitive/insensitive guanine nucleotide-binding proteins. *Eur. J. Pharmacol.* **289,** 343–351
14. Berg, K. A., Maayani, S., Goldfarb, J., Scaramellini, C., Leff, P., and Clarke, W. P. (1998) Effector pathway-dependent relative efficacy at serotonin type 2A and 2C receptors: evidence for agonist-directed trafficking of receptor stimulus. *Mol. Pharmacol.* **54,** 94–104
15. Kenakin, T. (1995) Agonist-receptor efficacy II: agonist trafficking of receptor signals. *Trends Pharmacol. Sci.* **16,** 199–205.

13

Identification and Quantitation of G-Protein α-Subunits

Ian Mullaney and Graeme Milligan

Summary

The demonstration that many intracellular signaling processes are mediated by a family of closely related guanine nucleotide binding proteins (G-proteins) has led to the development of specific techniques that can be used to identify which of these polypeptide(s) is involved on receptor activation by ligand. In addition, these methods can be used to probe the specificity of the interaction and to yield information about the stoichiometries involved.

Key Words

G-protein, immunoblotting, pertussis toxin.

1. Introduction

The classical G-proteins exist as heterotrimers comprising nonidentical α-, β-, and γ-subunits. Initial identification of G-proteins involved the ability of the α-subunit to act as a substrate for mono-ADP-ribosylation catalyzed by the ADP-ribosyltransferase activity of a number of bacterial exotoxins, a modification that functionally altered the involvement of the G-protein in signal transduction. The use of [^{32}P]NAD as a substrate allowed the visualization of ADP-ribosylated polypeptides following separation in sodium dodecyl sulfate-polyacrylamide gel electrophoresis (SDS-PAGE) gels and autoradiography. In this way G_s, the G-protein involved in the hormonal stimulation of adenylyl cyclase was identified as a substrate for ADP-ribosylation by cholera toxin. Similarly, members of the G_i family, the inhibitory G-proteins, were characterized by their ability to act as substrates for pertussis toxin. Although the use of toxins yields little further information about the molecular identity of the G-proteins involved, it can be useful in an initial investigation into G-protein function in a particular system. However, not all G-protein α-subunits are toxin

From: *Methods in Molecular Biology, vol. 259, Receptor Signal Transduction Protocols, 2nd ed.*
Edited by: G. B. Willars and R. A. J. Challiss © Humana Press Inc., Totowa, NJ

substrates: the ability of agonists to activate phospholipase Cβ and cause hydrolysis of inositol containing phospholipids has been shown to be unaffected by treatment with either pertussis or cholera toxins in the vast majority of cells and tissues studied. This implied the involvement of toxin-insensitive G-proteins that were subsequently identified as members of the G_q subfamily. Indeed the use of the polymerase chain reaction (PCR) based on conserved sequence domains across the G-protein family and the isolation of cDNAs has allowed identification more than 20 G-protein α-subunits including the members of the G_q and G_{12} subfamilies. Because these G-proteins do not have the conserved cysteine residue close to the C-terminus that is the hallmark of the G_i subfamily G-proteins they are not substrates for pertussis toxin. As a consequence of these studies, the primary amino acid sequences of all of the G-protein α-subunits has been deduced, allowing for the generation of antipeptide antisera for use as specific immunological tools. Although the G- protein superfamily is highly conserved, there are regions of sequence variation, particularly in the C-terminal region, that contain key receptor coupling sites that have been successfully used to produce specific polyclonal antipeptide antisera *(1)* (*see* **Table 1**). Although a variety of monoclonal antisera are available, polyclonal antisera are in widespread use and are generally produced by subcutaneous injection of peptide conjugated to carrier protein into rabbits or other suitable hosts. The titer and specificity of these antisera can be tested either in enzyme-linked immunosorbent assay (ELISA) assays against the antigen peptide or by immunoblotting using either purified or recombinant G-protein α-subunits.

The majority of G-protein assays utilize cell-free systems, relying on the production of plasma membrane containing fractions from either tissue or cultured cells. Immunoblotting of supernatant fractions produced during membrane production generally shows little immunoreactivity, reinforcing the concept that the G-protein α-subunits are located at membranes. All of the electrophoretic and immunoblotting techniques described herein use a crude plasma membrane preparation as starting material. All of the G-protein α-subunits have predicted molecular masses of between 39 and 45 kDa. The gel conditions and sample treatments described have been optimized to separate polypeptides within this molecular mass range. Treatment of the samples with *N*-ethylmaleimide differentially alkylates the α-subunits of the pertussis toxin-sensitive G-proteins G_{i1}, G_{i2}, and G_{i3} and the isoforms of G_o on accessible cysteine residues. This has the effect of altering the migration of these α-subunits in SDS-PAGE with the result that it is possible to obtain greater resolution of the G_o isoforms from the G_i-like G-proteins. If sample alkylation is performed in conjunction with resolution on a 12.5% acrylamide, 0.06% *bis*-acrylamide gel, the separation achieved can be dramatic. This technique is particularly useful

Table 1
Peptide Sequences That Have Been Widely Used to Generate Antipeptide Antisera Directed Against the α-Subunits of Various G-Proteins

Peptide used	G-protein sequence	Antiserum identifies
RMHLRQYELL[a]	$G_s\alpha$372–381	$G_s\alpha$
KENLKDCGLF[a]	$TD_1\alpha$ 341–350	$TD_1\alpha$, $TD_2\alpha$ $G_{i1}\alpha$, $G_{i2}\alpha$
LERIAQSDYI	$G_{i2}\alpha$ 160–169	$G_{i2}\alpha$
KNNLKECGLY[a]	$G_{i3}\alpha$ 345–354	$G_{i3}\alpha$
ANNLRGCGLY[a]	$G_o\alpha$ 345–354	$G_o\alpha$
GCTLSAEERAALERSK	$G_o\alpha$ 1–16	$G_o\alpha$
NLKEDGISAAKDVK	$G_o\alpha$ 22–35	$G_o\alpha$
QLNLKEYNLV[a]	$G_q\alpha$ 350–359	$G_q\alpha$, $G_{11}\alpha$ ($G_{14}\alpha$?)
QENLKDIMLQ[a]	$G_{12}\alpha$ 370–379	$G_{12}\alpha$
HDNLKQLMLQ[a]	$G_{13}\alpha$ 368–377	$G_{13}\alpha$
QLNLREFNLV[a]	$G_{14}\alpha$ 346–355	$G_{14}\alpha$, $G_q\alpha$ $G_{11}\alpha$
ARYLDEINLL[a]	$G_{15}\alpha$ 365–374	$G_{15}\alpha$, $G_{16}\alpha$
QNNLKYIGLC[a]	$G_z\alpha$ 346–355	$G_z\alpha$

[a]Denotes C-terminal sequences. Amino acids are represented using the one-letter code. TD, Transducin.

when trying to identify G-proteins with antisera that crossreact; a common example is antisera directed against the C-terminus of G_o that can crossreact with G_i3 because of the presence of an immunodominant tyrosine in the peptide sequence of these two polypeptides. To separate the isoforms of $G_o\alpha$ or G-proteins of the G_q family, the best strategy we know of is to resolve the membranes on 12.5% acrylamide gels containing 6 *M* urea. It is also possible to separate these proteins on SDS-PAGE gels that contain a 4–8 *M* urea gradient or on two-dimensional gels. However, both of these methods are technically more difficult and have little advantage over the 6 *M* urea SDS-PAGE gel system. This is useful, however, for parallel analysis of the expression of the two widely expressed members of the G_q family, $G_q\alpha$ and $G_{11}\alpha$. As they share an identical C-terminal decapeptide, most of the available antisera do not discriminate between them, and without resolution *(2)* only a composite of their expression levels is obtained.

Sustained exposure of cell surface receptors to agonist frequently results in downregulation of both the receptor and the activated G-protein. Pulse–chase assay techniques, in which cells are first incubated with [^{35}S]methionine-labeled medium are then chased in the presence or absence of a receptor agonist and the G-proteins subsequently immunoprecipitated with specific antisera, have revealed that the mechanism of downregulation is agonist-induced accelerated turnover of the α-subunit *(3)*.

The aim of this chapter is to describe the techniques, outlined in the preceding, which can be used to identify and quantitate those G-proteins involved in particular signaling events.

2. Materials

2.1. Equipment

1. Tissue culture facilities and shaking incubator.
2. Sonicator.
3. Benchtop centrifuge, microcentrifuge, and refrigerated ultracentrifuge.
4. ELISA plate reader.
5. Tight-fitting Teflon-on-glass homogenizer or Polytron.
6. 100°C heating block.
7. SDS-PAGE gel apparatus and power pack.
8. Electroblotting apparatus and power pack.
9. Rotating wheel.
10. γ-Counter or phosphoimager.
11. Imaging densitometer.
12. Gel drier and vacuum pump.
13. X-ray film developing apparatus.

2.2. Reagents

1. Keyhole limpet hemocyanin (KLH), Freund's complete and incomplete adjuvants.
2. 21 m*M* glutaraldehyde in water.
3. Phosphate-buffered saline (PBS): 0.2 g of KCl, 0.2 g of KH_2PO_4, 8 g of NaCl, 1.14 g of Na_2HPO_4 (anhydrous), pH 7.4, up to 1000 mL of water. If kept as a stock, check pH before use.
4. Peptide solution (10 μg/mL): dilute 1:100 from 1 mg of peptide in 1 mL of PBS. Stock can be stored at –20°C.
5. ELISA blocking solution: 1 g of dried milk in 100 mL of PBS.
6. PBS–Tween-20: 0.5 mL of Tween-20 in 1000 mL of PBS.
7. ELISA antibody carrier solution: 0.05 mL of Tween-20, 0.1 g of dried milk in 100 mL of PBS.
8. ELISA primary antibody solution: G-protein antisera diluted in ELISA antibody carrier solution.
9. ELISA secondary antibody solution: 1:1000 dilution of horseradish peroxidase-conjugated donkey anti-rabbit IgG diluted in ELISA antibody carrier solution.
10. Citrate phosphate buffer:
    a. Buffer A: 0.1 *M* citric acid; 21.01 g in 1000 mL of water.
    b. Buffer B: 0.2 *M* Na_2HPO_4 (anhydrous); 28.4 g in 1000 mL of water. Mix 17.9 mL of buffer A, 32.1 mL of buffer B, and 50 mL of water, adjust pH to 6.0. Buffers A and B can be stored at 4°C.

11. H_2O_2 solution: 10 μL of H_2O_2 in 10 mL of water.
12. OPD substrate solution: 4 mg of *o*-phenylenediamide dihydrochloride (OPD), 9 mL of citrate phosphate buffer, 1 mL of H_2O_2 solution.
13. L-broth: 10 g of tryptone, 5 g of yeast extract, 10 g of NaCl to 1000 mL with water (adjust pH to 7.0 with NaOH), sterilize in a pressure cooker (e.g., Prestige medical series 2100 clinical autoclave).
14. L-broth/glucose: 10 g of tryptone, 5 g of yeast extract, 10 g of NaCl, 3.6 g of glucose to 1000 mL with water (adjust pH to 7.0 with NaOH), sterilize in a pressure cooker.
15. Ampicillin (50 mg/mL stock): 0.5 g in 10 mL of water, filter sterilize, and store as 1-mL aliquots at –20°C.
16. L-broth-agar: 10 g of tryptone, 5 g of yeast extract, 10 g of NaCl, 15 g of agar to 1000 mL with water (adjust pH to 7.0 with NaOH), sterilize in a pressure cooker. When hand cool, add 2 mL of ampicillin stock, shake to mix, and pour plates (approx 10 mL per plate).
17. Isopropyl-β-D-thiogalactopyranoside (100 m*M*): 0.476 g with up to 20 mL of water, filter sterilize, and store as 2-mL aliquots at –20°C.
18. TE buffer (10 m*M* Tris-HCl, 0.1 m*M* EDTA, pH 7.5): 1.21 g of Tris, 37.2 mg of EDTA in 1000 mL of water; adjust pH to 7.5.
19. Protease inhibitor TE buffer containing 10 m*M* NaF, 100 μ*M* Na_3VO_4, 1 m*M* phenylmethanesulfonyl fluoride, 3 m*M* benzamidine, 0.1 μ*M* soybean trypsin inhibitor, 10 μ*M* leupeptin, 0.2 μ*M* aprotinin, 1.5 μ*M* antipain.
20. 2% (w/v) 7-deoxycholic acid, sodium salt: 2 g in 100 mL of water.
21. 24% (w/v) Trichloroacetic acid (TCA): 24 g in 100 mL of water.
22. 1 *M* Tris base: 121.1 g in 1000 mL of water. Do not adjust the pH.
23. Laemmli sample buffer: 0.605 g of Tris, 30 g of urea, 5 g of SDS, 6 g of dithiothreitol (DTT), 10 mg of bromophenol blue in 100 mL of water; adjust the pH to 8.0 with HCl. Aliquot in 2-mL volumes and store at –20°C.
24. 5% (w/v) SDS, 50 m*M* DTT: 5 g of SDS, 0.771 g of DTT in 100 mL of water.
25. 100 m*M* *N*-ethylmaleimide (NEM): 1.25 g in 100 mL of water.
26. Cholera toxin (Sigma Chemical Co. Ltd., UK, product no. C-3012): kept as a 1 mg/mL stock at 4°C in 50 m*M* Tris-HCl, pH 7.5, 200 m*M* NaCl, 3 m*M* NaN_3, 1 m*M* EDTA.
27. Pertussis toxin (Porton Products, Ltd., UK): stored at –20°C as a 0.44 mg/mL stock in 50 m*M* phosphate buffer, pH 7.2; 500 m*M* NaCl; 50% (v/v) glycerol.
28. Nicotinamide adenine dinucleotide, di(triethylammonium) salt, [adenylate-^{32}P] (product no. NEG-023), specific activity 10–50 Ci/mmol from NEN Research Products, Du Pont (UK) Ltd., Stevenage, UK.
29. 1.5 *M* sodium phosphate buffer: 53.25 g Na_2HPO_4 in 250 mL of water (solution 1) and 58.5 g $NaH_2PO_4 \cdot 2H_2O$ in 250 mL of water (solution 2). Titrate solution 2 into solution 1 until pH 7.0 is reached.
30. ADP-ribosylation cocktail stocks:
    a. Thymidine (0.2 *M*) 48.44 mg/mL of water.
    b. Arginine HCl (1 *M*): 210.7 mg/mL of water.

c. ATP (0.04 *M*): 22.04 mg/mL of water.
d. GTP (1 m*M*): 0.54 mg/mL of water.
Store in 100-µL aliquots at –20°C.

31. ADP-ribosylation cocktail mix (for 15 samples): 75 µL of thymidine, 30 µL of GTP, 125 µL of 1.5 *M* sodium phosphate buffer, 10 µL of ATP, 15 µL of arginine hydrochloride, 30 µCi of [^{32}P]NAD$^+$ to 300 µL of water.
32. ^{35}S labeling medium: methionine- and cysteine-free Dulbecco's modified Eagle's medium (ICN Biomedicals, Inc.) supplemented with 50 µCi/mL of Trans ^{35}S-label (ICN) and 1% (v/v), heat inactivated, dialyzed fetal bovine serum.
33. 10% (w/v) SDS: 100 g of SDS in 1000 mL of water. Keep at room temperature and use as the stock for all SDS solutions.
34. Solutions for 10% (w/v) SDS-PAGE gels: resolving gel
    a. Acrylamide: 30 g of acrylamide, 0.8 g of *bis*-acrylamide to 100 mL with water (*see* **Note 1**).
    b. Gel buffer 1: 18.17 g of Tris, 4 mL of 10% (w/v) SDS, pH 8.8, to 100 mL with water.
    c. Glycerol: 50% (v/v) in water.
    d. APS: 100 mg of ammonium persulfate per mL of water.
35. Solutions for 10% (w/v) SDS-PAGE gels: stacking gel—gel buffer 2: 6 g of Tris, 4 mL of 10% (w/v) SDS, pH 8.8, to 100 mL with water.
36. Solutions for 12.5% (w/v) SDS PAGE gels: resolving gel—acrylamide: 30 g of acrylamide, 0.15 g of *bis*-acrylamide to 100 mL with water.
37. Electrophoresis running buffer: 6 g of Tris, 28.8 g of glycine, 20 mL of 10% (w/v) SDS to 2000 mL with water. Do not adjust the pH.
38. 6 *M* urea: 36.03 g of urea up to 100 mL with water. Make fresh on the day of use.
39. Solutions for 12.5% (w/v) SDS-urea PAGE gels: resolving gel:
    a. Acrylamide: 30 g of acrylamide, 0.15 g of *bis*-acrylamide, 36.025 g of urea to 100 mL with water.
    b. Gel buffer 1: 18.17 g of Tris, 36.025 g of urea, 4 mL of 10% (w/v) SDS, pH 8.8, to 100 mL with water.
    c. Glycerol: 50% (v/v) in water.
    d. APS: 100 mg of ammonium persulfate/mL of water.
40. Coomassie stain buffer: dissolve 2.5 g of Coomassie brilliant blue R-250 into a solution containing 450 mL of methanol, 450 mL of water, and 100 mL of glacial acetic acid. Filter through Whatman No.1 filter paper. This solution can be reused.
41. Gel destain buffer: 450 mL of methanol, 450 mL of water, 100 mL of glacial acetic acid.
42. Blotting buffer: 15 g of Tris, 72 g of glycine, 1000 mL of methanol to 5000 mL with water. Do not adjust pH.
43. Ponceau S solution: 15 g of TCA in 500 mL of water; allow to dissolve and add 0.5 g of Ponceau S. Keep stock solution at room temperature and reuse.
44. Immunoblot blocking buffer: 5 g of gelatin in 100 mL of PBS.
45. PBS–NP40: 2 mL of Nonidet P-40 in 1000 mL of PBS.

46. Immunoblot first antiserum solution: appropriate anti-G-protein antiserum dilution in PBS-NP40 containing 1% (w/v) gelatin.
47. Immunoblot second antibody solution: 1:500 dilution of a commercial horseradish peroxidase-conjugated donkey anti-rabbit IgG in PBS–NP-40 containing 1% (w/v) gelatin.
48. *o*-Diansidine solution: 10 mg of *o*-diansidine hydrochloride solution in 1 mL of water (*see* **Note 2**).
49. Sodium azide solution: 1 g of NaN_3 in 100 mL of water (*see* **Note 3**).
50. Immunoprecipitation (IP) buffer: 6.06 g of Tris, 11.1 g of NaCl, 2.23 g of EDTA, 12.5 mL of Triton X-100, pH 7.5, to 1000 mL with water.
51. IP wash buffer: 80 mL of IP buffer, 20 mL of 1% (w/v) SDS.
52. IP final wash buffer: 0.606 g of Tris, pH 6.8, to 100 mL of water.
53. ^{125}I-labeled donkey anti-rabbit immunoglobulin (product no. IM 134), specific activity 750–3000 Ci/mmol (Amersham International PLC, Amersham, UK).
54. ^{125}I-overlay solution: 50 mL of PBS–NP-40 containing 1% (w/v) gelatin and spiked with 5 μCi of ^{125}I-labeled donkey anti-rabbit immunoglobulin.

3. Methods

3.1. Production of G-Protein-Specific Antisera

3.1.1. Immunization and Serum Collection

Antipeptide antisera are prepared according to the method described by Goldsmith et al. *(1)*, in commercially purchased New Zealand White rabbits. Preimmune blood samples should be taken from each of the animals prior to injection and the serum checked for any significant titre or immunological identification of cellular proteins.

1. Dissolve 10 mg of KLH and 3 mg of the particular peptide (*see* **Note 4**) required in 1 mL of 0.1 *M* Na phosphate buffer, pH 7.0. Add 0.5 mL of 21 m*M* glutaraldehyde dropwise with stirring and incubate the combined 1.5 mL overnight at room temperature.
2. Mix with an equal volume of Freund's complete adjuvant and sonicate for 20 s.
3. Immediately after sonication, inject the resultant emulsion in 0.2-mL aliquots into multiple subcutaneous sites in the rabbit. Immunizations are normally performed simultaneously into two rabbits to maximize successful antibody production.
4. After 2 wk each animal receives a booster immunization with material prepared identically except that one half as much peptide and KLH are injected in Freund's incomplete adjuvant.
5. Bleed the animals 4 wk after the booster injections. Collect the blood into glass universals and allow to clot overnight at 4°C (*see* **Note 5**).
6. Remove the straw-colored serum from the clot and centrifuge at 500*g* for 5 min on a benchtop centrifuge to remove any remaining traces of erythrocytes. Aliquot the serum in appropriate volumes and store at –20°C.

3.1.2. ELISA Protocol

1. Coat a 96-well Titertek ELISA plate with antigen peptide. Add 100 µL of 10 µg/mL peptide solution, cover the plate with cling film, and incubate overnight at 4°C (*see* **Note 6**).
2. Remove liquid and wash each well two times with PBS (*see* **Note 7**).
3. Blot dry, add 100 µL of ELISA blocking solution to each well, and incubate at 37°C for 1 h.
4. Remove the blocker, wash each well two times with PBS–Tween-20, and blot dry.
5. Add 100 µL of increasing dilutions of antiserum (from 1:10 to 1:100,000) to each well, cover the plate with cling film, and incubate overnight at 4°C (*see* **Note 8**).
6. Remove antiserum dilutions, wash each well two times with PBS–Tween-20, and blot dry.
7. Add 100 µL of secondary antibody solution to each well and incubate at 37°C for 1 h.
8. Remove secondary antibody solution, wash each well five times with PBS–Tween-20, and blot dry.
9. Add 100 mL of OPD substrate solution to each well; wrap the plate in tin foil; and incubate, in the dark, at room temperature for 15–20 min (*see* **Note 9**).
10. Stop the reaction by addition of 50 µL of 2 *M* H_2SO_4 to each well and read at 492 nm on any ELISA plate reader, for example, Titertek Multiscan.
11. Plot the absorbency reading against antiserum dilution to determine specificity (*see* **Note 10**).

3.2. High Level Expression of Mammalian G-Protein α-Subunits in *Escherichia coli*

The method described below uses plasmid pT7.7 into which the various Gα genes have been subcloned *(4)*. This expression system utilizes the promoter for bacteriophage T7 RNA polymerase in the plasmid vector pT7.7. This expression vector contains an initiation codon ATG and a ribosome binding site positioned downstream from the T7 promoter such that maximal expression is ensured. Expression constructs were transformed into the lysogen BL21 (DE3), which contains a single chromosomal copy of the gene for T7 RNA polymerase under control of the isopropyl-β-D-thiogalactopyranoside (IPTG) inducible *lac* UV5 promotor.

3.2.1. Preparation of Competent E. coli Strain BL21 DE3, Transformation of Cells With the Expression Vector pT7.7, and Expression of Mammalian G-Protein α-Subunits

1. Take 50 µL of *E. coli* stock, add 5 ng of plasmid DNA, and incubate on ice for 15 min.
2. Heat the cells at 42°C for exactly 90 s, then return to ice for 2 min.

3. Add 450 μL of L-broth/glucose and allow cells to recover by incubation at 37°C for 1 h in a shaking incubator.
4. Spread 100 μL of the transformants onto L-broth-agar plates containing 100 μg/mL of ampicillin and incubate overnight at 37°C.
5. Select single colonies from the plate and put into 10 mL of L-broth containing 100 μg/mL of ampicillin and incubate in a shaking incubator overnight at 37°C.
6. Take 0.5 mL of the overnight culture, inoculate into 50 mL of L-broth containing 100 μg/mL of ampicillin, and incubate in a shaking incubator at 37°C until an A_{550} of 0.3–0.5 absorbency units is reached.
7. Remove 1 mL of the cell suspension as a control, add 0.5 mL of 100 m*M* IPTG to the remainder, and incubate in a shaking incubator at 37°C for 4 h.
8. Remove 1 mL of cell suspension, centrifuge in a benchtop microcentrifuge for 10 min at 15,000*g*, and discard the supernatant.
9. Add 25 μL of Laemmli sample buffer to the pellet, heat to 90°C for 10 min, and resolve the whole-cell extracts by SDS-PAGE.

3.3. Sample Preparation for Analysis of G-Proteins

3.3.1. Production of Crude Plasma Membrane Fractions

Crude plasma membrane fractions are produced essentially as described by Koski and Klee *(5)*.

1. Gently remove cells off the surface of the flasks with a Pasteur pipet or rubber policeman, collect in a 50-mL conical centrifuge tube on ice, and centrifuge at 500*g* at 4°C for 5 min in a benchtop centrifuge.
2. Discard the supernatant, resuspend the cell pellet in 30 mL of ice-cold PBS, and centrifuge as before.
3. Repeat this procedure twice and store the resultant washed cell paste at –80°C until needed.
4. Thaw the frozen cell pastes and resuspend in 2 mL of ice-cold TE buffer and homogenize with 20 strokes of a Teflon-on-glass tissue grinder (*see* **Note 11**).
5. If preparing plasma membrane fractions from tissues, chop the tissue with scissors, rinse in two washes of PBS (*see* **Note 12**), and homogenize for 60 s in 10 volumes of TE buffer with a Polytron at setting no. 4.
6. Centrifuge the homogenates for 10 min at 500*g* in an ultracentrifuge (a Beckman L5-50B, for example).
7. Discard the pellet (*see* **Note 13**) and centrifuge the supernatant, which will contain the bulk of plasma membrane fraction, at 48,000*g* for 10 min using the same centrifuge.
8. Discard the supernatant, resuspend the pellet in 5 mL of TE buffer, and recentrifuge for 10 min at 48,000*g*.
9. Again, discard the supernatant and resuspend the pellet in TE buffer.
10. Triturate the resuspended pellet with a syringe with a fine-gage needle. Aliquot into appropriate volumes and store at –80°C until needed.

11. Protein concentrations are determined. Membranes should have protein concentrations of 1–2 mg/mL.

3.3.2. TCA–Deoxycholate Precipitation of Samples

1. Take an appropriate amount of the crude membrane preparation (between 25 μg and 150 μg of membranes depending on the experiment) and place on ice in a 1.5-mL microcentrifuge tube. Samples with a volume of >100 μL should be centrifuged for 5 min at 15,000*g* on a microcentrifuge, the supernatant carefully removed, and the pellet resuspended in 20 μL of TE buffer.
2. Add 6.5 μL of 2% (w/v) 7-deoxycholic acid, sodium salt to each tube followed by 750 μL of double distilled water, then 250 μL of 24% (w/v) TCA (*see* **Note 14**).
3. Vortex-mix each sample and in a benchtop microcentrifuge for 10 min at 12,000 rpm.
4. Carefully discard the supernatant (*see* **Note 15**) and bring the pellet to weakly alkaline pH by addition of 20 μL of 1 *M* Tris base.
5. Add 20 μL of Laemmli sample buffer (*see* **Note 16**) and load the sample onto the gel.

3.3.3. NEM Treatment

Alkylation of samples prior to electrophoresis may produce better resolution of G-proteins with very similar molecular masses.

1. Take an appropriate amount of the crude membrane preparation (between 25 μg and 150 μg of membranes depending on the experiment) and place on ice in a 1.5-mL microcentrifuge tube.
2. Centrifuge the samples at 15,000*g* for 5 min on a benchtop microcentrifuge.
3. Remove the supernatant and resuspend the pellet in 20 μL of TE buffer.
4. Add 10 μL of 5% (w/v) SDS and 50 m*M* DTT and incubate at 90°C for 5 min.
5. Cool the samples on ice, add 10 μL of freshly prepared 100 m*M* NEM to each tube, and leave at room temperature for 20 min.
6. Add 20 μL of Laemmli sample buffer and load the sample onto the gel.

3.3.4. Mono-ADP-Ribosylation of Membranes by Bacterial Toxins

Mono-ADP-ribosylation of G-proteins for further analysis by SDS-PAGE is derived from the method of Hudson and Johnson ***(6)***.

1. Both cholera and pertussis toxins must be preactivated before in vitro use. Add an equal volume of 100 m*M* DTT to the toxin, gently mix, and allow to stand at room temperature for up to 1 h (*see* **Note 17**).
2. Take an appropriate amount of the crude membrane preparation (between 25 μg and 50 μg of membranes in a final volume of 25 μL) and place on ice in a 1.5-mL microcentrifuge tube.
3. Add 20 μL of the ADP-ribosylation cocktail mix to each tube and start the incubation by adding 5 μL of the appropriate preactivated toxin (*see* **Note 18**). Incubate for up to 90 min in a 37°C water bath.

4. Place on ice and precipitate samples using the TCA–deoxycholate method (*see* **Subheading 3.3.2.**).
5. Add 20 μL Laemmli sample buffer and load the sample onto the gel (*see* **Note 19**).

3.3.5. [^{35}S]Trans Pulse-Chase Protocol

1. Seed cells into 75-cm^3 flasks or 6-well culture dishes.
2. When cells are approx 60% confluent replace growth medium with ^{35}S labeling medium (*see* **Note 20**).
3. Incubate cells in ^{35}S-label for 20–48 h *(pulse)*.
4. Wash cell monolayer twice with normal growth medium and leave in fresh growth medium (*see* **Note 21**) *(chase)*.
5. At appropriate times, gently wash cells off the surface of the flask with a Pasteur pipet or rubber policeman, collect in a 12.5-mL plastic centrifuge tube on ice, and centrifuge at 500*g* at 4°C for 5 min on a benchtop centrifuge.
6. Remove the medium, resuspend the cells in 100 μL of water, and add 100 μL of 2% (w/v) SDS (*see* **Note 22**). Transfer to 2-mL screw-cap Eppendorf tubes.
7. Tighten the screw-cap onto the tube and heat to 100°C for 20 min (*see* **Note 23**).
8. Transfer tube to ice, pulse spin to collect all moisture to the bottom of the tube, and proceed to the immunoprecipitation protocol (*see* **Note 24**).

3.4. SDS-PAGE

Gels are performed using the basic approach of Laemmli *(7)* on slab gels.

3.4.1. 10% Acrylamide (w/v) SDS-PAGE Gels

Gel plates (180 mm × 160 mm with spacers of 1.5 mm) are run as part of a Bio-Rad Protean I electrophoresis apparatus (Bio-Rad Laboratories Ltd., Watford, Herts.).

1. Recipe for one gel:
   Resolving gel (lower) solutions:

a. H_2O	8.2 mL
b. Buffer 1	6 mL
c. Acrylamide	8 mL
d. 50% Glycerol	1.6 mL
e. APS	90 μL
f. *N,N,N′,N′*-tetramethylethylenediamine (TEMED)	8 μL

2. Set up the gel apparatus according to the manufacturer's guidelines. Add all reagents in the order given into a 250-mL conical flask and mix gently. Cast gel using Pasteur pipet (*see* **Note 25**).
3. Carefully overlay cast gel with approx 1 mL of 0.1% (w/v) SDS and allow the gel to polymerize (between 1 and 2 h at room temperature).
4. After polymerization, remove the SDS overlay and prepare to add the stacker gel.
   Stacker gel (upper) solutions:

a. H_2O 9.75 mL
b. Buffer 2 3.75 mL
c. Acrylamide 1.5 mL
d. APS 150 µL
e. TEMED 8 µL

5. Add all reagents in the order given into a 100-mL conical flask and mix gently. Pour stacker gel on top of resolving gel and place the well forming comb in top of the gel, ensuring no air bubbles are trapped under the comb. Leave to polymerize (approx 1 h at room temperature).
6. After polymerization, remove the sample well comb and place the gel in the gel tank containing enough running buffer in the base to cover the bottom edge of the gel and add the remaining running buffer to the top.
7. Load the prepared samples in the preformed wells using a Hamilton syringe.
8. Run the gel overnight (approx 16 h) at 60 V and 15 mA per plate until the dye front reaches the bottom of the gel plates.

3.4.2. 12.5% Acrylamide (w/v) SDS-PAGE Gels

Gel plates 180 mm × 200 mm with spacers of 1.5 mm. When higher resolution of proteins within a narrow molecular weight range is needed, 12.5% (w/v) SDS-PAGE gels should be run instead of 10% (w/v) gels. The longer gels are able to maximize detection of molecular mass differences and are run as part of a Bio-Rad Protean II electrophoresis system 1.

1. Recipe for one gel:
   Resolving gel (lower) solutions:
   a. H_2O 11.6 mL
   b. Buffer 1 12 mL
   c. Acrylamide 20 mL
   d. 50% Glycerol 4 mL
   e. APS 180 µL
   f. TEMED 16 µL
2. Set up the gel apparatus according to the manufacturer's guidelines. Add all reagents in the order given into a 250-mL conical flask and mix gently. Cast gel using Pasteur pipet (*see* **Note 25**).
3. Carefully overlay cast gel with approx 1 mL of 0.1% (w/v) SDS and allow the gel to polymerize (between 1 and 2 h at room temperature).
4. After polymerization, remove SDS overlay and prepare to add stacker gel.
   Stacker gel (upper) solutions:
   a. H_2O 9.75 mL
   b. Buffer 2 3.75 mL
   c. Acrylamide 1.5 mL
   d. APS 150 µL
   e. TEMED 8 µL

5. Add all reagents in the order given into a 100-mL conical flask and mix gently. Pour stacker gel on top of resolving gel and place well forming comb in top of the gel, ensuring no air bubbles are trapped under the comb. Leave to polymerize (approx 1 h at room temperature).
6. After polymerization, remove the sample well comb and place the gel in the gel tank containing enough running buffer in the base to cover the bottom edge of the gel and add the remaining running buffer to the top.
7. Load the prepared samples in the preformed wells using a Hamilton syringe.
8. Run the gel overnight (approx 16 h) at 100 V and 15 mA per plate until the dye front reaches the bottom of the gel plates.

3.4.3. 12.5% Acrylamide (w/v) SDS-Urea-PAGE Gels

Resolution of many closely related G-protein α-subunits can be achieved using 12.5% SDS-PAGE gels in which 6 *M* urea has been added to the gel mixture (*see* **Note 26**). Gel plates 180 mm × 200 mm with spacers of 1.5 mm are run as a part of a Bio-Rad Protean II electrophoresis apparatus.

1. Recipe for one gel:
   Resolving gel (lower) solutions:

a. 6 *M* urea	11.6 mL
b. 6 *M* urea/buffer 1	12 mL
c. 6 *M* urea/acrylamide	20 mL
d. 50% Glycerol	4 mL
e. APS	180 µL
f. TEMED	16 µL

2. Set up the gel apparatus according to the manufacturer's guidelines. Add all reagents in the order given into a 250-mL conical flask and mix thoroughly. Cast gel using Pasteur pipet (*see* **Note 25**).
3. Carefully overlay cast gel with approx 1 mL of 0.1% (w/v) SDS and allow the gel to polymerize (between 3 and 4 h at room temperature) (*see* **Note 27**).
4. After polymerization, remove SDS overlay and prepare to add the stacker gel. Because there is no urea in the stacker gel, follow the procedure given for non-urea-containing gels.
5. After polymerization, remove the sample well comb and place the gel in the gel tank containing enough running buffer in the base to cover the bottom edge of the gel and add the remaining running buffer to the top.
6. Load the prepared samples in the preformed wells using a Hamilton syringe.
7. Run the gel for 18–20 h at 120 V and 50 mA per plate (*see* **Note 28**).

3.4.4. Autoradiography

1. Remove SDS-PAGE gel with resolved radiolabeled polypeptides from the electrophoresis apparatus and soak for 1 h in Coomassie stain buffer.
2. Remove stain, add destain solution, and leave for 2–3 h.

3. Remove stain and dry the gel onto Whatman 3-mm filter paper under suction from an electric vacuum pump attached to a gel drier (e.g., a Bio-Rad model 583) at 70°C for 2 h.
4. Transfer the dried gel to a Kodak X-omatic cassette with intensifying screens (or similar) containing Kodak X-omat S X-ray film and allow to autoradiograph at –80°C for an appropriate time (1–4 d).
5. Develop the film (e.g., on a Kodak X-omat developing machine) and quantitate the autoradiograph using an imaging densitometer.

3.4.5. Phosphoimaging

Follow **steps 1–3** from **Subheading 3.4.4.**

4. Transfer the dried gel to a phosphoimager plate and leave for an appropriate length of time.
5. Develop the image using a phosphoimager (e.g., Fujix BAS1000) and quantitate the resultant image.

3.5. Immunological Methods

Transfer of proteins from SDS-PAGE gels onto nitrocellulose and subsequent incubation with antisera essentially follows the strategies reported by Towbin *(8)*.

3.5.1. Electroblotting of Proteins Onto Nitrocellulose

Proteins that have been separated on SDS-PAGE gels are transferred onto nitrocellulose using an electroblotting apparatus (e.g., an LKB 2005 Transphor unit).

1. Soak the sponge pad in blotting buffer and place in the lower part of the transfer cassette. All subsequent loading procedures are done with the cassette totally submerged in blotting buffer to prevent formation of any air bubbles within the cassette, which prevents successful transfer of the proteins from the gel.
2. Place a piece of Whatman 3-mm chromatography filter paper with dimensions slightly larger than the gel on the sponge and put the gel on top.
3. Position a piece of nitrocellulose, cut approximately to the size of the gel, over the gel, then another piece of Whatman 3-mm filter paper, and finally another sponge to complete the sandwich.
4. Close the cassette, insert into the transfer apparatus, and electroblot toward the anode at 1.5 A for 2 h (*see* **Note 29**).
5. To check if transfer is complete, remove nitrocellulose from the blotting sandwich, place in a clean container, and cover with Ponceau S solution. Gently rock until protein banding appears.
6. Discard staining solution and wash the blot with blotting buffer from the electroblotting tank until the bands disappear.

3.5.2. Incubation of Nitrocellulose With Antisera

1. Transfer the electroblotted nitrocellulose sheet into a dish, cover with 100 mL of immunoblot blocking buffer, and incubate for 2 h at 30°C.
2. Remove the blocker, wash the nitrocellulose using copious amounts of double-distilled water, add the first antiserum solution (normally the specific anti-G-protein antiserum), and incubate overnight at 30°C (*see* **Note 30**).
3. Remove primary antiserum and wash the blots thoroughly with double-distilled water to remove all the unbound antiserum.
4. Wash the blot with PBS–NP-40 two times 10 min, then incubate at 30°C for 2 h in the second antiserum solution.
5. Thoroughly wash the blot with double-distilled water then with PBS–NP-40 for 2 × 10 min and finally with two washes each of 10 min with PBS.
6. Place blot in dish containing 40 mL of PBS. Add 1 mL of freshly prepared *o*-diansidine solution, then 10 µL of stock hydrogen peroxide.
7. Remove developer, terminate reaction by addition of sodium azide solution (*see* **Note 31**) and leave for 2 min. Pour off sodium azide and wash with water.

3.5.3. Immunoprecipitation Protocol

1. To each sample in a final volume of 50 µL, add 150 µL of 1.33% (w/v) SDS, containing the cocktail of protease inhibitors previously described, in 2-mL screwtop Eppendorf tubes (*see* **Note 32**).
2. Heat the samples to 100°C for 4 min, place on ice to cool, and pulse the samples to the bottom of the tubes by briefly centrifuging the samples at maximum speed in a microcentrifuge.
3. Add 0.8 mL of ice-cold IP buffer containing protease inhibitors to each sample and mix by inverting.
4. Centrifuge samples in a microcentrifuge at 15,000*g* for 10 min at 4°C and transfer the supernatant to a fresh tube.
5. Add an appropriate amount of antibody (between 2 and 20 µL depending on the antiserum) and incubate with rotation at 4°C overnight.
6. Add 50 µL of Pansorbin (or 20 µL of protein A–sepharose) (*see* **Note 33**) to each sample and incubate at 4°C with rotation for a minimum of 4 h.
7. Centrifuge samples in a microcentrifuge at 15,000*g* for 2 min at 4°C and wash with 3 × 1 mL IP wash buffer, pelleting the Pansorbin complex between each wash by centrifugation for 30 s at 15,000*g*.
8. After the final centrifugation, remove the supernatant and wash the pellet with 1 mL of IP final wash buffer.
9. Repeat the centrifugation step.
10. Remove the final wash buffer and add 50 µL of Laemmli buffer. Heat samples to 100°C for 10 min, centrifuge for 10 min at 15,000*g* to pellet Pansorbin, then load supernatant on gel.

3.6. Quantitation of G-Protein α-Subunits

This is achieved by immunoblotting various amounts (0–100 ng) of either *E. coli*-expressed G-protein (*see* **Subheading 3.2.**) or purified G-protein along with known amounts of the plasma membrane fractions. A standard curve can be constructed in the following ways and levels of G-protein in the sample can be assessed and expressed in terms of membrane protein, tissue amount or even cell number. Commercial preparations of partially purified G-protein α-subunits are available (*see* **Note 34**).

3.6.1. Densitometric Quantitation of Immunoblots

1. Place the developed immunoblot onto filter paper and allow to dry.
2. Scan the blot into an imaging densitometer (e.g., Bio-Rad GS-670) and quantitate.
3. Plot the standard curve showing amount of recombinant or purified G-protein against arbitrary densitometric values and interpolate the values for the unknowns from the curve.

3.6.2. ^{125}I-Labeled Donkey Anti-Rabbit Immunoglobulin Overlay Technique

1. Place the developed blots in a dish containing 50 mL of ^{125}I-overlay solution and incubate for 1 h at 30°C.
2. Remove the overlay solution, and wash the blots thoroughly with double-distilled water to remove all the unbound label and then with two washes each of 30 min with PBS.
3. Allow the blot to air-dry, excise the immunoreactive bands, and measure by γ-counting.
4. Plot the standard curve showing amount of recombinant or purified G-protein against the dpm obtained from the bound ^{125}I-overlay solution and extrapolate the values for the unknowns from the curve.

4. Notes

1. Care should be taken with acrylamide and bisacrylamide, as both have been reported to be neurotoxins.
2. Dissolve *o*-diansidine hydrochloride in water, as it is poorly soluble in PBS.
3. NaN_3 is toxic. The solution can be reused but tends to turn brown and discolor the nitrocellulose.
4. See **Table 1** for peptide sequences used in G-protein anti-peptide antisera production.
5. Do not use plastic universals. Glass universals allow the clot to shrink, enabling serum harvest.
6. It is advisable to test any antisera produced against other peptides, for example, other C-terminal G-protein sequences and nonrelated peptides to examine for nonspecific immunoreactivity.
7. For simplicity, fill water bottles with PBS or PBS–NP-40 and use these to wash the ELISA plates.

8. Dilute the antiserum simply by a series of 1:1 dilutions starting from 1:10, 20, 40, 80, 160, and so forth.
9. A positive reaction is purple which turns to brown on addition of H_2SO_4. If no color appears after 1 h, mix the remaining secondary antibody solution with the remaining OPD solution. If the solutions are correct, a purple color will appear instantaneously.
10. Typical half-maximal antiserum dilutions for useful antisera are 1:10,000–1:100,000.
11. Protease inhibitors should be included in all buffers.
12. The easiest way is to mince the tissue on a plastic Petri dish with curve ended scissors before disruption.
13. Depending on the tissue or cell line used, the pellet after the first centrifugation can be large. It is advisable to keep this pellet on ice until the plasma membrane fractionation is complete in case the original homogenization was unsuccessful.
14. The solutions must be added in the order 7-deoxycholate, water, then TCA. Solutions can be kept indefinitely at room temperature.
15. Remove the supernatant and blot dry onto tissue to eliminate as much liquid as possible. The pellet should be firmly sedimented to the bottom of the tube.
16. If the blue Laemmli buffer solution turns brown/orange then the sample is still acidic. Simply add 5-μL amounts of 1 *M* Tris-base until the sample turns blue.
17. To treat cells in vivo with pertussis toxin or cholera toxin, incubate in growth medium containing 100 ng/mL of toxin and harvest the cells as described in **Subheading 3.3.**
18. If a number of samples contain the ribosylation cocktail and the toxin, then mix the two solutions together in the ratio of 4:1 and add 25 μL to each sample. This will avoid sample variation caused by addition of small volumes.
19. As these samples are radioactive, load the gel using an automatic pipet and dispose of the tips along with the rest of the radioactive waste.
20. **Caution:** ^{35}S is volatile. Stocks should be opened and aliquoted in a fume hood. Store aliquots in 2-mL screwtop Eppendorf tubes and store at –80°C.
21. This is the chase element in pulse–chase protocols.
22. To ensure proper solubilization of the cell pellet always resuspend in water, then add the SDS. Do not resuspend the cell pellet directly in SDS.
23. If the sample is still viscous at this stage, pass it through a 25-gage needle and syringe and reboil as before for 10 min.
24. At this stage it is possible to freeze the sample at –20°C and store until further use.
25. There is no need to degas the solutions before pouring the gel.
26. Although we have reported that separation of certain G-protein α-subunits (such as $G_q\alpha$ from $G_{11}\alpha$) can best be achieved on 4–8 *M* urea gradient gels, our more recent experience suggests that there is little advantage over the 6 *M* urea gels, which are considerably easier to set up.
27. If the room temperature is on the cold side, there is the possibility that the urea will come out of solution as the gel is polymerizing. To avoid this allow the gel to set in warmer rooms, for example, tissue culture room, 30°C hot room, and so forth.

28. Polypeptides do not run normally on SDS-urea gels. They may appear to run at different molecular masses than in the absence of urea, making it more difficult to identify particular α-subunits. It is suggested that purified or recombinant α-subunits be first resolved on these gels and used as reference standards.
29. Be careful after blotting; the tank buffer is very hot, sometimes approaching 80–90°C.
30. Appropriate antibody dilutions can be determined by immunoblotting different amounts of plasma membrane preparations and selecting an amount that gives a strong immunoreactive response.
31. First and second antibody solutions can be stored at 4°C indefinitely if a small spatula amount of the antimicrobial agent thimerosol is added prior to each use. These solutions can be used three or four times.
32. If solubilizing membranes, take the appropriate amount, centrifuge on a benchtop microcentrifuge for 5 min at top speed, and remove the supernatant. Resuspend the membrane pellet in 100 µL of water and add 100 µL of 2% (w/v) SDS. Proceed with rest of the protocol.
33. Pansorbin is considerably less expensive than protein A–Sepharose and should be considered as first choice. However, better results may be obtained with protein A–Sepharose.
34. Calbiochem Novabiochem (UK) Ltd (Freepost, Nottingham, England, NG7 1BR) supply partially purified recombinant G-protein α-subunits (including G_{i1}, G_{i2}, G_{i3}, G_o, and G_s) for use as standards on immunoblots.

References

1. Goldsmith, P., Gierschik, P., Milligan, G., et al. (1987) Antibodies directed against synthetic peptides distinguish between GTP-binding proteins in neutophil and brain. *J. Biol. Chem.* **262,** 14,683–14,688.
2. Milligan, G. (1993) Regional distribution and quantitative measurement of the phosphoinositidase C-linked guanine nucleotide binding proteins $G_{11}\alpha$ and $G_q\alpha$ in rat brain. *J. Neurochem.* **61,** 845–851.
3. Wise, A., Lee, T. W., MacEwan, D. J., and Milligan, G. (1995) Degradation of $G_{11}\alpha/G_q\alpha$ is accelerated by agonist occupancy of $\alpha_{1A/D}$, α_{1B} and α_{1C} adrenergic receptors. *J. Biol. Chem.* **270,** 17,196–17,203.
4. Wise, A. and Milligan, G. (1994) High level expression of mammalian G-protein α subunit G_q subtypes in *Escherichia coli*. *Biochem. Soc. Trans.* **22,** 12S.
5. Koski, G. and Klee, W. A. (1981) Opiates inhibit adenylate cyclase by stimulating GTP hydrolysis. *Proc. Natl. Acad. Sci. USA* **78,** 4185–4189.
6. Hudson, T. H. and Johnson, G. L. (1980) Peptide mapping of adenylyl cyclase regulatory proteins that are cholera toxin substrates. *J. Biol. Chem.* **255,** 7480–7486.
7. Laemmli, U. K. (1970) Cleavage of structural proteins during the assembly of the head of bacteriophage T4. *Nature* **227,** 680–695.
8. Towbin, H., Staehelin, T., and Gordon, J. (1979) Electrophoretic transfer of proteins from polyacrylamide gels to nitrocellulose sheets: procedure and some applications. *Proc. Natl. Acad. Sci. USA* **76,** 4350–4354.

14

Analysis of Function of Receptor–G-Protein and Receptor–RGS Fusion Proteins

Richard J. Ward and Graeme Milligan

Summary

Fusion constructs between G-protein-coupled receptors and G-protein α-subunits have been used to examine a variety of aspects of the functioning of these signaling systems. Here we describe some of the various techniques and methods used in detail. The process of fusing the two components, the receptor and the G-protein α-subunit, result in a construct that acts as an agonist stimulated GTPase enzyme whose expression level can be accurately determined by radioligand binding. The effects of different ligands, mutations, and other signaling system components can thus be analyzed by a series of relatively simple assays. Recently we have begun to supplement this approach with the use of fusions between G-protein-coupled receptors and regulator of G-protein signaling proteins to examine the function of this important group of signaling molecules.

Key Words

GTPase, radiological binding, GTPγs.

1. Introduction

The family of G-protein-coupled receptors (GPCRs) is an extensive one. Its members act to transmit signals from various extracellular elements to the induction of cellular responses via heterotrimeric G-proteins. These G-proteins are made up of α-, β-, and γ-subunits, of which the α-subunit has a guanine nucleotide binding site. Upon activation of the GPCR, this site exchanges GDP for GTP and following the dissociation of the β/γ-complex from the α-subunit, both elements can interact with and regulate the activity of cellular effectors. The signal is terminated by the intrinsic GTPase activity of the α-subunit, which hydrolyzes the GTP to GDP and this allows reassociation of the heterotrimeric complex. The rate of the GTPase reaction and hence the rate of the

From: *Methods in Molecular Biology, vol. 259, Receptor Signal Transduction Protocols, 2nd ed.*
Edited by: G. B. Willars and R. A. J. Challiss © Humana Press Inc., Totowa, NJ

removal of the signal can be increased by GTPase activating proteins known as regulator of G-protein signaling (RGS) proteins. These are a family of proteins defined by a common, highly conserved RGS domain.

In recent years extensive use has been made of GPCR–G-protein α-subunit fusion constructs to examine the details and specificity of interactions between these proteins. Such constructs have several important characteristics that make them particularly valuable for quantitative analyses:

1. The GPCR and G-protein are present at 1:1 stoichiometry and in close proximity to one another. Thus, the interactions between them are well defined.
2. Such constructs function as agonist-dependent GTPases and basic enzyme kinetics can define the details of GPCR–G-protein interactions and their regulation by other proteins.
3. Because ligand binding studies can be used to quantitate accurately the expression level of the GPCR this also measures the amount of the G-protein fused to it.

Such fusion constructs can be expressed in mammalian cells in tissue culture. Membrane preparations generated from these cells provide appropriate material for analysis of function of the fusions.

Because of the ability of bacterially expressed RGS proteins to regulate the GTPase activity of a number of GPCR–G-protein fusions we have recently also begun to generate GPCR–RGS protein fusion constructs. These are useful in the analysis of the effects of posttranslational modifications, for example, phosphorylation and palmitoylation, on the activity of RGS proteins.

2. Materials

2.1. Equipment

1. Tissue culture facilities.
2. Water bath.
3. Refrigerated centrifuges.
4. Benchtop ultracentrifuge, rotor, and thick-walled polycarbonate tubes.
5. Plate reader.
6. Benchtop refrigerated centrifuge with rotor for 96-well plates.
7. Packard Topcounter NXT.
8. Packard filtermate harvester.
9. Rotating wheel.
10. Scintillation counter.
11. Heating block.
12. Polyacrylamide gel electrophoresis apparatus.

2.2. Reagents

1. pcDNA3 (Invitrogen Ltd., Paisley, UK).
2. Lipofectamine (Invitrogen Ltd., Paisley, UK).

3. HEK-293(T) cells (European Collection of Animal Cell Cultures, Porton Down, Salisbury, Wiltshire, UK).
4. HEK-293(T) media, Dulbecco's modified Eagle's medium (DMEM) supplemented with 10% newborn calf serum and 2 m*M* glutamine (Sigma-Aldrich Ltd., Gillingham, UK).
5. OptiMem, serum free media (Invitrogen Ltd., Paisley, UK).
6. 1X phosphate-buffered saline (PBS) buffer: 120 m*M* NaCl, 25 m*M* KCl, 10 m*M* Na_2HPO_4, 3 m*M* KH_2PO_4, pH 7.4. Dissolve 7.2 g of NaCl, 0.2 g of KCl, 1.48 g of Na_2HPO_4, 0.43 g of KH_2PO_4 in 1000 mL of water; adjust the pH to 7.4.
7. Pertussis toxin (PTX), supplied as 0.2 mg/mL and diluted to 50 ng/mL final (Sigma-Aldrich Ltd., Gillingham, UK).
8. Deepwell 96-well blocks, Eppendorf 1.2-mL 96-well plates (Helena Biosciences, Sunderland, UK) (*see* **Note 1**).
9. Sealing film, Linbro adhesive plate sealer (ICN Biomedicals Inc., Aurora, OH, USA).
10. 1X TE buffer: 10 m*M* tris[hydroxymethyl]aminomethane-HCl (Tris-HCl), 0.1 m*M* diaminoethanetetraacetic acid (EDTA), pH 7.5. Dissolve 1.21 g of Tris and 37.2 mg of EDTA in 1000 mL of water; adjust pH to 7.5.
11. Protein standards, bovine serum albumin (BSA) made up in water to 0–2.0 μg/μL in 0.2 μg/μL steps. Store as 100-μL aliquots at –20°C.
12. BCA reagent: 26 m*M* 2,2′-biquinoline-4,4′-dicarboxylic acid disodium salt (Fluka), 200 m*M* Na_2CO_3, 5 m*M* $C_4H_4KNaO_6$ sodium potassium tartrate, 100 m*M* NaOH, 100 m*M* $NaHCO_3$, pH 11.25. Dissolve 10 g of 2,2′-biquinoline-4,4′-dicarboxylic acid disodium salt, 20 g of Na_2CO_3, 1.6 g of $C_4H_4KNaO_6$, 4 g of NaOH, 9.5 g of $NaHCO_3$ 1000 mL of water; adjust the pH to 11.25 with 2 *M* NaOH. Immediately before use add 1 part in 50 of 160 m*M* $CuSO_4$ (40 g of $CuSO_4$ in 1000 mL of water).
13. 50–100 Ci/mmol of [*ethyl*-^{3}H]RS-79948-197 (cat. no. TRK1039). 75–95 Ci/mmol of [*o-methyl*-^{3}H]yohimbine (cat. no. TRK684, Amersham BioSciences UK Ltd.).
14. Idazoxan hydrochloride (cat. no. 0793, Tocris Cookson Ltd., Avonmouth, UK). For 1 m*M* stock, dissolve 2.41 mg of idazoxan hydrochloride in 10 mL of water. Store at –20°C as 1-mL aliquots.
15. 1X TEM buffer: 75 m*M* Tris-HCl, 5 m*M* EDTA, 12.5 m*M* $MgCl_2$, pH 7.4. Dissolve 9.1 g of Tris, 1.86 g of EDTA, 2.54 g of $MgCl_2$ in 1000 mL of water; adjust the pH to 7.4.
16. Unifilter 96 GF/C (Packard Bioscience).
17. Optiplates 96 white (Packard Bioscience).
18. 1X TE wash buffer: 75 m*M* Tris-HCl, 1 m*M* EDTA, pH 7.5. Dissolve 9.1 g of Tris, 0.37 g of EDTA in 1000 mL of water; adjust the pH to 7.4.
19. Microscint 20 (Packard Bioscience).
20. 1 m*M* Adrenaline, ([–]-epinephrine bitartrate, cat. no. E-104, Research Biochemicals International, MA, USA). Dissolve 0.67 g of (–)-epinephrine bitartrate in 20 mL of water to make 0.1 *M*, then dilute 100X in water to make 1 m*M*. Store at –20°C as 1-mL aliquots.

21. 2 m*M* Ascorbic acid: dissolve 35.2 mg in 100 mL of water. Store at –20°C as 5-mL aliquots.
22. 2.5 U/μL Creatine kinase (Roche Diagnostics, Lewes, UK, cat. no. 736-988. One milligram is equal to 800 U at 37°C). Dissolve 31.25 mg of creatine kinase in 10 mL of water. Store at –20°C as 200-μL aliquots.
23. 0.4 *M* Creatine phosphate (Roche Diagnostics, Lewes, UK, cat. no. 621 722). Dissolve 2.6 g of creatine phosphate in 20 mL of water. Store at –20°C as 250-μL aliquots.
24. 40 m*M* ATP, pH 7.5 (adenosine 5′-triphosphate disodium salt, cat. no. A3377, Sigma-Aldrich Ltd., Gillingham, UK). Dissolve 0.48 g of adenosine 5′-triphosphate disodium salt in 20 mL of water; adjust the pH to 7.5. Store at –20°C as 250-μL aliquots.
25. 40 m*M* App(NH)p (adenylyl-imidodiphosphate tetralithium salt, Roche Diagnostics, Lewes, UK, cat. no. 102 547). Dissolve 25 mg of adenylyl-imidodiphosphate tetralithium salt in 1.19 mL of water. Store at –20°C as 50-μL aliquots.
26. 10 m*M* Ouabain (Sigma-Aldrich Ltd, Gillingham, UK, cat. no. O3125). Dissolve 0.585 g ouabain in 100 mL of water. Store at –20°C as 5-mL aliquots (*see* **Note 4**).
27. 4 *M* NaCl: dissolve 23.4 g of NaCl in 100 mL of water.
28. 1 *M* $MgCl_2$: dissolve 20.3 g of $MgCl_2$ in 100 mL of water.
29. 0.1 *M* DTT: dissolve 0.31 g of dithiothreitol in 20 mL of water. Store at –20°C as 100-μL aliquots.
30. 20 m*M* EDTA, pH 7.5: dissolve 0.75 g of EDTA in 100 mL of water; adjust the pH to 7.5.
31. 2 *M* Tris-HCl, pH7.5: dissolve 24.2 g of Tris in 100 mL of water; adjust the pH to 7.5.
32. 1 m*M*/100 n*M* GTP, pH 7.5 (guanosine 5′-triphosphate lithium salt, cat. no. G5884 Sigma-Aldrich Ltd., Gillingham, UK). Dissolve 0.11 g of guanosine 5′-triphosphate lithium salt in 20 mL of water to make 0.01 *M*. Dilute 10X or 100X with water to make 1 m*M* or 100 n*M* respectively. Store at –20°C as 200-μL aliquots.
33. Guanosine 5′-triphosphate [γ^{32}P], [γ^{32}P]GTP): 30 Ci/mmol, 2 mCi/mL, cat. no. NEG004, Perkin Elmer Life Sciences.
34. Acidified charcoal: dissolve 50 g of activated charcoal, 1.2 mL concentrated P_3HO_4 in 1000 mL of water. Store at 4°C.
35. LB media: dissolve 10 g of tryptone, 5 g of yeast extract, 10 g of NaCl in 1 L of water. Adjust pH to 7.5 and sterilize by autoclaving. When cool, add ampicillin from a 100 mg/mL stock (0.22 μm sterile filtered) to a final concentration of 100 μg/μL.
36. 1 *M* Isopropyl β-D-1-thiogalactopyranoside (IPTG): dissolve 4.77 g in 25 mL of water. Sterilize by 0.22-μm filtration and store at –20°C as 1-mL aliquots.
37. Glutathione Sepharose 4B (cat. no. 17-0756-01, Amersham BioSciences UK Ltd.).
38. Glutathione elution buffer: 0.154 g of glutathione in 50 mL of 50 m*M* Tris-HCl, pH 8.0.
39. 24% Trichloroacetic acid (TCA): 24 g of trichloroacetic acid made up to 100 mL with water.

40. 1 m*M* GTPγS (1 m*M* guanosine 5′-[γ-thio]triphosphate tetralithium salt, cat. no. G8634 Sigma-Aldrich Ltd., Gillingham, UK). Dissolve 5 mg in 8.9 mL of water. Store as aliquots at –20°C.
41. 10X GTPγS binding assay stock buffer: 200 m*M* *N*-2-hydroxyethylpiperazine-*N*′-2-ethanesulfonic acid (HEPES), 30 m*M* $MgCl_2$, 1 *M* of NaCl, 2 m*M* of L-ascorbic acid. Dissolve 47.7 g of HEPES, 6.1 g $MgCl_2 \cdot 6H_2O$, 58.44 g of NaCl, 0.35 g of L-ascorbic acid in 1000 mL of water; adjust the pH to 7.4.
42. Guanosine 5′-(γ-thio) triphosphate [^{35}S], ([^{35}S]GTPγS), 1250 Ci/m*M* (cat. no. NEG030H, Perkin Elmer Life Sciences).
43. 40 μ*M* GDP (guanosine 5′-diphosphate sodium salt, cat. no. G7127 Sigma-Aldrich Ltd, Gillingham, UK). Dissolve 35.5 mg in 20 mL of water to give 4 m*M*, then dilute 1:100 in water to give 40 μ*M*. Store at –20°C as 1-mL aliquots.
44. Stop solution (1X GTPγS binding assay buffer with protease inhibitors). Dilute 10X GTPγS binding assay buffer 1:10 with water and add one protease inhibitor tablet/50 mL of solution (complete protease inhibitor cocktail, Roche Diagnostics, Lewes, UK, cat. no. 1697 498).
45. Solubilization buffer: 100 m*M* Tris-HCl, 200 m*M* NaCl, 1 m*M* EDTA, 1.25% Nonidet P-40, protease inhibitors, pH 7.4. Dissolve 12.1 g of Tris, 11.7 g of NaCl, 0.4 g of EDTA in 987.5 mL of water; adjust the pH to 7.4. Add 12.5 mL of Nonidet P-40 and one protease inhibitor tablet/50 mL solution (*see* **item 44**). Store at 4°C.
46. Normal rabbit serum, obtained from preimmune bleeds of New Zealand white rabbits. Allow whole blood to clot and centrifuge to remove red blood cells before storing at –80°C as 100-μL aliquots.
47. Protein G–Sepharose Fast Flow (cat. no. P3296 Sigma-Aldrich Ltd., Gillingham, UK).
48. Bead buffer: 2% bovine serum albumin, 1% NaN_2, protease inhibitors. Dissolve 2 g of bovine serum albumin, 0.1 g of NaN_2 in 100 mL of water. Add one protease inhibitor tablet/50 mL of solution. Store at 4°C.
49. Antisera should be obtained from the appropriate commercial sources or raised by the methods given in ***(1)***.
50. Scintillation fluid, Ultima Gold XR (Packhard Bioscience).
51. 2X RIPA: 100 m*M* HEPES, 300 m*M* NaCl, 2% Triton X-100, 1% sodium deoxycholate, 0.2% sodium dodecyl sulfate. Dissolve 5.96 g of HEPES, 4.38 g of NaCl, 5 g of Triton X-100, 2.5 g of sodium deoxycholate, 0.5 g of sodium dodecyl sulfate (SDS) in 250 mL of water. Store at 4°C.
52. 1X RIPA: 1X RIPA, 10 m*M* NaF, 5 m*M* EDTA, 10 m*M* sodium phosphate buffer, 5% ethylene glycol, protease inhibitors. 25 mL of 2X RIPA, 1 mL of 0.5 *M* NaF, 0.5 mL of 0.5 *M* EDTA, pH 8.0; 5 mL of 0.1 *M* sodium phosphate buffer, 2.5 mL of ethylene glycol, made up to 50 mL with water. Add one protease inhibitor tablet/50 mL of solution. Store at 4°C.
53. 2X Sample buffer: 0.126 *M* Tris-HCl, 20% glycerol, 0.1 *M* DTT, 0.16 *M* SDS, pH 6.8. Dissolve 0.76 g of Tris base, 10 mL of glycerol, 0.78 g of DTT in 30 mL of water; adjust the pH to 6.8. Add 2.3 g of SDS and a small amount of bro-

mophenol blue, then make volume up to 50 mL with water. Store at room temperature.

54. Poly-D-lysine: Poly-D-lysine (cat. no. P6407 Sigma-Aldrich Ltd., Gillingham, UK). 0.01% stock solution in sterile water.
55. 1 mCi/mL of [9,10(*n*)-^{3}H]palmitic acid (cat. no. TRK909, Amersham BioSciences UK Ltd.).
56. Dialyzed serum: newborn calf serum dialyzed at 4°C against Earle's salts for 2 d; dialysis buffer changed every 12 h.
57. Earle's salts: 120 m*M* NaCl, 1.2 m*M* KCl, 0.8 m*M* $MgSO_4$•$7H_2O$, 1 m*M* NaH_2PO_4, 5 m*M* glucose. Dissolve 6.8 g of NaCl, 0.1 g of KCl, 0.2 g of $MgSO_4$•$7H_2O$, 0.14 g of NaH_2PO_4, 1 g of glucose in 1000 mL of water.
58. DMEM–sodium pyruvate–ascorbic acid: DMEM, 5 m*M* sodium pyruvate, 0.1 m*M* L-ascorbic acid. Dissolve 0.55 g of sodium pyruvate, 18 mg of L-ascorbic acid in 1 L of DMEM.
59. 1% SDS: 1 g of SDS in 100 mL of water.
60. Kahn solubilization buffer: 1% Triton X-100, 10 m*M* EDTA, 100 m*M* NaH_2PO_4, 10 m*M* NaF, 50 m*M* HEPES, pH 7.2. Dissolve 12 g of HEPES, 3.72 g of EDTA, 14.2 g of NaH_2PO_4, 0.5 g of NaF in 990 mL of water; adjust the pH to 7.2 and add 10 mL of Triton X-100.
61. Pansorbin (cat. no. 507858, CN Biosciences [UK] Ltd., Nottingham).
62. Protein A–Sepharose (cat. no. P9424 Sigma-Aldrich Ltd., Gillingham, UK) (*see* **Note 5**).
63. Kahn immunoprecipitation wash buffer: 1% Triton X-100, 10 m*M* EDTA, 100 m*M* NaH_2PO_4, 10 m*M* NaF, 0.5% SDS, 50 m*M* HEPES, pH 7.2. Dissolve 12 g of HEPES, 3.72 g of EDTA, 14.2 g of NaH_2PO_4, 0.5g of NaF in 990 mL of water; adjust the pH to 7.2 and add 10 mL of Triton X-100 and 5 g of SDS.
64. PVDF membrane (Immobilin P, cat. no. IPVH00010, Millipore Corperation, USA).
65. EA-wax (Thistle Scientific, cat. no. TS-EA-10026).

3. Methods

3.1. Fusion Protein Constructs

The general strategy for production of GPCR–G-proteins fusion constructs has been described in detail in *(2)*. The same basic concept was used to generate the GPCR–RGS fusions. These were subcloned into the mammalian expression vector pcDNA3 using standard molecular biological techniques. Many of the fusion constructs have been constructed to incorporate epitope tags such as HA, myc, VSV-G, or FLAG to which commercially available antisera can be obtained.

3.2. Transfection

Transfections are carried out using the Lipofectamine reagent and HEK-293(T) cells. The cells are maintained according to standard cell culture procedures in a laminar flow cabinet and grown at 37°C in a humidified 5% CO_2

atmosphere. DNA for transfection is purified using standard QIAGEN maxi-prep kits, no need has been found to use kits specifically designed to remove endotoxin.

Constructs containing $G_i/G_o\alpha$-subunits, whether fused to a GPCR or cotransfected with a GPCR–RGS fusion construct, are mutated at position Cys^{351} to Ile. The Cys^{351}Ile mutation renders the $G_i/G_o\alpha$ resistant to pertussis toxin-catalyzed ADP-ribosylation and so allows the transfected cells to be treated with pertussis toxin to prevent GPCR activation of endogenously expressed forms of these G-proteins. The replacement of Cys^{351} of $G_{i1}\alpha$ with Ile, rather than any other amino acid, allows the greatest degree of activation of this G-protein by a coexpressed α_{2A}-adrenoceptor ***(3)***.

1. Seed HEK-293(T) cells into 10-cm dishes and grow to approx 70% confluence.
2. Dilute appropriate quantities of DNA stock to 0.1 µg/µL with sterile water.
3. Prepare a 15-mL disposable centrifuge tube for each plate of cells containing 5–10 µg of DNA, either as a single construct, or a combination of constructs for cotransfection. (Where an experiment requires both single and cotransfection the amount of DNA should be kept constant by adding vector DNA where necessary). Add serum-free media (OptiMem) to adjust the volume to 600 µL.
4. Add 600 µL of Lipofectamine/serum-free media (20 µL of Lipofectamine in 580 µL of OptiMem) dropwise to each tube.
5. Incubate at room temperature for 30 min.
6. During the incubation wash the 10-cm dishes of cells by removing the media and replace with 10 mL of serum-free media. Return the plates to the incubator until needed.
7. Add 4800 µL of serum-free media to each tube.
8. Replace the serum-free media on the plates with the DNA/OptiMem/Lipofectamine mix (6000 µL/plate).
9. Incubate the plates for 5 h at 37°C.
10. Add 10 mL of HEK-293(T) media to each plate and incubate overnight at 37°C.
11. Replace the media with 9 mL each of HEK-293(T) media (morning).
12. Add 1 mL to each of PTX media (late afternoon). Incubate at 37°C overnight. (PTX media added 10X concentrated to give 1X final concentration).
13. Harvest the cells by scraping and transfer the suspension to disposable centrifuge tubes.
14. Centrifuge at 3220*g* for 10 min at 4°C and completely remove the supernatant.
15. Wash the cells (twice) by resuspending in 10 mL of 1X PBS and centrifuge as described in **step 14**.
16. Store the cell pellets at –80°C.

3.3. Membrane Preparations

1. Add 1 mL of 1X TE to the frozen cell pellet obtained from the procedure described in **Subheading 3.2.** Allow to thaw on ice.

2. Resuspend the pellet using a 1-mL syringe and 25-gage needle.
3. Transfer the resuspended cells to a (cold) Teflon–glass homogenizer and homogenize for 1 min, on ice.
4. Return the suspension to the original tube and pass through the needle 10 times, ensuring that the suspension is kept ice-cold at all times.
5. Centrifuge at 200*g* for 5 min at 4°C.
6. Transfer the supernatant to 1-mL thick-walled polycarbonate centrifuge tubes, balance, and centrifuge at 90,000*g* for 30 min at 4°C.
7. Remove the supernatant and resuspend the pellet in an appropriate volume 1X TE.
8. Determine the protein concentration by BCA assay with reference to BSA standards. Add duplicate 10-µL aliquots of standards and samples (including sample dilutions) to a 96-well enzyme-linked immunosorbent assay (ELISA) plate and then dispense 200 µL of BCA reagent into each well. Seal the plate and incubate at 37°C for 20 min. Measure the absorbances at 492 nm and calculate the sample concentrations with reference to the standard curve. Dilute the membranes to 1 µg/µL.
9. Store the membranes as appropriate volume aliquots at –80°C. Membranes must not be subjected to more than one freeze–thaw cycle.

3.4. Ligand-Binding Studies

The level to which the fusion constructs are expressed can be determined by the use of ligand-binding experiments. In the case of α_{2A}-adrenoceptor based fusions we use the antagonist [*ethyl*-^{3}H]RS-79948-197 or the antagonist/inverse agonist [*o-methyl*-^{3}H]yohimbine. Nonspecific binding is determined by the addition of a high concentration of a competitive antagonist such as idazoxan.

The protocol is designed to be carried out in deepwell 96-well blocks using a Packard filtermate harvester and Topcounter, but can be easily modified to use other membrane harvesters and more conventional liquid scintillation counting. The assay can then be carried out in disposable glass or plastic tubes.

3.4.1. Binding of a Single Concentration of Ligand

If a single concentration of ligand in the region of 10 times greater than the previously determined K_d is used, an approximation of the levels of expression can be obtained. At 10X the K_d an antagonist ligand would be expected to occupy 91% of the total GPCR binding sites.

1. Dilute the stock [*ethyl*-^{3}H]RS-79948-197 to give a working concentration of 12 n*M* in 1X TEM buffer. Prepare a dilution of the membranes such that 1–2 µg of protein is added to each well.
2. Prepare 96-well deepwell blocks, on ice, by adding the following in order (*see* **Note 4**) to quadruplicates of wells: 20 µL of 12 n*M* [*ethyl*-^{3}H]RS-79948-197; 130 µL of 1X TEM; and 50 µL of membranes.

A second quadruplicate of wells should be prepared, to determine nonspecific binding, by adding the following: 20 μL of 12 n*M* [*ethyl*-^{3}H]RS-79948-197; 20 μL of 1 m*M* idazoxan; 110 μL of 1X TEM; and 50 μL of membranes.

3. Seal the blocks and vortex mix. Centrifuge briefly at 200*g* to collect the liquid at the bottom of the wells.
4. Incubate by floating the blocks in a water bath at 30°C for 30 min.
5. Harvest the membranes using a 96-well filtermate harvester and Unifilter GF/C filters. The filters should be washed three times in cold 1X TE wash buffer after harvesting.
6. Allow the filter plates to dry and seal the lower surface. Add 25 μL of Microscint to each filter and seal the upper surface of the plate. Count the plates on a topcounter.
7. Standards should be prepared by adding six aliquots of the ^{3}H-ligand to a 96-well white Optiplate. Add 200 μL of Microscint to the appropriate wells and seal the plate. Count as above.
8. Calculate the averages of the quadruplicates and subtract the nonspecific binding counts (those in the presence of idazoxan). Use the specific activity of the radioligand and the amount of protein added to the well to calculate a result in pmoles of radioligand bound per milligram of protein. The concentration of radioligand used can be confirmed by using the standard counts and the specific activity.

3.4.2. Saturation Analysis

The assay is performed similarly to that in **Subheading 3.4.1.** but with the use of a series of differing concentrations of the radioligand.

1. Set up a series of dilutions of [*ethyl*-^{3}H]RS-79948-197 from 40 n*M* to 1 n*M* in 1X TEM buffer.
2. Use these dilutions to set up a series of 12 binding experiments across the plate. Thus wells A1–H1 would contain the first dilution, A2–H2 the second, and so on. The other components of the binding assay should be added in the order described in **Subheading 3.4.1.**, with idazoxan going into four of the eight wells for each [*ethyl*-^{3}H]RS-79948-197 dilution.
3. Set up standards of each dilution of [*ethyl*-^{3}H]RS-79948-197 by adding four aliquots of each to a 96-well white Optiplate and then 200 μL of Microscint. The plate should then be sealed and counted.
4. Centrifuge the blocks briefly at 200*g* to ensure that all the liquid is at the bottom of the wells and then incubate at 30°C for 30 min. The membranes should then be harvested, the filter plates sealed, and scintillant added as described in **Subheading 3.4.1.**
5. Calculate the averages of the standards and use the specific activity to determine the amount of [*ethyl*-^{3}H]RS-79948-197 added in fmoles and as a nanomolar concentration for each concentration added.
6. Calculate the averages of the binding quadruplicates and subtract the nonspecific binding counts from each. Convert the corrected binding counts to fmol [*ethyl*-

^{3}H]RS-79948-197 bound and use the amount of protein added to calculate the binding in fmoles bound per milligram of protein. This can then be plotted against concentration of [*ethyl*-^{3}H]RS-79948-197 to obtain a saturation binding plot, from which the maximum concentration of [*ethyl*-^{3}H]RS-79948-197 bound (B_{max}) and the K_d for the ligand can be determined by nonlinear regression using range of commercially available packages, for example, Graphpad Prism **(Fig. 1)**. The B_{max} can be used in conjunction with the data from high-affinity GTPase assays (*see* **Subheading 3.6.**) to calculate agonist-induced turnover number of GTP (K_{cat}) for the fusion construct. It is also possible to display the data as a Scatchard plot (*see* **Note 5**) by plotting the radioligand bound (fmol/mg) against bound (fmol/mg) divided by free radioligand (n*M*), as shown in **Fig. 1B**.

3.5. High-Affinity GTPase Assay

This assay is designed to examine the effect of a receptor–ligand interaction on the GTPase activity of the Gα part of a receptor–G-protein fusion construct. As guanine nucleotide exchange on the G-protein (*see* **Subheading 3.7.**) is the rate-limiting step in the cycle of G-protein activation and deactivation, then the subsequent GTPase rate is also a useful monitor of G-protein activation.

The assay can be carried out as a saturation GTPase assay (*see* **Subheading 3.6.**), by setting up a series of experiments with increasing concentrations of unlabeled GTP. The maximum GTPase velocity of each can then be plotted against GTP concentration and nonlinear regression used to obtain values for maximum velocity (V_{max}) and K_m. When agonist is added an increase in V_{max} is observed but generally the K_m value remains the same. This is not consistent with the activity having been regulated by a GTPase activating protein *(4)*. This can be contrasted to the situation when RGS protein is also added to the assay, when a greater increase in V_{max} is seen together with an increase in K_m *(5)*. These characteristics are also seen for GPCR–RGS protein fusion constructs.

In this case the system is slightly different in that the GPCR is linked to an RGS-protein and this construct is cotransfected into the cell with a separate Gα construct. Thus the experiment examines both GPCR-mediated activation of a G-protein and the subsequent effect of the RGS on the activated G-protein.

The assay protocol is designed so that it can be carried out in deepwell 96-well blocks and then counted using a topcounter. It can be modified, however, by performing the experiment in individual 1.5-mL tubes. The supernatant from the last stage can be counted using a conventional liquid scintillation counter.

3.5.1. Reaction Mix

A 2X concentrated reaction mix prepared by mixing the following solutions on ice: 500 µL of 2 m*M* ascorbic acid, 250 µL of 0.4 *M* creatine phosphate,

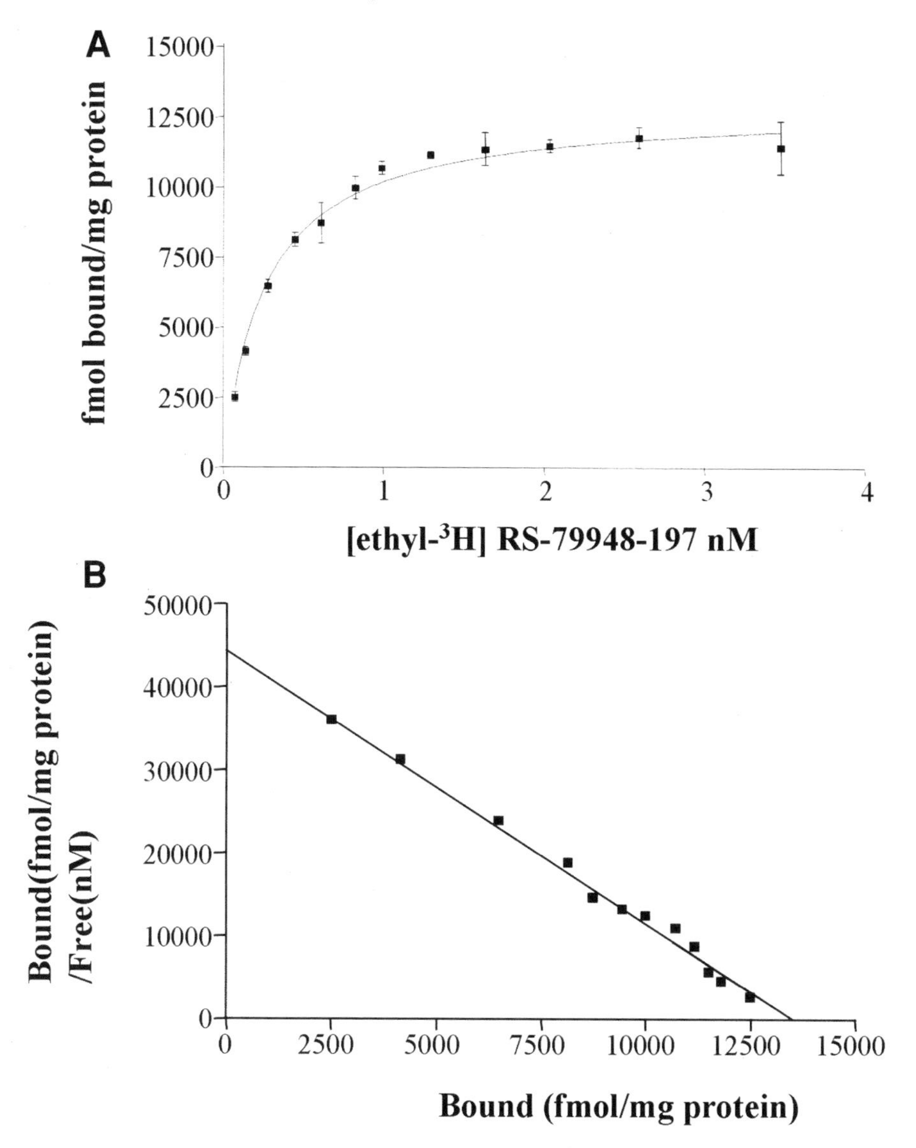

Fig. 1. Saturation binding of the radioligand [*ethyl*-^{3}H]RS-79948-197 to membranes containing the $\alpha_{2A}G_{i1}\alpha C^{351}I$ fusion construct. **(A)** Saturation plot of n*M* [*ethyl*-^{3}H]RS-79948-197 against fmol/mg of protein. The B_{max} (12,877 fmol/mg protein) and K_d (0.263 n*M*) are calculated by nonlinear regression analysis. **(B)** Scatchard plot of radioligand bound against radioligand bound divided by free radioligand. B_{max} is the *x*-intercept (13,470 fmol/mg) and the K_d is the reciprocal of the slope (0.303 n*M*).

200 μL of 2.5 U/μL of creatine kinase, 250 μL of 40 m*M* ATP, pH 7.5, 25 μL of 40 m*M* App(NH)p, 1000 μL of 10 m*M* ouabain, 250 μL of 4 *M* NaCl, 50 μL of 1 *M* $MgCl_2$, 200 μL of 1 *M* DTT, 50 μL of 20 m*M* EDTA, pH 7.5, 200 μL of 2 *M* Tris-HCl, pH 7.5, 50 μL of 0.1 m*M* GTP, --μL of [γ^{32}P] GTP (to add 50,000 cpm/well), ----μL of Water (up to 5000 μL final volume for 100 wells).

3.5.2. Standard Assay

High-affinity GTPase assays are traditionally performed with a single concentration of GTP that is usually in the region of 0.5 μ*M*. Using the mix described in **Subheading 3.5.1.**, GTP concentration in the assay will be 0.5 μ*M* with the radioactive GTP acting as a tracer. A series of "nonspecific" high GTP assays are also performed in the presence of 100 μ*M* GTP. Counts present in these assays are assumed to be unrelated to the actions of heterotrimeric G-proteins as these display a relatively high K_m (*see* **Subheading 3.6.**) for the nucleotide.

1. Dilute the membranes with 1X TE buffer such that 20 μL contains 5–10 μg of protein.
2. Prepare 96-well deepwell blocks, on ice, by adding the following to triplicate sets of wells:

Basal	High GTP	Ligand
20 μL of membranes	20 μL of membranes	20 μL of membranes
30 μL of water	20 μL of water	20 μL of water
	10 μL of 10^{-3} *M* GTP	10 μL of 10^{-3} *M* adrenaline
50 μL	50 μL	50 μL

3. Add 50 μL of 2X reaction mix to each well, seal the blocks, vortex-mix, and centrifuge briefly at 200*g* at 4°C to ensure that all the liquid is at the bottom of the wells.
4. Prepare standards by adding 50-μL aliquots of the reaction mix to white 96-well optiplates.
5. Incubate the blocks at 37°C for 30 min.
6. Stop the reaction by the addition of 900 μL of acidified charcoal to each well.
7. Reseal the blocks and centrifuge at 2500*g* for 15 min at 4°C.
8. Transfer 300 μL from each well into the equivalent well in an Optiplate 96-well plate.
9. The sample and standard plates should be Cherenkov counted using a topcounter.

3.5.3. Analysis of Results

Calculate the averages of the triplicates and subtract the high GTP counts (to correct for hydrolysis of [γ^{32}P]GTP not produced via heterotrimeric

G-proteins) from the basal and ligand added counts. Use the standard counts, the specific activity of the GTP, the length of incubation and the amount of protein added to calculate the results in pmoles of GTP hydrolyzed per milligram of protein per minute.

3.6. Saturation GTPase Assays

This is a modification of the standard high-affinity GTPase assay to allow the determination of V_{max} and K_m values. A series of experiments are set up with increasing concentrations of unlabeled GTP added (instead of that added to the 2X reaction mix), to give GTPase activity values at a series of specific activities of [γ^{32}P]GTP. These can then be used to calculate agonist-stimulated GTPase V_{max} and the K_m for GTP.

3.6.1. Reaction Mix

Prepared as in **Subheading 3.5.1.** but without the addition of 1×10^{-4} *M* GTP.

3.6.2. Assay

1. Prepare a series of 12–14 dilutions of GTP from 20 m*M* to 200 n*M*. These will provide an isotopic dilution curve for the [γ^{32}P]GTP.
2. Use these dilutions to set up a series of GTPase assays as described in **Subheading 3.5.2.** Thus for each GTP dilution a set of 12 wells should be prepared (4 basal, 4 high GTP, and 4 with the appropriate ligand) each with the addition of 10 µL of the GTP dilution replacing 10 µL of the water added in the method described in **Subheading 3.5.2.**
3. The assay should then be carried out as described in **Subheading 3.5.2.**
4. Two sets of standards must be set up, the first as described in **Subheading 3.5.2.** that are Cherenkov counted (cpm standards) and the second with the addition of scintillant (200 µL of Microscint) to be counted on the ^{32}P channel (dpm standards).

3.6.3. Analysis of Results

1. Calculate the concentration of GTP added as [γ^{32}P]GTP from the dpm standards. This, together with the counts from the cpm standards, can then be used to determine the specific activity of the label in cpm.
2. Determine the total amount of GTP in each well by adding the concentration of GTP (nonradioactive) to the concentration of GTP added as [γ^{32}P]GTP, calculated in **step 1**.
3. Calculate the averages of the triplicates and subtract the high GTP counts from those of the basal and ligand added triplicates (specific cpm values).
4. Divide the specific cpm values by the concentration of GTP (in n*M*) added as [γ^{32}P]GTP **(step 1)** and then multiply by the concentration of unlabeled GTP (in n*M*) from the dilution added to each triplicate. This gives corrected cpm values based on all the GTP present being labeled.

5. The pmoles of GTP hydrolyzed can then be determined by dividing the corrected cpm values by the specific activity of the label in cpm. These values can be expressed as pmoles hydrolyzed per milligram of protein per minute (velocity of the reaction) by taking account of the amount of protein present, the time of incubation, and the proportion of the reaction counted after pelleting the charcoal by centrifugation.
6. Plot the concentration of GTP (in n*M*) against velocity of the reaction (pmol/mg protein/min). The maximum velocity of the reaction and the K_m can then be determined by nonlinear regression analysis (**Fig. 2**).
7. For illustrative purposes it is often useful to present the data based on a linear transformation. We prefer the Eadie–Hofstee transformation. Divide the velocity of the reaction (pmol/mg/min) by the total amount of GTP added (*V/S*). Plot *V/S* against velocity *(V)* and determine maximum velocity from the *y*-intercept and the K_m for GTP from the slope.
8. Once the basal and agonist added maximum velocities have been determined, the turnover number (K_{cat}) for GTP can be calculated. This is the change in maximum velocity with the addition of agonist divided by the expression level.

$$\frac{V_{max} + \text{agonist (pmol/mg/min)} - V_{max} \text{ basal (pmol/mg/min)}}{\text{Expression (pmol/mg)}} = \text{turnover number (min}^{-1}\text{)}$$

9. It is important to use [γ^{32}P]GTP that has a sufficiently high specific activity to allow the addition of the desired number of counts without adding too much extra GTP. Because the K_m for GTP in the absence of an RGS is generally in the region of 200 n*M*, only a small amount of additional GTP can be added before too few data points remain below the K_m value.

3.6.4. Effects of a Recombinant RGS Protein on GTPase Activity

Generally, analysis of the effects of agonist ligands on the GTPase activity of GPCR–G-protein fusion proteins demonstrates an increase in V_{max} with the K_m for GTP being unaltered. This is not consistent with the G-protein element of the fusion being subjected to the GTPase activating protein activity of an RGS (*see* **ref. *4*** for details).

3.6.5. Saturation GTPase Assays in the Presence of Recombinant RGS Protein

The addition of recombinant RGS protein to a saturation GTPase assay requires only minor modifications to the protocol. The RGS must be added to the diluted membranes prior to setting up the assay so that they can be incubated together on ice for 30 min to allow association. The volume of RGS solution added should be subtracted from the volume of water added to maintain the final volume of the reaction.

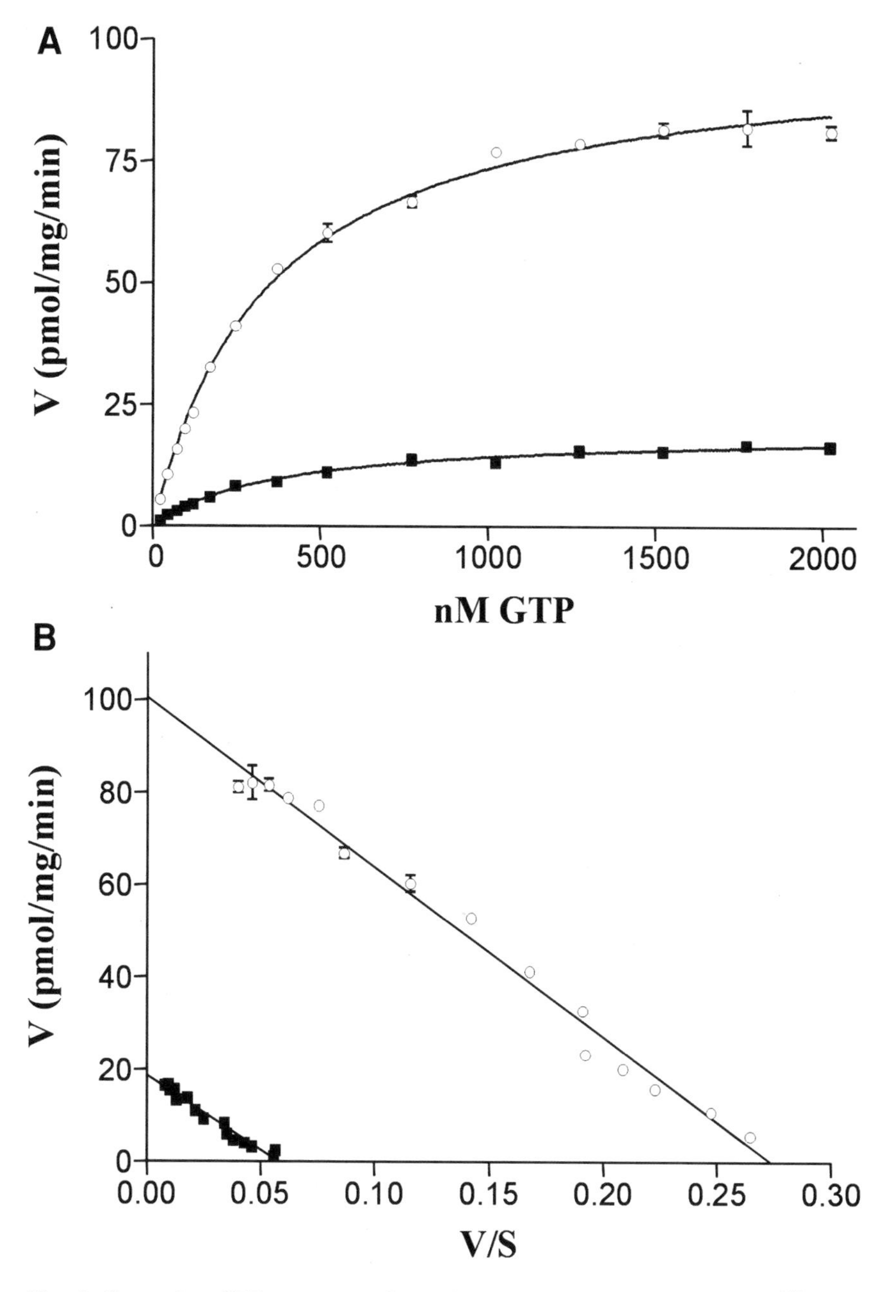

Fig. 2. Saturation GTPase assay of membranes containing the $\alpha_{2A}G_{i1}\alpha C^{351}I$ fusion construct. The addition of adrenaline results in an increase in V_{max}, while the K_m value remains unaltered. **(A)** Saturation plot of assays without *(closed squares)* and with adrenaline *(open circles)*. Values for V_{max} (19.6 pmol/mg/min and 99.0 pmol/mg/min) and K_m (373.5 n*M* and 348.2 n*M*) determined by linear regression. **(B)** Eadie–Hofstee plot of the above. The *y*-intercept gives the V_{max} values (18.6 pmol/mg/min and 100.5 pmol/mg/min) and the slope gives the K_m values (321.2 n*M* and 367.3 n*M*).

Recombinant RGS protein is expressed and purified using the glutathione-*S*-transferase (GST) fusion protein system (Amersham BioSciences). The RGS cDNA is subcloned, in frame, into one of the pGEX series of vectors and transformed into the *E. coli* strain BL21 DE3. The bacteria are grown up and the protein purified as follows, (some steps will require optimization with different constructs).

1. Prepare a 50-mL LB ampicillin overnight culture of the bacteria and use it to inoculate a 400-mL LB ampicillin culture. This is incubated, shaking, at 37°C until it reaches an optical density of 1.0 at 600 nm. (Samples should be taken for SDS-polyacrylamide gel electrophoresis [SDS-PAGE] before and after induction and also at all the subsequent steps.)
2. Induce by adding 400 µL of 1 *M* IPTG (final concentration 1m*M*) and incubate for a further 4 h.
3. Harvest the cells by centrifugation at 6000*g* for 15 min at 4°C. Discard the supernatant. (The bacterial pellets may be stored at this stage at –80°C).
4. Resuspend the pellet in 5 mL/g of Bugbuster reagent and add 1 µL/mL of benzonase (both Novagen). Incubate on a rotating wheel at 4°C for 1 h; brief sonication may be necessary if lysis has not occurred after this time.
5. Centrifuge at 30,000*g* for 20 min at 4°C and transfer the supernatant to a fresh 50-mL tube.
6. Add 1 *M* DTT to a final concentration of 5 m*M*.
7. Add 800 µL of a 50% slurry of glutathione Sepharose 4B beads, previously washed 3 times in 5 mL 1X PBS (*see* **Note 3**) and incubate on the rotary wheel for 30 min at 4°C.
8. Centrifuge at 500*g* for 5 min at 4°C. Remove the supernatant and store at –80°C for further extraction of GST fusion proteins.
9. Wash the beads three times with 5 mL of 1X PBS.
10. Add 1 mL of glutathione elution buffer, mix, and incubate at room temperature for 10 min.
11. Centrifuge at 500*g* for 5 min and transfer the supernatant to a fresh tube.
12. Repeat the elution three times.
13. Run the samples out on SDS-PAGE to determine yield and combine the fractions with the best yields of fusion construct.
14. Determine the protein concentration as in **Subheading 3.3.** To prevent the elution buffer affecting the protein assay the GST–RGS fusion must be precipitated with TCA. Take a 50-µL aliquot of the combined fractions and add 12.5 µL of 24% TCA. Incubate on ice for 25 min and centrifuge at 16,000*g* for 15 min at 4°C. Remove the supernatant and then briefly and carefully wash the pellet (without resuspending) with 200 µL of 1X TE. Resuspend the pellet in 400 µL 1X TE and carry out the protein assay as described in **Subheading 3.6.5.** The final concentration must be corrected for the dilution when redissolving the TCA precipitate.
15. Calculate molarity of the GST–RGS fusion protein to add known molar amounts to the GTPase assay.

3.6.6. GPCR–RGS Fusion Proteins

The GPCR–RGS fusion proteins are cotransfected with a Gα-subunit clone in equal proportions as described in **Subheading 3.2.** Membranes prepared from these cells can then be used in a GTPase assay in the same way as those of a GPCR–G-protein fusion construct.

These constructs usually show an increase in V_{max} with agonist accompanied by an increase in K_m in a similar manner to the GRCP–G-protein/recombinant RGS protein experiments **(Fig. 3)**.

3.7. [35S]GTPγS Binding Assays

GTPγS is a nonhydrolyzable analog of GTP, which is able to bind to an activiated G-protein in exchange for GDP, but then remains bound as it cannot subsequently be hydrolyzed to GDP by the intrinsic GTPase activity of the G-protein. [^{35}S]GTPγS binding can therefore be used as a measure of the level of activation of G-protein α-subunits.

This subheading describes a modification of standard [^{35}S]GTPγS binding assays that increases the signal-to-background ratio by incorporating an immunoprecipitation step *(6)*. This immunoprecipitation may be carried out with an antibody to part of the fusion protein sequence or to either a carboxy or amino terminal epitope tag. The assay is carried out in 1.5-mL centrifuge tubes rather then deepwell 96-well blocks.

3.7.1. The Assay

1. Prepare 2X concentrated assay mix by adding the following to a 7-mL bijou bottle: 1000 μL of 10X stock buffer, ---μL of [^{35}S] GTPγS (to add 50 nCi/tube), 250 μL of 40 μ*M* GDP, ----μL of water (up to 5000 μL final volume for 100 tubes).
2. Dilute the membranes in 1X TE buffer so that the addition of 25 μL per tube gives 5–10 μg of protein per tube.
3. For each membrane preparation to be assayed set up the following tubes on ice:

Tube	Membranes	Assay mix	GTP/agonist	Water
1–4	25 μL	50 μL		25 μL
5–8	25 μL	50 μL	10 μL of 1 m*M* GTPγS	15 μL
9–12	25 μL	50 μL	10 μL of 1 m*M* adrenaline	15 μL
13–16	25 μL	50 μL	10 μL of 1 m*M* GTPγS + 10 μL of 1 m*M* adrenaline	5 μL

4. Incubate at 30°C for 10 min using a water bath and appropriate tube racks.
5. Stop the reaction by adding 500 μL of 1X assay buffer (with protease inhibitors) to each tube.
6. Centrifuge at 20,000*g* for 10 min at 4°C.

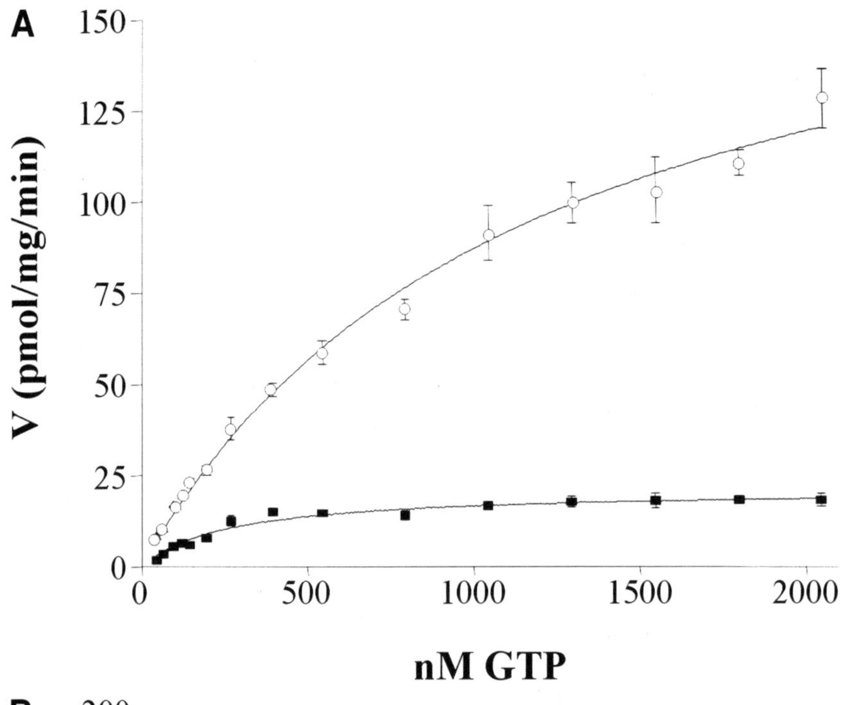
A
150
125
100
75
50
25
0
V (pmol/mg/min)
0
500
1000
1500
2000
nM GTP

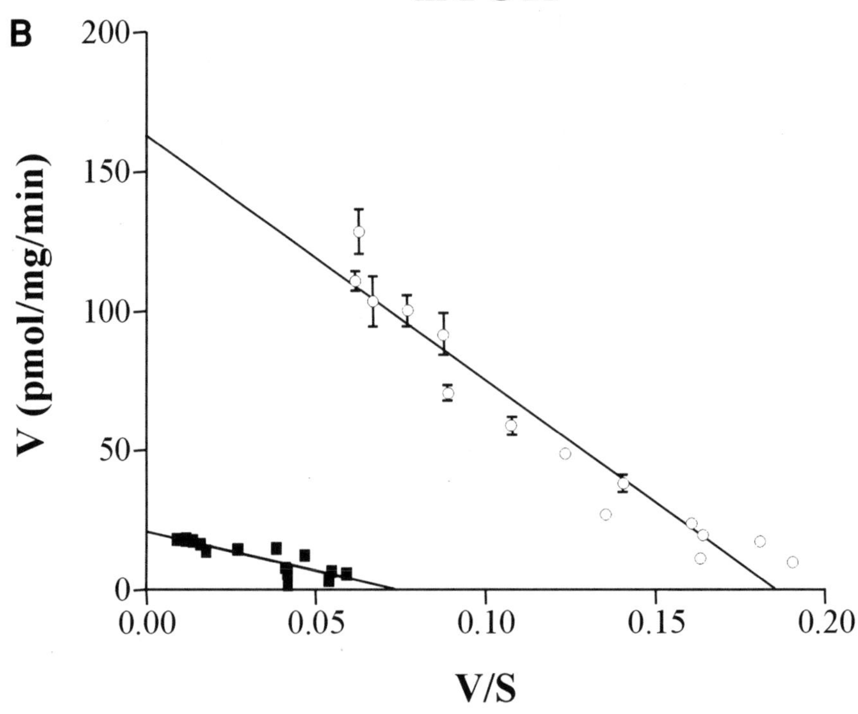
B
200
150
100
50
0
V (pmol/mg/min)
0.00
0.05
0.10
0.15
0.20
V/S

7. Remove and discard the supernatant. (Dispose of as radioactive waste.)
8. Add 50 µL of solubilization buffer (with SDS) to each tube and incubate at 4°C for 1 h on a rotating wheel.
9. Centrifuge at 20,000*g* for 10 min at 4°C.
10. Preclear the solublized membranes by transferring the supernatants to fresh tubes each containing 8 µL of normal rabbit serum (diluted 1:10 with solubilization buffer) and 52 µL of protein G–Sepharose suspension (washed and resuspended in bead buffer).

    For example, for 100 tubes, 60 µL per tube:
    60 µL of normal rabbit serum plus 740 µL of solubilization buffer
    added to
    600 µL of protein G–Sepharose plus 4600 µL of bead buffer.
11. Incubate on the rotating wheel at 4°C for 1 h.
12. Centrifuge at 1000*g* for 2 min at 4°C.
13. Transfer 90 µL of each supernatant to fresh tubes each containing diluted antibody and 40 µL of protein G–Sepharose suspension (washed and resuspended in bead buffer). (The actual quantity of antisera added must be determined experimentally for each antibody used.)

    For example, for 100 tubes, 50 µL per tube:
    25 µL of antisera plus 975 µL of solubilization buffer
    added to
    500 µL of protein G–Sepharose plus 3500 µL of bead buffer.
14. Incubate on the rotating wheel at 4°C overnight.
15. Centrifuge at 1000*g* for 2 min at 4°C.
16. Remove the supernatant and wash the beads with 500 µL of solubilization buffer, vortex mix, and centrifuge at 1000*g* for 2 min at 4°C. Remove the supernatant and repeat the wash.
17. Remove the supernatant and add 1 mL of scintillation fluid, vortex mix to resuspend the beads, and count using a scintillation counter.

Fig. 3. *(see facing page)* Saturation GTPase assay of membranes containing an α_{2A}-adrenoceptor–RGS16 fusion construct cotransfected with $G_{o1}\alpha C^{351}I$. The addition of adrenaline results in an increase in V_{max} and an increase in the K_m value. Although this system describes a fused RGS protein, the situation using recombinant RGS protein and membranes containing a GPCR–G-protein fusion would be similar. **(A)** Saturation plot of assays without *(closed squares)* and with adrenaline *(open circles)*. Values for V_{max} (21.0 pmol/mg/min and 190.8 pmol/mg/min) and K_m (262.5 n*M* and 1189 n*M*) determined by linear regression. **(B)** Eadie–Hofstee plot of the above. The *y*-intercept gives the V_{max} values (20.9 pmol/mg/min and 163.3 pmol/mg/min) and the slope gives the K_m values (283.5 n*M* and 881.4 n*M*).

3.7.2. Analysis of Results

1. Calculate the averages of each set of quadruplicates and subtract the counts for the tubes containing GTPγS (nos. 5–8)> from the basal tubes (nos. 1–4) and the counts from the tubes containing GTPγS and adrenaline (nos. 13–16) from those containing adrenaline alone (nos. 9–12).
2. The specific activity of the [^{35}S]GTPγS and the amount of protein added can then be used to calculate the amount of [^{35}S]GTPγS bound per milligram of protein.

3.8. Immunoprecipitation

Immunoprecipitations are carried out as described in this subheading. The samples must then be run out on SDS-PAGE and transfered to nitrocellulose membrane by standard Western blotting protocols. (We use the Novogen High Performance Pre-cast gel system and its associated Western blotting system, Invitrogen.) Detection should be carried out using an appropriate antibody and enzyme conjugated secondary antibody. (We find the conjugated horseradish peroxidase/enhanced chemiluminescence detection system to be the most satisfactory).

1. Transfect HEK-293(T) cells with the required constructs as described in **Subheading 3.2.** up to part 12. The cells should be grown on polylysine-coated plates prior to transfection as described in **Subheading 3.9.**
2. Remove the media from the plates of transfected cells and wash with 3 × 5 mL cold 1X PBS.
3. Remove the final wash and add 800 μL of 1X RIPA to each plate. Scrape the plates with cell scrapers to detach the cells and transfer the suspension to Eppendorf tubes.
4. Incubate on a rotating wheel at 4°C for 1 h.
5. Centrifuge at 20,000*g* for 10 min at 4°C. Transfer the supernatant to fresh Eppendorf tubes.
6. Determine the protein concentration by BCA assay as described in **Subheading 3.3.** The lysates should then be aliquoted and those aliquots not immediately required stored at –80°C.
7. To carry out the immunoprecipitation add sufficient lysate for 500 μg of protein to the required number of Eppendorf tubes and make up the volume to 800 μL with 1X RIPA.
8. Add 40 μL of a 1:1 suspension of washed protein G–Sepharose in 1X RIPA to each tube and incubate at 4°C on a rotating wheel for 1 h to preclear.
9. Centrifuge at 1000*g* at 4°C for 1 min. Transfer the supernatants to fresh tubes containing 40 μL of protein G–Sepharose suspension and 2 μg of the appropriate antibody. Incubate at 4°C on a rotating wheel overnight.
10. Centrifuge at 1000*g* for 1 min at 4°C and remove the supernatant.
11. Wash the beads with 3 × 1 mL 1X RIPA.

12. Add 60 µL of reducing sample buffer, resuspend the beads and incubate for 1 h at room temperature, vortex mixing every 15 min, then incubate at 85°C for 4 min. The samples can then be run on SDS-PAGE and Western blotted.

3.9. Acylation

GPCRs, G-protein α-subunits, and RGS proteins are all subject to post-translational acylation that is dynamic. This assay is used to detect the palmitoylation of a specific protein construct by ^{3}H-labeling and subsequent immunoprecipitation *(7)*. Enhanced Autoradiography-wax (EA-wax) is used to reduce the exposure times required to generate an image from the final blot.

1. To polylysine coat sufficient 6-cm tissue cultures dishes, add 2 mL of poly-D-lysine solution to the first plate and then remove the solution immediately. This should then be repeated for the remaining plates using the same 2 mL of poly-D-lysine. Repeat the process for all the plates and allow to dry.
2. Transfect cells with the appropriate constructs according to the protocol described in **Subheading 3.2.** to **step 10**. After overnight incubation, split the cells into two polylysine coated 6-cm plates for each construct. Incubate overnight at 37°C.
3. Add 5% final volume of media to the dried down [9,10*(n)*-^{3}H]palmitate, which should then be added to the DMEM–sodium pyruvate–ascorbic acid. Mix the media, split into two equal volumes, and add ligand to one half.
4. Remove the media from the plates and replace with 1 mL of the labeled media, (with and without ligand), onto two plates for each construct. Incubate at 37°C for the desired time. Time points should be arranged to finish at the same time.
5. Put the plates on ice, remove the media, and wash with 2 mL each ice-cold 1X PBS.
6. Remove the PBS and add 200 µL of 1% SDS. Scrape the plates and then pass the resuspended cells through 25-gage needles 10 times before transferring to 2-mL screw cap tubes. Heat the tubes to 100°C for 5 min, cool, and centrifuge briefly.
7. Transfer to Eppendorf tubes and add 800 µL of Kahn solubilization buffer and 100 µL of Pansorbin. Incubate at 4°C on the rotating wheel for 1–2 h.
8. Centrifuge at 20,000*g* for 3 min at 4°C and transfer the supernatant to fresh tubes.
9. Remove 10 µL of each and measure the protein concentration by BCA assay as described in **Subheading 3.3.** Dilute all samples to the same protein concentration with Kahn solubilization buffer.
10. Add 100 µL of protein A–Sepharose and 5–10 µL of the appropriate antiserum to each tube, incubate overnight at 4°C on a rotating wheel. (The actual amount of antiserum required will vary and so must be determined experimentally.)
11. Centrifuge at 20,000*g* for 1 min and remove the supernatants. Wash the pellets twice with 1 mL each Kahn immunoprecipitation wash buffer. Remove the last wash and add 40 µL each sample buffer. Mix and heat to 80°C for 3 min.
12. Run equal volumes (20–30 µL each) on 10% SDS-PAGE by standard techniques. Include prestained markers for size determination. It is important that all the gel solutions used contain low levels of reducing agents, that is, <20 m*M* DTT.

13. Transfer the separated proteins to a polyvinylidene fluoride (PVDF) membrane by semidry blotting using standard techniques.
14. Remove the blot, allow it to dry and treat with EA-wax as directed by the supplier.
15. Set up to autoradiograph with preflashed film. Exposure is likely to be 2–4 wk. Develop film by standard method.

4. Notes

1. We recommend the use of Eppendorf 1.2-mL deepwell 96-well blocks as these can be floated on the surface of a water bath to incubate the contents of the wells. Some other deepwell 96-well blocks need to be modified by the addition of holes in the sides to allow them to float correctly.
2. Ouabain; care should taken with this compound.
3. To wash and resuspend protein A– or G–Sepharose fast flow beads, take the required volume of suspension and centrifuge for 1 min at 500*g*. Remove the supernatant and replace with 1 mL of the required buffer. Centrifuge as before and remove the supernatant. The wash should then be repeated twice and the beads resuspended in the original volume of the required buffer.
4. It is important to maintain the order of addition of components to a binding assay as the results can be affected by the exposure of the membranes to a higher concentration of ligand then planned in the final volume. This could be the case if membranes were added first, followed by the ligand.
5. Although a description is given for methods to calculate Scatchard plots this should be used to illustrate results only, nonlinear regression analysis gives more accurate values.

References

1. Mullaney, I. and Milligan, G. (1997) Identification and quantitation of G proteins, in *Methods in Molecular Biology*, Vol. 83: *Receptor Signal Transduction Protocols* (Challiss, R. A. J., ed.), Humana Press, Totowa, NJ, pp. 159–177.
2. Wise, A. (2000) Construction and analysis of receptor-G protein fusion proteins, in *Signal Transduction, A Practical Approach*, 2nd ed. (Milligan, G., ed.), Oxford University Press, Oxford, pp. 103–137.
3. Bahia, D. S., Wise, A., Fanelli, F., Lee, M., Rees, S., and Milligan, G. (1998) Hydrophobicity of residue[351] of the G-protein $G_{i1}\alpha$ determines the extent of activation by the α_{2A}-adrenoceptor. *Biochemistry* **37,** 11,555–11,652.
4. Cavalli, A., Druey, K. M., and Milligan, G. (2000) The regulator of G protein signaling RGS4 selectively enhances α_{2A}-adrenoceptor stimulation of the GTPase activity of $G_{o1}\alpha$ and $G_{i2}\alpha$. *J. Biol. Chem.* **275,** 23,693–23,899.
5. Hoffmann, M., Ward, R. J., Cavalli, A., Carr, C., and Milligan, G. (2001) Differential capacities of the RGS1, RGS16 and RGS-GAIP regulators of G protein signaling to enhance α_{2A}-adrenoceptor agonist stimulated GTPase activity of $G_{o1}\alpha$. *J. Neurochem.* **78,** 797–806.
6. Lui, S., Carrillo, J. J., Pediani, D. J., and Milligan, G. (2002) Effective information transfer from the α_{1b}-adrenoceptor to $G_{11}\alpha$ requires both β/γ interactions and an

aromatic group four amino acids from the C terminus of the G protein. *J. Biol. Chem.* **277,** 25,707–25,714.

7. Stevens, P. A., Pediani, D. J., Carrillo, J. J., and Milligan, G. (2001) Coordinated agonist regulation of receptor and G-protein palmitoylation and functional rescue of palmitoylation-deficient mutants of the G-protein $G_{11}\alpha$ following fusion to the α_{1b}-adrenoceptor. *J. Biol. Chem.* **276,** 35,883–35,890.

15

Assessment of Receptor Internalization and Recycling

Jennifer A. Koenig

Summary

Internalization of G-protein-coupled receptors (GPCRs) occurs in response to agonist activation of the receptors and causes a redistribution of receptors away from the plasma membrane toward endosomes. Internalization of lower-affinity small molecule GPCRs such as muscarinic acetylcholine and adrenergic receptors has been measured using hydrophilic antagonist radioligands that are membrane impermeant. In contrast, internalization of peptide hormone receptors is assessed by measuring the internalization of a radiolabeled- or fluorescently labeled peptide hormone. More recently, the use of epitope-tagged receptors has allowed the measurement of changes in receptor subcellular distribution by the use of immunoassay and immunofluorescence confocal microscopy. This chapter describes each of these approaches to the measurement of receptor internalization and describes the advantages and disadvantages of each method.

Key Words

Clathrin, dynamin, endocytosis, endosomes, epitope tag, immunoassay, immunofluorescence, internalization, muscarinic acetylcholine receptor, peptide hormone endocytosis, quantitation, radioligand binding, recycling, sequestration, somatostatin.

1. Introduction

1.1. Mechanism of Internalization and Recycling

Agonist activation of G-protein-coupled receptors (GPCRs) results in a rapid redistribution of receptors from the plasma membrane to endosomes through receptor-mediated endocytosis (reviewed in **refs.** ***1–5***). Endocytosis involves phosphorylation of the receptor by G-protein receptor kinases, followed by binding of arrestin and clathrin and accumulation of receptors in clathrin-coated pits **(Fig. 1)**. Invagination of clathrin-coated pits results in the formation of clathrin-coated vesicles which then uncoat and fuse with endosomes. Smooth pits and caveolin-dependent mechanisms have also been described but are less

From: *Methods in Molecular Biology, vol. 259, Receptor Signal Transduction Protocols, 2nd ed.*
Edited by: G. B. Willars and R. A. J. Challiss © Humana Press Inc., Totowa, NJ

well understood *(6–8)*. The actual mechanism involved may depend on the cell type under study as well as the receptor type. Endosomes are a tubulovesicular compartment characterized by a relatively low internal pH and are comprised of early recycling endosomes as well as late multivesicular endosomes *(9–14)*. After internalization, GPCRs can either be recycled back to the plasma membrane or sent to lysosomes for degradation (**Fig. 1**). The recycling step is thought to be constitutive (i.e., occurring continuously and independent of the presence of agonist). The mechanisms involved in determining the fate of internalized receptors are beginning to be worked out and a number of motifs in the C-terminal tail of the receptors are thought to be important determinants in interacting with intracellular proteins that can direct protein sorting *(15–20)*.

1.2. Quantitative Analysis of Receptor Trafficking

It is important here to clarify the meaning of the terms internalization, sequestration, and endocytosis. All of these terms have been used to describe a decrease in plasma membrane receptor number. However, "endocytosis" is used more specifically to mean the movement of receptors from the plasma membrane to endosomes. Recycling refers to the movement of receptors from endosomes to the plasma membrane. Because recycling is constitutive, the net loss of plasma membrane receptors depends on both the rate of endocytosis and the rate of recycling. This is illustrated in **Fig. 2A**, which shows three curves resulting from endocytosis with the same rate constant ($k_e = 0.1$ min^{-1}; $t_{1/2}$ ~7 min) but with different rate constants for recycling. The faster the recycling (indicated by the higher value for k_r), the smaller decrease in surface receptor number (i.e., less apparent internalization).

To explore this idea further, consider unstimulated cells where most receptors reside at the plasma membrane. On agonist activation, endocytosis is induced and the rate of endocytosis is proportional to the number of surface receptors (R_s) with proportionality constant equal to the rate constant k_e. Immediately after agonist application, endocytosis is rapid, as the number of surface receptors is high. As the receptors reach endosomes, they are recycled continuously at a rate determined by the number of endosomal receptors (R_e) and the recycling rate constant (k_r). Eventually the system reaches a new steady state at

Fig. 1. *(see facing page)* Schematic diagram illustrating the main steps in the processes of receptor endocytosis and recycling. *(1)* Agonist (indicated by D) activated receptors are rapidly phosphorylated on their intracellular loops and/or C-terminal tail by G-protein receptor kinases (and by second messenger kinases). For many receptors this step is thought to be irreversible or only slowly reversible at the plasma membrane. *(2)* Arrestin binds to the phosphorylated receptor (P) and is thought to cause

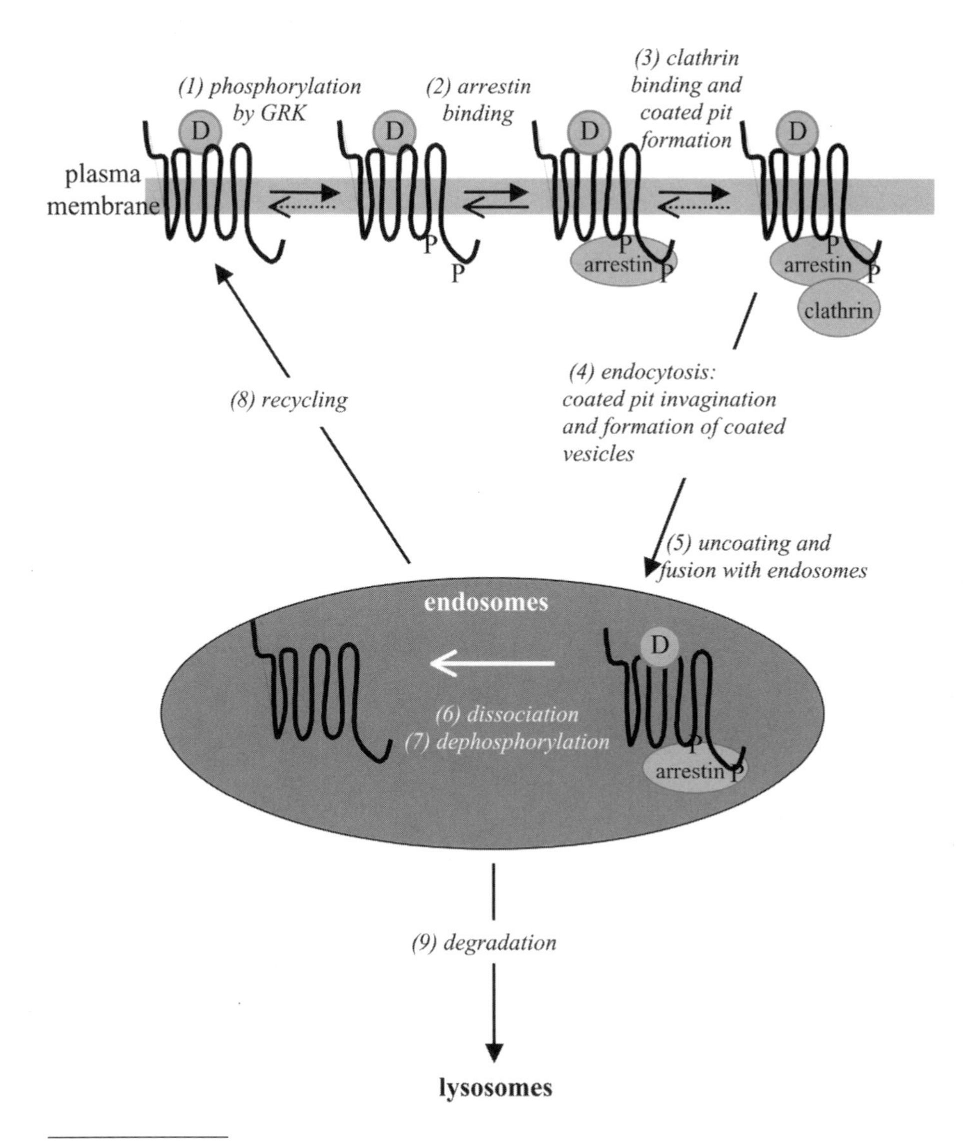

Fig. 1. *(continued)* desensitization by preventing G-protein binding. *(3)* Arrestin also binds clathrin as well as other endocytic and signaling components. *(4)* Receptors accumulate in clathrin-coated pits, which invaginate to form clathrin-coated vesicles. *(5)* These vesicles uncoat and fuse with endosomes. Endosomes are a tubulovesicular intracellular compartment characterized by a relatively low internal pH. *(6)* The low pH environment promotes dissociation of any internalized agonist. Some receptors colocalize with arrestin in endosomes while others do not and this appears to dictate their recycling characteristics ***(68,69)***. *(7)* Dephosphorylation occurs in endosomes and there appears to be some sorting between recycling and lysosomal pathways determined by motifs in the C terminal tail of the receptor. *(8)* Most receptors are recycled back to the plasma membrane. *(9)* Some receptors are sent to lysosomes for degradation.

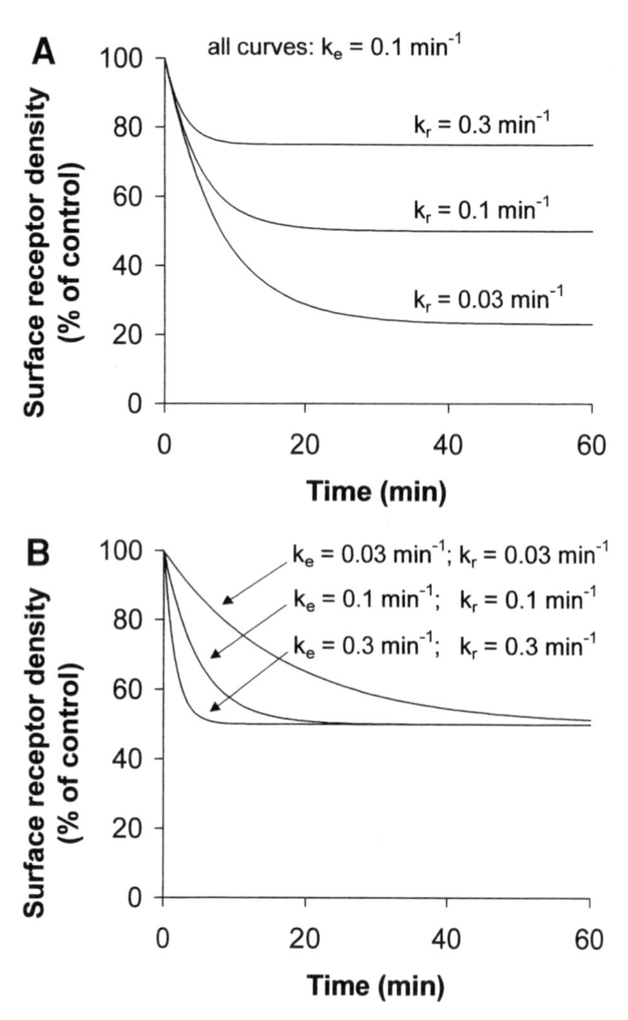

Fig. 2. Changes in cell surface receptor density on agonist activation. This model starts with all of the receptors at the plasma membrane in unstimulated cells. On agonist activation at $t = 0$, the rate constants for endocytosis (k_e) and recycling (k_r) are as shown. **(A)** Effect of different recycling rate constants on the net change in surface receptor number. All curves have the same endocytosis rate constant, $k_e = 0.1$ min^{-1} (corresponds to $t_{1/2} \sim 7$ min) but have different recycling rate constants, $k_r = 0.3$, 0.1, and 0.03 min^{-1} (corresponding to $t_{1/2} \sim 2.3$, 7, and 23 min, respectively). The graph shows that the height of the plateau (the number of receptors at steady state) depends on k_r as well as k_e and faster recycling (indicated by a higher value for k_r) means fewer internalized receptors at steady state but faster turnover of receptors at steady state. **(B)** The time taken to reach steady state depends on both k_e and k_r. Higher values for k_e and k_r result in faster internalization and faster turnover during steady state. However the number of receptors at steady state is the same for all three curves because the ratio $k_e/(k_e + k_r)$ is the same.

which the rate of endocytosis equals the rate of recycling and there is no net change in the number of surface receptors. The rate of internalization (defined as the rate of change in number of surface receptors) is the rate of receptors arriving at the plasma membrane by recycling less the rate of receptors leaving the plasma membrane by endocytosis.

$$\text{Rate of endocytosis} = k_e \times R_s$$

$$\text{Rate of recycling} = k_r \times R_e$$

$$\text{Rate of internalization} = k_r \times R_e - k_e \times R_s$$

If the number of plasma membrane and endosomal receptors at the beginning of the experiment is R_{s0} and R_{e0} respectively, then the number of surface receptors at any time is given by:

$$R_s = \frac{1}{k_e + k_r}\left[k_r(R_{s0} + R_{e0}) + (k_e R_{s0} - k_r R_{e0})e^{-(k_e + k_r)t}\right] \quad (1)$$

There are two important consequences from this derivation. The first is that the rate constant for change of surface receptor number (i.e., internalization) $= k_e + k_r$. Furthermore, the steady-state number of surface receptors ($R_{s,ss}$) and the steady-state number of endosomal receptors ($R_{e,ss}$) are given by:

$$R_{s,ss} = \frac{k_r\,(R_{s0} + R_{e0})}{k_e + k_r} \qquad R_{e,ss} = \frac{k_e\,(R_{s0} + R_{e0})}{k_e + k_r}$$

This is illustrated in **Fig. 2B**, which shows three curves where the extent of internalization at steady state is the same, as in all cases $k_e = k_r$ and so $k_r/(k_e + k_r) = 0.5$. However, the curve with higher values for k_e (0.3 min^{-1}) and k_r (0.3 min^{-1}) reaches steady state much more quickly than the curve with lower values for k_e (0.03 min^{-1}) and k_r (0.03 min^{-1}). Furthermore the rate of endocytosis at steady state (~ 60 min) is $k_e \times R_s$ which is $0.3 \times 50 = 15\%$ of receptors per minute for the fastest endocytosing receptors and $0.03 \times 50 = 1.5\%$ of receptors per minute for the slowest endocytosing receptors. A more detailed description of this derivation is given in *(2,21)*.

1.3. Inhibitors of Receptor Internalization and Recycling

Endocytosis and recycling are both strongly temperature and energy dependent. Depletion of ATP with either sodium azide *(22)* or antimycin and deoxyglucose *(23)* reduces somatostatin-14 internalization by >85%. Recycling of somatostatin-14 is less ATP dependent and is reduced by only 50% *(23)*. Incubation at 15°C reduces internalization of somatostatin by 90% and recycling by 70% while incubation at 4°C reduces internalization by 97% and recycling by 80% *(23)*. Another approach is to prevent internalization by treatment with agents such as concanavalin A, phenylarsine oxide, or hyperosmolar sucrose or by depletion of intracellular potassium *(24–26)*. The mechanism of action of concanavalin A is unclear although it may exert its effect by crosslinking sugar residues. It works well with some receptors (e.g., β-adrenergic receptors *[27]*) but is less effective at other receptors (e.g., muscarinic acetylcholine receptors *[28]*). Phenylarsine oxide has been used to inhibit internalization successfully by a number of groups (e.g., ***29,30***) but is known to have many effects on cellular function through interactions with sulfhydryl groups (e.g., interactions with phosphatases and depletion of ATP) *(31)*. I have had variable results with phenylarsine oxide but have noticed a considerable problem with decreased adherence of cells to plastic dishes. Hypertonic media (such as 0.45 *M* sucrose) or potassium depletion are thought to prevent endocytosis by preventing clathrin coated pit formation *(32)*. Hyperosmolar sucrose is quick (5-min preincubation required) and effective and shown to prevent internalization in a number of receptor types *(24,33,34)*. However, it does have nonspecific effects on measurement of signaling pathways such as fluorometric determination of intracellular calcium and it also reduces receptor recycling (L. A. Frawley, S. L. R. Ball, and J. A. Koenig, *unpublished observations*).

An alternative approach to inhibiting endocytosis is the use of dominant negative mutants of dynamin, an important protein in clathrin-mediated endocytosis *(14,35)*. The most common dominant-negative mutant used is the K44A mutant. Whereas the endocytosis of many GPCRs is sensitive to expression of this mutant, others are not and this has been interpreted as a dynamin-insensitive mechanism (e.g., *[36]*). However, for some receptors another isoform of dynamin (dynamin 2) appears to be important *(37)*. Furthermore, receptors that were previously thought to use dynamin-insensitive pathways have now been shown to be dynamin sensitive through the use of other dynamin dominant-negative mutants *(38)*.

Receptor recycling is inhibited by agents such as monensin and nigericin that raise the pH within endosomes *(27,39,40)*, although these inhibitors have other effects and do not necessarily inhibit recycling completely *(28)*. Dominant-negative mutants of the small GTPases such as Rab4 and Rab5 have been used to inhibit recycling and endocytosis respectively, and this approach may prove to be more specific and perhaps more successful *(41,42)*.

1.4. Techniques for Measurement of Receptor Internalization and Recycling

Early experiments investigating receptor internalization relied on the measurement of plasma membrane receptor density using membrane-impermeant radioligand binding assays (e.g., [^{3}H]CGP 12177 for β_2-adrenergic receptors *[43]* and [^{3}H]*N*-methylscopolamine for muscarinic acetylcholine receptors *[44]*). These assays are straightforward and reliable but the main disadvantage is that any agonist used to initiate internalization must be thoroughly washed off before the radioligand binding assay. For rapidly dissociating agonists such as the muscarinic and adrenergic agonists this is not generally a problem but it may become problematic for higher affinity ligands. Total cellular receptor number could be measured with membrane-permeant radioligands although issues of access to intracellular receptors and the low pH of internal compartments as well as extremely long incubations required to reach equilibrium meant that these assays were less popular ***(21)***. Alternatively, 0.2% digitonin can be used to permeabilize cells and allow access of membrane-impermeant ligands ***(43)***.

More recently, epitope tagging has led to the development of assays for plasma membrane receptor density based on an immunoassay detecting an N-terminal epitope tag such as HA, FLAG, or myc (*see* Chapter 5). An immunoassay has the advantage that the agonist or antagonist ligands do not need to be washed off before the assay. Furthermore, they can be used for receptors where hydrophilic antagonist radioligands are not available, as is the case for many peptide hormone receptors. Fluorescence-linked immunoassays are commonly followed by flow cytometry to quantitate changes in fluorescence ***(43)***. Enzyme-linked and radioimmunoassays have also been used (*see* **Subheading 2.2.**). Whichever detection system is used, it is important that the cells are fixed in formaldehyde (or similar fixative) before application of antibodies, as in some cases, incubation with the epitope-tag antibody can initiate receptor internalization (J. A. Koenig, *unpublished observations* with HA-tagged somatostatin$_2$ receptors) and ***(45,46)***. Alternatively, appropriate controls must be used. Antibody-feeding assays have been used to demonstrate continuous endocytosis and recycling while the total number of plasma membrane receptors remains unchanged ***(47,48)***. In these assays, the antibody is added to live cells and removal of antibody from the cell surface is measured. Although these are valuable assays, it is important to design the experiment carefully and include the appropriate controls.

Although the above methods for measuring receptor internalization are quantitative and based on the study of cell populations, an important alternative method is the analysis of single cells by confocal microscopy. This approach relies on either immunofluorescence staining or on the generation of a GFP-labeled receptor. GFP-receptor internalization has been reviewed recently and is not discussed further here ***(49–51)***. Confocal microscopy is extremely valuable in determining the intracellular location of internalized receptors by colocalization with markers for endosomes such as antibodies to early endosome antigen1 (EEA1) ***(52,53)***, transferrin receptor, or indeed fluorescently labeled transferrin ***(39)***. The main drawback with confocal microscopy is that it is extremely difficult to quantitate ***(54)***. Confocal immunofluorescence micro-

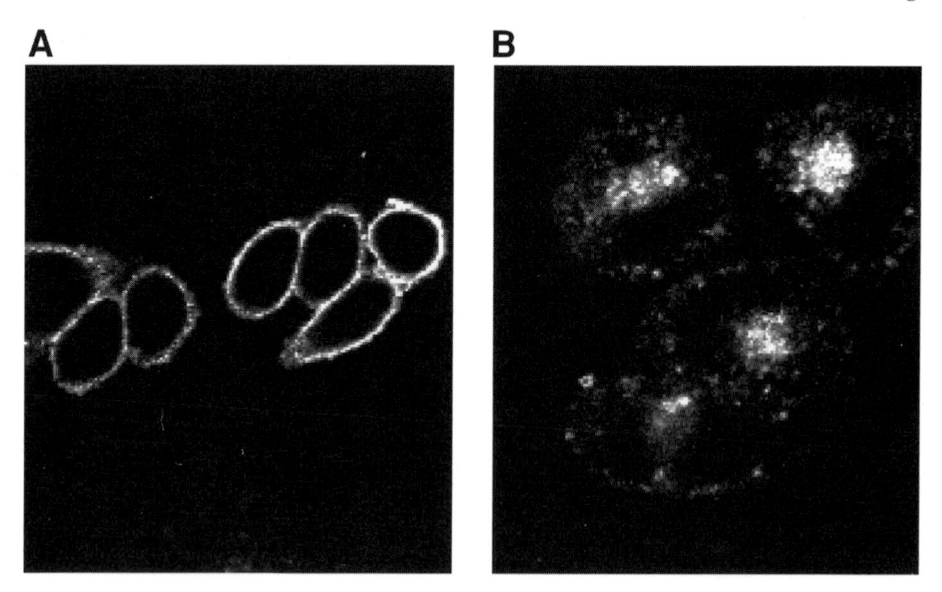

Fig. 3. Immunofluorescence confocal microscopy of HA-tagged somatostatin receptors (sstr2). **(A)** Unstimulated cells, **(B)** after 60 min of incubation with 100 n*M* somatostatin-14. Cells were fixed, permeabilized, and incubated with anti-HA and then fluorescein–anti-mouse IgG. Radioimmunoassay and ELISA experiments detecting the cell surface HA-tag showed a reduction in surface receptor density of 20–30% after 60 min of incubation with 100 n*M* somatostatin-14.

scopy of somatostatin-induced internalization of sstr2 somatostatin receptors in CHO-K1 after 60 min suggests that the extent of internalization is substantial **(Fig. 3)**. However, radioimmunoassay and enzyme-linked immunosorbent assay (ELISA) data of exactly the same system, the only difference being the secondary antibody, gave a reduction of only 20–30% of surface receptor number after 60-min incubation with 100 n*M* somatostatin-14. Quantitation of confocal images is possible, however, with careful experimental design and analysis but such a large topic is outside the scope of this chapter (as a starting point *see* **refs. *55–64***).

1.5. Techniques for Measurement of Agonist Internalization and Recycling

Receptor internalization has also been measured by monitoring the internalization of a radiolabeled or fluorescently labeled peptide agonist. In general, the agonists that have been used successfully have been of relatively high affinity (at least nanomolar) and it is generally assumed that lower affinity agonists such as those for muscarinic acetylcholine and β-adrenergic receptors will not show significant internalization. It is worth noting, however, that a much higher concentration of ^{3}H-radiolabeled muscarinic receptor agonist oxotremorine would be required to reach receptor occupancy equivalent to that of ^{125}I-radiolabeled somatostatin agonist and that such concentrations of [^{3}H]oxotremorine would be impractical.

Measurement of agonist internalization relies on being able to distinguish between surface-bound ligand and internalized ligand. Typically, cells are incubated at 37°C for up to 1 h to allow binding and internalization, and then washed rapidly to remove free ligand. This leaves both surface-bound and internalized ligand. Further washing (usually with low pH buffer) then removes the surface-bound ligand, which can then be quantitated with a γ-counter or scintillation counter. Alternatively, cells are incubated with radioligand at 4°C to allow binding but prevent internalization, washed to remove free radioligand, then warmed to 37°C to allow both dissociation of agonist from cell surface receptors (which can be measured by collecting the extracellular medium) and internalization (which is measured by solubilizing the cells).

The most difficult aspect of this procedure is determining the correct washing procedures to remove free ligand and separate surface from internalized ligand. Very little work has been published describing the rate constants for dissociation for peptide hormone agonists under close to physiological conditions. Dissociation half-times for ^{125}I-somatostatin-14 in cell membranes vary dramatically depending on the assay buffer and temperature ***(33,65)***. For somatostatin-14, estimates of the dissociation half-time in intact cells vary from 20 to 40 s and are comparable to the dissociation half-time measured in cell membranes including GTP in physiological saline but very much shorter than dissociation half-times measured in low ionic strength buffer without GTP. Another consideration when determining the best washing procedure is cell viability. Very low pH washes (below pH 5) significantly reduce cell viability, which is important if the recycling of agonist or receptor is to be measured.

Internalization of agonists has also been demonstrated using fluorescently labeled peptides. The choice of fluorophore is important because fluorescein derivatives are susceptible to quenching in the low pH environment of endosomes and may show photobleaching upon extended illumination. Some of the Alexa Fluor™, BODIPY (Molecular Probes), and Cy-dyes (Amersham International) do not have this problem. Small fluorescent peptides often are not retained on fixation with formaldehyde, and this precludes colocalization experiments with antibodies for intracellular marker proteins or receptors.

For both fluorescent- and radiolabeled peptides the issue of peptide integrity is an important one. I have found that the extent of somatostatin degradation varies with cell type and cell density and occurs more so after internalization than in the extracellular medium *(23)*. Many authors have shown that fluorescence is detectable in lysosomes after internalization of fluoropeptides and thus concluded that the peptides are degraded. However, with confocal microscopy, it is extremely difficult to quantify the images sufficiently well to determine what proportion of internalized fluoropeptide is degraded and to detect any peptide that is recycled and diffuses into the extracellular medium.

Successful measurement of agonist recycling requires that any recycled agonist is not able to be reendocytosed. This can be achieved by including an excess concentration of agonist or antagonist in the extracellular medium during the recycling step or by washing extensively throughout recycling. This prevents the recycled agonist from rebinding to its receptor and being reendocytosed. In the case of ^{125}I-somatostatin at the somatostatin$_2$ receptor, recycling of ^{125}I-somatostatin was detectable only if reendocytosis was prevented *(23)*. Furthermore, after incubation with saturating concentrations of somatostatin (100 n*M*) to induce desensitization, sufficient agonist was recycled to reactivate the somatostatin$_2$ receptors and cause inhibition of adenylyl cyclase *(23)*.

2. Materials

2.1. Measuring Internalization and Recycling With Antagonist Radioligands: l-[N-methyl-^{3}H]-Scopolamine Methyl Chloride ([^{3}H]NMS) Binding to Muscarinic Receptors in CHO Cells

1. DMEM–HEPES: use Dulbecco's modified Eagle medium (DMEM) powder and make up in 10 m*M* 4-(2-hydroxyehtyl)piperazine-1-ethanesulfonic acid (HEPES), pH 7.4.
2. PBS–Ca–Mg: 137 m*M* NaCl, 2.7 m*M* KCl, 8.1 m*M* Na_2HPO_4, 1.5 m*M* KH_2PO_4, 0.9 m*M* $CaCl_2$, 0.5 m*M* $MgCl_2$.
3. l-[*N-methyl*-^{3}H]scopolamine methyl chloride ([^{3}H]NMS) is available at a specific activity of 60–90 Ci/mmol from Amersham International and Perkin Elmer Life Sciences (*see* **Note 1**).
4. *Optional:* propylbenzilylcholine mustard (PrBCM, obtained from Dr. J. M. Young, Department of Pharmacology, University of Cambridge, Tennis Court Road, Cambridge CB2 1PD, UK there are no commercial suppliers). Dissolve in ethanol at 0.01 *M* and store indefinitely at –20°C. On the day of assay, dilute to 0.1 m*M* in PBS–Ca–Mg, pH 7.2, and leave to cyclize to the aziridinium ion for 1 h at room temperature, then dilute to 10 n*M*.
5. Carbachol and *N*-methyl atropine (Sigma) dissolve readily in water or assay buffer, can be stored in aliquots at –20°C.
6. Triton-X-100: 0.5% (v/v) in water.
7. Scintillant: Optiphase Safe or similar scintillant that is water miscible at a water to scintillant volume ratio of 1:8.

8. Scintillation counter.
9. Hemocytometer or Coulter counter.

2.2. Measuring Surface Receptor Number by Radioimmunoassay for an Epitope Tag: HA-Tagged Somatostatin Receptors in CHO-K1 Cells

1. DMEM–HEPES: use DMEM powder and make up in 10 m*M* HEPES, pH 7.4.
2. Phosphate-buffered saline (PBS): 137 m*M* NaCl, 2.7 m*M* KCl, 8.1 m*M* Na_2HPO_4, 1.5 m*M* KH_2PO_4.
3. 3% Paraformaldehyde in PBS (weigh and heat gently to dissolve, cool and filter through standard laboratory filter paper in fume hood).
4. PBS–BSA–HS: PBS containing 1% bovine serum albumin (BSA) and 1% horse serum.
5. Somatostatin (Bachem): dissolve in PBS to 1 m*M* and store frozen aliquots at –20°C.
6. Triton-X-100, 1% (v/v) in water.
7. Anti-HA from Babco through Cambridge Bioscience diluted to approx 4 μg/mL (1:500–1:1000) in PBS–BSA–HS.
8. ^{125}I-anti-mouse IgG $F(ab')_2$ fragment (from sheep) available from Perkin Elmer Life Sciences (*see* **Note 2**). The whole antibody could also be used. Aliquot upon arrival and store at –20°C. On the day of assay, dilute in PBS–BSA–HS to give a final concentration of approx 100,000 dpm/well.
9. γ-Counter.
10. Appropriate lead acetate safety shielding for ^{125}I and Geiger counter for monitoring working area (*see* **Note 2**).
11. Vacuum pump.

2.3. Immunofluorescence Confocal Microscopy

1. Cover slips, thickness no. 1 (should be 170 ± 5 μm; *see* **refs. *54*** and ***66***).
2. DMEM–HEPES: use DMEM powder and make up in 10 m*M* HEPES, pH 7.4.
3. PBS: 137 m*M* NaCl, 2.7 m*M* KCl, 8.1 m*M* Na_2HPO_4, 1.5 m*M* KH_2PO_4.
4. 3% Paraformaldehyde in PBS. (Use in a fume hood. Weigh and heat gently to dissolve, cool and filter through standard laboratory filter paper.)
5. Triton X-100: 0.2% (v/v) in PBS, filtered through 0.2-μm filter disks.
6. PBS–BSA–HS: PBS containing 1% BSA and 1% horse serum. Dissolve the BSA by scattering the powder on the surface of the PBS and leaving it to soak in for about 20 min. This gives fewer small particles of undissolved BSA in suspension. Filter through 0.22-μm filter disks and keep in covered bottles to avoid accumulation of dust.
7. Anti-HA from Babco through Cambridge Bioscience diluted to approx 4 μg/mL (1:500–1:1000) in PBS–BSA–HS. Centrifuge for a few minutes at approx 13,000*g* in a bench centrifuge to pellet any aggregates or particulates. Take the sample from the supernatant for dilution.

8. Alexa Fluor™ 488: anti-mouse IgG from Molecular Probes. (Use at the manufacturer's recommended dilution in PBS–BSA–HS (*see* **Note 3**). Keep the stock antibody solution in the dark at 4°C and centrifuge for a few minutes at approx 13,000*g* in a benchtop centrifuge to pellet any aggregates or particulates. Take the sample from the supernatant for dilution. Keep in the dark.
9. Microscope slides: clean by spraying with ethanol and air-drying. Avoid accumulation of dust.
10. Nail varnish: any type or color.
11. Mounting medium: Vectashield, Vector Labs.

2.4. Radiolabeled Agonist Binding, Internalization, and Recycling: ^{125}I-Somatostatin-14

1. ^{125}I-Somatostatin-14 is available from Perkin Elmer Life Sciences and Amersham International. It arrives as a powder and should be dissolved in water (25 μCi/mL) and frozen at –20°C in aliquots. It should not be refrozen (*see* **Note 4**). Dilute to 5 n*M* in DMEM–HEPES, pH 7.2.
2. Somatostatin (Bachem), dissolve in PBS to 1 m*M* and store frozen aliquots at –20°C.
3. 10 m*M* DMEM–HEPES, pH 7.4.
4. Acid wash buffer: 10 m*M* DMEM–MES, pH 5.0. Use DMEM powder and make up in 10 m*M* 2-(*N*-morpholine)ethanesulfonic acid (MES), pH 5.0.
5. 1% Triton X-100 in water.

2.5. Fluorescently Labeled Agonist Binding and Internalization

1. Phenol red-free DMEM powder made up in 10 m*M* HEPES, pH 7.2.
2. Fluo-somatostatin (*see* **Note 5**) diluted to 50 n*M* in phenol red-free DMEM–HEPES, pH 7.2.
3. Confocal microscope with inverted microscope (Zeiss LSM510 or similar) with heated stage and preferably a drug delivery system (by injection) (*see* **Note 6**).
4. Glass cover slips (thickness 1) to fit the heated stage on the microscope.

3. Methods

3.1. Measuring Internalization and Recycling With Antagonist Radioligands: [^{3}H]NMS Binding to Muscarinic Receptors in CHO Cells

3.1.1. Internalization

1. Seed cells at 10,000 cells/well (*see* **Note 7**) in 24-well multiwell plates. Incubate overnight.
2. Remove from incubator and wash twice with 1 mL of DMEM–HEPES at 37°C (*see* **Note 8**).
3. Remove DMEM–HEPES wash and add 0.45 mL of DMEM–HEPES at 37°C and incubate for 10 min at 37°C.
4. Initiate internalization by adding 0.05 mL of 5 m*M* carbachol or vehicle and incubate for varying times (from 5 to 120 min) at 37°C. Use three wells for each

time point. Include three sets of three wells with vehicle only on each plate. One set will be used to measure total binding, one set will be used to measure nonspecific binding and one set will be used to measure the cell count.

5. Wash twice with 1 mL of PBS–Ca–Mg to remove carbachol.
6. Incubate at 4°C overnight with 0.25 mL 0.2 n*M* [^{3}H]NMS (this measures the total binding) or with 0.25 mL of 0.2 n*M* [^{3}H]NMS containing 1 μ*M* *N*-methylatropine for the nonspecific binding. Leave three wells on each plate without radiolabeled drug to be used for cell counting.
7. Rinse each well rapidly twice with 1 mL of PBS–Ca–Mg to remove unbound radioligand. Leave to dry.
8. Add 0.25 mL of 0.5% Triton X-100 to wells with radioactivity. Leave for at least 1 h to solubilize.
9. Add 0.25 mL of 1 m*M* PBS–EDTA–trypsin (0.5 mg/mL) to wells containing cells for counting and leave for 10 min. Dislodge cells by taking them up and down in a pipet tip and count using a hemocytometer or Coulter counter (*see* **Note 9**).
10. Transfer solubilized radioactivity from wells to scintillation vials. Rinse with 0.25 mL of water and add this to the scintillation vial also. Add 4 mL of scintillant, mix well, and transfer to a scintillation counter.
11. Subtract the nonspecific binding from the total binding to obtain the specific binding. The vehicle-only specific binding should be the same (within experimental error) for each plate.
12. Data analysis: *see* **Subheading 3.7.**

3.1.2. Recycling

1–3. As for **Subheading 3.1.1., steps 1–3**

4. Initiate internalization by adding 0.05 mL of 5 m*M* carbachol or vehicle and incubate for 60 min at 37°C. Use three wells for each recycling time point. Include three sets of three wells with vehicle only on each plate. One set will be used to measure total binding, one set will be used to measure nonspecific binding and one set will be used to measure the cell count.
5. Wash twice with 1 mL of PBS–Ca–Mg to remove carbachol.
6. Incubate with 10^{-8} *M* PrBCM for 20 min at room temperature (*see* **Note 10**).
7. Remove PrBCM, wash twice with 1 mL of DMEM–HEPES and incubate in 1 mL of DMEM–HEPES at 37°C for varying times (0–120 min) (*see* **Note 11**).
8. Wash twice with 1 mL PBS–Ca–Mg to remove medium.
9. Measure [^{3}H]NMS binding as described in **Subheading 3.1.1., steps 6–11**.
10. Data analysis: *see* **Subheading 3.7.**

3.2. Measuring Surface Receptor Number by Radioimmunoassay for an Epitope Tag: HA-Tagged Somatostatin Receptors in CHO-K1 Cells

3.2.1. Internalization

1. Seed cells at 10,000 cells/well (*see* **Notes 7** and **12**) in 24-well multiwell plates. Incubate overnight.

2. Remove from incubator and wash twice with 1 mL of DMEM–HEPES at 37°C (*see* **Note 8**).
3. Remove DMEM–HEPES, wash and add 0.45 mL of DMEM–HEPES at 37°C and incubate for 10 min at 37°C.
4. Initiate internalization by adding 0.05 mL of 1 μ*M* somatostatin-14 or vehicle and incubate for varying times (from 5 to 120 min) at 37°C. Use three to four wells for each time point. Include three sets of three wells with vehicle only on each plate. One set will be used to measure total binding and one set will be used to measure nonspecific binding (*see* **Note 13**). One set will be used to determine the number of cells; these cells should not be treated with paraformaldehyde but instead determine the number of cells as described in **Subheading 3.1.1., step 9**.
5. Fix with 3% paraformaldehyde in 0.5 mL of PBS for 15 min at room temperature.
6. Wash with 3 × 2 mL pf PBS then 1 × 1 mL PBS–BSA–HS.
7. Incubate with 0.25 mL of anti-HA (~4 μg/mL) in PBS–BSA–HS for 1 h at room temperature.
8. Wash 3 × 1 mL PBS–BSA–HS then 3 × 2 mL PBS.
9. Incubate with 0.25 mL ^{125}I-anti-mouse IgG Fab fragment at 4°C overnight (*see* **Note 14**).
10. Wash 3 × 1 mL PBS.
11. Solubilize in 0.5 mL of 1% Triton-X-100.
12. Transfer to vials, rinse with 0.5 mL of water and combine rinse with original sample.
13. Quantitate in a γ-counter.
14. Subtract the nonspecific binding from the total to obtain specific binding and express as a percentage of the control (vehicle only). The specific binding for the control should be the same (within experimental error) for each plate.
15. Data analysis: *see* **Subheading 3.7.**

3.3. Immunofluorescence Confocal Microscopy of HA-Tagged Somatostatin Receptors in CHO-K1 Cells

1. Sterilize the glass cover slips by dipping in 70% ethanol in water, then air-drying. Waving through a Bunsen flame is optional.
2. Place the cover slips one per well in a 6-well plate.
3. Harvest cells with trypsin–EDTA (or by the usual method) and resuspend in growth medium.
4. Place a 20–50-μL drop of cell suspension onto the middle of the cover slip and leave for 15 min (in a humidified incubator).
5. Gently cover with medium (2 mL per well) and incubate in a cell culture incubator. On the day of the experiment, the cells should be subconfluent so that individual cells can be seen under the microscope.
6. Rinse with DMEM–HEPES, pH 7.2.
7. Incubate with saturating concentrations of agonist (100 n*M* somatostatin-14) at 37°C for varying times (0–120 min) to initiate internalization.

8. Wash, 3X PBS.
9. Fix with 3% paraformaldehyde in 1 mL of PBS for 15 min at room temperature.
10. Wash, 2X PBS.
11. Permeabilize with 1 mL of 0.2% Triton X-100, 10 min at room temperature.
12. Wash with 3 × 2 mL of PBS, then 1 × 1 mL PBS–BSA–HS.
13. Incubate with 0.25 mL of anti-HA (~4 μg/mL) in PBS–BSA–HS for 1 h at room temperature (in a humidified atmosphere to prevent drying out). *See* **Note 15**.
14. Wash 3 × 1 mL of PBS–BSA–HS, then 3 × 2 mL of PBS.
15. Incubate with 0.25 mL of Alexa Fluor™ 488 anti-mouse IgG (*see* **Note 15**) for 1 h at room temperature in the dark.
16. Wash 3 × 1 mL with PBS.
17. Place a 20-μL drop of mounting medium (Vectashield) on a clean microscope slide.
18. Carefully pick up the cover slip from the dish with forceps, dip into distilled water, and gently clean the back of the cover slip with a wet cotton bud. *See* **Note 16**. Repeat, then dab the edge of the cover slip on a tissue to remove excess water. Invert onto the mounting medium, carefully avoiding any bubbles.
19. Seal the edges of the cover slip onto the microscope slide with nail varnish, then air-dry.

3.4. Radiolabeled Agonist Binding and Internalization: ^{125}I-Somatostatin-14

1 and 2. As for **Subheading 3.1.1., steps 1** and **2**.

3. Remove DMEM–HEPES wash and add 0.2 mL of DMEM–HEPES at 37°C and incubate for 10 min at 37°C (*see* **Note 17**).
4. Add 0.05 mL of 1 n*M* ^{125}I-somatostatin-14 to define total binding and internalization or 1 n*M* ^{125}I-somatostatin-14 containing 5 μ*M* somatostatin-14 to define nonspecific binding and internalization. Leave three wells on each plate with no radioactivity to use for the cell count as described in **Subheading 3.1.1., step 9**.
5. Incubate for varying times (5–120 min) at 37°C.
6. Terminate the incubation by washing rapidly twice with 1 mL of ice-cold DMEM–HEPES pH 7.4.
7. At this point there are two alternatives:
   a. To measure both surface-bound and internalized ligand, leave the wells dry, and go on to step 8, or
   b. To measure internalized ligand, incubate with 2 mL of ice-cold acid wash buffer for 10 min, then remove and leave the wells dry (*see* **Note 18**).
8. Solubilize the cells in 1% Triton X-100 for at least 1 h.
9. Transfer to vials and count in a γ-counter.
10. The surface-bound ligand is obtained by subtracting the dpm for the acid-washed wells **(step 7b)** from the dpm values for the rapid pH 7-washed wells **(step 7a)**. *See* **Note 19**.
11. Data analysis: *see* **Subheading 3.7.**

3.5. Measuring Radiolabeled Agonist Recycling: ^{125}I-Somatostatin-14

1–4. As for **Subheading 3.4., steps 1–4**.
5. Incubate for 60 min at 37°C.
6. Terminate the incubation by washing rapidly twice with 1 mL of ice-cold DMEM–HEPES pH 7.4, then add 1 mL ice-cold acid wash buffer and leave for 10 min on ice. Remove acid wash buffer and add 0.5 mL of 1 μ*M* somatostatin-14 in DMEM–HEPES, pH 7.4, warmed at 37°C (*see* **Note 20**).
7. Incubate at 37°C for varying times (0–120 min).
8. Remove the extracellular medium to vials for counting (this gives the amount of recycled radioactivity).
9. Wash twice with 1 mL of DMEM–HEPES, pH 7, and leave wells dry.
10. Solubilize the cells in 1% Triton X-100 for at least 1 h and transfer to vials for counting (this gives the remaining internalized radioactivity).
11. Count in a γ-counter.
12. Subtract nonspecific from total to obtain specific internalization and recycling (*see* **Note 21**).
13. Data analysis: *see* **Subheading 3.7.**

3.6. Fluorescently Labeled Agonist Binding, Internalization, Dissociation, and Recycling

1–6. Prepare cells on cover slips as for **Subheading 3.3.**
7. Place in the heated stage and cover with 0.4 mL of phenol red-free DMEM–HEPES, pH 7.2.
8. Take a number of images with the confocal microscope before adding fluoropeptide (0.1 mL of 50 n*M* fluo-somatostatin) and then continue to image at regular intervals (*see* **Note 22**).
9. After observing binding of fluorophore at the plasma membrane followed by internalization, wash and include excess unlabeled agonist to observe agonist dissociation and recycling (*see* **Note 23**).

3.7. Data Analysis

Equation 1 is written in such a way as to be useful for data fitting of both receptor internalization and recycling experiments.

3.7.1. Internalization Experiments

Perform nonlinear curve fitting using the general **Eq. 1**. Programs such as GraphPad Prism allow input of user-defined equations. Commonly available spreadsheet programs can also be used for nonlinear curve fitting (e.g., using the Solver function in Microsoft Excel) ***(67)***. For many receptors where there are effectively no endosomal receptors in unstimulated cells, $R_{e0} = 0$ and R_{s0} is normalized to 100%: then **Eq. 1** reduces to:

$$R_s = \frac{R_{s0}}{k_e + k_r}\,[k_r + k_e e^{-(k_e+k_r)t}].$$

3.7.2. Recycling Experiments

In the PrBCM experiments where recycling is measured, $R_{s0} = 0$ and we can assume $k_e \sim 0$, as agonist is not present and there is no significant constitutive activity. In this case, **Eq. 1** reduces to $R_s = R_{e0}(1 - e^{-k_r)t})$. The recycling data can be fitted to find the best-fit values for R_{e0} (which is now the number of endosomal receptors at the beginning of the recycling step) and k_r. If PrBCM or a similar ligand is not available, then R_{s0} is the number of surface receptors at the beginning of the recycling step and, depending on the evidence available, it may be safe to assume that $k_e \sim 0$, as agonist is no longer present.

Data analysis for agonist internalization experiments is often more complex. There may be significant agonist degradation during the course of the experiment, which would make the effective free concentration vary with time. There is evidence for at least the somatostatin$_2$ receptor, that not all of the somatostatin agonist is recycled after endocytosis *(23)*. Equation (1) is derived from a two-compartment model. Bringing in the agonist as well makes it a three-compartment model where (if the experiment is designed correctly) the amount of agonist is in great excess over the number of receptors at the cell surface. It is possible to use simple exponential rise equations (such as $I = I_{max} \cdot e^{(-kt)}$) where I is the amount of internalized ligand at time t, I_{max} is the maximum amount of internalized ligand, and k is the rate constant for internalization. Numerical simulation modeling may be a better method for curve fitting of more complex models.

4. Notes

1. Relevant safety procedures should be in place for storage, use, and disposal of ^{3}H-ligands.
2. When using ^{125}I-radioligands, relevant safety precautions should be taken and local procedures for storage, use and disposal should be followed. From **step 9** onwards, all manipulations should be performed behind lead acetate screens. Hold the vacuum line with a clamp to increase the distance from the radioactivity. Note that >90% of the radioactivity is removed with the incubation buffer at the end of the incubation. Therefore, shielding the waste container from the working area decreases exposure significantly. The radioactivity can be added to the cells in a small volume (10 μL) with a repeating pipet to decrease exposure times. Include an in-line 0.22-μm filter to help prevent accidental contamination of the pump by overflow of the waste container.
3. The choice of fluorophore depends greatly on the type of confocal microscope being used and which laser lines are available. I have found Alexa Fluor™ 488 to be superior to fluorescein, FITC or Cy2 if using the 488-nm line of an Argon or argon–krypton laser but have also used Cy3, which is excited by the 543-nm line of a HeNe laser. Some authors use a biotinylated secondary antibody followed by a fluorescent avidin but I have found this extra step to be unnecessary.

4. I have found that a significant proportion (sometimes up to 40%) of the radioactivity adheres to the tubes (we use screw-cap Eppendorf tubes) and this was reduced somewhat by the inclusion of bacitracin (0.2 mg/mL).
5. Fluorosomatostatin was available from Advanced Bioconcept and then NEN Life Sciences. At the time of writing, Perkin Elmer Life Sciences (who incorporated NEN Life Sciences) do not sell fluorosomatostatin but will consider a custom synthesis. Molecular Probes also sells a variety of fluorescently labeled peptide hormones. This situation has been repeated with many fluoropeptides, as there are often manufacturing difficulties. Ensure that you check the activity of the fluoropeptide (by radioligand binding competition experiment or a functional assay) and then buy sufficient for a whole series of experiments. If possible choose a fluorophore that is not pH sensitive (e.g., Cy3).
6. The confocal microscope is extremely sensitive to small fluctuations in temperature, as these will cause sufficient focal drift to take the images out of focus during the course of an experiment. This is a major technical problem and has been discussed in detail in confocal microscopy textbooks (e.g., ***66***) and on various microscopy websites (e.g., the confocal microscopy listserver archives at http://listserv.acsu.buffalo.edu/archives/confocal.html).
7. I use 24-well plates because they hold enough cells to give a reasonable signal: 96-well plates are difficult to wash well without dislodging the cells. The cell seeding density required depends on the level of receptor expression. If there are too many receptors in the well, a significant proportion of the added radioactivity is bound to the receptors and the free concentration of radioligand is depleted. In practice, the assay should be set up so that the level of free radioligand depletion is $<10\%$. To check this, count a sample of the radioligand used to incubate cells; this will typically be about 100,000 dpm/well. The amount of total binding to receptors should be no more than 10,000 dpm. In CHO cells, the muscarinic receptors are overexpressed at approx 10^6 receptors/cell whereas in cell lines endogenously expressing the receptor (e.g., NG-108 or SH-SY5Y) there are likely to be 10^4 receptors per cell. Therefore we use a lower seeding density for the CHO cells (subconfluent approx 10^4 cells/well) compared to NG108 or SH-SY5Y (confluent, approx 10^6 cells/well). The volume of [^{3}H]NMS can be increased to 0.5 or even 1 mL if ligand depletion becomes a problem and the budget allows this.
8. We generally incubate the plates at 37°C by floating the plates in a water bath. Alternatively we have used a hot room or incubator for long incubation times. To wash, tilt the plate and remove the medium with a 1-mL pipet tip connected to a vacuum line. Add the wash buffer gently down the side of the well using a Gilson pipet tip with the end cut off. Neuronal cell lines generally do not adhere too well and require very gentle treatment. Some cells adhere better after pretreating the plate with polylysine (add 0.5 mL of 50 µg/mL poly-L-lysine in PBS, leave for at least 1 h or overnight, and remove before adding cells). CHO cells adhere very strongly and will tolerate washing with a repeating pipet.

9. Even the gentlest of washing procedures causes a loss of cells. Therefore it is important to count the number of cells in wells that have been through all of the same washing procedures as the ones that have been used for radioactivity.
10. PrBCM alkylates specifically 95–100% of all cell surface muscarinic receptors under these conditions ***(44)***. The PrBCM treatment step could be omitted if such a ligand is not available for the receptor of interest.
11. Recovery of receptor number after removal of agonist is due to both newly synthesized receptors and recycling of internalized receptors. Inclusion of 20 μg/mL of cycloheximide 30 min before and throughout the agonist treatment and recycling steps (but not the [^{3}H]NMS binding assay) inhibits new synthesis of receptors and thus enables the investigation of only the receptor recycling step.
12. A calibration curve was constructed with each experiment where the absorbance was measured from 4000, 5000, 6000, 8000, and 10,000 cells/well. Absorbance showed a linear relationship with increasing cell number up to 10,000 cells/well. At more than 20,000 cells/well the relationship was no longer linear, perhaps because the antibody concentrations became limiting or because the cells started to become confluent and the number of receptors per cell decreased or the accessibility of the antibody decreased.
13. The "nonspecific binding" is best determined by plating a similar number of cells that do not contain the epitope tag—either CHO-K1 or wild-type receptor-expressing CHO-K1 cells. An alternative measure of nonspecific binding is just to leave out the primary antibody. However, with high cell densities, we have found a significant difference between the two approaches indicating some nonspecific binding of the primary antibody.
14. An alternative to a radioimmunoassay (RIA) is an ELISA. However, the RIA gives the more reproducible results. This seemed to be mainly due to batch-to-batch variation in the affinity of the secondary enzyme-conjugated antibody—alkaline phosphatase-conjugated anti-mouse IgG (1:150, Vector Labs or 1:1000 Boehringer Mannheim). After washing (6 × 2 mL of Tris-buffered saline) the signal is developed by incubation with *p*-nitrophenyl phosphate substrate (Vector Labs) in 100 m*M* sodium bicarbonate pH 10 for 10–20 min. Absorbance is measured at 405 nm and the absorbance of the substrate alone subtracted.
15. There are at least two methods; the choice is a matter of personal preference. (1) Dry around the cover slip with the vacuum line, then carefully drop 100 μL of antibody solution on the cover slip. Keep a wet tissue inside the plate and keep the lid on to achieve a humidified atmosphere. (2) Place a drop (~25 μL) of antibody solution onto a strip of Parafilm or similar. Pick up the cover slip with forceps, dab the edge gently onto a tissue to remove most of the water, and invert onto the antibody droplet so the cells are facing downwards. At the end of the incubation, gently lift up the cover slip by injecting a little PBS underneath the cover slip with a syringe and needle. The cover slip can then be picked up with forceps and transferred back to the multiwell plate for washing.

16. The purpose of this step is to remove any salts and dust that might have accumulated on the cover slip. It is very easy to break them at this stage; I prepare duplicate cover slips just in case.
17. Many authors include protease inhibitors before and during the somatostatin incubation. I have used a simple reverse phase chromatography method for measuring ^{125}I-somatostatin-14 breakdown to ^{125}I-tyrosine and found that there is very little degradation in the CHO-cell experiments but that there was significant degradation when I used Neuro2A cells ***(33)***. The Neuro2A cells were used at a much higher cell density and this probably meant there were just more proteases around. Bacitracin was the only inhibitor found to prevent most of the degradation: leupeptin, phosphoramidon, amastatin, and soybean trypsin inhibitor were without effect ***(33)***.
18. The pH of the acid wash buffer should be chosen so as to enhance the rate of dissociation of the radioligand but not reduce cell viability. This is particularly important if radioligand recycling is to be measured after internalization. I have shown that a pH 5 wash was the best compromise for these cells, as a pH 5 wash at 15°C could dissociate 90% of the bound radioactivity in membranes. Also, cells washed in this manner retained the ability to exclude Trypan blue. Using a lower pH wash reduced cell viability by >10%.
19. The ability to measure surface bound radioligand is critically dependent on being able to wash free from bound radioligand quickly enough to reduce nonspecific binding but not lose a significant amount of specific binding. The most reliable way of checking the washing procedure is to determine the dissociation rate constant and show that the wash time is significantly shorter than the half-time for dissociation. Very few authors have demonstrated this and most assume that an acid wash will suffice. Measurement of the dissociation rate constant of fluoro-somatostatin (*see* **Subheading 3.6.**) showed that the half-time for dissociation was approx 20 s which was of a similar time scale to the wash duration (~10 s). This meant that it was extremely difficult to get good data for surface bound radioligand and therefore all my published data has been of internalized radioligand.
20. Excess unlabeled somatostatin-14 is included in the recycling medium to prevent reendocytosis of recycled radioligand ***(23)***. The same effect could be achieved by the use of an antagonist or by continuous washing during the recycling process.
21. At least three total internalization and three nonspecific internalization wells should be included for all time points as the nonspecific binding can sometimes vary with internalization time. The extended incubation times and washing procedures for the recycling experiments can lead to washing off of cells and so cell counts should be performed routinely on several different time points to check for this.
22. The method of addition of drug depends on the equipment available. A Gilson pipette is sufficient but a dedicated drug delivery or perfusion system would be better.
23. Being able to observe binding of fluoropeptide at the cell surface depends on the brightness of the fluorophore and the receptor density. I have observed cases

where surface binding was not detectable, but after 30 min internalized ligand accumulated in sufficient concentrations in endosomes to become detectable.

References

1. Bohm, S. K., Grady, E. F., and Bunnett, N. W. (1997) Regulatory mechanisms that modulate signalling by G protein-coupled receptors. *Biochem. J.* **322,** 1–18.
2. Koenig, J. A. and Edwardson, J. M. (1997) Endocytosis and recycling of G protein-coupled receptors. *Trends Pharmacol. Sci.* **18,** 276–287.
3. Lefkowitz, R. J. (1998) G protein coupled receptors: III new roles for receptor kinases and β-arrestins in receptor signaling and desensitization. *J. Biol. Chem.* **273,** 18,677–18,680.
4. Carman, C. V. and Benovic, J. L. (1998) G Protein-coupled receptors: turn-ons and turn-offs. *Curr. Opin. Neurobiol.* **8,** 335–344.
5. Ferguson, S. S. G. (2001) Evolving concepts in G protein-coupled receptor endocytosis: the role in receptor desensitization and signaling. *Pharmacol. Rev.* **53,** 1–24.
6. Ceresa, B. P. and Schmid, S. L. (2000) Regulation of signal transduction by endocytosis. *Curr. Opin. Cell Biol.* **12,** 204–210.
7. Zajchowski, L. D. and Robbins, S. M. (2002) Lipid rafts and little caves. *Eur. J. Biochem.* **269,** 737–752.
8. Ostrom, R. S. (2002) New determinants of receptor-effector coupling: trafficking and compartmentation in membrane microdomains. *Mol. Pharmacol.* **61,** 473–476.
9. Mellman, I. (1996) Endocytosis and molecular sorting. *Annu. Rev. Cell Dev. Biol.* **12,** 575–625.
10. Gruenberg, J. and Maxfield, F. R. (1995) Membrane transport in the endocytic pathway. *Curr. Opin. Cell Biol.* **7,** 552–563.
11. Vallee, R. B. and Okamoto, P. M. (1995) The regulation of endocytosis: identifying dynamin's binding partners. *Trends Cell Biol.* **5,** 43–47.
12. Geli, M. I. and Riezman, H. (1998) Endocytic internalization in yeast and animal cells: similar and different. *J. Cell Sci.* **111,** 1031–1037.
13. Clague, M. J. (1998) Molecular aspects of the endocytic pathway. *Biochem. J.* **336,** 271–282.
14. Marsh, M. and McMahon, H. T. (1999) The structural era of endocytosis. *Science* **285,** 215–220.
15. Cao, T. T., Deacon, H. W., Reczek, D., Bretscher, A., and von Zastrow, M. (1999) A kinase-regulated PDZ domain interaction controls endocytic sorting of the β2-adrenergic receptor. *Nature* **401,** 286–290.
16. Gage, R. M., Kim, K. A., Cao, T. T., and von Zastrow, M. (2001) A transplantable sorting signal that is sufficient to mediate rapid recycling of G protein coupled receptors. *J. Biol. Chem.* **276,** 44,712–44,720.
17. Seachrist, J. L., Anborgh, P. H., and Ferguson, S. S. G. (2000) β2-adrenergic receptor internalization, endosomal sorting and plasma membrane recycling are regulated by rab GTPases. *J. Biol. Chem.* **275,** 27,221–27,228.

18. Trejo, J. and Coughlin, S. R. (1999) The cytoplasmic tails of protease-activated receptor-1 and substance P receptor specify sorting to lysosomes versus recycling. *J. Biol. Chem.* **274,** 2216–2224.
19. Tsao, P. I. and von Zastrow, M. (2000) Type-specific sorting of G protein-coupled receptors after endocytosis. *J. Biol. Chem.* **275,** 11,130–11,140.
20. Tsao, P., Cao, T., and von Zastrow, M. (2001) Role of endocytosis in mediating downregulation of G protein-coupled receptors. *Trends Pharmacol. Sci.* **22,** 91–96.
21. Koenig, J. A. and Edwardson, J. M. (1994) Kinetic analysis of the trafficking of muscarinic acetylcholine receptors between the plasma membrane and intracellular compartments. *J. Biol. Chem.* **269,** 17,174–17,182.
22. von Zastrow, M. and Kobilka, B. K. (1994) Antagonist-dependent and independent steps in the mechanism of adrenergic receptor internalization. *J. Biol. Chem.* **269,** 18,448–18,452.
23. Koenig, J. A., Kaur, R., Dodgeon, I., Edwardson, J. M., and Humphrey, P. P. A. (1998) Fates of endocytosed somatostatin receptors and associated agonists. *Biochem. J.* **336,** 291–298.
24. Roettger, B. F., Rentsch, R. U., Pinon, D., et al. (1995) Dual pathways of internalization of the cholecystokinin receptor. *J. Cell Biol.* **128,** 1029–1041.
25. Slowiejko, D. M., McEwen, E. L., Ernst, S. A., and Fisher, S. K. (1996) Muscarinic receptor sequestration in SH-SY5Y neuroblastoma cells is inhibited when clathrin distribution is perturbed. *J. Neurochem.* **66,** 186–196.
26. Hunyady, L., Merelli, F., Baukal, A. J., Balla, T., and Catt, K. J. (1991) Agonist-induced endocytosis and signal generation in adrenal glomerulosa cells. *J. Biol. Chem.* **266,** 2783–2788.
27. Pippig, S., Andexinger, S., and Lohse, M. J. (1995) Sequestration and recycling of β2-adrenergic receptors permit receptor resensitization. *Mol. Pharmacol.* **47,** 666–676.
28. Szekeres, P. G., Koenig, J. A., and Edwardson, J. M. (1998) Involvement of receptor cycling and receptor reserve in resensitization of muscarinic responses in SH-SY5Y human neuroblastoma cells. *J. Neurochem.* **70,** 1694–1703.
29. Garland, E. M., Grady, E. F., Payan, D. G., Vigna, S. R., and Bunnett, N. W. (1994) Agonist induced internalization of the substance P (NK1) receptor expressed in epithelial cells. *Biochem. J.* **383,** 177–186.
30. Hertel, C., Coulter, S. J., and Perkins, J. P. (1985) A comparison of catecholamine-induced internalization of β-adrenergic receptors and receptor-mediated endocytosis of epidermal growth factor in human astrocytoma cells. *J. Biol. Chem.* **260,** 12,547–12,553.
31. Singh, S. and Aggarwal, B. B. (1995) Protein tyrosine phosphatase inhibitors block tumour necrosis factor dependent activation of the nuclear transcription factor NF-KB. *J. Biol. Chem.* **270,** 10,631–10,639.
32. Heuser, J. E. and Anderson, R. G. W. (1989) Hypertonic media inhibit receptor-mediated endocytosis by blocking clathrin-coated pit formation. *J. Cell Biol.* **108,** 389–400.

33. Koenig, J. A., Edwardson, J. M., and Humphrey, P. P. A. (1997) Somatostatin receptors in Neuro-2a cells: 2, ligand internalisation. *Br. J. Pharmacol.* **120,** 52–59.
34. Moore, R. H., Sadovnikoff, N., Hoffenberg, S., et al. (1995) Ligand stimulated β2-adrenergic receptor internalization via the constitutive endocytic pathway into rab5 containing endosomes. *J. Cell Sci.* **108,** 2983–2991.
35. vanderBliek, A. M. (1999) Functional diversity in the dynamin family. *Trends Cell Biol.* **9,** 96–102.
36. Vogler, O., Baogatkewitsch, G. S., Wriske, C., Krummererl, P., Jakobs, K. H., and van Koppen, C. J. (1998) Receptor subtype-specific regulation of muscarinic acetylcholine receptor sequestration by dynamin. *J. Biol. Chem.* **273,** 12,155–12,160.
37. Gaborik, Z., Szaszak, M., Szidonya, L., et al. (2001) β-Arrestin and dynamin-dependent endocytosis of the AT1 angiotensin receptor. *Mol. Pharmacol.* **59,** 239–247.
38. Werbonat, Y., Kleutges, N., Jakobs, K. H., and vanKoppen, C. J. (2000) Essential role of dynamin in internalization of M2 muscarinic acetylcholine and angiotensin AT1A receptors. *J. Biol. Chem.* **275,** 21,969–21,974.
39. Gicquiaux, H., Lecat, S., Gaire, M., et al. (2002) Rapid internalization and recycling of the human neuropeptide YY1 receptor. *J. Biol. Chem.* **277,** 6645–6655.
40. Pless, D. D. and Wellner, R. B. (1996) In vitro fusion of endocytic vesicles: effects of reagents that alter endosomal pH. *J. Cell Biochem.* **62,** 27–39.
41. Volpicelli, L. A., Lah, J. J., Fang, G., Godenring, J. R., and Levey, A. I. (2002) Rab11a and myosin Vb regulate recycling of the M4 muscarinic acetylcholine receptor. *J. Neurosci.* **22,** 9776–9784.
42. Volpicelli, L. A., Lah, J. J., and Levey, A. I. (2001) Rab5 dependent trafficking of the m4 muscarinic acetylcholine receptor to the plasma membrane, early endosomes and multivesicular bodies. *J. Biol. Chem.* **2746,** 47,590–47,598.
43. Clark, R. B. and Knoll, B. J. (2002) Measurement of receptor desensitization and internalization in intact cells. *Methods Enzymol.* **343,** 506–529.
44. Koenig, J. A. and Edwardson, J. M. (1994) Routes of delivery of muscarinic acetylcholine receptors to the plasma membrane in NG108-15 cells. *Br. J. Pharmacol.* **111,** 1023–1028.
45. Petrou, C., Chen, L., and Tashjian, A. H. (1997) A receptor-G protein coupling independent step in the internalization of the thyrotropin releasing hormone receptor. *J. Biol. Chem.* **272,** 2326–2333.
46. Tolbert, L. M. and Lameh, J. (1998) Antibody to epitope tag induces internalization of human muscarinic subtype 1 receptor. *J. Neurochem.* **70,** 113–119.
47. Signoret, N., Pelchen-Matthes, A., Mack, M., Proudfoot, A. E. I., and Marsh, M. (2000) Endocytosis and recycling of the HIV coreceptor CCR5. *J. Cell Biol.* **151,** 1281–1293.
48. Hein, L., Meinel, L., Pratt, R. E., Dzau, V. J., and Kobilka, B. K. (1997) Intracellular trafficking of angiotensin II and its AT1 and AT2 receptors: evidence for selective sorting of receptor and ligand. *Mol. Endocrinol.* **11,** 1266–1277.

49. Rizzuto, R., Carrington, W., and Tuft, R. A. (1998) Digital imaging microscopy of living cells. *Trends Cell Biol.* **8,** 288–291.
50. White, J. and Stelzer, E. (1999) Photobleaching GFP reveals protein dynamics inside live cells. *Trends Cell Biol.* **9,** 61–65.
51. Kallal, L. and Benovic, J. L. (2000) Using green fluorescent proteins to study G protein-coupled receptor localization and trafficking. *Trends Pharmacol. Sci.* **21,** 175–180.
52. Corvera, S. and Czech, M. P. (1998) Direct targets of phosphoinositide 3-kinase products in membrane traffic and signal transduction. *Trends Cell Biol.* **8,** 442–446.
53. Simonsen, A., Lippe, R., Christoforidis, S., et al. (1998) EEA1 links PI(3)K function to Rab5 regulation of endosome fusion. *Nature* **394,** 494–498.
54. Pawley, J. (2000) The 39 steps: a cautionary tale of quantitative 3-D fluorescence microscopy. *Biotechniques* **28,** 884.
55. Sakai, T., Mizuno, T., Miyamoto, H., and Kawasaki, K. (1998) Two distinct kinds of tubular organelles involved in the rapid recycling and slow processing of endocytosed transferrin. *Biochem. Biophys. Res. Commun.* **242,** 151–157.
56. Patterson, G. H., Knobel, S. M., Sharif, W. D., Kain, S. R., and Piston, D. W. (1997) Use of the green fluorescent protein and its mutants in quantitative fluorescence microscopy. *Biophys. J.* **73,** 2782–2790.
57. Van der Voort, H. T. M. and Strasters, K. C. (1995) Restoration of confocal images for quantitative image analysis. *J. Microsc.* **178,** 165–181.
58. Go, W. Y., Roettger, B. F., Holicky, E. L., Hadac, E. M., and Miller, L. J. (1997) Quantitative dynamic multicompartmental analysis of cholecystokinin receptor movement in a living cell using dual fluorophores and reconstruction of confocal images. *Anal. Biochem.* **247,** 210–215.
59. Niswender, K. D., Blackman, S. M., Rohde, L., Magnuson, M. A., and Piston, D. W. (1995) Quantitative imaging of green fluorescent protein in cultured cells: comparison of microscopic techniques, use in fusion proteins and detection limits. *J. Microsc.* **180,** 109–116.
60. Ghosh, R. N., Mallet, W. G., Soe, T. T., McGraw, T. E., and Maxfield, F. R. (1998) An endocytosed TGN38 chimeric protein is delivered to the TGN after trafficking through the endocytic recycling compartment in CHO cells. *J. Cell Biol.* **142,** 923–936.
61. Conway, B. R., Minor, L. K., Xu, J. Z., D'Andrea, M. R., Ghosh, R. N., and Demarest, K. T. (2001) Quantitative analysis of agonist-dependent parathyroid hormone receptor trafficking in whole cells using a functional green fluorescent protein conjugate. *J. Cell Physiol.* **189,** 341–355.
62. Becker, P. L. (1996) Quantitative fluorescence measurements, in *Fluorescence Imaging Spectroscopy and Microscopy* (Wang, X. F. and Herman, B., eds.), John Wiley & Sons, New York, pp. 1–29.
63. Mallard, F., Antony, C., Tenza, D., Salamero, J., Goud, B., and Johannes, L. (1998) Direct pathway from early/recycling endosomes to the Golgi apparatus revealed through the study of shiga toxin B-fragment transport. *J. Cell Biol.* **143,** 973–990.

64. Janecki, A. J., Janecki, M., Akhter, S., and Donowitz, M. (2000) Quantitation of plasma membrane expression of a fusion protein of Na/H exchanger NHE3 and green fluorescence protein (GFP) in living PS120 fibroblasts. *J. Histochem. Cytochem.* **48,** 1479–1491.
65. Koenig, J. A., Edwardson, J. M., and Humphrey, P. P. A. (1997) Somatostatin receptors in Neuro-2a cells: 1, operational characteristics. *Br. J. Pharmacol.* **120,** 45–51.
66. Pawley, J. B. (1995) *Handbook of Biological Confocal Microscopy*, 2nd ed. Plenum Press, New York.
67. Bowen, W. P. and Jerman, J. C. (1995) Nonlinear regression using spreadsheets. *Trends Pharmacol. Sci.* **16,** 413–417.
68. Oakley, R. H., Laporte, S. A., Holt, J. A., Barak, L. S., and Caron, M. G. (1999) Association of β-arrestin with G protein-coupled receptors during clathrin-mediated endocytosis dictates the profile of receptor resensitization. *J. Biol. Chem.* **274,** 32,248–32,257.
69. Oakley, R. H., Laporte, S. A., Holt, J. A., Barak, L. S., and Caron, M. G. (2001) Molecular determinants underlying the formation of stable intracellular G protein-coupled receptor-β-arrestin complexes after receptor endocytosis. *J. Biol. Chem.* **276,** 19,452–19,460.

16

G-Protein-Coupled Receptor Phosphorylation and Palmitoylation

Andrew B. Tobin and Mark Wheatley

Summary

It is now clear that nearly all G-protein-coupled receptors (GPCRs) are phosphorylated and palmitolyated. The process of receptor phosphorylation has been extensively studied because it offers a regulatory mechanism that is both rapid and dynamic. However, it has recently become clear that palmitoyaltion of GPCRs at C-terminal cysteine residues may also offer dynamic receptor modification. A growing number of GPCRs have been demonstrated to undergo rapid agonist-mediated changes in their palmitoylation status with functional implications to receptor signaling. This chapter aims to outline the methods we have used to investigate agonist-mediated changes in GPCR phosphorylation and palmitoylation.

Key Words

Antibodies, G-protein-coupled receptor, immunoprecipitation, ligand binding, orthophosphate, palmitic acid, palmitoylation, phosphorylation, posttranslational modification, SDS-PAGE, solubilization.

1. Introduction

G-protein-coupled receptors (GPCRs) are known to undergo a number of posttranslationally modifications. Most GPCRs have at least one extracellular residue that is glycosylated, usually *N*-linked oligosaccharide. In addition, specific intracellular residues can be modified by phosphorylation or acylation by palmitate. These intracellular modifications have been the focus of intensive research in recent years as changes in the phosphorylation and/or acylation status have been shown to have profound effects on GPCR signaling capability ***(1,2)***.

Reversible phosphorylation of key components is a common regulatory phenomenon for a wide range of biological processes and cell signaling is no exception ***(3,4)***. Phosphorylation of GPCRs can be mediated by both second

From: *Methods in Molecular Biology, vol. 259, Receptor Signal Transduction Protocols, 2nd ed.*
Edited by: G. B. Willars and R. A. J. Challiss © Humana Press Inc., Totowa, NJ

messenger-activated kinases and GPCR kinases (GRKs), with GRKs phosphorylating only agonist-occupied receptors *(1,5,6)*. The phosphorylated receptors initiate a multimolecular desensitization process involving recruitment of β-arrestins to the cell membrane, uncoupling of receptor–G-protein complexes and GPCR internalization via clathrin-coated pits. Subsequent dephosphorylation of internalised GPCRs can precede their recycling back to the cell surface *(5,6)*.

Many, but not all, GPCRs have one or more cysteines located distal to the seventh transmembrane helix that is modified by the C_{16} fatty acid palmitate. This palmitoylation anchors the C-terminus of the receptor to the plasma membrane, thereby imposing conformational constraints on this region of the GPCR. Unlike other acylations, such as myristoylation and prenylation, palmitoylation can undergo rapid turnover with potential for palmitoylation–depalmitoylation cycles *(2,7)*. Given the structural importance of palmitoylation to the conformation of the C-terminal region of GPCRs, this could have important ramifications with respect to intracellular signaling. Our work on the V_{1a} vasopressin receptor has established that palmitoylation of this receptor is regulated by agonist stimulation and that the palmitoylation status of the receptor affects the ability of the receptor to be phosphorylated *(2)*. Studies by others on different GPCRs have also identified a relationship between the phosphorylation and palmitoylation status of receptors, suggesting that these two forms of post-translational modification may be commonly used by GPCRs in a regulatory function *(7,8)*.

In this chapter, we present appropriate methodologies for studying phosphorylation and palmitoylation of GPCRs.

2. Materials

1. [^{32}P]orthophosphate. We use [^{32}P]orthophosphate from AmershamPharmacia (cat. no. PBS-11), 10 mCi/mL stock concentration.
2. Phosphate-free Krebs–HEPES: 10 m*M* 4-(2-hydroxyethyl)piperazine-1-ethanesulfonic acid (HEPES), pH 7.4, 118 m*M* NaCl, 4.3 m*M* KCl, 1.17 m*M* $MgSO_4$, 1.3 m*M* $CaCl_2$, 25.0 m*M* $NaHCO_3$, 11.7 m*M* glucose.
3. Krebs–HEPES: 10 m*M* HEPES, pH 7.4, 118 m*M* NaCl, 4.3 m*M* KCl, 1.17 m*M* $MgSO_4$, 1.3 m*M* $CaCl_2$, 25.0 m*M* $NaHCO_3$, 1.18 m*M* KH_2PO_4, 11.7 m*M* glucose.
4. Solubilization buffer: 10 m*M* Tris-HCl, pH 7.4, 10 m*M* EDTA, 500 m*M* NaCl, 1% (v/v) Nonidet P-40 (NP-40) (*see* **Note 1**).
5. TE buffer: 10 m*M* Tris-HCl, pH 7.4; 10 m*M* EDTA.
6. TE+P: 10 m*M* Tris-HCl, pH 7.4, 10 m*M* EDTA, 1 m*M* phenylmethylsulfonyl fluoride, 1 μg/mL of pepstatin A, 1 μg/mL of leupeptin, 10 μg/mL of soybean trypsin inhibitor, 25 m*M* glycerophosphate.
7. [9,10-^{3}H]Palmitic acid, 52 Ci/mmol stock concentration (NEN).
8. Laemmli buffer: 125 m*M* Tris-HCl, pH 6.8; 4% sodium dodecyl sulfate (SDS), 20% glycerol, 10 m*M* mercaptoethanol; 0.5% bromophenol blue.

9. Coomassie stain: 40% methanol, 10% acetic acid, 0.2% Coomassie brilliant blue R 250.
10. Destain: 40% methanol, 10% acetic acid.

3. Methods

3.1. Whole Cell ^{32}P-Labeling and Stimulation

The protocol that is described here has been used in our laboratory in studies of the M_3-muscarinic receptor *(3,4)* and V_{1a} vasopressin receptor *(2)* expressed as recombinant proteins in Chinese hamster ovary (CHO) cells. In the case of the M_3-muscarinic receptor we have developed our own receptor-specific antibodies to a fusion protein containing the third intracellular loop of the receptor. In studies on the V_{1a} vasopressin receptor we have used a commercial antibody that recognizes a hemaglutinin tag (monoclonal antibody 12CA5) that was engineered into the N-terminus of the receptor.

1. Cells are prepared on 6-well culture dishes (*see* **Note 2**).
2. Wash each well with 1 mL of phosphate-free Krebs.
3. Add 1 mL of phosphate-free Krebs containing 50 μCi/mL of [^{32}P]orthophosphate.
4. Incubate for 60–90 min at 37°C.
5. Stimulate cells by directly adding 10 μL of agonist made up as a 100X stock in phosphate-free Krebs.
6. Stop stimulation by removing medium (we use a suction line) and adding 1 mL of ice-cold solubilization buffer.

3.2. Receptor Solubilization and Immunoprecipitation

1. Allow the cells to solubilize for 10 min on ice. All subsequent steps should be carried out with ice-cold buffers and centrifuges set at 4°C.
2. Pipet up and down to ensure all cellular material is in suspension.
3. Transfer to a 1.5-mL microfuge tube and clear by centrifuging in a benchtop centrifuge maximum speed (~20,000*g* for 5 min). Sealed screw-top microfuge tubes can be used here to reduce contamination of the microfuge.
4. Transfer 800 μL of supernatant to a fresh tube and add anti-receptor antibody (1–5 μg).
5. Vortex-mix and leave on ice for 60–90 min.
6. Add 180 μL of a protein A–Sepharose slurry (*see* **Note 3**). Mix by vortexing and place on a roller at 4°C for 15 min. To prevent accidental radioactive contamination it is recommended that the microfuge tubes are placed in a large 50-mL Falcon tube while rolling.
7. Pellet the protein A–Sepharose beads by a brief 20-s pulse in a benchtop centrifuge (*see* **Note 4**).
8. Remove supernatant. It is very important that none of the protein A beads are inadvertently removed during this step. We usually use a suction line and remove sufficient supernatant so that there is approx 100 μL left in the tube. It is better to have more wash steps than to lose sample by sucking up the protein A pellet.

9. Add 1 mL of TE buffer and vortex thoroughly.
10. Pellet the protein A beads and remove the supernatant as before.
11. Add 1 mL of solubilization buffer, vortex-mix thoroughly, centrifuge protein A pellet, and remove the supernatant as before.
12. Wash protein A pellet twice more with TE buffer.
13. On the final wash, all of the supernatant has to be removed. To do this we remove the supernatant with a fine-tipped gel loading pipet (*see* **Note 5**).
14. Add 20 µL of Laemmli buffer to the protein A pellet.
15. Mix the sample by flicking the tube and then heat at 50–60°C for 2 min.
16. Centrifuge the sample briefly in a benchtop centrifuge and apply to a sodium dodecyl sulfate-polyacrylamide gel electrophoresis (SDS-PAGE) gel.
17. To check for equivalent antibody loading, the gel is stained with Coomassie blue for 10 min and then destained for 30 min before being dried and exposed for autoradiography.

3.3. Increasing the Sensitivity by Making Micromembranes

If background problems are encountered, the signal-to-noise ratio could be improved by immunoprecipitation of the receptor from a membrane preparation derived from cells that have been labeled with [^{32}P]orthophosphate. However, the samples generated from whole-cell phosphorylation are highly radioactive, which means that techniques such as sonication and ultracentrifugation, which can be employed to make membrane preparations, are not generally practical as this will result in large general purpose laboratory equipment being contaminated with ^{32}P.

The method we use for the preparation of membranes from radiolabeled cells is described in **Subheading 3.4.** The receptor contained in these membranes can then be used for immunoprecipitation.

3.4. Making a Membrane Preparation from ^{32}P-Labeled Cells

1. Label cells with [^{32}P]orthophosphate at 37°C as described in **Subheading 3.1.**
2. Stimulate as in **Subheading 3.1.**
3. Stop the reaction using 1 mL of ice-cold TE+P buffer.
4. Leave cells to swell on crushed ice for 15 min.
5. Pipet to bring cells into suspension.
6. Pass six times through a fine-bored needle connected to a 1-mL syringe.
7. Centrifuge the sample in a cooled (4°C) benchtop centrifuge set at maximum speed (~20,000*g*) for 5 min.
8. *Optional.* Remove supernatant and wash the pellet with ice-cold TE+P buffer containing 500 m*M* NaCl.
9. Resuspend the membrane pellet in 500 µL of ice-cold solubilization buffer and leave to solubilize for 10 min.
10. Clear the sample by centrifugation (20,000*g* for 5 min).
11. Immunoprecipitate as described in **Subheading 3.1.**

3.5. Identification of GPCR Palmitoylation

Our studies on the V_{1a} vasopressin receptor have identified that the receptor is palmitoylated on at least two cysteine residues.

The method for measuring palmitoylation of GPCRs is essentially similar to that of phosphorylation in that the cells are labeled with the radiolabel (in this case [^{3}H]palmitic acid), the receptors solubilized, immunoprecipitated, and run out on an SDS-PAGE gel. The major difference is that ^{3}H is a very low energy β-emitter compared to ^{32}P and therefore gels have to be impregnated with scintillant to improve the efficiency of the autoradiography. Nevertheless, exposure times are typically several weeks (usually 3–7 wk). The turnover between experiments is therefore rather slow.

3.6. Labeling Cells With [^{3}H]Palmitic Acid, Solubilization, and Immunoprecipitation

1. Plate out CHO cells expressing the receptor of interest onto 6-well dishes. Allow to adhere overnight.
2. Wash cells with phosphate-buffered saline (PBS) and add 1 mL of of Dulbecco's modified Eagle medium (DMEM) plus 1% foetal calf serum containing 400 μCi of [^{3}H]palmitic acid/well (*see* **Note 6**) and incubate at 37°C for 4–16 h (*see* **Note 7**).
3. Remove the medium and wash with Krebs–HEPES buffer.
4. Leave to equilibrate in 1 mL of Krebs–HEPES buffer for 10 min.
5. Stimulate the cells.
6. Stop the reaction by removing medium and adding of 1 mL of ice-cold solubilization buffer.
7. Solubilize receptors and immunoprecipitate as described in **Subheading 3.2.**
8. Once the SDS-PAGE gel has run, first fix the gel using 40% methanol–10% acetic acid for 15 min. Then impregnate the gel with scintillant (Enhance™ from Packard or Amplify™ from Amersham Pharmacia). Impregnate for 15 min. Dry the gel and expose at –70°C to Kodak X-Omat film for 3–7 wk.

4. Notes

1. To increase the effectiveness of this buffer, 0.1% (w/v) SDS can be added. However, the SDS may cause the genomic DNA from the solubilized cells to precipitate. This can cause serious problems with background contamination. We have found with the M_3-muscarinic receptor, which is known to be particularly resistant to solubilization *(3)*, that the addition of SDS has still not been necessary. We would recommend, therefore, that solubilization buffer not containing SDS should be tried first.
2. We have conducted this procedure on receptors stably expressed in CHO and transiently transfected in CHO, COS-7, and HEK-293 cells. In the case of HEK-293 we routinely treat the dishes with poly-L-lysine to ensure that they adhere well since there are a number of wash steps. Also remember to include extra wells for the binding experiment to check for equal loading of transfected receptors.

3. Protein A–Sepharose purchased from AmershamPharmacia (CL-4B protein A–Sepharose) comes as a 1.5-g pack. To prepare this for use reconstitute the 1.5-g in 50 mL of TE buffer. After equilibrating for at least 1 h (more usually overnight) on a roller at 4°C the protein A–Sepharose made up in this way is stable for approx 3 mo. Sodium azide can be added as a preservative.
4. Contamination with cellular DNA can be a source of major background problems. We have found that cells of human origin (e.g., HEK-293, SH-SY5Y) are particularly problematic. If there is a problem with DNA contamination it is usually evident at this stage, with the protein A–Sepharose not forming a uniformed pellet but instead "smearing" up the side of the tube. If this appears to be a problem, then extensive vortexing at this stage and throughout the subsequent wash steps usually solves the problem. Reducing the number (i.e., confluency) of the cells at the start of the experiment will also reduce this problem.
5. If two different cell lines are being compared for their ability to phosphorylate receptors, then it is important to load the same number of receptors to each well. It is at this final wash step that we adjust for equal receptor loading. We generally use cells transfected with the M_3-muscarinic receptor, the levels of which are easily determined by ligand binding studies. When the cells are split into 6-well dishes for the phosphorylation study, three extra wells are included for the ligand binding assays. Two wells are used to determine the total binding and one the nonspecific binding. To adjust for equal receptor loading, the protein A pellet in the final wash step is resuspended in 1 mL of ice-cold TE. Then a volume is removed that contains equivalent numbers of receptors. This is then spun down and the protein A pellet resuspended in 20 μL of Laemmli buffer ready for loading on to the SDS-PAGE gel.
6. Preparation of [^{3}H]palmitic acid. Remove solvent by drying under a stream of N_2. A Pasteur pipet on the end of a length of tubing from the gas cylinder is convenient for this. Resuspend in 50 μL of dimethyl sulfoxide and then make up in DMEM (1% v/v final concentration).
7. Avoid very long incubation periods to minimize metabolic incorporation of ^{3}H label into the cellular amino acid pool as this could result in nonspecific labeling of proteins.

References

1. Tobin, A. B. (2002) Are we β-arking up the wrong tree? Casein kinase 1α provides an additional pathway for GPCR phosphorylation. *Trends Pharmacol. Sci.* **7,** 337–343.
2. Hawtin, S. R., Tobin, A. B., Patel, S., and Wheatley, M. (2001) Palmitoylation of the vasopressin V_{1a} receptor reveals different conformational requirements for signaling, agonist-induced receptor phosphorylation and sequestration. *J. Biol. Chem.* **276,** 38,139–38,146.
3. Tobin, A. B. and Nahorski, S. R. (1993) Rapid agonist-mediated phosphorylation of m3-muscarinic receptors revealed by immunoprecipitation. *J. Biol. Chem.* **268,** 9817–9823.

4. Budd, D. C., Willars, G. B., McDonald, J. E., and Tobin, A. B. (2001) Phosphorylation of the $G_{q/11}$-coupled m3-muscarinic receptor is involved in receptor activation of the ERK-1/2 mitogen activated protein kinase pathway. *J. Biol. Chem.* **276,** 4581–4587.
5. Ferguson, S. S. G. (2001) Evolving concepts in G protein-coupled receptor endocytosis: the role in receptor desensitization and signaling. *Pharmacol. Rev.* **53,** 1–24.
6. Pitcher, J. A., Freeman, N. J., and Lefkowitz, R. J. (1998) G protein-coupled receptor kinases. *Annu. Rev. Biochem.* **67,** 653–692.
7. Moffett, S., Rousseau, G., Lagace, M., and Bouvier M. (2001) The palmitoylation state of the β_2-adrenergic receptor regulates the synergistic action of cyclic AMP-dependent protein kinase and β-adrenergic receptor kinase involved in its phosphorylation and desensitisation. *J. Neurochem.* **76,** 269–279.
8. Sadeghi, H. M, Innamorati, G., Dagarag, M., and Birnbaumer, M. (1997) Palmitoylation of the V_2-vasopressin receptor. *Mol. Pharmacol.* **52,** 21–29.

17

Identification of G-Protein-Coupled Receptor Phosphorylation Sites by 2D Phosphopeptide Mapping

Andree Blaukat

Summary

Reversible phosphorylation is important for G-protein-coupled receptor (GPCR) signaling, desensitization, and endocytosis, yet the precise location and role of in vivo phosphorylation sites is unknown for most receptors. This chapter describes a powerful analytical method for the direct identification of GPCR phosphorylation sites by two-dimensional (2D) phosphopeptide mapping. The GPCR of interest is isolated from ^{32}P-labeled cells by immunoprecipitation and transferred to nitrocellulose membranes. *In situ* cleavage by trypsin releases phosphopeptides that are separated by a combination of high-voltage electrophoresis and chromatography. Phosphoamino acid analysis and Edman sequencing of isolated phosphopeptides reveals information that can lead to the direct identification of GPCR phosphorylation sites. Furthermore, the 2D phosphopeptide mapping technique allows the analysis of temporal and positional changes in the GPCR phosphorylation pattern under different physiological conditions.

Key Words

Desensitization, Edman sequencing, G-protein-coupled receptor, immunoprecipitation, kinase, phosphopeptide mapping, phosphorylation, radiolabelling, thin-layer chromatography, tryptic digest.

1. Introduction

G-protein-coupled receptors (GPCRs) constitute the largest family of transmembrane receptors and are involved in the control of virtually all cellular functions. On repeated or prolonged agonist stimulation GPCR signaling is attenuated by negative feedback loops. Although this desensitization of agonist/receptor/G-protein/effector systems in general involves perturbations of all components, GPCR phosphorylation resulting in uncoupling of the receptor from further G-protein activation appears to be the most important. Furthermore, there is evidence for a role of GPCR phosphorylation in regulating

From: *Methods in Molecular Biology, vol. 259, Receptor Signal Transduction Protocols, 2nd ed.*
Edited by: G. B. Willars and R. A. J. Challiss

internalization and receptor trafficking *(1)*. More recently, receptor phosphorylation has been shown to redirect the G-protein-coupling profile of certain GPCRs and balance activation of distinct signaling pathways *(2)*. GPCRs are phosphorylated by GPCR kinases that seem to specifically recognize activated receptors, by second messenger-activated kinases, such as protein kinase A and C and by casein kinases *(1,3)*. Despite its importance in regulation of GPCR function, the precise location and role of in vivo phosphorylation sites is unknown for most receptors.

This chapter describes a powerful analytical method for the direct identification of GPCR phosphorylation sites by two-dimensional (2D) phosphopeptide mapping *(4)*. Cells expressing the GPCR of interest, endogenously or introduced by transfection, are metabolically labeled with [^{32}P]orthophosphate. Following treatment of choice (e.g., agonist stimulation) cells are lysed and the GPCR is isolated by immunoprecipitation. After transfer to nitrocellulose membranes, *in situ* cleavage by trypsin releases phosphopeptides that are separated on thin-layer chromatography (TLC) plates by a combination of high-voltage electrophoresis and chromatography.

Phosphoamino acid analysis and Edman sequencing of isolated phosphopeptides reveals information that can lead to the direct identification of GPCR phosphorylation sites. In addition, the 2D phosphopeptide mapping technique allows the analysis of temporal and positional changes in the GPCR phosphorylation pattern under different physiological conditions. A workflow of the experimental procedure is shown in **Fig. 1**.

The technique of 2D phosphopeptide mapping was originally developed in the laboratory of Tony Hunter *(4)* and first applied to GPCRs in a rather descriptive way by Ozcelebi and Miller in 1995 to visualize phosphorylation patterns of cholecystokinin receptors *(5)*. However, only recently the method was combined with Edman sequencing to identify individual cellular phosphorylation sites of a GPCR *(6)*.

2. Materials

1. [^{32}P]H_3PO_4, HCl- and carrier-free.
2. Phosphate-free medium, for example, 4-(2-hydroxyehtyl)piperazine-1-ethanesulfonic acid (HEPES)-buffered Dulbecco's modified Eagle's medium (DMEM). If no phosphate-free medium is available, a balanced phosphate-free salt solution may also work.
3. A cell culture quality balanced salt solution, for example, phosphate- (PBS) or HEPES-buffered saline (HBSS).
4. RIPA lysis buffer: 1% Nonidet P-40 (NP-40), 0.5% deoxycholate, 0.1% sodium dodecyl sulfate (SDS), 50 m*M* Tris-HCl, 150 m*M* NaCl, 2 m*M* EDTA, 25 m*M* NaF, pH 7.5. Phosphatase and protease inhibitors should be added just before the experiment: 1 m*M* sodium vanadate (*see* **ref. *7*** for preparation), 0.1 m*M* Pefabloc

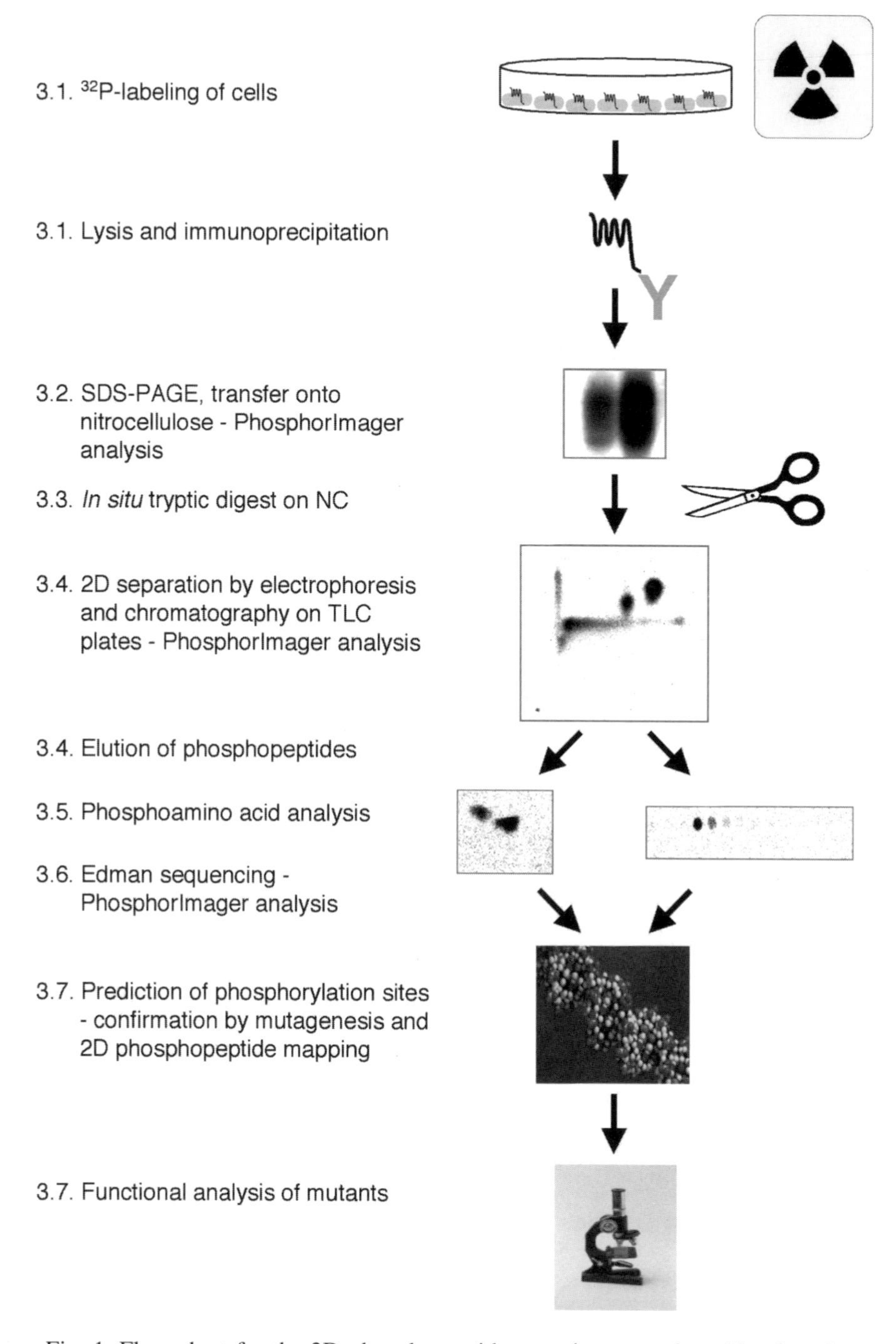

Fig. 1. Flow chart for the 2D phosphopeptide mapping procedure. Numbered steps correspond to **Subheading 3.**

(or 1 m*M* phenylmethylsulfonyl fluoride [PMSF]), 10 μg/mL aprotinin, leupeptin, and pepstatin A. Other, relatively high stringency lysis buffers may be used as well.

5. Precipitating antibody against the target protein or a fused epitope tag (*see* **Note 1**).
6. Protein A–Sepharose for rabbit, protein G–Sepharose for mouse, rat, goat, and sheep antibodies.
7. SDS sample buffer: 62.5 m*M* Tris-HCl, pH 6.8, 2% SDS, 30% (v/v) glycerol, 0.01% bromophenol blue, 25 m*M* dithiothreitol (DTT).
8. Nitrocellulose blotting membranes (e.g., from Schleicher & Schuell).
9. Ammonium bicarbonate, 50 m*M* in water, freshly prepared from 1 *M* stock that is tightly locked; stable for 1–2 mo at 4°C.
10. Modified sequencing grade trypsin (e.g., from Promega). The trypsin should be dissolved in 1 m*M* HCl to 0.1 μg/μL and stored at –20°C. Repeated thaw–freeze cycles (more than three times) are not recommended.
11. Chemicals: acetic acid, *n*-butanol, formic acid, hydrogen peroxide, isobutyric acid, ninhydrin, polyvinylpyrrolidon K30 (PVP), pyridine.
12. Cellulose TLC plates, 20 × 20 cm (e.g., from Merck).
13. Equipment: facilities to handle necessary ^{32}P quantities of 5–50 mCi (e.g., appropriate shielding, Plexiglas boxes and containers for tubes, protected membrane pump), filter tips to avoid contamination of pipettes, 1.5 to 2-mL screw cap tubes with rubber or silicon sealing, end-over-end shaker at 4°C, cooled microcentrifuge, electrophoresis and blotting instruments, vacuum centrifuge, Hunter thin-layer peptide mapping electrophoresis unit (from CBS Scientific Inc., Del Mar, CA, USA) including power supply, chromatography tank, PhosphorImager (autoradiography may also work but is usually not sensitive enough, in particular for analysis of Edman sequencing cycles).

3. Methods

3.1. ^{32}P-Labeling and Immunoprecipitation

1. Cells expressing the GPCR of interest are grown to 75–100% confluence during 1–3 d of culture (*see* **Note 2**). Wash cells once with phosphate-free medium and incubate them in the same medium for 30–120 min to (partially) deplete intracellular phosphate pools. Common antibiotics, 0.1% bovine serum albumin (BSA) or 0.5% dialyzed fetal calf serum may be added to the medium, if necessary for survival of cells.
2. Remove the medium and label cells with 0.2–2 mCi/mL of [^{32}P]orthophosphate in phosphate-free medium for 6–12 h (*see* **Notes 2** and **3**).
3. Stimulate with a GPCR ligand of choice and remove medium either using a pipet with filter tips or a membrane pump (*see* **Note 4**). For a pilot experiment, stimulation with a high ligand concentration (e.g., 10 times the EC_{50} determined in functional assays) for 5–15 min is recommended. Later, concentration-response and kinetic analysis can be performed and may reveal interesting details of the receptor phosphorylation pattern *(6)*.
4. Transfer cells to ice and wash twice with an ice-cold balanced salt solution (e.g., PBS or HBSS) and add 1 mL of ice-cold RIPA buffer (*see* **Note 5**). Scrape

the cells with a disposable spatula, suspend them using a filter tip and transfer cell suspensions to chilled screw-cap tubes (regular Eppendorf tubes frequently leak during the procedure!). Incubate lysates for 45 min at 4°C with gentle rocking/shaking.

5. Centrifuge tubes in a benchtop centrifuge at maximum speed for 20 min at 4°C. Carefully transfer the supernatants to fresh, chilled tubes with 2–10 μL of antiserum (or 2–20 μg of purified antibody) and 35 μL of protein A– or protein G–Sepharose suspension. Incubate lysates for 1–3 h at 4°C with gentle rocking/shaking. However, in some cases a short incubation of 15–30 min at room temperature may result in less background (*see* **Note 6**).
6. Centrifuge the samples briefly, remove the supernatant, and wash three times with RIPA buffer and once with a low-salt buffer (e.g., 20 m*M* HEPES, pH 7.5) or water. A prolonged washing step for 5–15 min, the use of RIPA with 0.1–1% BSA or 500 m*M* LiCl may help to reduce background (*see* **Note 6**).
7. After the last washing step and careful drying of the Sepharose beads (e.g., with a small-diameter syringe) 30 μL of SDS sample buffer are added and the samples boiled for 2–5 min. However, some GPCRs may aggregate under these conditions. To avoid this, sample buffer containing 6 *M* urea (instead of glycerol) may be used and samples should not be boiled but treated for 30–60 min at 40–45°C. Alkylation of reduced SH-groups should be avoided, as it may interfere with the following procedure.

3.2. SDS-PAGE and Transfer

1. Briefly centrifuge the samples and apply supernatants on an appropriate SDS-polyacrylamide gel (usually 8% or 10% for GPCRs). The protein A–Sepharose beads still contain significant amounts of precipitated protein that can be recovered by boiling them in another 30 μL of sample buffer. Run the gel with the usual voltage but avoid leakage of the bromophenol blue front that may contain radioactive protein decay products, nucleotides and phospholipids. Therefore, the front of the gel should be cut and disposed of before transfer. The running buffer will be slightly contaminated with ^{32}P and should be handled according to the local guidelines while the electrophoresis equipment is usually not significantly contaminated. In general nothing argues against using small 8 × 10 cm and 0.75 to 1-mm thick gels. The higher transfer yields obtained with these small and thin gels may improve results.
2. Perform a regular transfer of proteins onto nitrocellulose membranes using a semi-dry or wet blot device. Optimal transfer conditions should be established in non-radioactive pilot experiments. Ponceau S staining and the use of polyvinylidene fluoride (PVDF) that may affect the following procedure should be avoided. After transfer, the gel may be analyzed using a PhosphorImager to monitor transfer yield. The filter papers in direct contact with the gel and the membrane will be contaminated with ^{32}P while the transfer equipment is not usually contaminated.
3. Wrap the filter in thin plastic foil and expose on X-ray film or phosporimager (*see* **Note 7**). The filter should not dry as this may affect the tryptic digest. An

overnight film exposure or 2–4 h on a PhosphorImager should be sufficient to obtain satisfying results.

3.3. Tryptic Digest

1. Exactly overlay the autoradiogram or a 1 : 1 print of the PhosphorImager analysis with the nitrocellulose membrane and cut out the bands of interest. Using a slightly overexposed picture on which the contours of the membrane are visible will help to hit the bands exactly. Alternatively, radioactive ink (regular ink plus a few μL of a ^{14}C source) or a commercial phosphorescent ruler may be used as a guide. Contours of bands may be marked using a needle or a pen. Precision of the cuts should be verified by reexposing the nitrocellulose membrane on X-ray film or phosporimager.
2. Remove residues of plastic foil, transfer membrane pieces to 200 μL of 0.5% PVP in 0.6% acetic acid, and incubate at 37°C for 30 min. This will block protein binding sites on the nitrocellulose that may otherwise capture trypsin. A protein-free blocking procedure is crucial because proteins would compete with the radiolabeled GPCR for active trypsin. During or after blocking, radioactivity on the nitrocellulose pieces should be determined in a β-counter using a Cerenkov program.
3. Aspirate blocking solution and wash the nitrocellulose pieces three times with water.
4. Add 200 μL of freshly prepared 50 m*M* ammonium bicarbonate solution (a 1 *M* stock solution, tightly sealed and stored at 4°C is stable for 1–2 mo) containing 1 μg of modified sequencing grade trypsin and incubate overnight at 37°C. Proteins attached to the nitrocellulose will be cleaved and peptides will be released from the nitrocellulose membrane.
5. Transfer the supernatants to fresh tubes, wash filters twice with water for 5–15 min, and pool the supernatants. Measure the free and residual membrane-bound radioactivity in a β-counter using a Cerenkov program. About 75–90% of ^{32}P should be released from the nitrocellulose. If this is not the case, membranes can be incubated with trypsin a second time. The amount of radioactivity in phosphopeptides that is needed to obtain high-quality 2D maps depends on the individual protein, however, about 1000 cpm can be considered as a good starting point (*see* **Note 8**).
6. Dry samples in a vacuum centrifuge (2–6 h).
7. Prepare a fresh performic acid solution by mixing 9 parts formic acid with 1 part 30% hydrogen peroxide solution and incubate for 1 h at room temperature. Add 50 μL of performic acid to chilled samples and incubate for 1 h on ice. This treatment will oxidize cysteine and methionine residues and improve yields of a second trypsin cleavage step.
8. Stop the reaction by diluting samples with 450 μL of water and freeze them at –80°C.
9. Dry samples in a vacuum centrifuge (6–12 h). At this step, a high and very stable vacuum is necessary to avoid irreversible oxidative damage of samples (*see* **Note 9**)!

10. Add 50 μL of freshly prepared 50 m*M* ammonium bicarbonate solution and sonicate samples in a water bath for 5 min to dissolve peptides. Add 1 μg of modified sequencing grade trypsin, mix, and incubate overnight at 37°C. After this second cleavage step, particular peptides may be isolated by an affinity-purification (e.g., with antibodies against a particular peptide of the target protein, with antibodies against phosphotyrosine to enrich tyrosine phosphorylated peptides or with peptide binding domain fusion proteins, such as SH2-domain GST fusions) (*see* **Note 10**).
11. Add 140 μL of pH 1.9 electrophoresis buffer (formic acid–acetic acid–water 44 : 156 : 1800 v/v/v) mix and centrifuge for 5 min. Transfer 180 μL of the supernatant to fresh tubes, carefully avoiding insoluble material and vacuum dry. The remaining 20 μL and eventually insoluble material may be used for a phosphoamino acid analysis to determine the overall phosphoamino acid composition of the protein (*see* **Subheading 3.5.**).

3.4. 2D Separation of Tryptic Peptides

1. Dissolve phosphopeptides in 5–10 μL of electrophoresis buffer, mix intensively, and centrifuge for 1 min.
2. Apply supernatants in minute portions onto cellulose TLC plates on the smallest possible spot area. A fan with cold air may be used to accelerate the procedure, but hot air that would "bake" peptides to the plate must be avoided. The spot should be at a 3-cm distance from an edge of the plate in a central position (suggested for an initial experiment) or shifted 2–5 cm to the left (*see* **Note 11**).
3. Wet the plate with pH 1.9 electrophoresis buffer using a Whatman paper with a circular hole at the position of the phosphopeptide spot. On contact of the paper with the application spot, phosphopeptides may be lost by absorption to the paper.
4. Connect the anode to the left and the cathode to the right side and separate samples by electrophoresis at 2000 V using pH 1.9 electrophoresis buffer for 30–40 min. Other buffer systems (e.g., 1% ammonium carbonate, pH 8.9) can be used, but the pH 1.9 buffer usually gives superior results (*see* **Note 11**).
5. Extensively air-dry TLC plates after electrophoresis.
6. Run an ascending chromatography step in isobutyric acid chromatography buffer (isobutyric acid–*n*-butanol–pyridine–acetic acid–water 1250 : 38 : 96 : 58 : 558, by vol) for at least 12 h. **Caution:** the buffer has a terrible and persistent smell! It is possible to shorten the separation time and distance without loss of quality by scraping the TLC plate 3 cm from the top. Other buffer systems may be used *(4)*, but the relatively hydrophilic isobutyric acid buffer usually results in better phosphopeptide maps (L. Rönnstrand and C. Wernstedt, personal communication).
7. Extensively air-dry the TLC plates, label the top corners with some radioactive ink, wrap in thin plastic foil and expose to a PhosphorImager for 1–3 d. Film exposure, even with enhancer screens, is usually only sensitive enough to detect major phosphopeptides.
8. Overlay a 1 : 1 print of the PhosphorImager analysis with the TLC plate using the radioactive ink spots as guides. Mark spots on the TLC plate with a soft pencil,

scrape the cellulose containing phosphopeptides of interest using a blade and transfer to an Eppendorf tube. This should be done in a calm area, as draughts may whirl up scraped cellulose powder containing phosphopeptides.

9. Measure radioactivity of the cellulose-bound phosphopeptides in a β-counter and proceed according to the results. Approximately 25–50 cpm are necessary for a phosphoamino acid analysis and about 50–500 are needed for Edman sequencing.
10. Several buffers and procedures can be applied to isolate phosphopeptides from the cellulose. In most cases three extractions with 250 μL of 20% (v/v) acetonitrile with 5 min sonication in a water bath will result in highest recovery rates (70–95%). Alternatively, electrophoresis or chromatography buffer can be used. If sufficient radioactivity is isolated, split samples and use 25–50 cpm or 10–20% for a phosphoamino acid analysis and the remainder for Edman sequencing. Eventually, 100–500 cpm may be saved for a secondary digest with another protease (*see* **Note 12**).
11. Vacuum dry samples for 2–4 h and proceed with **Subheading 3.5.** and/or **3.6.**

3.5. Phosphoamino Acid Analysis by 2D Electrophoresis

1. Dry the samples, transfer to a screw-cap tube (Eppendorf tubes may burst!), add 200 μL of 6 *M* HCl, tightly seal the tube, and incubate at 110°C for 1 h to hydrolyze proteins/peptides.
2. Let the tubes cool down to room temperature, add 300 μL of water, mix, and vacuum dry for 4–6 h.
3. Add 500 μL of water, mix, and centrifuge for 5 min. Take 450 μL of the supernatant, carefully avoiding any insoluble material, transfer to a fresh tube, and vacuum dry again.
4. Dissolve the samples in 5–10 μL of pH 1.9 electrophoresis buffer and add each 1.5 μg of nonradioactive phosphoserine, phosphothreonine, and phosphotyrosine standards (prepared as concentrated stock solution in pH 1.9 electrophoresis buffer stable at –20°C for several years). Apply supernatants in minute portions onto cellulose TLC plates on the smallest possible spot area. A fan with cold air may be used to accelerate the procedure, but hot air that would "bake" phosphoamino acids to the plate must be avoided. Four samples can be simultaneously analyzed on a 20 × 20 cm cellulose TLC plate. Consult the manual of the Hunter thin-layer peptide mapping electrophoresis system *(8)* or Boyle et al. (1991) *(4)* for detailed instructions and outlines of TLC plates.
5. Wet the plate with pH 1.9 electrophoresis buffer using a Whatman paper with circular holes at positions corresponding to the samples.
6. Connect the anode to the left and the cathode to the right side and run the first dimension in pH 1.9 electrophoresis buffer at 2000 V for 20 min.
7. Air-dry the TLC plate extensively.
8. Wet the plate with a Whatman paper soaked in pH 3.5 electrophoresis buffer (acetic acid–pyridine–water 100:10:1794 v/v/v), but leave out areas where phosphopeptides have been separated in the first dimension.

9. Turn the plate 90° left and run the second dimension in pH 3.5 electrophoresis buffer at 1800 V for 20 min.
10. Dry the plate with hot air (fan or oven) and spray with 0.25% ninhydrin (w/v) in acetone. Dry the plate with hot air again to develop the stains of the phosphoamino acid standards. Distinct purple spots will appear that correspond to the position of phosphoserine (top left), phosphothreonine (middle), and phosphotyrosine (bottom).
11. Expose to a PhosphorImager for 1–3 d.
12. Identify phosphorylated amino acid(s) by overlaying a 1 : 1 PhosphorImager print with the TLC plate and comparison with the nonradioactive, ninhydrin-stained standards. Because the hydrolysis is only partial, residual phosphopeptides (above the starting point) and free phosphate (top left corner) are usually seen.

3.6. Edman Sequencing

1. Quantities of phosphopeptides are usually not sufficient for direct Edman sequencing. Instead, a solid-phase sequencer (e.g., ABI 477) can be customized such that cleavage cycles are directly collected without further chemical modification as necessary for the common Edman sequencing procedure. Routinely 20 cleavage cycles are performed, but if sufficient radioactivity is incorporated in the peptide, up to 40 cycles may be possible.
2. Vacuum dry fractions corresponding to the individual cleavage cycles, dissolve in 5–10 µL of trifluoracetic acid (*caution:* toxic and corrosive!) and apply to a cellulose TLC plate with preformed circular holes.
3. Expose for 2–3 d on a PhosphorImager to visualize cleavage cycles containing ^{32}P-labeled amino acids.

3.7. Analysis, Interpretation, and Validation of Data

1. Several pieces of information are obtained by the 2D phosphopeptide mapping procedure (*see* **Fig. 2** for a typical set of results):
   a. The change of total GPCR phosphorylation as consequence of treatment of cells (e.g., agonist stimulation).
   b. The overall phosphoamino acid composition of the GPCR.
   c. The number and physicochemical properties of (major) phosphopeptides.
   d. The phosphoamino acid composition of (major) phosphopeptides.
   e. The position of phosphoamino acids in tryptic peptides of the GPCR.
2. Perform a theoretical tryptic cleavage of the particular GPCR protein sequence, for example, by using the peptide mass tool on the ExPASy Molecular Biology Server (http://www.expasy.org/tools/peptide-mass.html).
3. Potential phosphorylation sites in tryptic peptides are identified by their match to information (d) and (e) (e.g., if phosphoserine [d] in position four of a tryptic peptide [e] is found, search peptides in the GPCR sequence that fit to these criteria).
4. If there is more than one peptide matching to information (d) and (e), theoretical charge and chromatographic mobility based on their amino acid composition can

be calculated to narrow down the number of candidate peptides and potential phosphorylation sites *(4)*.

5. Another possible way to gain more information about a particular phosphopeptide is to perform a secondary digest with another enzyme (e.g., endoprotease Glu-C; *see* **Note 12**) after extraction from the TLC plate or an affinity purification of particular tryptic peptides (*see* **Note 10**).
6. If the appropriate instrumentation is available, a mass spectrometry analysis of extracted phosphopeptides, eventually even including direct sequencing by post-source decay techniques, may be performed *(9)*. This would combine the power of both of the methods, 2D phosphopeptide mapping and mass spectrometry, in a perfect way.
7. To validate the identification of a phosphorylation site, it is crucial to mutate the predicted amino acid to alanine (serine and threonine) or phenylalanine (tyrosine) and to repeat the whole procedure with this GPCR variant. If the information obtained from the wild-type receptor was correct the corresponding phosphopeptide spot should disappear in the mutant GPCR.
8. Finally, GPCR phosphorylation site mutants can be used to correlate phosphorylation events with functions and cellular responses to receptor activation.

3.8. Troubleshooting

1. Highly charged, very hydrophobic and large peptides (>30 amino acids) may not sufficiently resolve in electrophoresis and/or chromatography. This can result in radioactive "clouds" or smears at different positions of the TLC plate. A change of buffer system or alternative cleavage methods, eventually in combination with trypsin, may improve results. Furthermore, tryptic cleavage sites in strategic positions may be introduced by mutagenesis to generate shorter peptides.
2. Phosphorylation of a second amino acid in a peptide will shift the corresponding spot slightly down and to the left side as compared to the monophosphorylated peptide (in the described buffer system). This may explain a surprising disappearance of spots in samples with increases in total phosphorylation.
3. Poor chromatographic resolution may be due to evaporation of buffer components. Many plates can be rescued by rechromatography in freshly prepared buffer.
4. Increases in tyrosine phosphorylation are often masked by high-stoichiometry serine/threonine posphorylation of proteins. Isolation of tryptic peptides by immunoprecipitation with an anti-phosphotyrosine antibody (e.g., pY99 from Santa Cruz Biotechnology, Inc.) may improve recovery of tyrosine-phosphorylated peptides (*see* **Note 10**).
5. For solid-phase sequencing, peptides are covalently coupled to a disc via their C-terminal carboxyl group and cleaved starting from their N-terminus using the Edman reagent. However, peptides may also couple through carboxyl side chains, which will result in a release of the complete peptide in the corresponding cleavage cycle and eventually appearance of a false signal in the PhosphorImager analysis. These signals are usually less intensive than "real" signals and may even help to identify particular phosphopeptides.

4. Notes

1. Precipitating antibodies for GPCRs are rare and respective promises by commercial suppliers and academic groups are not always fulfilled. As an alternative, the GPCR can be fused with an epitope tag, transfected into suitable cells, and isolated using anti-tag antibodies for immunoprecipitation. Particularly good experience has been made with FLAG and hemagglutinin tags (*see* Chapter 5).
2. The amount of protein needed for the 2D mapping procedure is frequently overestimated while there is often an intuitive inhibition to use sufficiently high ^{32}P concentrations. ^{32}P labeling of the cellular phosphate pool, ^{32}P incorporation in the protein of interest, and quality of the antibody used for immunoprecipitation are the most important parameters. Therefore, an increase of the ^{32}P concentration, a decrease of cell number and an optimization of the immunoprecipitation procedure are highly recommended. For a pilot experiment, a 35-mm plate (or a well of a 6-well plate) and 1 mCi of ^{32}P/mL are suggested. For endogenously expressed GPCRs and Edman sequencing a 100-mm plate may be used. It should be noted that there is no need to prepare and apply the ^{32}P-containing medium to cells under sterile conditions. In particular, when antibiotics are included, no contamination of culture by microorganisms should be expected during the 6–12 h of labeling.
3. When handling mCi amounts of ^{32}P, appropriate shielding against the high-energy β- and secondary γ/X-ray irradiation is absolutely essential. The workplace, the first suction bottle (when using a pump), and the waste container should be shielded by 15- to 25-mm Plexiglas covered with a 1–5 mm layer of lead. To avoid unnecessary bremsstrahlung, the lead should be at the outer side of the Plexiglas, that is, not directly exposed to the source of radiation. Alternatively, double shielding using conventional and lead-containing Plexiglas that is available from some commercial providers (e.g., Amersham) can be used. During ^{32}P-labeling and transport from the bench to the incubator, cells should be kept in a Plexiglas box. In general, exposure to mCi amounts of ^{32}P must be kept as low as possible by minimizing handling times and by maximally increasing distance from the source of radiation. It is suggested that an initial "faked" test experiment is performed without ^{32}P for training purposes and to identify potential shortcoming in the procedure and the equipment.
4. To minimize the risk of radioactive contamination of the pump it is recommended to use two suction bottles in series. The second bottle may contain an absorbent to bind radioactive liquids accidentally carried over. A few drop of commercial antifoam or a higher aliphatic alcohol (e.g., octanol) in the suction bottle reduce generation of radioactive foam that may be carried over to the pump.
5. Cellular reactions may also be stopped by dipping the whole plate in liquid nitrogen. Frozen cell culture plates can be stored for 1–3 d at –20°C with appropriate shielding. To inhibit phosphatases more efficiently, highly potent phosphatase inhibitors, such as okadaic acid or calyculin A may be added to the RIPA buffer at 0.1–1 μ*M*. Furthermore, cells may be preincubated with these inhibitors and/or with 100 μ*M* sodium vanadate 30–50 min prior to agonist stimulation.

6. To optimize the immunoprecipitation procedure (amount of antibody, cell lysis, incubation times, washing conditions) it is suggested that pilot experiments are performed with cells metabolically labeled with [^{35}S]methionine (analogous procedure to the ^{32}P-labeling but using 50–100 μCi/mL [^{35}S]methionine in sulfur-free medium).
7. Although exposure to film results in better pictures with higher resolution, phosporimager analysis is preferable as it is more sensitive, has a broader linear range and allows precise quantification of the levels of radioactivity that may reveal interesting details of GPCR phosphorylation (*see* **Fig. 2** and **ref. *6***).
8. The levels of incorporation of ^{32}P required to identify phosphorylation sites depends on the number of tryptic phosphopeptides generated. For PhosphorImager detection, <25 cpm/peptide may be enough, whilst for phosphoamino acid analysis at least 25–50 cpm/peptide are necessary. For Edman sequencing, approx 50–500 cpm/peptide are needed. In general it is recommended that the fate of phosphopeptides during the whole procedure is followed by Cerenkov counting.
9. To reach a sufficient vacuum, an oil pump and freeze trap (also to protect the pump) between the centrifuge and pump are necessary. The vacuum can be tested simply by drying 500 μL of frozen water that should not thaw during drying or, in other words, it should sublimate. If the vacuum is not stable, the oxidization reaction can be stopped using 450 μL of a 20% (v/v) ammonia solution. This makes

---

Fig. 2. *(see facing page)* Example of the successful identification of GPCR phosphorylation (from **ref. *6***): mapping of serine 339, 346, and 348 as the major phosphorylation sites of the bradykinin B_2 receptor. **(A)** HF-15 human foreskin fibroblasts grown on 100-mm plates were labeled with [^{32}P]orthophosphate and left untreated or stimulated with 1 μ*M* bradykinin (Bk) for 10 min. Following immunoprecipitation and 10% SDS-PAGE, isolated proteins were transferred onto nitrocellulose membranes and analyzed using a PhosphorImager (BAS2000, Fuji). The ^{32}P-labeled bradykinin B_2 receptor was digested *in situ* with trypsin and the resulting peptides were separated on TLC plates by high-voltage electrophoresis and ascending chromatography. A *cross* illustrates where samples were applied; "+" and "–" indicate the polarity during electrophoresis. Thereafter, phosphopeptides were localized by PhosphorImager analysis. **(B)** Peptides were eluted and a fraction was hydrolyzed, and subjected to phosphoamino acid analysis followed by 2D electrophoretic separation on TLC plates and phosphorimager analysis. Phosphorylated amino acids were identified using commercial standards (locations indicated by *dashed circles*). **(C)** Major fractions of phosphopeptides were subjected to 20 cycles of Edman degradation and cleaved amino acids were collected and analyzed using a PhosphorImager to locate the position of the phosphorylation site(s) as exemplified for peptide 1. The content of ^{32}P radioactivity of each sequencing cycle was quantified and expressed in arbitrary units *(AU)*. **(D)** Phosphopeptides on TLC plates were quantified using a PhosphorImager and the relative contributions of individual phosphopeptides to total basal (light) as well as bradykinin-mediated (dark) phosphorylation were determined.

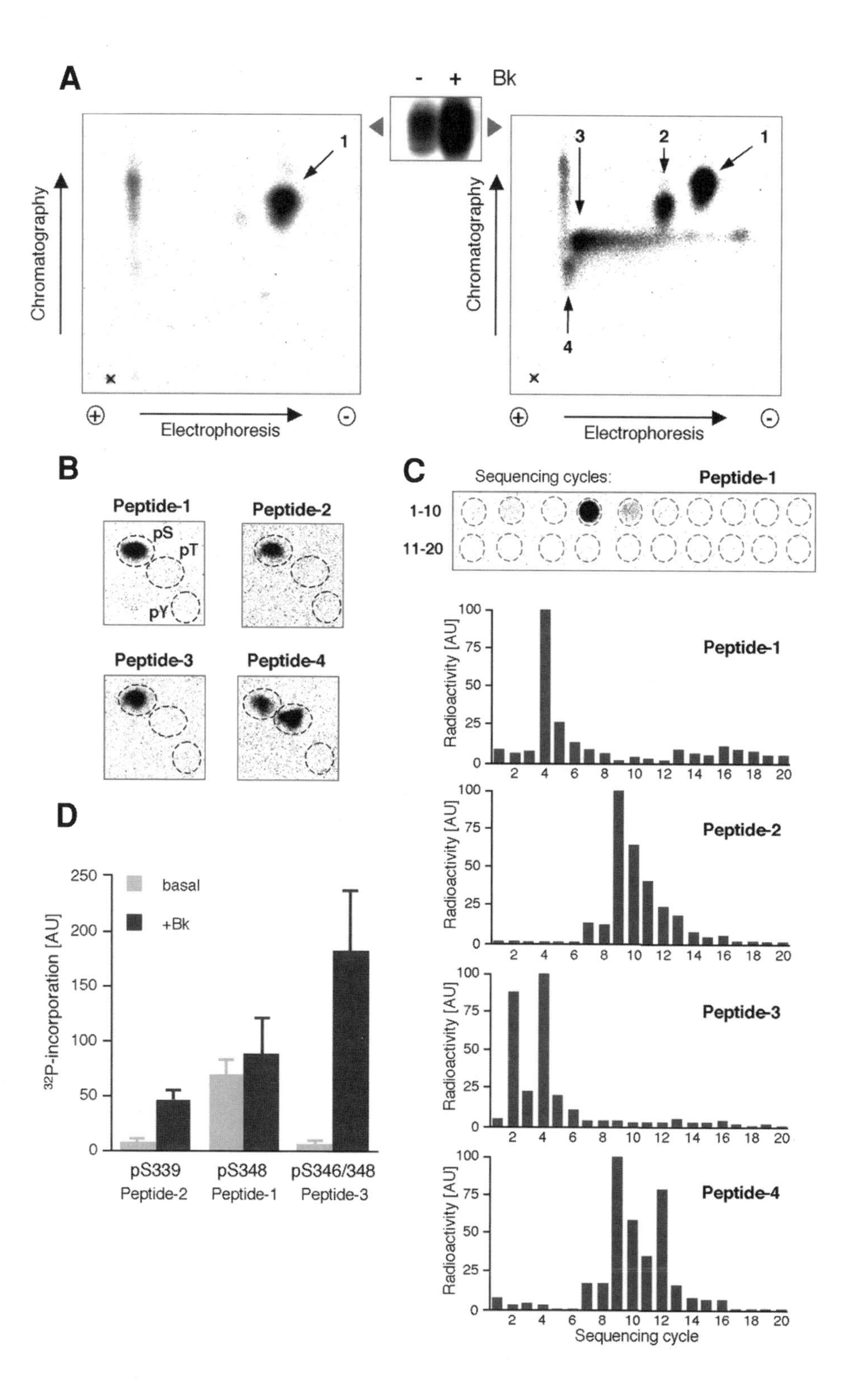
A
-
+
Bk
Chromatography
Electrophoresis
1
2
3
4
B
Peptide-1
Peptide-2
Peptide-3
Peptide-4
pS
pT
pY
C
Sequencing cycles:
Peptide-1
1-10
11-20
Radioactivity [AU]
Sequencing cycle
D
basal
+Bk
^{32}P-incorporation [AU]
pS339
Peptide-2
pS348
Peptide-1
pS346/348
Peptide-3

the vacuum a less critical parameter, but will leave an ammonium formate pellet that must be removed by repeated cycles of addition of water and vacuum drying. Salt affects trypsin cleavage and electrophoretic separation of phosphopeptides.

10. For affinity-purification of particular phosphopeptides, remove the trypsin by incubation with 50 µL of soybean trypsin inhibitor Sepharose slurry (e.g., from SIGMA) for 1 h at 4°C with gentle shaking. Remove the supernatant, wash the beads with 200 µL of 50 m*M* ammonium bicarbonate, and pool the supernatants. Add the bait (e.g., 2 µL of antiserum or 2–5 µg of purified antibody or fusion protein) and incubate for 2 h at 4°C with gentle shaking. Add 35 µL of protein A– or G–Sepharose slurry (if antibodies were used) and incubate for another 30 min at 4°C. Wash the beads three times with 100 m*M* ammonium bicarbonate, 1% (v/v) Triton X-100, and three times with water. Elute twice with 250 µL of 1% (v/v) diethylamine. Vacuum dry the samples, add 200 µL of water, dry again, and subject to 2D separation. It is recommended to follow the fate of phosphopeptides by Cerenkov counting.
11. Detailed instruction and outlines for TLC plates are given by Boyle et al. (1991) *(4)* and in the manual of the Hunter thin layer peptide mapping electrophoresis system from CBS Scientific (no. HTLE-7002) *(8)*. Currently, there does not seem to be any commercial alternative for this system, which performs very well. In the described setting, most phosphopeptides will be positively charged or neutral and migrate to the cathode on the right side. Free phosphate (intense spot on the left side approx one third way of the chromatography distance) and highly negatively charged peptide will migrate to the anode at the left side.
12. To gain information about phosphopeptides that contain glutamate residues, an endoprotease Glu–C cleavage of phosphopeptides extracted from TLC plates can be performed. Samples are dried, dissolved in water, and vacuum dried twice to evaporate residual organic solvent that may disturb the enzymatic reaction. Peptides are dissolved in 50 µL of 50 m*M* ammonimum bicarbonate, mixed, and sonicated for 5 min in a water bath. Add 0.2 µg of endoprotease Glu–C (prepare stock solution in 50 m*M* ammonimum bicarbonate and store aliquots at –20°C), incubate for 2 h at 37°C, add another 0.2 µg of enzyme, incubate for 2 h, and repeat this procedure five times in total. Thereafter the mixture is incubated overnight at 37°C and analysis of endoprotease Glu–C cleavage products is continued as outlined in **Subheadings 3.4.–3.7.**

Acknowledgments

I am grateful to Lars Rönnstrand and Christer Wernstedt from the Ludwig Institute for Cancer Research, Uppsala, Sweden who introduced me to the 2D phosphopeptide mapping technique and generously shared their knowledge with me.

References

1. Krupnick, J. G. and Benovic, J. L. (1998) The role of receptor kinases and arrestins in G protein-coupled receptor regulation. *Annu. Rev. Pharmacol. Toxicol.* **38,** 289–319.

2. Hall, R. A., Premont, R. T., and Lefkowitz, R. J. (1999) Heptahelical receptor signaling: beyond the G protein paradigm. *J. Cell Biol.* **145,** 927–932.
3. Tobin, A. B. (2002) Are we β-ARKing up the wrong tree? Casein kinase 1 alpha provides an additional pathway for GPCR phosphorylation. *Trends Pharmacol. Sci.* **23,** 337–343.
4. Boyle, W. J., van der Geer, P., and Hunter, T. (1991) Phosphopeptide mapping and phosphoamino acid analysis by two-dimensional separation on thin-layer cellulose plates. *Methods Enzymol.* **201,** 110–149.
5. Ozcelebi, F. and Miller, L. J. (1995) Phosphopeptide mapping of cholecystokinin receptors on agonist-stimulated native pancreatic acinar cells. *J. Biol. Chem.* **270,** 3435–3441.
6. Blaukat, A., Pizard, A., Wernstedt, C., Alhenc-Gelas, F., Müller-Esterl, W., and Dikic, I. (2001) Determination of the bradykinin B_2 receptor *in vivo* phosphorylation sites and their role in the regulation of receptor function. *J. Biol. Chem.* **276,** 40,431–40,440.
7. Gordon, J. A. (1991) Use of vanadate as protein–phosphotyrosine phosphatase inhibitor. *Methods Enzymol.* **201,** 477–482.
8. Bolye, W. J. and Hunter, T. Instruction manual: Hunter Thin Layer Peptide Mapping Electrophoresis System, C.B.S. Scientific Company, Inc., Del Mar, CA, USA.
9. Affolter, M., Watts, J. D., Krebs, D. L., and Aebersold, R. (1994) Evaluation of two-dimensional phosphopeptide maps by electrospray ionisation mass spectrometry of recovered peptides. *Anal. Biochem.* **223,** 74–81.

18

Ubiquitination of G-Protein-Coupled Receptors

Adriano Marchese and Jeffrey L. Benovic

Summary

In this chapter we describe methods for detecting the ubiquitination state of G-protein-coupled receptors (GPCRs). This involves coexpression of a GPCR with an epitope-tagged ubiquitin construct in a heterologous expression system. Modification by ubiquitin of the GPCR resulting from agonist activation is detected by immunoprecipation and subsequent immunoblotting for the epitope-tagged ubiquitin. We use here the chemokine receptor CXCR4 as the model receptor; however, this could be easily modified to detect the ubiquitination state of any GPCR.

Key Words

Agonist, CXCR4, degradation, G-protein-coupled receptor, immunoblot, immunoprecipitation, lysosome, sorting, ubiquitin.

1. Introduction

Many agonist-activated G-protein-coupled receptors (GPCRs) are subject to a complex series of regulatory events involving phosphorylation, interaction with nonvisual arrestins, recruitment to clathrin-coated pits, and internalization into endocytic vesicles *(1)*. Internalized receptors are subject to a series of sorting events directing them to follow one of two routes from the early endosomal compartment. One route leads to recycling of the receptor back to the plasma membrane, allowing for a process known as resensitization. The other route targets the receptor to lysosomes where it is degraded, a process known as downregulation. Very little is known about the molecular mechanisms mediating the sorting events linked to the postinternalization trafficking of GPCRs. However, recently we have shown that agonist-promoted ubiquitination of the chemokine receptor CXCR4 is important for targeting the receptor to lysosomes *(2)*. A receptor mutant in which three lysine residues in the C-terminus were changed to arginines internalized normally but failed to undergo agonist-

From: *Methods in Molecular Biology, vol. 259, Receptor Signal Transduction Protocols, 2nd ed.*
Edited by: G. B. Willars and R. A. J. Challiss © Humana Press Inc., Totowa, NJ

promoted degradation. The fact that this receptor mutant was not ubiquitinated suggests that the ubiquitin moiety serves as a sorting signal in an endosomal compartment *(2)*. Similar results have been observed for the β_2-adrenergic receptor *(3)*.

Ubiquitin is a highly conserved 76-amino-acid polypeptide that is covalently attached to proteins through the formation of an isopeptide bond between the C-terminal glycine residue of ubiquitin and the ε-amino group of lysine side chains of target proteins *(4)*. Through internal lysine residues ubiquitin can then be modified by itself to form polyubiquitin chains. Generally, polyubiquitination serves to target the protein for proteolysis by the proteasome. However, recent developments have revealed a nonproteasomal function of ubiquitin when attached to substrate proteins *(5,6)*. In particular, ubiquitin appears to play an integral role in the endosomal trafficking of a subset of cellular proteins. Ubiquitin is involved in the internalization of cell surface proteins *(7)* and the sorting of membrane proteins into invaginating domains of endosomes *(8–10)*. Membrane proteins destined for degradation are concentrated in these invaginating domains, which bud away from the cytosol and pinch off into the lumen of the endosome *(11)*. As the endosome matures, it accumulates internal vesicles and forms a structure known as the multivesicular body (MVB) *(12)*. The limiting membrane of MVBs then fuses with lysosomes where protein degradation eventually occurs. Recent studies have suggested that the ubiquitin moiety on cargo molecules is recognized by a subset of proteins of the transport machinery located on endosomes that harbor a specialized domain that binds ubiquitin, thus providing the framework for a network of ubiquitin-dependent interactions that ultimately leads to the proper sorting and degradation of ubiquitinated cargo in lysosomes *(8–10,13)*.

The most effective means to detect the ubiquitination status of a protein is by immunoblotting. The covalent attachment of ubiquitin will add at least 8 kDa to the size of a protein, therefore leading to slower migration when subjected to sodium dodecyl sulfate-polyacrylamide gel electrophoresis (SDS-PAGE) analysis. The ubiquitinated forms can then be detected by immunoblotting using an antibody that recognizes the protein of interest with ubiquitinated forms migrating at approx 8-kDa intervals above the unmodified version. Because the ubiquitinated state of a GPCR may represent a small fraction of the total membrane pool, antibodies raised against GPCRs are often not sensitive enough to detect ubiquitinated forms. Therefore, a more sensitive means of detecting the ubiquitinated status of a GPCR is to concentrate the receptor by immunoprecipitation followed by immunoblotting with an antibody against ubiquitin. We describe here a procedure that enables the detection of the ubiquitinated chemokine receptor CXCR4. This procedure can easily be adapted and applied to detect the ubiquitinated state of other members of the GPCR family.

2. Materials

1. Human embryonic kidney cells (HEK-293; American Tissue Type Collection).
2. DNA expression constructs: HA-tagged CXCR4 in pcDNA3; His-tagged dynamin-K44A; 3X FLAG-tagged ubiquitin (*see* **Note 1**).
3. Antibodies: anti-FLAG M2 monoclonal antibody (Sigma, St. Louis, MO); anti-mouse IgG conjugated with horseradish peroxidase (Vector Laboratories, Burlingame, CA); anti-HA monoclonal and polyclonal antibodies (Covance, Richmond, CA).
4. Lysis buffer: 50 m*M* Tris-HCl, pH 8, 150 m*M* NaCl, 5 m*M* EDTA, 0.5% sodium deoxycholate (w/v), 1% Nonidet P-40 (NP-40; v/v), 0.1% SDS (w/v), 10 m*M* sodium fluoride, 10 m*M* sodium orthovanadate (*see* **Note 2**).
5. Wash buffer: same as lysis buffer.
6. SDS-PAGE gels: we typically use 7% gels (*see* **Note 3**).
7. Denaturation solution for immunoblots: 62.5 m*M* Tris-HCl, pH 6.7, 100 m*M* β-mercaptoethanol, 2% SDS (*see* **Note 4**).
8. Tris-buffered saline (50 m*M* Tris-HCl, pH 7.4; 150 m*M* NaCl) containing 0.05% Tween-20 (TBST). Also, TBST containing 5% (w/v) nonfat dry milk (TBST–5% milk).
9. Phosphate-buffered saline (PBS): 137 m*M* NaCl, 2.7 m*M* KCl, 10 m*M* Na_2HPO_4, 1.4 m*M* KH_2PO_4, pH 7.4.
10. Protease inhibitors are added fresh each time with the following final concentrations: 10 μg/mL of aprotinin, 10 μg/mL of leupeptin, 0.2 mg/mL of benzamidine, and 1 μg/mL of pepstatin-A.
11. *N*-Ethylmaleimide (NEM) is added fresh each time to a final concentration of 20 n*M* (*see* **Note 5**).
12. 2X Sample buffer: 0.0375 *M* Tris-HCl, pH 6.5, 8% SDS, 10% glycerol, 5% β-mercaptoethanol, 0.003% bromophenol blue.
13. SuperSignal® chemiluminescent substrate (Pierce, Rockford, IL).
14. FuGene6 transfection reagent and protein A–agarose beads (Roche, Indianapolis, IN).
15. Dulbecco's modified Eagle's medium (DMEM; Mediatech, Herndon, VA) and fetal bovine serum (FBS; GIBCO, Grand Island, NY).
16. Stromal cell derived growth factor-1α (SDF) (PeproTech, Rocky Hill, NJ) dissolved in DMEM containing 0.5% FBS.

3. Methods

1. HEK-293 cells are maintained in DMEM supplemented with 10% FBS.
2. The day before transfection, passage cells onto 6-cm dishes such that they are 50–60% confluent the next day (*see* **Note 6**).

3. Cells are then transiently transfected using Fugene 6 transfection reagent with 1 μg each of DNA encoding HA-tagged CXCR4, 3X FLAG-tagged ubiquitin, and dynamin dominant negative K44A mutant, according to the manufacturer's instructions (*see* **Note 7**).
4. After 48 h, wash cells once with 2 mL of warm DMEM supplemented with 20 m*M* HEPES, pH 7.4.
5. Then incubate cells in the same media in the presence or absence of 100 n*M* SDF (a CXCR4 agonist) for 30 min (*see* **Note 8**).
6. Place dishes on ice and wash once with 2 mL of ice-cold PBS.
7. Add 1 mL of lysis buffer, scrape cells, and transfer to microcentrifuge tubes. Incubate tubes on ice for approx 10 min to allow for complete solubilization (*see* **Note 9**).
8. To pellet cellular debris, centrifuge samples at 30,000 rpm (55,000*g*) for 30 min at 4°C in a TLA-45 rotor in an Optima™ TL Ultracentrifuge (Beckman).
9. Carefully transfer 500 μL of supernatant to a fresh microcentrifuge tube. Add 2.5 μL of polyclonal anti-HA antibody and incubate for 1 h while rocking at 4°C (*see* **Note 10**).
10. Add 50 μL of protein A–agarose (prepared by diluting 1:1 [w/v] with lysis buffer) and continue incubation for an additional 1 h while rocking at 4°C.
11. Collect receptor–protein A–agarose complexes by centrifugation in a microcentrifuge at 12,000*g* for 5 s.
12. Carefully remove supernatant, resuspend beads in 750 μL of lysis buffer, and incubate for 10 min at 4°C while rocking.
13. Repeat **steps 11** and **12**.
14. Collect complexes as in **step 11** and carefully remove the last traces of lysis buffer.
15. Elute proteins from agarose beads by adding 30 μL of 2X SDS-gel sample buffer and incubating for 30 min at room temperature (*see* **Note 11**).
16. Resolve proteins by SDS-PAGE and transfer to nitrocellulose membranes using a standard Western blot protocol ***(14)***.
17. Treat blots in approx 40 mL of denaturation solution for 30 min at 60°C, followed by washing with TBST for 10 min while rocking at room temperature. Repeat this washing step two more times with fresh TBST.
18. The membrane is then blocked for 30 min in 10 mL of TBST containing 5% (w/v) nonfat dry milk.
19. The nitrocellulose is next incubated with 10 mL of TBST–5% milk containing 5 μg/mL of mouse monoclonal anti-FLAG M2 antibody for 1 h at room temperature.
20. Wash the nitrocellulose three times for 10 min each in TBST.
21. Incubate the nitrocellulose for 30 min with 10 mL of TBST–5% milk containing goat anti-mouse IgG conjugated to horseradish peroxidase at a dilution of 1:3000.
22. Wash the membrane five times for 10 min each in TBST.
23. Overlay the nitrocellulose with 1–2 mL of Supersignal Chemiluminescence reagent for approx 5 min, allow the blot to drip dry, wrap in plastic wrap, and visualize on X-ray film.

24. At this point, blots can be treated with the denaturation solution to remove bound antibody and reprobed with the monoclonal HA antibody to detect receptor levels.

4. Notes

1. Although commercial anti-ubiquitin antibodies are available, we used an epitope-tagged version of ubiquitin to increase the sensitivity of detection. The presence of the tag does not appear to affect conjugation to substrate proteins *(15)*. The version of the ubiquitin construct we use carries a triple FLAG sequence at the N-terminus *(2)*. Hemagglutinin *(16)* and myc-tagged *(15)* versions have also been used successfully. A commercially available anti-ubiquitin monoclonal antibody from Zymed Laboratories (San Francisco, CA) appears to recognize monoubiquitin attached to substrate proteins very well *(8)*.
2. The precise recipe for immunoprecipitation buffers will vary depending on the receptor being immunoprecipitated. Ideally, high-stringency immunoprecipitation and washing conditions are preferred in order to reduce the likelihood of immunoprecipitating nonspecific proteins that may confound the interpretation of ubiquitin immunoblots.
3. We typically use 7% gels because it helps resolve ubiquitinated species of receptor.
4. We typically use nitrocellulose membranes for our immunoblots. Even though proteins subjected to SDS-PAGE are significantly denatured we have found that treatment of blots in a solution that further denatures proteins before immunoblotting significantly enhances the ability to detect ubiquitinated receptor.
5. Ubiquitin attached to proteins is removed by a specialized set of serine proteases known as deubiquitinating enzymes *(17)*. A function of these enzymes is to recycle ubiquitin. Therefore the ubiquitinated state of a protein is determined by the relative rates of ubiquitin conjugation and deconjugation. Because the activity of these enzymes can compromise the ability to detect ubiquitinated proteins, it is critical that they are inhibited when a cell lysate is prepared. Deubiquitinating enzymes contain active sufhydryl groups that can be blocked by alkylating agents such as iodoacetamide and NEM. Iodoacetamide can be used at a final concentration of 1 mg/mL. We have used NEM at a final concentration as high as 20 n*M*. These agents should be added fresh to the lysis buffer just before use.
6. We typically perform 48-h transfections. Therefore, cells should be passaged at an appropriate density such that they are not overgrown the day of the experiment.
7. For the yeast α-mating factor receptor the ability to detect ubiquitinated receptor is enhanced in a yeast strain that is defective in clathrin-mediated endocytosis *(18)*. This is due in part because the ubiquitination reaction occurs at the plasma membrane, therefore increasing the proportion of receptors that are modified by ubiquitin, but also because it prevents the receptor from entering a compartment where it is deubiquitinated *(19)*. We reasoned that perhaps a similar scenario is occurring with mammalian GPCRs; therefore we performed our ubiquitination experiments in the presence of the dominant negative mutant of dynamin–K44A, which inhibits internalization of CXCR4. Indeed we found that the ability to

detect monoubiquitinated CXCR4 was dramatically increased in cells in which dynamin–K44A was coexpressed *(2)*. Whether this will be true for other GPCRs remains to be determined.

8. We initially performed a time course to determine the optimal length of stimulation that resulted in maximal ubiquitination of CXCR4. We found that 30 min of stimulation led to the greatest ubiquitination under our conditions; however, this will have to be determined initially for other GPCRs.
9. These conditions have been optimized for our experimental conditions such that there is complete lysis and solubilization of membranes. To ensure complete lysis, cell lysates may be subjected to sonication. Longer incubations while rocking at 4°C may also be performed to allow for complete membrane solubilization.
10. We found that these conditions were optimal for immunoprecipitating CXCR4 and subsequent detection of its ubiquitination state. This will vary depending on the GPCR. An important consideration to make is that incubation times after the creation of the cell lysate should be kept to a minimum owing to the high activity of deubiquitinating enzymes present in the lysate.
11. It is critical that samples are not boiled to elute receptors off of the beads. Boiling will cause receptors to aggregate and impede their migration on SDS-PAGE.

Acknowledgments

A. Marchese was supported by postdoctoral fellowships from the Canadian Institutes of Health Research and the American Heart Association.

References

1. Claing, A., Laporte, S. A., Caron, M. G., and Lefkowitz, R. J. (2002) Endocytosis of G protein-coupled receptors: roles of G protein-coupled receptor kinases and β-arrestin proteins. *Prog. Neurobiol.* **66,** 61–79.
2. Marchese, A. and Benovic, J. L. (2001) Agonist-promoted ubiquitination of the G protein-coupled receptor CXCR4 mediates lysosomal sorting. *J. Biol. Chem.* **276,** 45,509–45,512.
3. Shenoy, S. K., McDonald, P. H., Kohout, T. A., and Lefkowitz, R. J. (2001) Regulation of receptor fate by ubiquitination of activated β_2-adrenergic receptor and β-arrestin. *Science* **294,** 1307–1313.
4. Weissman, A. M. (2001) Themes and variations on ubiquitylation. *Nat. Rev. Mol. Cell Biol.* **2,** 169–178.
5. Hicke, L. (2001) Protein regulation by monoubiquitin. *Nat. Rev. Mol. Cell Biol.* **2,** 195–201.
6. Bonifacino, J. S. and Weissman, A. M. (1998) Ubiquitin and the control of protein fate in the secretory and endocytic pathways. *Annu. Rev. Cell Dev. Biol.* **14,** 19–57.
7. Hicke, L. (1997) Ubiquitin-dependent internalization and down-regulation of plasma membrane proteins. *FASEB J.* **11,** 1215–1226.
8. Katzmann, D. J., Babst, M., and Emr, S. D. (2001) Ubiquitin-dependent sorting into the multivesicular body pathway requires the function of a conserved endosomal protein sorting complex, ESCRT-I. *Cell* **106,** 145–155.

9. Reggiori, F. and Pelham, H. R. (2001) Sorting of proteins into multivesicular bodies: ubiquitin-dependent and -independent targeting. *EMBO J.* **20,** 5176–5186.
10. Urbanowski, J. L. and Piper, R. C. (2001) Ubiquitin sorts proteins into the intralumenal degradative compartment of the late-endosome/vacuole. *Traffic* **2,** 622–630.
11. Seto, E. S., Bellen, H. J., and Lloyd, T. E. (2002) When cell biology meets development: endocytic regulation of signaling pathways. *Genes Dev.* **16,** 1314–1336.
12. Piper, R. C. and Luzio, J. P. (2001) Late endosomes: sorting and partitioning in multivesicular bodies. *Traffic* **2,** 612–621.
13. Raiborg, C., Bache, K. G., Gillooly, D. J., Madshus, I. H., Stang, E., and Stenmark, H. (2002) Hrs sorts ubiquitinated proteins into clathrin-coated microdomains of early endosomes. *Nat. Cell Biol.* **4,** 394–398.
14. Mundell, S. J., Orsini, M. J., and Benovic, J. L. (2002) Characterization of arrestin expression and function. *Methods Enzymol.* **343,** 600–611.
15. Ellison, M. J. and Hochstrasser, M. (1991) Epitope-tagged ubiquitin. A new probe for analyzing ubiquitin function. *J. Biol. Chem.* **266,** 21,150–21,157.
16. Courbard, J. R., Fiore, F., Adelaide, J., Borg, J. P., Birnbaum, D., and Ollendorff, V. (2002) Interaction between two ubiquitin-protein isopeptide ligases of different classes, CBLC and AIP4/ITCH. *J. Biol. Chem.* **277,** 45,267–45,275.
17. Wilkinson, K. D. (2000) Ubiquitination and deubiquitination: targeting of proteins for degradation by the proteasome. *Semin. Cell Dev. Biol.* **11,** 141–148.
18. Hicke, L., Zanolari, B., and Riezman, H. (1998) Cytoplasmic tail phosphorylation of the α-factor receptor is required for its ubiquitination and internalization. *J. Cell Biol.* **141,** 349–358.
19. Amerik, A. Y., Nowak, J., Swaminathan, S., and Hochstrasser, M. (2000) The Doa4 deubiquitinating enzyme is functionally linked to the vacuolar protein-sorting and endocytic pathways. *Mol. Biol. Cell* **11,** 3365–3380.

19

Receptor Mutagenesis Strategies for Examination of Structure–Function Relationships

Marion Blomenröhr, Henry F. Vischer, and Jan Bogerd

Summary

This chapter describes three different strategies of receptor mutagenesis with their advantages, disadvantages, and limitations. Oligonucleotide-directed mutagenesis using either the Altered Sites® II in vitro mutagenesis system or the GeneTailor® site-directed mutagenesis system can generate base substitutions/deletions/insertions that yield single/multiple amino acid substitutions/deletions/insertions and/or N- or C-terminal truncations in GPCRs. Polymerase chain reaction-based mutagenesis strategies allow substitutions/deletions/insertions of larger domains within GPCRs, creating truncated receptors or receptor chimeras.

In addition, some guidelines are given and examples are provided to facilitate design and interpretation of mutational experiments.

Key Words

Domain swapping, G-protein-coupled receptor, molecular model, mutagenic oligonucleotide, polymerase chain reaction, receptor chimera, receptor–ligand complex, site-directed mutagenesis.

1. Introduction

Direct structural data of G-protein-coupled receptors (GPCRs) is limited to the low-resolution structures of bovine and frog rhodopsin *(1,2)* and the 2.8-Å resolution X-ray structure of bovine rhodopsin *(3)*. Site-directed mutagenesis of GPCRs is an indirect approach to study the involvement of particular amino acid residues/domains of GPCRs in specific functions, such as direct interaction with ligands, receptor activation, subsequent coupling to downstream signaling events, or regulation of receptor expression at the cell surface. To test the relative contribution of each amino acid residue in these receptor functions, wild-type and mutant receptors are expressed in heterologous cell systems. Subsequently, they are tested for changes in specific characteristics, such as

From: *Methods in Molecular Biology, vol. 259, Receptor Signal Transduction Protocols, 2nd ed.*
Edited by: G. B. Willars and R. A. J. Challiss © Humana Press Inc., Totowa, NJ

ligand binding, coupling to different signaling pathways (e.g., inositol phosphate or cAMP production), their phosphorylation state, or their localization within the cell, depending on the question one addresses. The methods used to express (mutant) GPCRs in cell systems are described in Chapters 8, 9, and 10. The function of such receptors can of course be assessed by many techniques and described throughout this book are a wide variety that concentrate directly on the receptor or events immediately downstream of the receptor.

Because of the pitfalls and limitations of mutational experiments, their design and interpretation must occur carefully. The following are some guidelines, followed in our laboratories, as well as some examples.

1. As a first "scan," use progressive deletions of the domain of interest at the N- or C-terminal ends *(4)* or exchange whole domains between related GPCRs from one species (*see* **Fig. 1**) *(5)* or between homologous GPCRs from different species *(6)* before precisely characterizing the role of individual residues by single amino acid mutations.
2. Consider the nature of the specific amino acid substitution (i.e., charge, hydrogen bonding potential, hydrophobicity, volume and shape of the side chain) and use multiple mutations at a locus designed to test a hypothesis (*see* **Fig. 2**) *(7)*.

Fig. 1. *(see facing page)* Intracellular cAMP production in response to increasing concentrations of follicle-stimulating hormone (FSH, ■) and chorionic gonadotropin (hCG, ○) in HEK-293(T) cells *(upper panel)*, transiently transfected with the wild-type human FSH receptor **(A)**, a chimeric receptor construct encoding the extracellular N-terminal tail of the human luteinizing hormone (LH) receptor fused to the transmembrane domain of the human FSH receptor **(B)** or the wild-type LH receptor constructs **(C)** and cotransfected with a plasmid containing a β-galactosidase gene under control of a promoter containing five cAMP-response elements. Receptor constructs are schematically depicted in the *lower panel*, illustrating the N-terminal extracellular domain, the seven transmembrane helices, and the C-terminal intracellular tail. *Arrows* indicate the nine β-strands in each receptor that together form the concave surface of the horseshoe-shaped hormone-binding domain. FSH- and LH-receptor amino acids are indicated by *open* and *filled circles*, respectively. Results are means ± SEM for triplicate observations from a single representative receptor stimulation experiment. To compensate for interassay variations based on the reporter gene's transfection/expression efficiencies, reporter gene activities are normalized in each experiment to that observed in the presence of 10 μ*M* forskolin. Hence, the cAMP-mediated reporter gene activities are presented as arbitrary units (AU). The FSH receptor **(A)** is approx 700-fold less responsive to hCG than the LH receptor **(C)**. However, replacing the exodomain of the FSH receptor with the corresponding domain of the LH receptor conferred an approx 700-fold increased hCG responsiveness to the chimearic receptor. Thus, hormone selectivity of the FSH- and LH-receptor is exclusively determined by its N-terminal exodomain *(5)*. (From **ref.** *7*; reproduced with permission © the Biochemical Society.)

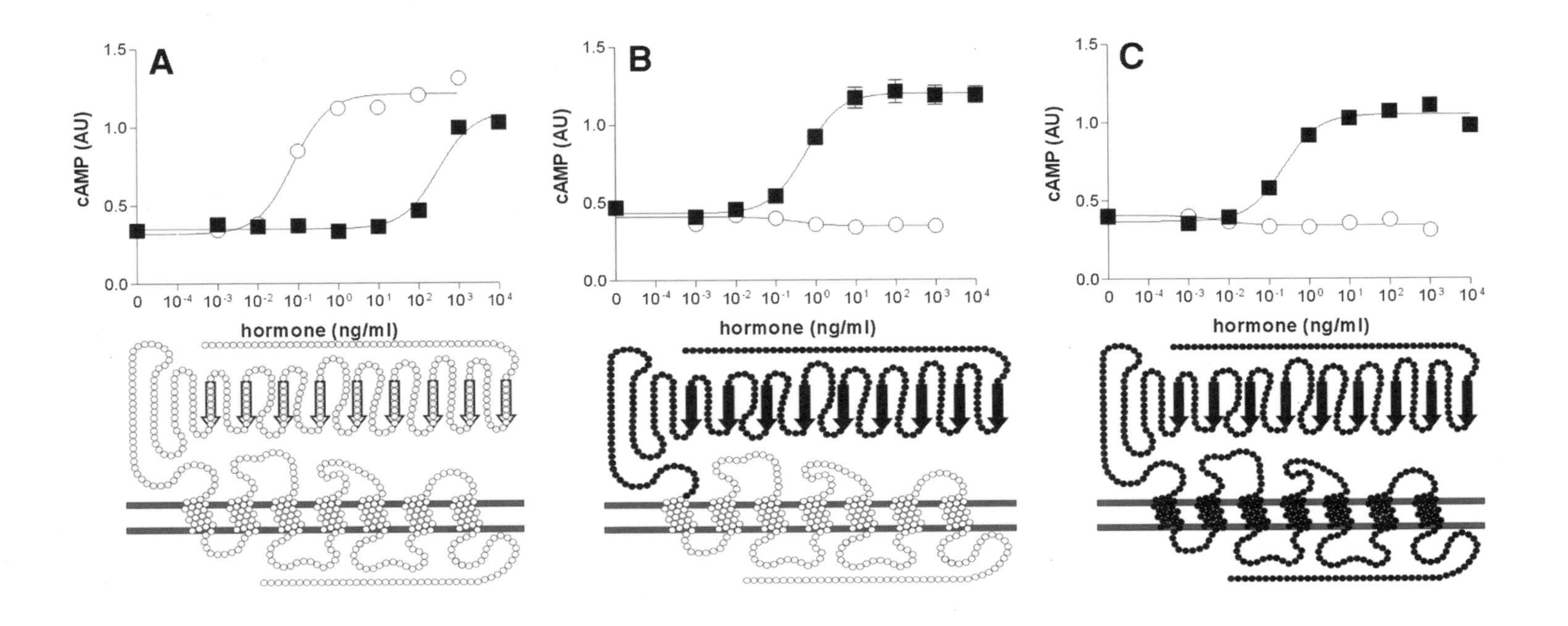

A
B
C
cAMP (AU)
1.5
1.0
0.5
0.0
0
10^{-4}
10^{-3}
10^{-2}
10^{-1}
10^{0}
10^{1}
10^{2}
10^{3}
10^{4}
hormone (ng/ml)

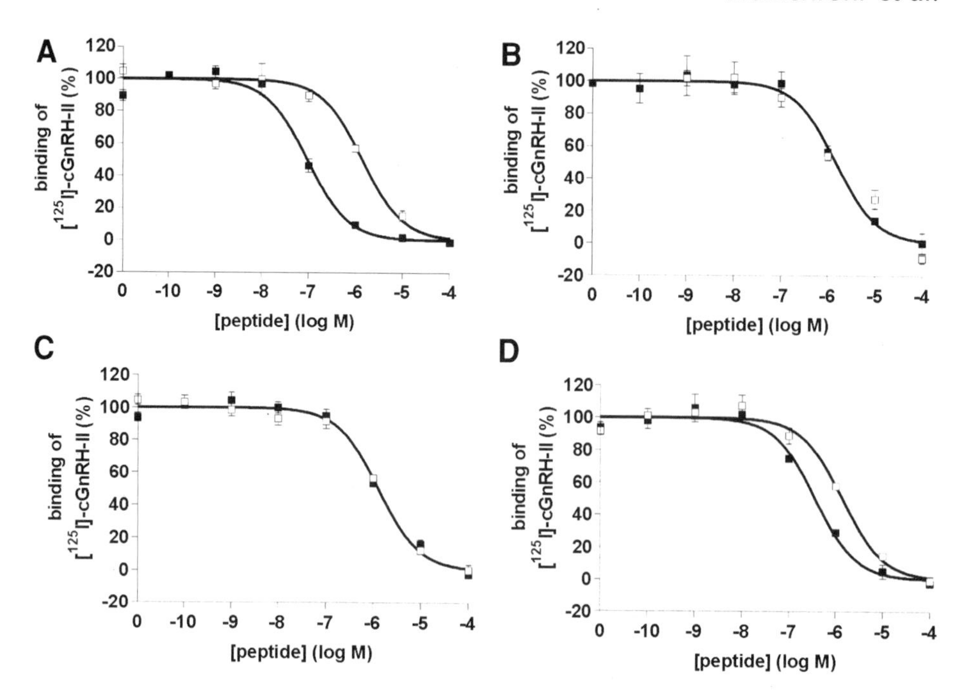

Fig. 2. Ligand binding of [His^5,Leu^7,Asn^8]gonadotropin-releasing hormone (catfish GnRH, □) and [His^5,Leu^7,Arg^8]gonadotropin-releasing hormone (■) to membranes prepared from HEK-293(T) cells transiently expressing wild-type **(A)**, Asp^{304}Ala mutant **(B)**, Asp^{304}Asn mutant **(C)**, or Asp^{304}Glu mutant **(D)** catfish gonadotropin-releasing hormone (GnRH) receptors. Results are means ± SEM for triplicate observations from a single representative competition binding experiment in which binding of a constant amount of ^{125}I-chicken GnRH-II to the receptor-expressing membranes is replaced by increasing concentrations of unlabeled catfish GnRH or [His^5,Leu^7,Arg^8]GnRH, respectively. The lower the concentration of the unlabeled ligand necessary to replace half of the bound ^{125}I-chicken GnRH-II ligand (IC_{50} value), the higher the affinity of the receptor for the unlabeled ligand. Thus, the affinity of the catfish GnRH receptor for catfish GnRH can be improved by replacing Asn^8 by an arginine residue because the IC_{50} value for [His^5,Leu^7,Arg^8]GnRH is lower than that of catfish GnRH. Moreover, replacement of the negatively charged Asp^{304} residue in the receptor with uncharged (alanine or asparagine) residues abolished the preference of this receptor for ligands with Arg^8. This demonstrates that Asp^{304} of the catfish GnRH receptor is able to mediate the recognition of [Arg^8]GnRH peptides. (Reprinted with permission from **ref.** *7* © 2002 The Biochemical Society.)

3. To test intramolecular networks of interactions perform single and reciprocal/compensating mutations. In cases in which a second mutation restores the function disrupted by a first mutation this is an indication that these residues share the

same microenvironment and that they have a related rather than an independent contribution to the function analyzed *(8,9)*.

4. A mutation can either disrupt or restore a direct interaction with a ligand or G-protein, it can result in the pertubation of a local conformation or disturb the overall integrity of the receptor. Therefore it is necessary to test mutant GPCR constructs in different functional assays in order to analyze the nature of the induced effect *(4)*.
5. The basis for the rational design of novel peptic or nonpeptic ligands of GPCRs is a precise knowledge regarding the ligand–receptor complex. It is necessary not only to characterize the ligand-binding pocket of the receptor by site-directed mutagenesis but also to use multiple selected receptor agonists and antagonists to evaluate specific hypotheses about the molecular basis of the ligand–receptor interaction (*see* **Fig. 2**) *(7)*.
6. Computational three-dimensional molecular models of the ligand–receptor complex and molecular dynamic simulation are extremely valuable for designing and interpreting mutagenesis studies *(7,9)*. Mutational experiments and the 2.8-Å resolution crystal structure of bovine rhodopsin *(3)*, on the other hand, help fine-tune the computational models.

In this chapter, we describe three different strategies of receptor mutagenesis:

1. Oligonucleotide-directed mutagenesis using the Altered Sites II in vitro mutagenesis system (*see* **Note 1** and **Fig. 3**) involves annealing of an antibiotic repair oligonucleotide and the mutagenic oligonucleotide to alkaline-denatured pALTER® II DNA template containing the GPCR insert, followed by synthesis of the mutant strand with T4 DNA polymerase and T4 DNA ligase. The heteroduplex DNA is then transformed into the repair minus *E. coli* strain ES1301 *mutS* (*see* **Note 2**) *(10)* and grown in selective medium to isolate clones containing the mutant plasmid. Antibiotic-resistant plasmids are isolated and transformed into the competent final host strain. Restriction analysis or direct sequencing of the plasmid DNA allows confirmation of the introduction of the mutation(s). This strategy is very efficient for single/multiple amino acid substitutions/deletions/insertions and/or N- or C-terminal truncations of GPCRs (by mutating start or stop codons), but it is inappropriate for substituting/deleting/inserting larger domains within GPCRs. The yield of mutants is high owing to the use of the repair minus *E. coli* strain ES1301 *mutS*. Moreover, analysis of mutants is very quick because of the preselection by antibiotics. We have never observed any additional spontaneous mutations. A disadvantage of this mutagenesis system is that it requires in vitro digestion and purification steps to clone the GPCR insert into the pALTER II vector before mutagenesis, and back into an appropriate expression vector such as pcDNA3 after mutagenesis.
2. Recent developments in mutagenesis technology now allow oligonucleotide-directed mutagenesis directly on your GPCR construct of interest. This eliminates additional subcloning steps and transformation into a specific *E. coli* strain. For example, the GeneTailor site-directed mutagenesis system (*see* **Note 3** and **Fig. 4**) involves methylation of plasmid DNA, harboring the GPCR insert, after which

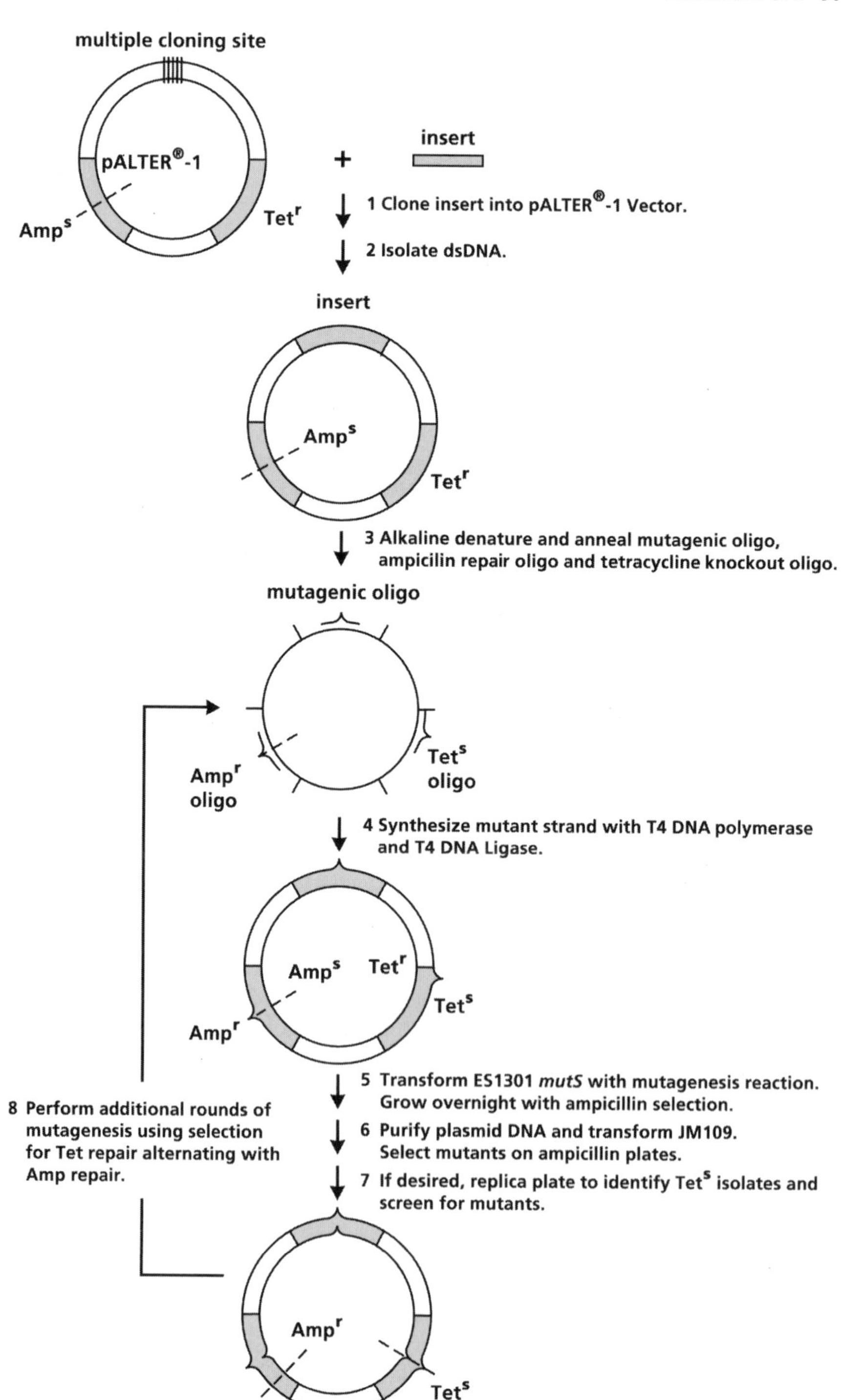
multiple cloning site
pALTER®-1
Amp^s
Tet^r
+
insert
1 Clone insert into pALTER®-1 Vector.
2 Isolate dsDNA.
insert
Amp^s
Tet^r
3 Alkaline denature and anneal mutagenic oligo, ampicilin repair oligo and tetracycline knockout oligo.
mutagenic oligo
Amp^r oligo
Tet^s oligo
4 Synthesize mutant strand with T4 DNA polymerase and T4 DNA Ligase.
Amp^s
Tet^r
Tet^s
Amp^r
5 Transform ES1301 mutS with mutagenesis reaction. Grow overnight with ampicillin selection.
6 Purify plasmid DNA and transform JM109. Select mutants on ampicillin plates.
7 If desired, replica plate to identify Tet^s isolates and screen for mutants.
8 Perform additional rounds of mutagenesis using selection for Tet repair alternating with Amp repair.
Amp^r
Tet^s

the plasmid DNA is amplified in a mutagenesis reaction with two overlapping primers, of which one contains the target mutation(s). Next, the amplified plasmid is transformed into wild-type *E. coli* in which the endogenous *Mcr*BC endonuclease digests the methylated template DNA, leaving only the unmethylated, mutated amplification product. This system enables fast and high-throughput mutagenesis in just three easy steps, generating multiple mutants for functional studies.

3. The polymerase chain reaction (PCR)-based mutagenesis strategy requires the use of four oligonucleotide primers to generate one chimeric GPCR (*see* **Note 4** and **Fig. 5**). Two oligonucleotide primers overlap in the chimeric junction region, one being antisense (primer C) and the other being sense (primer B), and both encoding the chimeric junction between the domains derived from two different GPCRs (e.g., GPCR1 and GPCR2; *see* **Fig. 5**). The other two oligonucleotide primers are located upstream ([sense] primer A) and downstream ([antisense] primer D) of the chimeric junction region, and are either flanking the multiple cloning site, or perfectly complementary to the 5′- and 3′-end coding sequences, of GPCR1 and GPCR2, respectively (*see* **Note 5**). Two PCR fragments (between primers A and C, and between primers B and D, respectively) that each have an overlapping end at the primer B/C region are generated in high-fidelity PCRs (*see* **Note 6**). Next, the A/C and B/D fragments are fused in a self-primed fusion PCR, in which the 3′ ends of the complementary sequences of primers B and C are allowed to anneal, and serve as primers for elongation by DNA polymerase. Subsequently, the fusion products are PCR amplified using the primers A and D. Next, the resulting chimeric GPCR PCR product (A/D) is either TA subcloned into an appropriate expression vector (e.g., pcDNA3.1/V5-His) or subcloned using the restriction endonuclease sites present in the multiple cloning site, and entirely sequenced.

   This strategy is especially useful to generate GPCR chimeras, swap receptor domains, and to introduce or delete large GPCR domains (*see* **Note 4**). However, a major disadvantage of this strategy is that the entire PCR product has to be sequenced because of the spontaneous occurrence of nucleotide misincorporation(s) during PCR.

2. Materials

1. Altered Sites II in vitro mutagenesis system and ES1301 *mutS* Competent Cells (*see* **Note 1**).
2. GPCR insert cloned into pALTER II vector (*see* **Note 7**).
3. Phosphorylated mutagenic oligonucleotide (*see* **Note 8**).
4. TE buffer: 10 m*M* Tris-HCl, pH 8.0, 1 m*M* EDTA.

Fig. 3. *(see opposite page)* Schematic diagram of the oligonucleotide-directed mutagenesis using the Altered Sites II in vitro Mutagenesis System. (Adapted from the manufacturer's technical manual.)

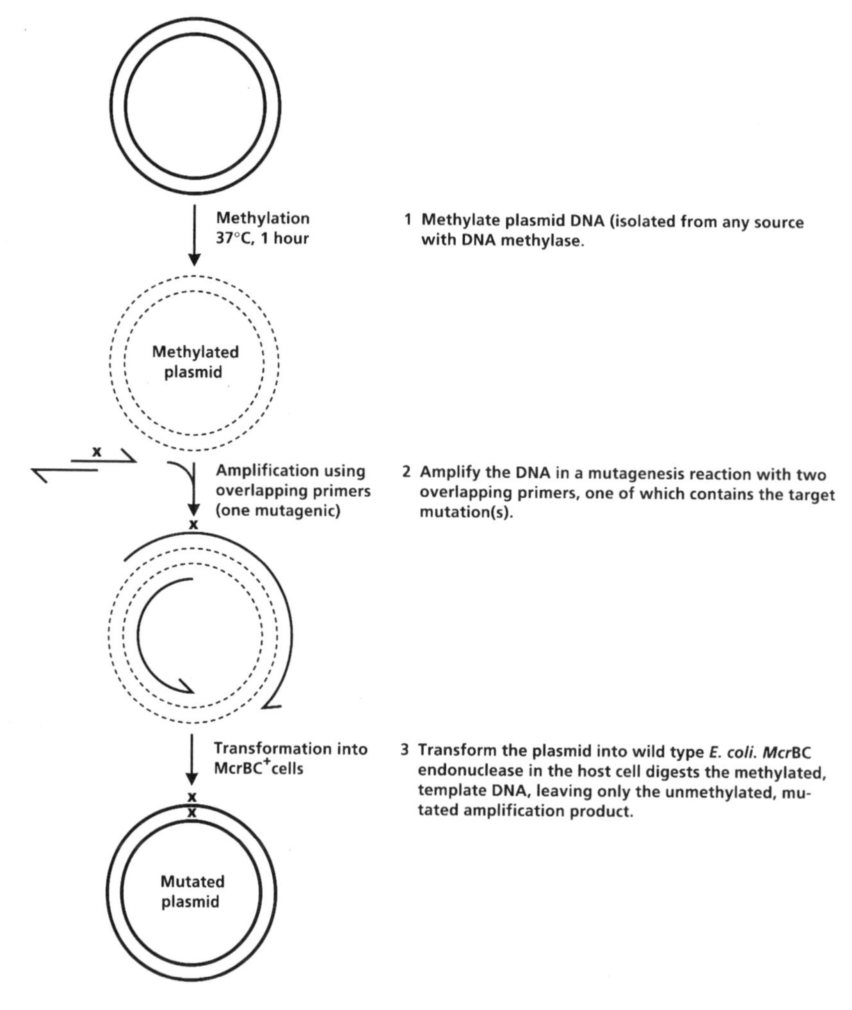

Fig. 4. Schematic diagram of the GeneTailor Site-Directed Mutagenesis System. (Adapted from the manufacturer's technical manual.)

5. 10X Annealing buffer: 200 m*M* Tris-HCl, pH 7.5; 100 m*M* $MgCl_2$; 500 m*M* NaCl.
6. 10X Synthesis buffer: 100 m*M* Tris-HCl, pH 7.5; 5 m*M* dNTPs; 10 m*M* ATP; 20 m*M* dithiothreitol (DTT).
7. 5X KCM: 0.5 *M* KCl, 0.15 *M* $CaCl_2$; 0.25 *M* $MgCl_2$.
8. GPCR cDNA insert(s) cloned in an expression vector(s).
9. Sense and antisense open-reading frame oligonucleotide primers or primers flanking the cloning site.
10. Sense and antisense chimeric overlapping oligonucleotide primers.
11. Advantage®-High Fidelity (HF) PCR kit (*see* **Note 6**).

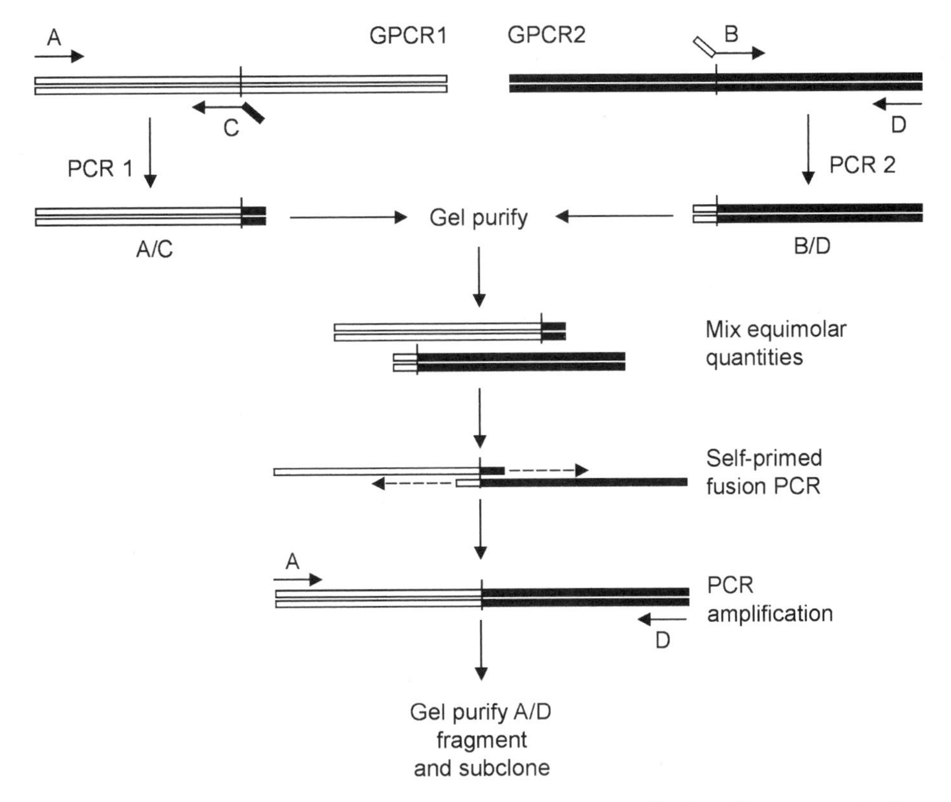

Fig. 5. Construction of a chimeric receptor (A/D) by PCR-based mutagenesis (*see* **Subheadings 3.8.** and **3.9.** for a detailed description).

12. pcDNA3.1/V5-His TOPO® TA expression kit and One Shot® TOP10 chemically competent *E. coli* (Invitrogen; *see* **Note 9**).
13. SOC medium: 2% tryptone, 0.5% yeast extract, 10 m*M* NaCl, 2.5 m*M* KCl, 10 m*M* $MgCl_2$, 10 m*M* $MgSO_4$, 20 m*M* glucose.
14. LB medium: 1% tryptone, 0.5% yeast extract, 100 m*M* NaCl, pH 7.5.
15. Ampicillin, tetracycline, kanamycin.
16. JM109 bacterial strain.

3. Methods

3.1. Oligonucleotide-Directed Mutagenesis: Denaturation of Double-Stranded DNA Template

1. Set up the following alkaline denaturation reaction and incubate for 5 min at room temperature: 0.05 pmol (*see* **Note 10**) of dsDNA template, 2 µL of 2 *M* freshly prepared NaOH, and double-distilled water (ddH_2O) up to 20 µL final volume.

2. Add 2 µL of 2 *M* NH_4Ac, pH 4.6, adjusted with acetic acid; 10–20 µg of linear polyacrylamide (*see* **Note 11**), and 75 µL of 96% ethanol. Incubate for 10 min at room temperature.
3. Precipitate the DNA by centrifugation for 25 min at top speed in a microcentrifuge, drain, and wash the pellet with 200 µL of 96% ethanol. Then drain and dissolve the pellet in 10 µL of TE, pH 8.0.

3.2. Oligonucleotide-Directed Mutagenesis: Annealing Reaction and Mutant Strand Synthesis

1. Prepare the annealing reaction on ice: 10 µL (0.05 pmol) of alkaline-denaturated dsDNA, 1.25 pmol of mutagenic oligonucleotide (*see* **Note 12**), 1 µL (0.25 pmol) of ampicillin repair oligonucleotide, 1 µL (0.25 pmol) of tetracycline knock-out oligonucleotide, 2 µL of 10X annealing buffer, and ddH_2O up to 20 µL final volume.
2. Heat annealing reactions at 75°C for 5 min and allow them to cool slowly to room temperature by placing heating block or beaker containing 300 mL of water for 30 min at room temperature and for another 10–15 min on ice.
3. Place the annealing reactions on ice and add these components in the order listed: 3 µL of 10X synthesis buffer, 1 µL of T4 DNA polymerase, 1 µL of T4 DNA ligase, and 5 µL ddH_2O to reach 30 µL final volume.
4. Incubate the reaction at 37°C for 90 min to perform mutant strand synthesis and ligation.

3.3. Oligonucleotide-Directed Mutagenesis: Transformation of ES1301 mutS Competent Cells

1. Thaw competent ES1301*mutS* cells (*see* **Note 2**) quickly at room temperature, then keep on ice until use.
2. Prepare the transformation reactions on ice in the order listed: 65 µL of ddH_2O, 20 µL of 5X KCM, 15 µL of DNA from **Subheading 3.2., step 4**, 100 µL of ES1301 *mutS* cells. Then incubate for 20 min on ice and for another 10 min at room temperature.
3. Transfer the transformation reaction into a tube containing 4 mL of LB medium without antibiotics and incubate at 37°C for 3 h with shaking (~225 rpm).
4. Add 125 µg/mL of ampicillin and incubate overnight at 37°C with shaking (~225 rpm).
5. The next day use this culture for the plasmid miniprep (*see* **Note 13**) and transformation in competent JM109 cells.

3.4. Oligonucleotide-Directed Mutagenesis: Preparation of JM109 Competent Cells

1. Vortex-mix overnight culture of JM109 cells, transfer 1 mL of cell culture to 50 mL of LB medium, and leave for 2.5 h at 37°C with shaking.
2. Pellet the cells by centrifugation at 1500*g* for 4 min at 4°C, discard the supernatant, resuspend the pellet carefully in 25 mL of cold 50 m*M* $CaCl_2$, and leave for 20 min on ice.

3. Pellet the cells from **Subheading 3.4., step 2**, by centrifugation at 1500*g* for 4 min at 4°C, discard the supernatant, resuspend the pellet carefully in 5 mL of cold 50 m*M* $CaCl_2$, and leave for at least 1 h on ice before usage. These competent cells can be stored frozen at –80°C.

3.5. Oligonucleotide-Directed Mutagenesis: Transformation Into JM109 Competent Cells

1. Transfer 150 µL of JM109 competent cells from step **Subheading 3.4., step 3**, into a tube and add 5–10 ng of plasmid DNA from **Subheading 3.3., step 5**, moving the pipet tip through the cells while dispensing.
2. Immediately place the tube on ice for 1 h. Quickly flick the tube several times during the hour of incubation to keep the cells in suspension.
3. Heat-shock the cells in a water bath at exactly 42°C for 1 min; do not shake and immediately place the tube back on ice.
4. Add 200 µL of LB medium to each transformation reaction and incubate at 37°C for 1 h with shaking to let the cells recover.
5. Plate all 350 µL of each transformation reaction on a LB plate containing 125 µg/mL of ampicillin and leave the plates overnight at 37°C so that mutant colonies can grow.

3.6. Oligonucleotide-Directed Mutagenesis: Analysis of Transformants

1. Pick colonies and plate them in a grid format on grouped plates containing 10 µg/mL of tetracycline, 50 µg/mL of kanamycin, or 125 µg/mL of ampicillin, respectively. Pick each colony with a sterile toothpick and inoculate the three plates in sequence.
2. Incubate the inoculated plates overnight at 37°C.
3. Select $amp^R tet^S kan^S$ colonies (*see* **Note 14**), grow them overnight at 37°C in LB medium containing 125 µg/mL of ampicillin, and miniprep the DNA.
4. Screen for the desired mutation(s) in the GPCR insert by sequencing the region of the desired mutation(s) or do an in vitro digestion analysis (*see* **Note 15**).

3.7. Oligonucleotide-Directed Mutagenesis: The GeneTailor Site-Directed Mutagenesis System (Invitrogen, Carlsbad, CA)

For detailed information about this new oligonucleotide-directed mutagenesis technique (which we have not tested yet in the laboratory), see instruction manual: http://www.invitrogen.com/content/sfs/manuals/genetailor_man.pdf.

3.8. PCR-Based Mutagenesis: Primer Design (see Fig. 5)

1. Design sense and antisense chimeric primers (B and C, respectively) of which 24–30 nucleotides of their 3′ end are complementary to the GPCR1 or GPCR2 sequences, respectively. These primers are juxtaposed to the chimeric junction, having a calculated melting temperature (T_m) of >60°C (*see* **Note 16**). The 5′-ends of primers B and C consist of approx 10 nucleotides that encode for

GPCR2 or GPCR1, respectively. The overlap between the chimeric primers must have a calculated T_m values of 55–60°C.

2. Design perfectly matching sense and antisense primers (A and D, respectively) considering the mentioned (*see* **Note 16**) critical parameters, and having T_m values approx 10°C higher than the T_m values of the matching 3′-end of the chimeric primers.

3.9. PCR-Based Mutagenesis: PCR (see Fig. 5)

1. Set up a first PCR using primers A and C and GPCR1 cDNA as template as follows: 10 ng of GPCR1, 15 pmol of each primer A and C, 5 µL of 10X HF PCR reaction buffer, 5 µL of 10X HF dNTP mix, 1 µL of Advantage-HF polymerase mix, and ddH_2O to 50 µL final volume (*see* **Note 6**). Initiate PCR (*see* **Note 17**) with denaturation at 94°C for 30 s, followed by 10 cycles of 94°C for 8 s, *XX*°C (*see* **Note 18**) for 30 s, 68°C for *YY* s (*see* **Note 18**), then 25 cycles of 94°C for 8 s and 68°C for *YY* s, and finally 5-min extension at 68°C.
2. Perform a second PCR similar to the first PCR, but this time use primers B and D and GPCR2 cDNA as template.
3. Gel-purify the A/C and B/D PCR products.
4. Perform a self-primed fusion PCR, by mixing equimolar amounts of the A/C and B/D fragments (use 10 ng of the smallest fragment) and 5 µL of 10X HF PCR reaction buffer, 5 µL of 10X HF dNTP mix, 1 µL of Advantage-HF polymerase mix, and ddH_2O to 50 µL final volume. (**Note:** In this step, no primers are present in the PCR.) Anneal and extend the A/C- and B/D-overlaps by using the following PCR cycling: 94°C for 30 s, followed by 20 cycles: 94°C for 15 s, ZZ°C (*see* **Note 19**) for 4 min and 68°C for YY min.
5. Dilute the resulting fusion-PCR product 50X in ddH_2O, and subsequently use 2 µL as template for PCR amplification with the A and D primers using PCR cycling parameters that have been optimized for these primers.
6. Gel-purify the resulting A/D PCR product and dissolve DNA in 20 µL of ddH_2O final volume.

3.10. PCR-Based Mutagenesis: TOPO®-Cloning and Transformation

1. Mix 4 µL A/D PCR product, 1 µL of salt solution (1.2 *M* NaCl and 0.06 *M* $MgCl_2$) and 1 µL of pcDNA3.1/V5-His-TOPO solution and incubate for 20 min at room temperature (*see* **Note 20**).
2. Place the TOPO cloning reaction on ice, and subsequently add 2 µL into a vial of One Shot® TOP10 chemically competent *E. coli*, mix gently, and incubate on ice for 15 min.
3. Heat-shock the cells for 30 s at 42°C and immediately place the vial on ice.
4. Add 250 µL of SOC medium to each transformation reaction and shake the vial at 37°C for 1 h.
5. Plate 75 and 175 µL of each transformation reaction on LB plates containing 50 µg/mL of ampicillin and incubate overnight at 37°C.

3.11. PCR-Based Mutagenesis: Analysis of Transformants

1. Pick colonies with a sterile toothpick and resuspend them individually in 50 µL of ddH_2O.
2. Use 2.5 µL of these picked colonies as templates for PCR amplification (*see* **Note 21**) using a gene-specific sense primer in combination with a plasmid antisense primer to determine the presence of an A/D insertion as well as the right insert orientation in the plasmids.
3. Identify positive clones on gel, and inoculate 4 mL of LB medium containing 50 µg/mL of ampicillin with 35 µL of the resuspended colonies from **step 1** and grow them overnight at 37°C while shaking (~200 rpm).
4. Miniprep the plasmid DNA using a plasmid isolation kit and sequence the entire open-reading frame of the insert.

4. Notes

1. Altered Sites II in vitro Mutagenesis System and ES1301 *mutS* Competent Cells can be obtained from Promega (WI, USA).
2. The ES1301 *mutS* strain ***(10)*** suppresses in vivo mismatch repair. It is used for the initial round of transformation to decrease the chance that the antibiotic repair mismatch or the mutagenic mismatch will be repaired. We succeed in getting transformants only when using the highly competent (>10^7 cfu/µg) cells purchased from Promega.
3. The GeneTailor site-directed mutagenesis system can be obtained from Invitrogen (CA, USA).
4. We succesfully used the PCR-based mutagenesis method to introduce the nucleotide sequence coding for the HA-epitope tag between the sequences coding for the human follicle-stimulating hormone receptor (hFSH-R) signal peptide and the mature receptor protein as well as to perform site-directed mutagenesis.
5. Alternatively, when unique endonuclease restriction sites that flank the mutated/chimeric GPCR domain are present, perfectly complementary primers that are distally located from these endonuclease restriction sites can be used in combination with the mutation/chimeric primers, instead of the primers that are located at the 5′- and 3′-ends of the mutation/chimeric receptor construct. The mutated/chimeric PCR fragment can then replace the corresponding wild-type fragment via their unique restriction sites; this reduces the length of the PCR fragment and therefore decreases the incidence of spontaneous occuring PCR errors. Moreover, it avoids sequencing the entire receptor-coding sequences.
6. High-fidelity polymerase is composed of a mixture of *Taq* polymerase (variant) and a proofreading polymerase to decrease the incidence of PCR artifacts.
7. Oligonucleotide-directed mutagenesis can be performed on any vector containing the GPCR insert. However, we inserted our GPCR into the pALTER II vector to increase the yield of mutants using antibiotic selection (*see* **Note 1**).
8. The sythetic 5′-phosphorylated oligonucleotide must be complementary to the single-stranded template except for the region of mismatch (containing the desired

mutation) near the center. The size of the oligomer synthesized depends on the number of base substitutions required; a 17–20-oligonucleotide is sufficient for a single base change. For mutations involving two or more mismatches, oligomers 25 bases or longer are needed to allow 12–15 perfectly matched nucleotides on either side of the mismatch. Phosphorylation of the oligonucleotides at their 5′-end significantly increases the number of mutants.

9. The pcDNA3.1/V5-His-TOPO vector is supplied linearized with single 3′ thymidine overhangs and a topoisomerase covalently bound to the vector. PCR inserts are efficiently ligated into the vector via their single 3′ deoxyadenosine overhangs that are added during PCR amplification by most *Taq* polymerases in a template-independent fashion. To ensure the presence of 3′ deoxyadenosine overhangs on the PCR product the high-fidelity polymerase mixture must contain at least 10 times more *Taq* polymerase than proofreading polymerase.
10. Rule of the thumb: ng of dsDNA ≈ pmol of dsDNA × 0.66 × N; where N = length of dsDNA in basepairs.
11. Without polyacrylamide as carrier we never succeeded in precipitating the denaturated DNA. Prepare a 5% acrylamide solution (without *bis*-acrylamide) in 40 m*M* Tris-HCl, 20 m*M* NaAc, 1 m*M* EDTA, pH 7.8. Add 1/100 volume of 10% ammonium persulfate and 1/1000 volume of TEMED, and allow to polymerize for 30 min. When the solution has become viscous, precipitate the polymer with 2.5 volumes of ethanol, centrifuge, and redissolve the pellet in 20 volumes of ddH_2O by shaking overnight. The 0.25% linear polyacrylamide solution obtained can be stored at –20°C for several years ***(11)***.
12. We successfully used up to three different mutagenic oligonucleotides at the same time.
13. We purify plasmid DNA out of transformed ES1301 *mutS* cell culture following the procedure as described in the manual of QIAGEN Spin Plasmid Kit (QIAGEN, CA, USA). The column is eluted with 100 µL 10 m*M* Tris-HCl, pH 8.5.
14. If the transformants do not grow on the kanamycin plate you can be sure that transfer to JM109 strain occurred. If you only have $amp^R tet^R kan^S$ colonies proceed with them, as there is still a chance that they contain the aimed mutation in the GPCR insert. The only disadvantage of $amp^R tet^R kan^S$ cells is that they cannot be used for a second round of mutagenesis in which ampicillin resistance should be knocked out and tetracycline resistance should be restored. However, in our hands this second round of mutagenensis never worked.
15. Often restriction sites can be incorporated into the mutagenic primers without altering the amino acid sequence. These sites can provide a quick screen to identify those clones containing the desired mutation.
16. The efficiency of PCR-based mutagenesis is critically dependent on the design of the four primers. Generally, the part of the primer that will actually hybridize to the DNA template (i.e., the entire A and D primers and the matching 3′-ends of the B and C primers) must be designed considering the following: the length should be between 18 and 30 nucleotides long; the melting temperature should

be >60°C; 3′-end sequences that are complementary between primer pairs should be avoided. The entire primer sequence should be checked for self-complementary primer sequences to avoid intraprimer hairpin loop formation. We routinely optimize our primers using primer analysis software (Primer Express; Applied Biosystems, CA).

17. The cycling parameters have been optimized for a Perkin Elmer GeneAmp PCR system 2400, and may vary with different thermal cyclers.
18. Set the primer-template annealing temperature "*XX*" 5°C below the T_m values of the matching 3′-ends of the B and C primers during the initial PCR cycles. The annealing temperature is then increased to 68°C to increase specific PCR amplification. Elongation time "*YY*" depends on the length of the template to be amplified (~1 min/kb).
19. Set self-primed annealing temperature "*ZZ*" 7°C below the T_m value of the overlapping ends as calculated in **Subheading 3.8., step 1**.
20. Alternatively, digest the PCR-amplified chimeric/mutated GPCR fragment that is flanked by unique endonuclease restriction sites (*see* **Note 5**), as well as an expression vector containing the wild-type GPCR cDNA, with appropriate enzymes. Gel-purify the desired fragments, and ligate the chimeric/mutated GPCR fragment into the wild-type GPCR/vector construct in a 6:1 ratio using T4 DNA ligase in 20 μL final volume for 5 min at room temperature. Add 4 μL of ligation mix to One Shot® TOP10 chemically competent *E. coli*.
21. Alternatively, culture the picked colonies overnight in LB medium containing 50 μg/mL of ampicillin and isolate plasmid DNA the next day. Determine the presence of an A/D insertion as well as its orientation in the plasmids by restriction analysis.

References

1. Baldwin, J., Schertler, G. F. X., and Unger, V. M. (1997) An alpha-carbon template for the transmembrane helices in the rhodopsin family of G-protein coupled receptors. *J. Mol. Biol.* **272,** 144–154.
2. Unger, V. M. and Schertler, G. F. X. (1995) Low resolution structure of bovine rhodopsin determined by electron cryo-microscopy. *Biophys. J.* **68,** 1776–1786.
3. Palczewski, K., Kumasaka, T., Hori, T., et al. (2000) Crystal structure of rhodopsin: a G-protein coupled receptor. *Science* **289,** 739–745.
4. Blomenröhr, M., Heding, A., Sellar, R., et al. (1999) Pivotal role for the cytoplasmic carboxyl-terminal tail of a nonmammalian gonadotropin-releasing hormone receptor in cell surface expression, ligand binding, and receptor phosphorylation and internalization. *Mol. Pharmacol.* **56,** 1229–1237.
5. Vischer, H. F., Granneman, J. C. M., Noordam, M. J., Mosselman, S., and Bogerd, J. (2003) Ligand selectivity of gonadotropin receptors. The role of the β-strands of extracellular leucine-rich repeats 3 and 6 of the human luteinizing hormone receptor. *J. Biol. Chem.* **10,** 15,505–15,513.

6. Willars, G. B., Heding, A., Vrecl, M., et al. (1999) Lack of a C-terminal tail in the mammalian gonadotropin-releasing hormone receptor confers resistance to agonist-dependent phosphorylation and rapid desensitization. *J. Biol. Chem.* **274,** 30,146–30,153.
7. Blomenröhr, M., Ter Laak, T., Kühne, R., et al. (2002) Chimaeric gonadotropin-releasing hormone (GnRH) peptides with improved affinity for the catfish *(Clarias gariepinus)* GnRH receptor. *Biochem. J.* **361,** 515–523.
8. Blomenröhr, M., Bogerd, J., Leurs, R., et al. (1997) Differences in structure–function relations between nonmammalian and mammalian gonadotropin-releasing hormone receptors. *Biochem. Biophys. Res. Commun.* **238,** 517–522.
9. Blomenröhr, M., Kühne, R., Hund, E., Leurs, R., Bogerd, J., and Ter Laak, T. (2001) Proper receptor signalling in a mutant catfish gonadotropin-releasing hormone receptor lacking the highly conserved Asp^{90} residue. *FEBS Lett.* **501,** 131–134.
10. Siegel, E. C., Wain, S. L., Meltzer, S. F., Binion, M. L., and Steinberg, J. L. (1982) Mutator mutations in Escherichia coli induced by the insertion of phage mu and the transposable resistance elements Tn5 and Tn10. *Mutat. Res.* **93,** 25–33.
11. Gaillard, C. and Strauss, F. (1990) Ethanol precipitation of DNA with linear polyacrylamide as carrier. *Nucleic Acids Res.* **1,** 378.

20

Study of G-Protein-Coupled Receptor–Protein Interactions by Bioluminescence Resonance Energy Transfer

Karen M. Kroeger and Karin A. Eidne

Summary

Complex networks of protein–protein interactions are key determinants of cellular function, including those regulated by G-protein-coupled receptors (GPCRs). Formation of either stable or transitory complexes are involved in regulating all aspects of receptor function, from ligand binding through to signal transduction, desensitization, resensitization and downregulation. Today, 50% of all recently launched drugs are targeted against GPCRs. This particular class of proteins is extremely useful as a drug target because the receptors are partly located outside the cell, simplifying bioavailability and delivery of drugs directed against them. However, being located within the cell membrane causes difficulties for the study of GPCR function and bioluminescence resonance energy transfer (BRET), a naturally occurring phenomenon, represents a newly emerging, powerful tool with which to investigate and monitor dynamic interactions involving this receptor class. BRET is a noninvasive, highly sensitive technique, performed as a simple homogeneous assay, involving the proximity-dependent transfer of energy from an energy donor to acceptor resulting in the emission of light. This technology has several advantages over alternative approaches as the detection occurs within live cells, in real time, and is not restricted to a particular cellular compartment. The use of such biophysical techniques as BRET, will not only increase our understanding of the nature of GPCR regulation and the protein complexes involved, but could also potentially lead to the development of novel therapeutics that modulate these interactions.

Key Words

Bioluminescence resonance energy transfer, dimerization, G-protein-coupled receptor, protein–protein interaction.

From: *Methods in Molecular Biology, vol. 259, Receptor Signal Transduction Protocols, 2nd ed.*
Edited by: G. B. Willars and R. A. J. Challiss © Humana Press Inc., Totowa, NJ

1. Introduction

Complex networks of protein–protein interactions underpin G-protein-coupled receptor (GPCR) function and the identification and characterization of such dynamically regulated molecular interactions would undoubtedly aid in the design of more effective therapeutic strategies. The application of a novel biophysical system that would allow the measurement of protein–protein interactions involving GPCRs and their complex networks is described in this chapter. This system relies on the measurement of bioluminescence resonance energy transfer (BRET) and has been successfully applied to the study of GPCR homo- and heterodimerization as well as receptor/β-arrestin interactions involved in receptor desensitization and trafficking in mammalian cells *(1)*. The main advantage of BRET is that it allows protein–protein interactions to be monitored in real time, in their correct location, in live cells. Owing to the highly hydrophobic nature and cellular localization of GPCRs, conventional techniques for measuring protein–protein interactions (i.e., coimmunoprecipitation and yeast two-hybrid screening) have significant drawbacks. BRET is a simple, rapid homogeneous assay that may overcome some of these limitations.

BRET is a naturally occurring phenomenon, an example of which is the nonradiative transfer of energy occurring between aequorin and green fluorescent protein (GFP), in the jellyfish *Aequoria*. BRET interactions can be studied using fusion proteins that are tagged with either a bioluminescent donor such as the luciferase enzyme (Rluc) from the sea pansy *Renilla* or a fluorescent acceptor, such as enhanced green or enhanced yellow fluorescent protein (EGFP or EYFP). When the interaction partners are coexpressed in cells, energy is transferred in the presence of the substrate, coelenterazine, from Rluc to EYFP and light emitted only if the proteins are in close enough proximity (within 100 Å) *(2)*. The critical distance dependence between donor and acceptor molecules makes BRET an ideal technique to study receptor–protein interactions involved in all aspects of GPCR function and their modulation by agonists and antagonists. Similarly, fluorescence resonance energy transfer (FRET), which involves the transfer of energy between fluorescent donor and acceptor molecules, also represents a tool for the detection of GPCR–protein interactions and has been used in conjunction with imaging techniques to monitor GPCR interactions both spatially and temporally, and this is discussed in Chapter 21. Unlike its derivative FRET, BRET avoids the need for excitation, thus circumventing the problems associated with autofluorescence, photobleaching, and cell damage. Furthermore, the lower background fluorescence associated with BRET makes it a sensitive technique enabling the detection of low-level or weak protein interactions *(3)*. However, BRET is potentially limited because it cannot determine the cellular localization of the protein interaction. Single-cell BRET imaging would overcome this problem and would represent a significant advancement in the study of protein–protein interactions.

To apply BRET to the study of receptor interactions, fusion proteins are coexpressed in cells and if the two fusion molecules are in close enough proximity, then light is transferred from Rluc (480 nm) to EYFP. This results in an excitation and emission of light at the characteristic wavelength of 530 nm, with the degree of BRET quantified as a ratio of light emitted at 530 nm over 480 nm (**Fig. 1**). Initially BRET was performed in *E. coli* to study the interaction of light-sensitive circadian clock proteins *(2)*. Several other groups have now used BRET to study protein interactions between GPCRs in mammalian cells, including receptor homodimerization ***(4–9)*** and heterodimerization ***(10–13)***. Interactions between GPCRs and intracellular proteins required for receptor function have also been monitored using BRET, such as receptor–β-arrestin interactions involved in receptor desensitization and internalization ***(4,5,13)***. Thus, in theory BRET could represent a powerful tool with which to study the complex array of receptor interactions critical for GPCR functions ranging from ligand binding to receptor signaling, desensitization, and trafficking.

2. Materials

1. pRL-CMV (Promega) contains the gene encoding the *Renilla reniformis* luciferase enzyme (Rluc).
2. pEYFP (Clontech) contains the gene encoding the red-shifted variant of *Aequorea* green fluorescent protein, enhanced yellow fluorescent protein (EYFP).
3. Polyfect transfection reagent (QIAGEN).
4. BRET assay buffer: phosphate-buffered saline (PBS).
5. Coelenterazine, 500 μ*M* stock solution in methanol, diluted to 50 μ*M* in PBS just prior to use (*see* **Note 1**).

3. Methods

Performing a BRET assay to investigate a potential protein–protein interaction involves several steps: (1) generation of the two proteins of interest genetically fused with either Rluc or EYFP at either the N- or C-terminus; (2) coexpression of the two BRET fusion proteins in mammalian cells, and (3) detection of the BRET signal. This method can be applied to study any protein–protein interaction in mammalian cells and is not limited to investigation of interactions involving GPCRs. Here we have used the homo-oligomeric interaction between thyrotropin releasing hormone receptors (TRHRs) as an

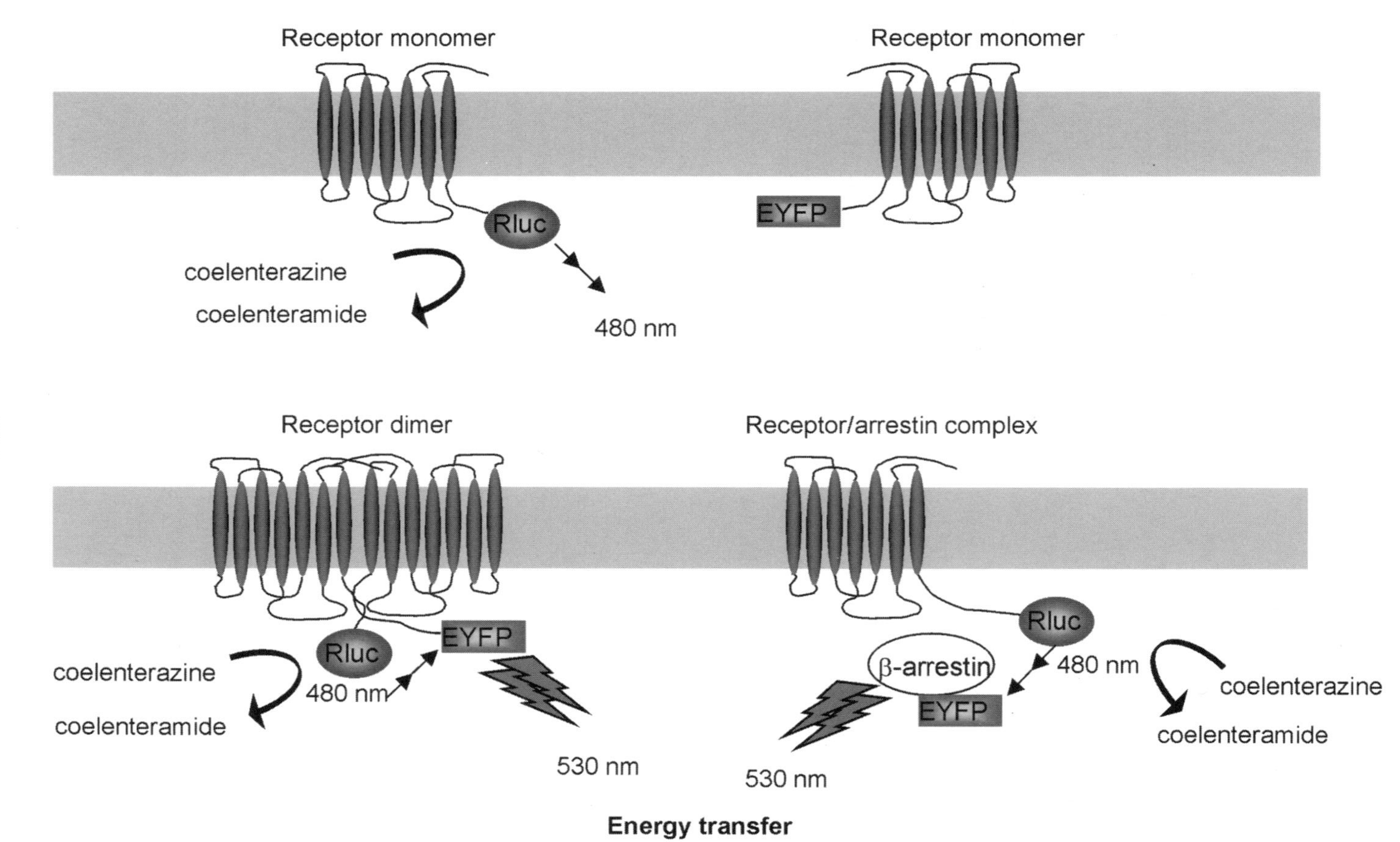
Receptor monomer
Rluc
coelenterazine
coelenteramide
480 nm
Receptor monomer
EYFP
Receptor dimer
EYFP
Rluc
coelenterazine
480 nm
coelenteramide
530 nm
Receptor/arrestin complex
Rluc
β-arrestin
480 nm
EYFP
coelenterazine
coelenteramide
530 nm
Energy transfer

example to demonstrate how the BRET technique can be applied to monitor protein–protein interactions involving GPCRs in live cells **(Fig. 2)**.

3.1. Generation of BRET Fusion Constructs

The following constructs encoding BRET fusion proteins were made:

1. Positive control BRET fusion construct for expression of the Rluc–EYFP fusion protein (*see* **Note 2**).
2. Negative control BRET constructs, pRluc, pEYFP for expression of Rluc and EYFP alone, and GnRHR–Rluc, β_2-adrenergic (β_2AR)–Rluc, GnRHR–EYFP, and β_2AR–EYFP for expression of other GPCRs (*see* **Note 3**). Gonadotropin releasing hormone (GnRH) and $\beta_2$adrenergic receptors were C-terminally tagged with either Rluc or EYFP (*see* **Note 3**).
3. BRET donor construct TRHR–Rluc for expression of the TRHR that was C-terminally tagged with Rluc (*see* **Notes 4–6**).
4. BRET acceptor construct TRHR–EYFP for expression of the TRHR that was C-terminally tagged with EYFP (*see* **Notes 4–6**).

3.2 Transfection and Coexpression of BRET Fusion Constructs

1. Transfect COS-1 cells (plated out the day prior to transfection at a density of 2×10^5 cells/well per 6-well plate) using transfection reagent according to the manufacturer's instructions (*see* **Note 7**).
2. Transfect cells with TRHR–Rluc fusion cDNA alone, or cotransfect with TRHR–Rluc fusion and TRHR–EYFP fusion cDNA (*see* **Note 8**). Rluc and Rluc-EYFP BRET control cDNA should also be transfected as a positive control for the experiment. Negative control transfections should also be included, for example TRHR–Rluc + EYFP, TRHR–EYFP + GnRHR–EYFP, TRHR–Rluc + β_2AR–EYFP, Rluc + TRHR–EYFP, and GnRHR–Rluc + TRHR–EYFP (*see* **Notes 3** and **8** and **Fig. 2**).

Fig. 1. *(see opposite page)* Schematic representation of BRET applications to detect GPCR protein–protein interactions. **(Upper panel)** Fusion proteins of the two proteins of interest are generated and coexpressed in mammalian cells. Following the addition of the cell-permeable Rluc substrate, coelenterazine, if no interaction and hence no energy transfer occurs then light is only emitted from Rluc at its peak wavelength of 480 nm. **(Lower panel)** If an interaction occurs between GPCRs (homo- or heterodimerization) or between receptor and intracellular protein (e.g., β-arrestin), Rluc and EYFP are brought into close enough proximity (<50–100 Å) to allow energy transfer to occur from donor (Rluc) to acceptor (EYFP), resulting in an additional emission of light at the wavelength characteristic of EYFP, 530 nm. The modulation of this interaction by various agonists or antagonists can then be investigated by monitoring the effect (increase, decrease, or no change) on the BRET signal.

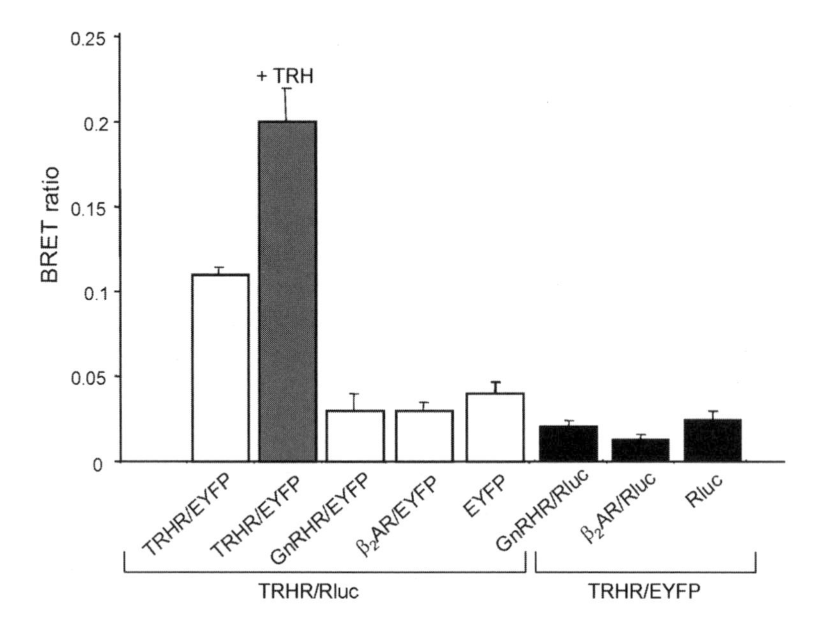

Fig. 2. Detection of dynamic receptor interactions involving thyrotropin releasing hormone receptor (TRHR) using BRET. **(A)** COS-1 cells were cotransfected with TRHR–Rluc and TRHR–EYFP, or as controls with GnRHR–EYFP, β_2AR–EYFP, or EYFP. Additional control transfections, TRHR–EYFP with either GnRHR–Rluc, β_2AR–Rluc, or Rluc were also performed. Following the addition of coelenterazine, light emission is immediately measured over two wavelength ranges and the BRET ratio determined demonstrating that a specific BRET signal, indicative of a homodimerization/oligomerization event, occurred between TRHRs. To determine the effect of ligand on the TRHR BRET, cells expressing Rluc and EYFP tagged TRHRs were preincubated for 15 min at 37°C with TRH (10^{-6} *M*) before the addition of coelenterazine.

3.3. BRET Assay

1. At 48 h posttransfection, cells are detached using PBS–0.05% trypsin, washed twice in PBS, and resuspended in 500 µL of PBS. For each transfection approx 100,000 cells are analyzed by flow cytometry for EYFP expression. For BRET, 50 µL of cell suspension (approx 20,000 cells) is distributed into each well of a 96-well plate.
2. Add 5 µL of coelenterazine (freshly diluted to 50 µ*M*) to each well to obtain a final concentration of 5 µ*M* and immediately read using a BRET plate reader (Mithras LB940, Berthold Technologies, Inc., Bad Wildbad, Germany) (*see* **Notes 9** and **10**).
3. To test the effect of an agonist or antagonist (or another reagent/chemical/treatment) on the interaction and hence on the BRET signal, the reagents can be preincubated prior to the addition of coelenterazine (*see* **Fig. 2**). Alternatively,

coelenterazine can be added, a reading taken, then the reagent added and the effect on the BRET ratio assessed over time.

4. Integrated readings are taken for 10 s collecting light filtered through two filters, each with a different wavelength range. The BRET ratio is then calculated from these two reads according to the following calculation:

$$\text{BRET ratio} = \frac{(\text{emission at 510–590 nm}) - (\text{emission at 440–500 nm}) \times cf}{(\text{emission at 440–500 nm})}$$

cf corresponds to (emission 510–590 nm /emission 440–500 nm)

5. Data are then represented as a normalized BRET ratio, which is defined as the BRET ratio for coexpressed Rluc and EYFP constructs normalized against the BRET ratio for the Rluc expression construct alone in the same experiment.

4. Notes

1. The form of coelenterazine used in the BRET assay is important. Several different forms of coelenterazine exist, each resulting in peak light emissions of slightly different wavelengths. For energy transfer to occur, the emission spectrum of the donor (Rluc) has to overlap significantly with the excitation/absorption spectrum of the energy acceptor (EYFP). In addition, the emission spectra of donor and acceptor have to be distinct enough to allow separate measurement of emission from each molecule with minimal overlap. With these considerations in mind, and using Rluc and EYFP as donor and acceptor molecules, respectively, the *h* form of coelenterazine was employed in the BRET assays. Coelenterazine is dissolved in methanol to prepare a stock solution (500 μ*M*) that is stored at –20°C protected from light. Just before use, the coelenterazine is diluted in BRET assay buffer to a final concentration of 50 μ*M* and wrapped in foil to protect it from light.
2. To confirm that a BRET signal can be obtained under the experimental conditions used, a BRET positive control should be used. A direct fusion of the Rluc and EYFP proteins separated by an amino acid linker is a good BRET positive control. It is constructed by cloning the Rluc coding region without its stop codon upstream and in-frame with the EYFP coding sequence. Following addition of coelenterazine to cells expressing the Rluc-EYFP fusion, light emitted from Rluc is transferred directly to EYFP, resulting in a high BRET signal compared to the BRET ratio obtained for Rluc and EYFP or Rluc alone *(1)*.
3. It is important to include negative controls in BRET assays when monitoring receptor–receptor or receptor–protein interactions. These include the expression of receptors tagged with either Rluc or EYFP with untagged EYFP or Rluc, respectively, to determine if the BRET signal is merely due to overexpression of Rluc and EYFP in the same cell and thus resulting from nonspecific interactions. To assess the specificity of the receptor–protein BRET signal, additional negative controls using other tagged receptors should also be performed, with these control tagged receptors expressed at similar levels to the receptors of interest (*see* **Fig. 2**). In addition, untagged proteins can be cotransfected along with the donor and

acceptor BRET fusions to disrupt the interaction, thereby reducing the BRET signal. This approach was adopted to assess the specificity of the BRET signal obtained on coexpression of TRHRs tagged with either Rluc or EYFP *(5)*.

4. To determine if an interaction occurred between TRHRs (homo-oligomerization), the TRHR was C-terminally tagged with either Rluc or EYFP to generate TRHR–Rluc or TRHR–EYFP *(5)*.
5. The absence of a BRET signal does not necessarily imply the lack of an interaction, but instead could mean that the donor and acceptor tags are not in an orientation that would favor energy transfer. This highlights the importance of performing BRET with Rluc or EYFP fusion partner proteins of interest constructed with tags at either the N- or C-terminus. Using this approach, the combination that allows the interaction of the two partner proteins with their respective donor and acceptor fusion molecules with the optimal orientation/distance to give a sensitive BRET signal, can be determined. The insertion of an amino acid linker sequence between the protein of interest and the fusion EYFP or Rluc may also be useful. When performing BRET to detect receptor–protein interactions, receptors are usually tagged at the C-terminus rather than the N-terminus to increase the chance of correct folding and membrane localization of the receptor and to reduce interference with ligand binding.
6. The functionality of BRET fusion proteins compared to the untagged proteins should be assessed, when possible, prior to their use in a BRET assay. Addition of either the Rluc or EYFP to the receptor or protein of interest may disrupt the expression, folding, localization, and/or function of the protein. Therefore, before assessing the relevance of potential receptor interactions it is important to ascertain whether the function of tagged proteins is compromised or not. To assess the functionality of BRET receptor fusion constructs, receptor binding and signaling assays were performed on the receptor–Rluc and receptor–EYFP fusions in comparison with untagged receptor *(5)*. The identity of the receptor interacting protein (e.g., β-arrestin) will determine what assays should be performed to assess whether function is retained for a particular protein of interest. In the case in which β-arrestin was the protein of interest, functionality of β-arrestin–EYFP or Rluc fusions were assessed using (1) receptor internalization assays to measure their ability to promote TRHR internalization and (2) confocal microscopy to monitor ligand-dependent translocation of β-arrestin–EYFP *(5)*. At 48 h following transfection, levels of expressed Rluc tagged protein can be assessed by measuring luminescence using a luminometer, while levels of expressed EYFP tagged protein can be determined using either a fluorescence plate reader or by fluorescence-activated cell sorting (FACS).
7. COS-1 cells have been used in this BRET assay protocol; however, BRET can be performed in any cell type able to undergo transfection. A variety of transfection reagents and techniques can be used to transfect cells with cDNA constructs. In our hands, consistent results were obtained using Polyfect transfection reagent (QIAGEN). However, the transfection reagent used may vary with the cell type chosen.

8. When performing BRET assays, the levels of Rluc compared to EYFP fusion proteins expressed is important. Different amounts of Rluc and EYFP fusion cDNAs should be transfected and assayed to determine the conditions necessary to obtain an optimal BRET signal. Extreme overexpression of fusion proteins is not recommended, as this may increase the risk of nonspecific BRET signals; hence negative controls such as other Rluc and EYFP tagged proteins expressed at similar levels to the proteins of interest should be included. Certain interactions, however (i.e., low-affinity interactions), may require higher levels of protein expression to allow detection. In addition to the total amount of cDNA transfected for protein expression, the ratio of donor to acceptor protein expressed is important and the optimal ratio needs to be determined empirically for each protein–protein interaction studied. By titrating the quantity of EYFP fusion protein expression, while keeping the expression of the Rluc fusion constant, the $BRET_{50}$ (the EYFP fusion protein expression as measured by degree of fluorescence, at which the BRET value is half-maximal) can be determined as a relative approximate measure of interaction affinities ***(14)***. Overexpression of GPCRs in heterologous systems can potentially lead to the generation of artefactual aggregation as a result of the highly hydrophobic nature of these seven transmembrane receptors and nonspecific BRET signals ***(10)***. Performing BRET at low levels of receptor expression, at or below physiological levels of receptor, may help to prevent nonspecific BRET signals. However, protein–protein interactions may vary in their relative affinities and thus different levels of protein expression, and different ratios of donor and acceptor molecules may be required to detect a BRET signal indicative of an interaction.
9. The main feature of an instrument that can detect BRET signals will include the ability to filter sequentially, then measure, light emission over two separate wavelength ranges. Several multifunctional plate readers are currently available that can perform this task; however, they can suffer from compromised sensitivity (compared to a dedicated luminometer) owing to their multifunctional nature. We have collaborated with Berthold Technologies in the development of a 96-well plate reader with the ability to measure BRET signals with high sensitivity, the Mithras LB940 (Berthold Technologies, Inc.). Owing to the above-mentioned problems encountered with extreme overexpression of proteins, it is important that BRET can be performed and signals detected using low levels of protein expression, that is, near or below physiological levels. Therefore, an instrument capable of detecting low BRET signals to reduce the presence of nonspecific BRET signals resulting from high expression levels of BRET fusion proteins is highly desirable.
10. On addition of coelenterazine to the cell suspension, light emission should be immediately measured as the luciferase reaction displays rapid decay kinetics. This can be performed manually by adding coelenterazine to each sample one at a time, reading immediately after addition, or by injection of substrate if this function is available on the instrument being used. Repeated reads are taken on all samples, as the BRET signal takes several seconds to equilibrate following the

addition of coelenterazine. Also, by performing repeated reads following the addition of substrate, the stability of the interaction can be monitored over time. This may be particularly important when monitoring the effect that various agonists or antagonists may have on receptor–protein interactions. By performing repeated reads, a BRET kinetic analysis is being performed, essentially allowing an interaction to be monitored in real time. Although the actual amount of light emitted decreases over time, the BRET ratio of the positive control Rluc-EYFP BRET fusion remains constant for at least 30 min after the addition of coelenterazine.

References

1. Eidne, K. A., Kroeger, K. M., and Hanyaloglu, A. C. (2002) Applications of novel resonance energy transfer techniques to study dynamic hormone receptor interactions in living cells. *Trends Endocrinol. Metab.* **13,** 415–421.
2. Xu, Y., Piston, D. W., and Johnson, C. H. (1999) A bioluminescence resonance energy transfer (BRET) system: application to interacting circadian clock proteins. *Proc. Natl. Acad. Sci. USA* **96,** 151–156.
3. Zacharias, D. A., Baird, G. S., and Tsien, R. Y. (2000) Recent advances in technology for measuring and manipulating cell signals. *Curr. Opin. Neurobiol.* **10,** 416–421.
4. Angers, S., Salahpour, A., Joly, E., et al. (2000) Detection of beta 2-adrenergic receptor dimerization in living cells using bioluminescence resonance energy transfer (BRET). *Proc. Natl. Acad. Sci. USA* **97,** 3684–3689.
5. Kroeger, K. M., Hanyaloglu, A. C., Seeber, R. M., Miles, L. E., and Eidne, K. A. (2001) Constitutive and agonist-dependent homooligomerization of the thyrotropin-releasing hormone receptor. Detection in living cells using bioluminescence resonance energy transfer. *J. Biol. Chem.* **276,** 12,736–12,743.
6. McVey, M., Ramsay, D., Kellett, E., et al. (2001) Monitoring receptor oligomerization using time-resolved fluorescence resonance energy transfer and bioluminescence resonance energy transfer. The human delta-opioid receptor displays constitutive oligomerization at the cell surface, which is not regulated by receptor occupancy. *J. Biol. Chem.* **276,** 14,092–14,099.
7. Cheng, Z. Y. and Miller, L. J. (2001) Agonist-dependent dissociation of oligomeric complexes of G protein-coupled cholecystokinin receptors demonstrated in living cells using bioluminescence resonance energy transfer. *J. Biol. Chem.* **276,** 48,040–48,047.
8. Ayoub, M. A., Couturier, C., Lucas-Meunier, E., et al. (2002) Monitoring of ligand-independent dimerization and ligand-induced conformational changes of melatonin receptors in living cells by bioluminescence resonance energy transfer. *J. Biol. Chem.* **277,** 21,522–21,528.
9. Issafras, H., Angers, S., Bulenger, S., et al. (2002) Constitutive agonist-independent CCR5 oligomerization and antibody-mediated clustering occurring at physiological levels of receptors. *J. Biol. Chem.* **277,** 34,666–34,673.
10. Ramsay, D., Kellett, E., McVey, M., Rees, S., and Milligan, G. (2002) Homo- and hetero-oligomeric interactions between G-protein-coupled receptors in living cells

monitored by two variants of bioluminescence resonance energy transfer (BRET): hetero-oligomers between receptor subtypes form more efficiently than between less closely related sequences. *Biochem. J.* **365,** 429–440.

11. Yoshioka, K., Saitoh, O., and Nakata, H. (2002) Agonist-promoted heteromeric oligomerization between adenosine A(1) and P2Y(1) receptors in living cells. *FEBS Lett.* **523,** 147–151.
12. Lavoie, C., Mercier, J. F., Salahpour, A., et al. (2002) Beta 1/beta 2-adrenergic receptor heterodimerization regulates beta 2-adrenergic receptor internalization and ERK signaling efficacy. *J. Biol. Chem.* **277,** 35,402–35,410.
13. Hanyaloglu, A. C., Seeber, R. M., Kohout, T. A., Lefkowitz, R. J., and Eidne, K. A. (2002) Homo- and hetero-oligomerization of thyrotropin-releasing hormone (TRH) receptor subtypes. Differential regulation of beta-arrestins 1 and 2. *J. Biol. Chem.* **277,** 50,422–50,430.
14. Mercier, J. F., Salahpour, A., Angers, S., Breit, A., and Bouvier, M. (2002) Quantitative assessment of beta 1- and beta 2-adrenergic receptor homo- and hetero-dimerization by bioluminescence resonance energy transfer. *J. Biol. Chem.* **277,** 44,925–44,931.

21

Fluorescence Resonance Energy Transfer to Study Receptor Dimerization in Living Cells

Jürgen E. Bader and Annette G. Beck-Sickinger

Summary

The versatility, sensitivity, and feasibility of fluorescence methods are very attractive to study protein–protein interaction at low levels of protein expression. However, one of the most severe limits in protein chemistry has been the difficulty of introducing site-specific fluorescent labels. The development of genetically encoded fluorescent probes, that is, green fluorescent protein (GFP) and its variants therefore opened up a broad field of novel applications. To characterize protein–protein interactions and determine detailed spatio–temporal dynamics of partners that are molecularly well characterized, fluorescence energy transfer methods are excellent nondestructive tools in living cells. Cellular responses to external factors are extensively based on direct molecular interaction and especially G-protein-coupled receptors (GPCRs) have been shown to interact with an unexpected level of complexity. Classical models of signal transduction describe GPCRs as monomeric proteins, while recent studies using fluorescence resonance energy transfer (FRET) and other methods show that GPCRs can also function as homo- or heterodimers. Theoretical background information on FRET technology and its diverse applications are summarized here. A detailed description of a spectroscopic method for FRET studies in the field of GPCR interaction is presented to facilitate and propagate studies to increase our understanding of protein–protein interactions involving GPCRs.

Key Words

Fluorescence microscopy, fluorescence resonance energy transfer (FRET), fluorescence spectroscopy, G-protein-coupled receptors, green fluorescent protein (GFP), GFP variants, homodimerization, neuropeptide Y (NPY) receptors, protein–protein interaction, red fluorescent protein, DsRed.

1. Introduction

Fluorescence-based methods are shedding new light on various biological processes from gene expression to protein–protein interaction, second-messenger

From: *Methods in Molecular Biology, vol. 259, Receptor Signal Transduction Protocols, 2nd ed.*
Edited by: G. B. Willars and R. A. J. Challiss © Humana Press Inc., Totowa, NJ

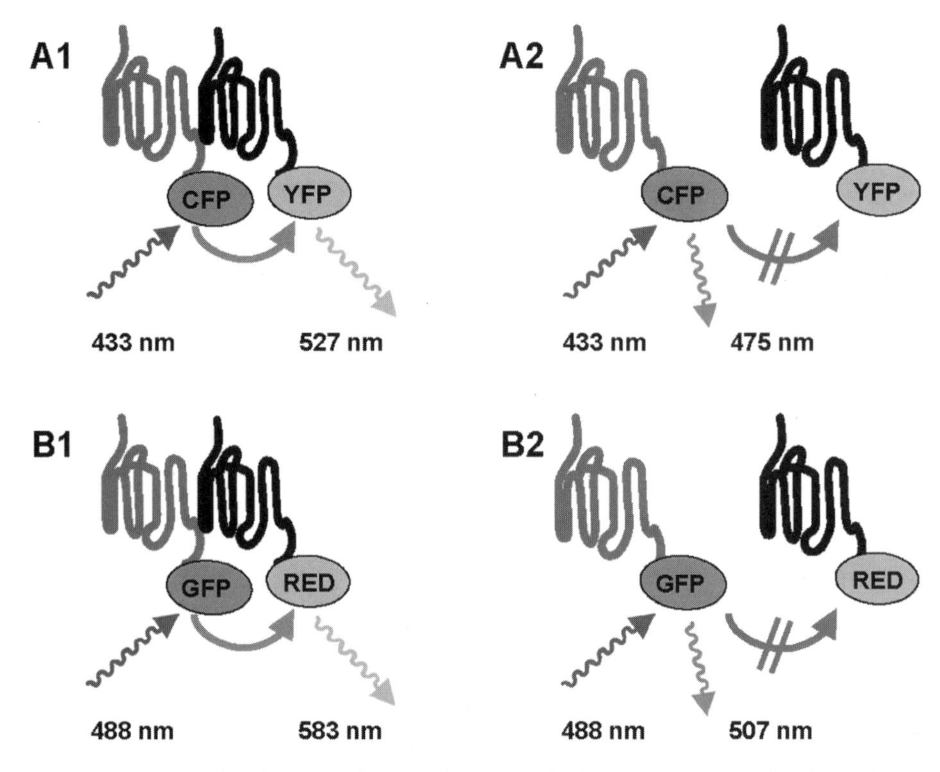

Fig. 1. Schematic diagram of FRET between GPCRs. Energy transfer form donor (CFP, GFP) to the acceptor molecule (YFP, DsRed) can occur only when the distance is lower than approx 10 nm (**A1**, **B1**). Otherwise no FRET is measurable (**A2**, **B2**).

cascades, and intra- as well as intercellular signaling. Numerous examples of applications of fluorescence in measurement and visualization both in vitro and in vivo can be found in the literature (as reviewed in **refs.** ***1*** and *2*). To monitor more accurately protein–protein interaction by using non-invasive methods in living cells, fluorescence resonance energy transfer (FRET) has proven to be a superior method and is becoming more and more popular *(3–5)*. It is a helpful tool not only to trace localization, but also to determine whether proteins make specific contacts or merely are colocalized. The idea behind this technique is to label two moieties of interest on the same molecule or two interacting molecules with different dyes and to measure molecular interaction via spectroscopic interaction as shown in **Fig. 1**. One moiety or molecule is called the "donor" and must be fluorescent, and the other is called the "acceptor" and is often but not necessarily fluorescent.

1.1. The FRET Effect

FRET is a spectroscopic process known for more than 50 yr by which energy is passed nonradiatively between molecules, over distances in the order of common macromolecular dimensions. This process of energy transfer by means of long-range dipole–dipole coupling occurs only under the following conditions: the donor molecule is in close proximity to the acceptor molecule (generally 1–10 nm) and is favorably oriented; there is an adequate overlap between the emission and excitation spectra of the donor and the acceptor molecules, respectively; and the quantum yield of the donor and the absorption coefficient of the acceptor are sufficiently high. The transferred energy leads to a reduction in the donor's fluorescence intensity and excited state lifetime and an increase of the acceptor's emission intensity (sensitized emission).

The efficiency E of this process can be defined as the fraction of donor molecules deexcited via energy transfer to the acceptor:

$$E = (1 - I_{DA} / I_D) = 1 - \tau_{DA} / \tau_D \qquad (1)$$

where I_{DA} and I_D are the emission intensities of the donor in the presence or absence of acceptor respectively, and τ_{DA} and τ_D are the respective lifetimes. Looking at the increase in fluorescence of the acceptor on the other hand, energy transfer efficiency can also be measured as:

$$E = (I_{AD} / I_A - 1)(\varepsilon_A / \varepsilon_D) \qquad (2)$$

where I_{AD} and I_A are the emission intensities of the acceptor in the presence of the donor (consisting of fluorescence arising from energy transfer and from direct excitation) or in the absence of the donor respectively. ε_A and ε_D are the molar absorption coefficients (formerly called extinction coefficients) of the acceptor and donor, respectively.

Förster ***(6,7)*** showed the strong distance dependency of this effect, setting up an equation in which the efficiency depends on the inverse sixth power of the distance r between donor and acceptor:

$$E = 1 / [\ 1 + (r / r_0)^6] \qquad (3)$$

where r_0 is the critical distance at which half of the energy is transferred. Depending on the spectral characteristics of the dyes and their relative orientation, the so-called "Förster distance" r_0 can be calculated ***(8)*** and enables FRET to be used in principle for measuring distances smaller than the resolution limit of a light microscope. Using a pair of dyes with r_0 equal to the distance to be measured, the FRET effect can theoretically be used as the often quoted "spectroscopic ruler" ***(9)***.

On the other hand, changes in the orientation and mobility of donor and acceptor fluorophores can significantly affect FRET efficiency and are included in calculations as a dipole coupling orientation factor *(3,10)*. It may be allowed for free fluorophores in solution to assume their relative orientation to be randomly distributed. However, interpreting results of such calculations in terms of absolute molecular distance can prove to be difficult for fluorophores attached to interacting proteins. Measurement of an approximate or *apparent* FRET efficiency with affordable techniques and interpretation of signal ratios provide suitable means to investigate protein interactions when absolute distance calculations are not essential. In this case Förster distances are approximate measures for the protein–protein distances.

1.2. Methods for FRET Measurement

Efficiency of energy transfer can be measured in numerous ways. Decay kinetic-based methods use the measurement of decreased excited-state lifetime of the donor or the change in lifetime of sensitized emission of the acceptor *(11–13)*. The most popular intensity-based method, in which the only required instrumentation is a steady-state fluorimeter, employs the detection of donor emission quenching and/or increased acceptor emission (in the case where the acceptor molecule is a fluorophore as well). For example, the steady-state donor fluorescence intensity is measured from a sample containing only the donor and from a corresponding sample containing donor and acceptor. After normalization to the corresponding sample concentrations, the fractional decrease in the donor fluorescence in the presence of the acceptor is equal to the efficiency of energy transfer as given in **Eq. 1**.

When only a small fraction of the sample contributes to energy transfer and in cases in which the samples are inhomogeneous or information on exact sample concentrations is missing as in living cells, measuring sensitized emission can be particularly useful. Whereas donor quenching can arise from several trivial sources, the increase in acceptor emission is an important confirmation that energy transfer is taking place because it can arise only from it. Furthermore, applying **Eq. 2** to measure sensitized emission of the acceptor with the same steady-state fluorimeter is possible but needs additional adjustment.

In a typical FRET experiment, the observed fluorescence emission spectrum is composed of the donor emission, the acceptor emission due to direct excitation, and the acceptor emission due to FRET. For reliable quantification of acceptor emission due only to FRET, adjustments for concentration differences are required. Even more essential is a precise determination of the proportion of the total emission signal that is due to direct excitation of the acceptor and that which is due to overlap of donor and acceptor emission spectra. This is normally achieved by measuring three samples expressing donor only, acceptor only, and both donor and acceptor at the excitation wavelengths of the donor and acceptor ***(14,15)***. In either case the calculation of energy transfer efficiency via donor quenching should agree with that calculated by sensitized emission. An example using this method is explained further in **Subheading 3.**

Digital imaging provides another possibility to quantify and compare fluorescence signals accurately and to observe, in parallel, the subcellular localization of FRET ***(5)***. However, this requires more elaborate equipment. When using fluorescence microscopy with different-color fluorophores, in addition to a microscope with at least 200–300 nm optical resolution, it is important to select filter combinations that reduce the spectral "bleed through." A charge-coupled device (CCD) camera detecting intensity uniformly over the corresponding range of light wavelengths is crucial, as is a fast readout speed to reduce exposure time and therefore photobleaching. Grayscale images are commonly acquired through acceptor, FRET, and donor channels from cells grown on either well plates or glass cover slips, which are a better option. Intensity values are corrected for spectral crosstalk and filter bleed-through by mathematical approaches ***(16,17)*** to extract the FRET signal. To avoid the introduction of errors during data processing, precise and reproducible measurements under the same conditions are essential to obtain correction coefficients. Calculations of corrected FRET ($FRET_c$) are carried out on a pixel-by-pixel basis for the entire image using the following equation:

$$FRET_c = FRET - (\text{Donor bleed through} \times \text{Donor}) - (\text{Acceptor bleed through} \times \text{Acceptor}) \quad (4)$$

$FRET_c$ images are reasonably displayed as intensity-modulated images in pseudocolor mode. Quantification of FRET values is performed by selecting a single cell or an area of special interest.

Another way to measure energy transfer that is particularly well suited to a microscope, because the required high light intensities are readily accessible, is based on the photobleachability of the donor or acceptor (pbFRET). The selective photobleaching of, for example, the acceptor fluorophore, eliminates or reduces FRET. Consequently in the regions of the cell where FRET occurred before, there will be an enhancement in the donor emission. Another kinetic-based approach in microscopy uses the fact that energy transfer is an alternative pathway for the excited state to emit energy and that the donor will photobleach more slowly if energy transfer is occurring to an acceptor. To confirm reliably that FRET has occurred, both quenching of the donor signal and the sensitized acceptor emission should be measured. The technique of acceptor photobleaching provides a method for including both measurements.

It can be shown that the fractional change in the photobleaching time constant by energy transfer is the same as the fractional change in the fluorescence lifetime of the donor. The fluorescence lifetime, a fixed property of individual fluorescent molecules and independent of the local concentration, but influenced by intermolecular events can be directly measured by fluorescence lifetime imaging microscopy (FLIM). However, specialized instrumentation is required because fluorescence lifetimes are measured in picoseconds or nanoseconds and the techniques in the field of time-resolved fluorescence methods are still developing rapidly ***(13,18)***.

Bleaching with an arc lamp is comparably slow but the bleaching speed can be increased with lasers available on confocal microscopes. In light microscopy, tissue above and below the plane of focus is illuminated and this also contributes to the image causing it to be diffuse and with reduced contrast. The confocal principle implies that illumination and detection is focused at the same small region within the tissue or cell. Fluorescent signals that arise from outside the sampled area are rejected by a spatial filter, such as a pinhole. An image is built up by scanning the sample across the focal plane, for example, with photomultiplier tubes. This method has the advantage that a higher resolution is achieved. Therefore it is possible to investigate the localization, the concentration, or, by using FRET, the interaction of molecules in single cells and even in subcellular structures. There is also less unfavorable photobleaching and phototoxicity found in the surrounding tissue. On the other hand the temporal resolution is limited by the time of the high-speed scanning process, and the sensitivity is reduced because the image is acquired from a single scanning point. Again, more specialized and costly equipment is required but the combination of more and more sophisticated imaging technologies will lead to revolutionary advances in measuring FRET in vivo ***(19–21)***.

1.3. Application of Green Fluorescent Protein and Variants

The development of genetically encoded fluorescent probes, for example, the green fluorescent protein (GFP), led to a dramatic revolution in biochemistry and cell biology. Discovered in the jellyfish *Aequorea victoria* by Shimomura et al. ***(22,23)*** and cloned by Prasher et al. ***(24)***, GFP has become one of the most widely used protein markers. GFP is an 11-stranded β-barrel with an α-helix running up the axis of the cylinder (so-called β-can structure) to which the tripeptide chromophore is attached. The gene contains all information necessary for the posttranslational synthesis of the chromophore. Fundamental research and extensive mutagenesis projects have produced a growing family of fluorescent protein variants with codons conforming to mammalian systems; improved folding properties at 37°C; reduced aggregation tendency; increased fluorescence quantum yields; and widely varying spectra with emissions ranging through blue, cyan, green, and yellow ***(25)***. Recently a red-emitting GFP-like fluorescent protein (drFP583) was cloned from the reef coral *Discosoma*

Table 1
Features of Fluorescent Protein Variants

	CFP	YFP	GFP	DsRed
Excitation maximum at λ (nm)	433	513	488	558
Emission maximum at λ (nm)	475	527	507	583
Absorption coefficient ε (cm^{-1} M^{-1})	26,000	84,000	55,000	22,500
Relative quantum yield	0.66	1.00	0.98	0.29
Photobleaching sensitivity	Low	Moderate	Low	Low
Förster distance (nm)	4.92 ± 0.10		4.73 ± 0.09	

Excitation maxima, emission maxima, and absorption coefficients at the excitation maxima, are listed as given in the literature ***(59)***. Fusion proteins may have different spectra and these should be determined. The quantum yields relative to YFP are calculated based on the literature ***(59)***. Förster distances are listed as calculated in the literature ***(8)*** assuming randomly distributed and freely oriented fluorophores.

sp. by Matz et al. ***(26)*** with a conserved β-can fold but overall only 26–30% sequence identity with GFP. The spectral properties of this so-called DsRed are ideal for dual-color experiments with GFP. Although improved variants of this exist and are commercially available they still have several drawbacks, including slow chromophore maturation, complex absorption spectrum, poor solubility, and a preferred tetramerization of the DsRed domain that can perturb function and localization of fusion proteins ***(27–29)***. To have minimum cross-excitation or cross-detection, fluorophores for FRET should have sharp and well-resolved excitation and emission spectra. DsRed is therefore an effective acceptor for GFP, facilitating detection of sensitized emission. However, despite improvements ***(30)*** the drawbacks have not been overcome and CFP and YFP will remain the most commonly used and best characterized proteins for FRET studies ***(31)***. The spectral features of the two FRET pairs CFP-YFP and GFP-DsRed are summarized in **Table 1**.

1.4. Oligomerization of G-Protein-Coupled Receptors

G-protein-coupled receptors (GPCRs) represent one of the largest families encoded in the human genome and are responsible for the cellular recognition of hormones, transmitters, light, and odorants. Studies on protein interactions can provide more insights into the molecular interactions of ligand and receptor or receptor and components of intracellular signal transduction cascades. The investigation of receptor–receptor interaction offers another rewarding application. Frequently applied methods for the investigation of receptor oligomerization are immunoprecipitation, photoaffinity labeling and crosslinking studies, size-separation chromatography, and Western blot analysis. As mentioned earlier, a number of new GFP-based experimental approaches, together with technical advances in microscopy and spectroscopy, provide the possibility of replacing these denaturing techniques *(32,33)* and make protein interaction studies accessible in vivo.

Classical models of signal transduction describe GPCRs as monomeric proteins characterized by seven transmembrane α-helices that interact with a family of heterotrimeric GTP-binding proteins, referred to as G-proteins. Although evidence for oligomerization arose many years ago and is well known in other receptor systems, only recent structural and biochemical studies show that GPCRs interact to form oligomers in vivo *(34,35)* and function as homo- or heterodimers as well *(36–38)*. It is still unknown which GPCRs can self-associate and/or interact with other GPCRs. The conditions required for oligomerization and the molecular principles that control these specific interactions remain to be resolved. The consequences for the binding site, ligand interaction, and signal transduction as well as receptor regulation have yet to be explored. Studies of oligomerization may also help to elucidate further the often unclear pharmacological profiles of GPCRs *(39,40)*.

The existence of homodimers has been shown so far for several GPCRs including the β_2-adrenergic receptor *(41,42)*, somatostatin receptor 5 *(43)*, δ- and κ-opioid receptors *(38,44–46)*, metabotropic glutamate receptor 5 *(47)*, and others *(48)*. Whereas dimerization of the somatostatin receptor 5, the δ-opioid receptor, and the β_2-adrenergic receptor are agonist mediated, dimerization of the κ-opioid receptor is agonist independent.

1.5. Neuropeptide Y Receptors

As an example, we describe here the successful application of FRET techniques to investigate the oligomerization of neuropeptide Y (NPY) receptors in living cells *(49)*. NPY is a 36-amino-acid peptide amide that was first isolated in 1982 *(50)*. It belongs to the family of pancreatic polypeptides and is one of the most abundant neurohormones in the mammalian central and peripheral nervous systems. NPY is involved in a great variety of physiological and pathophysiological processes such as cardiovascular and renal functions *(51)*, regulation of feeding *(52)*, nociception, memory *(53)*, anxiety, seizures *(54)*, circadian rhythm, and intestinal motility. These effects are mediated by at least five distinct GPCRs (Y_1, Y_2, Y_4, Y_5, and y_6) that have been cloned and partially characterized *(55,56)*. Besides inhibitory pathways via adenylyl cyclase ($G_{i/o}$), the effectors of signaling and the mechanisms of receptor regulation and interplay are still largely unknown. Initial speculation of receptor dimerization arose from studies with truncated NPY analogs and homodimeric, peptidergic agonists *(57,58)*. We investigated the human NPY Y_1, Y_2, and Y_5 receptor subtypes for their ability to form homodimers by means of direct FRET of receptor–GFP fusion proteins. For this purpose, we generated fusion proteins of NPY receptors tagged at their c-terminus with GFP or the cyan, yellow, and red spectral variants (CFP, YFP, and DsRed, respectively) that can be used as FRET pairs. Both of the FRET techniques that were applied (fluorescence microscopy and fluorescence spectroscopy) clearly showed that these NPY receptor subtypes are able to form homodimers. Dimerization was receptor subtype dependent in a range of 26–44% compared to tandem constructs as positive controls where the FRET signal was set to 100%. However, the dimerization was not influenced by either ligand or by G_α-protein binding.

2. Materials

1. hY_1-, hY_2-, and hY_5-receptor sequences cloned into the pcDNA3 vector (Invitrogen, Karlsruhe, Germany) and the pCR2.1-TOPO vector were used to clone the fusion proteins.
2. pECFP-N1, pEYFP-N1, pEGFP-N1, and pDsRed-N1 vectors (Clontech, Heidelberg, Germany) containing the genes encoding the spectral variants of the fluorescent protein labels were chosen as the expression vectors.
3. LipofectAMINE 2000 (Invitrogen) was used as the transfection reagent.
4. All materials for cell culture were supplied by Invitrogen. For baby hamster kidney (BHK) or Chinese hamster ovary (CHO) cell lines, standard growth medium was Dulbecco's modified Eagle's medium containing 10% fetal calf serum or 50% Dulbecco's modified Eagle's medium–50% nutrient mix Ham's F-12 with 10% fetal calf serum, respectively.
5. The buffer for FRET experiments was phosphate-buffered saline (PBS) containing protease inhibitors (4 m*M* pefabloc, 1 m*M* pepstatin, 28 μ*M* E-64) supplied by Fluka or Sigma (Taufkirchen, Germany).

3. Methods

Described below are general prerequisites and detailed experimental protocols for performing FRET analysis of GPCRs in living cells in a conventional fluorimeter (*see* **Note 1**).

3.1. Design of Tagged Receptors

To investigate potential protein–protein interactions by FRET, the generation of fusion proteins is required. The appropriate expression vectors for human NPY Y_1, Y_2, and Y_5 receptors, tagged at their C-termini to the green-, cyan-, yellow- or red-fluorescent protein (GFP, CFP, YFP, DsRed), respectively, were created using standard cloning techniques as previously described ***(49)***. As FRET donor constructs, the receptor sequences were fused in frame to the GFP or CFP sequence, while receptor sequences fused to DsRed or YFP were used as FRET acceptor constructs. As a positive control for FRET imaging and measurement, two different fusion proteins of hY_1- and hY_2-receptor were constructed, with GFP and DsRed or YFP and CFP, respectively, C-terminally fused as a tandem construct (*see* **Notes 2–4**).

3.2. Cell Culture and Functional Studies

BHK or CHO cells that do not express endogenous NPY-receptors were used and cultured prior to analysis according to standard protocols using the recommended media and culture conditions *(49)*.

One day before transient transfection, approx 2 million cells were seeded per 25-cm^2 culture flask to reach 90% confluency for optimal transfection conditions. According to the manufacturer's instructions and individual optimization steps for our equipment, 26 μL of LipofectAMINE 2000 (Invitrogen) and 13 μg of plasmid DNA were generally used. Coexpression of the FRET pair fusion proteins CFP/YFP or GFP/DSRed, respectively, was achieved by transfection with equal amounts of the corresponding receptor subtype fusion constructs (6.5 μg each). Before further analysis, cells were kept for another 24 h under normal growth conditions for sufficient expression and maturation of the fluorescent protein.

To prove whether the constructed receptor–GFP fusion proteins still are natively localized and have the activity and function of their untagged versions, competition binding and cAMP assays were performed after expression of the constructs in mammalian cells. The results showed that these constructs bind the natural ligand with comparable IC_{50} values to wild-type receptors, indicating that they have the correct conformation. The functional assays confirmed that all fusion proteins also inhibit adenylyl cyclase after stimulation with NPY. As another useful tool to quantify receptor expression and to correlate the number of binding sites to the fluorescence emission, a BHK cell line that stably expresses the hY_2–GFP fusion protein was established *(49)* (*see* **Notes 5–7**).

3.3. Fluorescence Spectroscopy

Most modern scanning fluorimeters should be sufficient to detect FRET between GPCRs at normal expression levels in mammalian cells. We carry out our spectrofluorimetric studies using a Fluorolog-3 spectrofluorimeter (Jobin Yvon Spex, Longjumeau, France) equipped with a 450-W xenon lamp and Datamax for Windows (Version 2.20) for data acquisition and processing. Further analysis and calculations are carried out with Microsoft Excel (Version 9.0). Typically our measurements are performed with an increment wavelength of 1 nm and an increment time of 0.75 s. Spectra of nontransfected cells, cells expressing only the donor or only the acceptor fusion protein, and cells expressing donor and acceptor pairs should be recorded using the following procedures (*see* **Notes 8** and **9**):

1. Wash cells with PBS and detach the grown monolayer by treatment with EDTA. Resuspend the cells to a concentration of 1.0×10^6 cells/mL using a buffer containing protease inhibitors (*see* **Note 10**).
2. Record a fluorescence emission spectrum from nontransfected control cells excited at both the donor and at the acceptor wavelengths. These values represent the autofluorescence of the cells and equipment and have to be subtracted from the corresponding sample spectra in each of the following steps (*see* **Note 11**).
3. Use samples that contain the same number of cells and record initially the donor emission spectrum (e.g., between 460 and 540 nm for CFP) of cells expressing only the donor fusion protein by irradiating this sample at the λ_{max} for the donor (e.g., CFP, 433 nm).
4. While still exciting at the λ_{max} for the donor (e.g., CFP, 433 nm) record the emission spectrum of a sample of cells coexpressing donor and acceptor fusion proteins (the FRET sample, for example, CFP and YFP fusion proteins, between 520 and 560 nm). After background subtraction this spectrum is a composite of donor emission and acceptor emission owing to direct excitation and to energy transfer.
5. Irradiate the same FRET sample at the λ_{max} for the acceptor (e.g., YFP, 510 nm) and record an acceptor-only emission spectrum of the coexpressed fusion proteins.
6. Record an acceptor emission spectrum of cells expressing only the acceptor fusion protein while exciting this sample at the λ_{max} for the acceptor (e.g., YFP, 510 nm, spectrum between 520 and 560 nm).
7. Lastly, record the emission spectrum for the same sample of cells expressing only the acceptor fusion protein while irradiating at the λ_{max} for the donor again (e.g., CFP, 433 nm, spectrum between 460 and 540 nm). This spectrum represents the acceptor emission owing to the direct excitation at the donor excitation wavelength.
8. Normalize the donor emission spectrum acquired in **step 3** to the peak height of the FRET spectrum of **step 4** and subtract it from this FRET spectrum. The resulting spectrum is composed of acceptor emission owing to direct excitation and to energy transfer (*see* **Notes 12** and **13**).
9. Divide the peak maximum of the acceptor emission spectrum generated in **step 6** by the one of the acceptor emission spectrum acquired with the FRET sample in **step 5**. This ratio of the two spectra can be used for scaling the acceptor emission resulting from direct excitation.
10. Normalize the acceptor emission spectrum generated in **step 7** with the ratio of **step 9** to gain the spectral emission resulting from direct excitation of the acceptor (spectral crosstalk) in the FRET spectrum and subtract it from the remaining spectrum of **step 8**. Finally this resulting spectrum is solely composed of acceptor emission owing to energy transfer and can be used for further calculations and comparative steps (*see* **Notes 14** and **15**).

As a positive control the preceding steps should be performed with samples containing donor–acceptor tandem fusion constructs ensuring a FRET effect. Cells containing the donor or acceptor fusion product and expressing colocalized but noninteracting free donor or acceptor labels, respectively, can be used as a negative control.

Calculated data are usually presented as ratios of fluorescence, comparing diverse samples under different conditions and setting the positive control for example as maximal FRET.

4. Notes

1. The limited ability to measure absolute distances is perhaps the most important drawback of FRET, but the method is valuable in measuring relative distances, namely, whether two points are closer together under condition A than condition B. Another point to consider when interpreting FRET data is that the efficiency of energy transfer depends not only on the distance between the donor and the acceptor, but on the relative orientation of the dyes as well. This is a factor that is often not precisely known and can sometimes be misinterpreted in ligand binding studies. A loss of FRET can be a result of conformational changes that lead to a disfavorable dipole orientation without indicating a loss of protein interaction. Interpreting the measurement of a significant FRET effect as a result of molecular interaction is in general not questionable. A lack of FRET on the other hand is not necessarily a clear sign of the absence of molecular interaction. Apart from these restrictions it becomes clear that practically every process during ligand–receptor interaction, signal transduction or the final intracellular responses can be investigated selectively in vivo and under microscopic observation, which is the major advantage and unique to FRET methods.
2. The cDNA encoding the target protein is fused in-frame with the cDNA for GFP or one of its variants. In general, fusion proteins can be engineered at either the N- or C-terminus of the host protein (*see also* Chapter 5 for epitope tagging applications). It might be necessary to splice GFP into non-critical loops of the host protein. This is possible because the N- and C-termini of its core domain are not far apart. We have successfully fused fluorescent protein variants at the C-terminus of human NPY receptors, but for other GPCRs it may be necessary to analyze several chimeras fused at various points of the sequence. Using a linker between receptor and fluorescent protein often improves the chances of proper folding, ensuring fluorescence and unaltered function.
3. GFP requires many transcripts for an easily detectable signal and, depending on the chosen variant, sufficient time for protein folding and fluorophore maturation is needed. In the meantime there are improved variants commercially available as described in the introduction and the usage of the latest enhanced variants can be helpful (eGFP, eCFP, and eYFP).
4. Selection of fluorescent proteins appropriate not only for FRET but also for the detection system to be used is important. The aim is to reduce the spectral crosstalk and optimize signal-to-noise ratio for each fluorophore and therefore spectroscopic properties of the donor and acceptor GFPs should be considered. Reasonable separation in emission spectra between donor and acceptor allows the selective stim-

ulation of the donor. Sufficient separation in excitation spectra is required to allow measurement of fluorescence independently. The combination of GFP and DsRed was successfully used for FRET at least as a second approach to reproduce experiments with CFP and YFP that are well suited for most applications.

5. In every case and regardless of the employed fusion strategy it is essential to verify that the fusion protein functionally resembles its native counterpart by biochemical and biophysical methods, and to show that the tagged protein adopts the same subcellular localization. This prerequisite for FRET studies with fluorescent proteins is one of the major disadvantages. It is recommended to test each construct for biological function. Sufficient time after transfection for adequate expression and fluorophore maturation is important.
6. Depending on the applied FRET measurement methods it is also advised to test the viability of the cells under the conditions used for the experiments. We used BHK and CHO cell lines for the experiments that do not natively express the probed receptors. Other strategies apply knock-out organisms to rescue and simultaneously prove the function of the constructs (*see* Chapter 24).
7. The transfection method and cell culture conditions have proved to be important for obtaining efficient expression. Our cells are cultured according to standard protocols and with standard media. Transfection is carried out with Lipofectamine 2000 as this was shown to be optimal and consistent under the given conditions. The amount of reagent and vector was optimized in several steps according to the manufacturer's instructions. For other studies, different transfection reagents should be tested and optimal conditions developed in order to reach constant and reliable expression (*see* Chapter 8). Seeking an expression ratio of donor and acceptor of nearly 1 : 1 is recommended for optimal FRET measurement.
8. We also established a cell line that stably expresses the hY_2-GFP fusion protein to quantify receptor expression by saturation binding experiments. Correlation of the number of binding sites to the fluorescence emission intensity allowed us to calculate receptor density on the cell surface of transiently transfected cells. By comparing the resulting distances to the appropriate Förster distances we could show that the measured FRET effect is not simply an effect of receptor overexpression. This issue should generally be addressed in all expression systems, alternatively by using cells that produce more physiological levels of receptor fusion protein.
9. To investigate a ligand-induced effect on FRET measurement, cells were incubated with different concentrations of NPY for different time periods. To obtain different numbers of occupied NPY receptors, the applied ligand concentrations were chosen based on the K_D value. A nearly complete saturation and with more diluted preparations accordingly less receptor occupation was achieved. However, in our experiments all the spectra recorded revealed no significant difference and no increase or decrease of the FRET effect was found. The influence of GTPγS binding was tested as another functional FRET assay and again no changes were measured leading to the speculation that NPY receptors are not transported in the single form to the cell membrane, but rather assemble to dimeric units in the

endoplasmatic reticulum already and therefore do not form dimers only after ligand stimulation.

10. While harvesting the cells and preparing the samples for measuring, the lysis of some cells cannot be prevented. To protect the extracellular domains and therefore the native constitution and conformation of the membrane proteins the activity of the released proteases is diminished by adding protease inhibitors. The protease inhibitor cocktail has to be optimized corresponding to the experimental design. The buffer for the experiments itself (e.g., PBS, 0.9% NaCl solution, HEPES, or Krebs–HEPES-type buffers) should closely resemble the conditions and requirements for binding of the ligand and/or cellular signaling, but should contain the minimal number of components that may interfere with the measured signal.
11. The qualities of steady-state measurement with a fluorimeter are not only its simplicity but also the advantage that relatively small amounts of energy transfer can be measured. If the optical density of the sample is kept sufficiently low so that no absorption of the donor fluorescence takes place, with care a 5% decrease in fluorescence is measurable. Using whole cells or other heterogeneous samples, special care should be taken that the samples are gently agitated during or at least between single measurement steps to ensure that sedimentation has not occurred.
12. On one side, a significant overlap of emission spectrum of the donor and excitation spectrum of the acceptor is required for energy transfer. In contrast, ideally the emission spectrum of the acceptor should be completely resolved, but in practice the emission spectra of donor and acceptor partially overlap. The greater the overlap in the donor–acceptor spectra, the greater the background signal from which, for example, the weak sensitized acceptor emission must be extracted. Therefore we recommend optimization of the applied donor excitation wavelength. This can be done by exciting the acceptor labeled samples at a variety of wavelengths near the λ_{max} for donor excitation, recording the acceptor emission spectra and selecting the wavelength for sample excitation where the acceptor emission reaches a minimum.
13. Nevertheless it is critical to determine the spectral crosstalk. If the steady-state emission spectra of a donor–acceptor-labeled sample and a donor-only sample are taken, the donor emission has to be removed from the donor–acceptor emission spectrum by subtracting the normalized donor-only emission. By normalizing the donor-only spectrum solely the shape of it is used, so this calculation is independent of the sample concentration and will result in the fluorescence spectrum of the acceptor comprising direct excitation and energy transfer. Note that the measured FRET spectrum itself should not be normalized because this would lead to an incorrect determination of *apparent* FRET efficiency. To finally attain a resulting spectrum due solely to FRET, the subtraction of a calculated spectrum consisting of acceptor emission due to direct excitation is required. For this purpose, two additional spectra of acceptor emission while irradiating the acceptor of the FRET sample or an acceptor-only sample are recorded and used as a scaling factor to level out concentration differences. The acceptor emission spectrum of an

acceptor-only sample recorded while exciting at the donor wavelength is thus easily scaled for subtraction.

14. In cases where FRET under different conditions will be compared and donor emission only marginally interferes with acceptor emission, another perhaps easier approach can be used for calculation. The first step, that is, adjustment of donor fluorescence emission described above is left out; only the FRET and acceptor spectra are recorded and normalized to a fluorescence value (often the maximum) of a selected emission spectrum, or vice versa. In either case, the resultant ratio spectrum is normalized for quantum yield of acceptor, for concentration of total molecules, and for incomplete acceptor labeling and can be used for relative comparisons. Positive controls are set to 100% FRET and the examined samples are estimated ratios of this maximum FRET. In all the above techniques it is important to subtract background arising from cells, equipment, or other sources; the best way to do this is to prepare a sample identical to the fluorescent sample, but without the attached dyes.
15. There are other techniques applying resonance energy transfer that represent an alternative method to study discrete protein–protein interactions in a nondestructive manner. Bioluminescence resonance energy transfer (BRET), a newly developed technique related to FRET, measures the transfer of energy between a luminescent donor (e.g., luciferase) and a fluorescent protein (e.g., YFP; *see* Chapter 20). While this method is free from the complications that arise from the need of excitation (spectral crosstalk and bleed through), it is similarly limited by the problems of autofluorescence, photobleaching, and cell damage. In addition BRET cannot provide detailed insight into the subcellular localization of the interacting partners, a unique feature of FRET as mentioned earlier.

References

1. van Roessel, P. and Brand, A. H. (2002) Imaging into the future: visualizing gene expression and protein interactions with fluorescent proteins. *Nat. Cell Biol.* **4,** E15–E20.
2. Hovius, R., Vallotton, P., Wohland, T., and Vogel, H. (2000) Fluorescence techniques: shedding light on ligand–receptor interactions. *Trends Pharmacol. Sci.* **21,** 266–273.
3. Wu, P. and Brand, L. (1994) Resonance energy transfer: methods and applications. *Anal. Biochem.* **218,** 1–13.
4. Selvin, P. R. (1995) Fluorescence resonance energy transfer. *Methods Enzymol.* **246,** 300–334.
5. Day, R. N., Periasamy, A., and Schaufele, F. (2001) Fluorescence resonance energy transfer microscopy of localized protein interactions in the living cell nucleus. *Methods* **25,** 4–18.
6. Förster, T. (1946) Energy transfer and fluorescence (in German). *Naturwissenschaften* **6,** 166–175.
7. Förster, T. (1948) Intermolecular energy transfer and fluorescence (in German). *Ann. Physik* **2,** 55–75.

8. Patterson, G. H., Piston, D. W., and Barisas, B. G. (2000) Forster distances between green fluorescent protein pairs. *Anal. Biochem.* **284,** 438–440.
9. Clegg, R. M. (1995) Fluorescence resonance energy transfer. *Curr. Opin. Biotechnol.* **6,** 103–110.
10. Overton, M. C. and Blumer, K. J. (2002) Use of fluorescence resonance energy transfer to analyze oligomerization of G-protein-coupled receptors expressed in yeast. *Methods* **27,** 324–332.
11. Patel, R. C., Lange, D. C., and Patel, Y. C. (2002) Photobleaching fluorescence resonance energy transfer reveals ligand-induced oligomer formation of human somatostatin receptor subtypes. *Methods* **27,** 340–348.
12. Harpur, A. G., Wouters, F. S., and Bastiaens, P. I. (2001) Imaging FRET between spectrally similar GFP molecules in single cells. *Nat. Biotechnol.* **19,** 167–169.
13. Wouters, F. S. and Bastiaens, P. I. (1999) Fluorescence lifetime imaging of receptor tyrosine kinase activity in cells. *Curr. Biol.* **9,** 1127–1130.
14. Clegg, R. M. (1992) Fluorescence resonance energy transfer and nucleic acids. *Methods Enzymol.* **211,** 353–388.
15. Cornea, A., Janovick, J. A., Maya-Nunez, G., and Conn, P. M. (2001) Gonadotropin-releasing hormone receptor microaggregation. Rate monitored by fluorescence resonance energy transfer. *J. Biol. Chem.* **276,** 2153–2158.
16. Gordon, G. W., Berry, G., Liang, X. H., Levine, B., and Herman, B. (1998) Quantitative fluorescence resonance energy transfer measurements using fluorescence microscopy. *Biophys. J.* **74,** 2702–2713.
17. Sorkin, A., McClure, M., Huang, F., and Carter, R. (2000) Interaction of EGF receptor and grb2 in living cells visualized by fluorescence resonance energy transfer (FRET) microscopy. *Curr. Biol.* **10,** 1395–1398.
18. Tadrous, P. J. (2000) Methods for imaging the structure and function of living tissues and cells: 2. Fluorescence lifetime imaging. *J. Pathol.* **191,** 229–234.
19. Emptage, N. J. (2001) Fluorescent imaging in living systems. *Curr. Opin. Pharmacol.* **1,** 521–525.
20. Tadrous, P. J. (2000) Methods for imaging the structure and function of living tissues and cells: 3. Confocal microscopy and micro-radiology. *J. Pathol.* **191,** 345–354.
21. Haraguchi, T., Shimi, T., Koujin, T., Hashiguchi, N., and Hiraoka, Y. (2002) Spectral imaging fluorescence microscopy. *Genes Cells* **7,** 881–887.
22. Shimomura, O., Johnson, F. H., and Saiga, Y. (1962) Extraction, purification and properties of aequorin, a bioluminescent protein from the luminous hydromedusan, *Aequorea. J. Cell. Comp. Physiol.* **59,** 223–239.
23. Shimomura, O., Johnson, F. H., and Saiga, Y. (1963) Further data on the bioluminescent protein, aequorin. *J. Cell. Physiol.* **62,** 1–8.
24. Prasher, D. C., Eckenrode, V. K., Ward, W. W., Prendergast, F. G., and Cormier, M. J. (1992) Primary structure of the *Aequorea victoria* green-fluorescent protein. *Gene* **111,** 229–233.
25. Tsien, R. Y. (1998) The green fluorescent protein. *Annu. Rev. Biochem.* **67,** 509–544.

26. Matz, M. V., Fradkov, A. F., Labas, Y. A., et al. (1999) Fluorescent proteins from nonbioluminescent *Anthozoa* species. *Nat. Biotechnol.* **17,** 969–973.
27. Wiehler, J., von Hummel, J., and Steipe, B. (2001) Mutants of *Discosoma* red fluorescent protein with a GFP-like chromophore. *FEBS Lett.* **487,** 384–389.
28. Bevis, B. J. and Glick, B. S. (2002) Rapidly maturing variants of the *Discosoma* red fluorescent protein (DsRed). *Nat. Biotechnol.* **20,** 83–87.
29. Mizuno, H., Sawano, A., Eli, P., Hama, H., and Miyawaki, A. (2001) Red fluorescent protein from *Discosoma* as a fusion tag and a partner for fluorescence resonance energy transfer. *Biochemistry* **40,** 2502–2510.
30. Campbell, R. E., Tour, O., Palmer, A. E., et al. (2002) A monomeric red fluorescent protein. *Proc. Natl. Acad. Sci. USA* **99,** 7877–7882.
31. Pollok, B. A. and Heim, R. (1999) Using GFP in FRET-based applications. *Trends Cell. Biol.* **9,** 57–60.
32. Devi, L. A. (2000) G-protein-coupled receptor dimers in the lime light. *Trends Pharmacol. Sci.* **21,** 324–326.
33. Eidne, K. A., Kroeger, K. M., and Hanyaloglu, A. C. (2002) Applications of novel resonance energy transfer techniques to study dynamic hormone receptor interactions in living cells. *Trends Endocrinol. Metabol.* **13,** 415–421.
34. Dean, M. K., Higgs, C., Smith, R. E., et al. (2001) Dimerization of G-protein-coupled receptors. *J. Med. Chem.* **44,** 4595–4614.
35. Overton, M. C. and Blumer, K. J. (2000) G-protein-coupled receptors function as oligomers in vivo. *Curr. Biol.* **10,** 341–344.
36. Devi, L. A. (2001) Heterodimerization of G-protein-coupled receptors: pharmacology, signaling and trafficking. *Trends Pharmacol. Sci.* **22,** 532–537.
37. Maggio, R., Barbier, P., Colelli, A., Salvadori, F., Demontis, G., and Corsini, G. U. (1999) G protein-linked receptors: pharmacological evidence for the formation of heterodimers. *J. Pharmacol. Exp. Ther.* **291,** 251–257.
38. Jordan, B. A. and Devi, L. A. (1999) G-protein-coupled receptor heterodimerization modulates receptor function. *Nature* **399,** 697–700.
39. Milligan, G. (2001) Oligomerisation of G-protein-coupled receptors. *J. Cell. Sci.* **114,** 1265–1271.
40. Edwards, S. W., Tan, C. M., and L. E., L. (2000) Localisation of G-protein-coupled receptors in health and disease. *Trends Pharmacol. Sci.* **21,** 304–308.
41. Angers, S., Salahpour, A., Joly, E., Hilairet, S., Chelsky, D., Dennis, M., and Bouvier, M. (2000) Detection of β 2-adrenergic receptor dimerization in living cells using bioluminescence resonance energy transfer (BRET). *Proc. Natl. Acad. Sci. USA* **97,** 3684–3689.
42. Hebert, T. E., Moffett, S., Morello, J. P., et al. (1996) A peptide derived from a beta2-adrenergic receptor transmembrane domain inhibits both receptor dimerization and activation. *J. Biol. Chem.* **271,** 16,384–16,392.
43. Rocheville, M., Lange, D. C., Kumar, U., Sasi, R., Patel, R. C., and Patel, Y. C. (2000) Subtypes of the somatostatin receptor assemble as functional homo- and heterodimers. *J. Biol. Chem.* **275,** 7862–7869.

44. Cvejic, S. and Devi, L. A. (1997) Dimerization of the delta opioid receptor: implication for a role in receptor internalization. *J. Biol. Chem.* **272,** 26,959–26,964.
45. Gomes, I., Jordan, B. A., Gupta, A., Trapaidze, N., Nagy, V., and Devi, L. A. (2000) Heterodimerization of μ and δ opioid receptors: a role in opiate synergy. *J. Neurosci.* **20,** RC110.
46. Jordan, B. A., Trapaidze, N., Gomes, I., Nivarthi, R., and Devi, L. A. (2001) Oligomerization of opioid receptors with β 2-adrenergic receptors: a role in trafficking and mitogen-activated protein kinase activation. *Proc. Natl. Acad. Sci. USA* **98,** 343–348.
47. Robbins, M. J., Ciruela, F., Rhodes, A., and McIlhinney, R. A. (1999) Characterization of the dimerization of metabotropic glutamate receptors using an N-terminal truncation of mGluR1α. *J. Neurochem.* **72,** 2539–2547.
48. Cornea, A. and Michael Conn, P. (2002) Measurement of changes in fluorescence resonance energy transfer between gonadotropin-releasing hormone receptors in response to agonists. *Methods* **27,** 333–339.
49. Dinger, M. C., Bader, J. E., Kobor, A. D., Kretzschmar, A. K., and Beck-Sickinger, A. G. (2003) Homodimerization of neuropeptide Y receptors investigated by fluorescence resonance energy transfer in living cells. *J. Biol. Chem.* **278,** 10,362–10,572.
50. Tatemoto, K., Carlquist, M., and Mutt, V. (1982) Neuropeptide Y—a novel brain peptide with structural similarities to peptide YY and pancreatic polypeptide. *Nature* **296,** 659–660.
51. Ganten, D., Paul, M., and Lang, R. E. (1991) The role of neuropeptides in cardiovascular regulation. *Cardiovasc. Drugs Ther.* **5,** 119–130.
52. Gerald, C., Walker, M. W., Criscione, L., et al. (1996) A receptor subtype involved in neuropeptide-Y-induced food intake. *Nature* **382,** 168–171.
53. Flood, J. F. and Morley, J. E. (1989) Dissociation of the effects of neuropeptide Y on feeding and memory: evidence for pre- and postsynaptic mediation. *Peptides* **10,** 963–966.
54. Vezzani, A., Civenni, G., Rizzi, M., Monno, A., Messali, S., and Samanin, R. (1994) Enhanced neuropeptide Y release in the hippocampus is associated with chronic seizure susceptibility in kainic acid treated rats. *Brain Res.* **660,** 138–143.
55. Michel, M. C., Beck-Sickinger, A., Cox, H., et al. (1998) XVI. International Union of Pharmacology recommendations for the nomenclature of neuropeptide Y, peptide YY, and pancreatic polypeptide receptors. *Pharmacol. Rev.* **50,** 143–150.
56. Cabrele, C. and Beck-Sickinger, A. G. (2000) Molecular characterization of the ligand–receptor interaction of the neuropeptide Y family. *J. Pept. Sci.* **6,** 97–122.
57. Daniels, A. J., Matthews, J. E., Slepetis, R. J., et al. (1995) High-affinity neuropeptide Y receptor antagonists. *Proc. Natl. Acad. Sci. USA* **92,** 9067–9071.
58. Matthews, J. E., Jansen, M., Lyerly, D., et al. (1997) Pharmacological characterization and selectivity of the NPY antagonist GR231118 (1229U91) for different NPY receptors. *Regul. Pept.* **72,** 113–119.
59. CLONTECH (2003) Features of Living Colors™ vectors. http:/www.clontech.com/.

22

Identification of Protein Interactions by Yeast Two-Hybrid Screening and Coimmunoprecipitation

Michael Tanowitz and Mark von Zastrow

Summary

Many protein interactions with G-protein-coupled receptors (GPCRs) appear to influence receptor signaling and functional regulation. There is great interest therefore in methods for the identification of novel or unanticipated GPCR binding proteins. A proven method for identifying such protein interactions is the yeast two-hybrid screen, which involves screening the protein products of a cDNA library with a selected domain derived from a GPCR. Once it is established that a candidate protein produces a specific positive interaction within the yeast two-hybrid system, it is important to demonstrate further that this interaction is likely to occur in vivo. Coimmunoprecipitation, in which proteins of interest are copurified with the receptor under study, is a good way to address this important issue. Together, the yeast two-hybrid screen and coimmunoprecipitation are a useful way to identify and sort through candidate GPCR-interacting proteins prior to analysis in physiological studies.

Key Words

GAL4 system, GPCR binding protein, GPCR-protein interactions, immunoprecipitation optimization, interaction cloning, receptor–protein crosslinking, receptor-protein solubilization.

1. Introduction

A number of cellular proteins, distinct from heterotrimeric G-proteins and arrestins, have been found to interact with certain G-protein-coupled receptors (GPCRs). In many cases these protein interactions were not anticipated from functional studies, which has led to increased interest in methods for detecting protein interactions independent of physiological function **(Fig. 1)**. One of the most popular of these methods is the yeast two-hybrid system, which allows one to screen for interactions between a protein domain isolated from a GPCR and a large number of cellular proteins represented in a cDNA library. Yeast

From: *Methods in Molecular Biology, vol. 259, Receptor Signal Transduction Protocols, 2nd ed.*
Edited by: G. B. Willars and R. A. J. Challiss © Humana Press Inc., Totowa, NJ

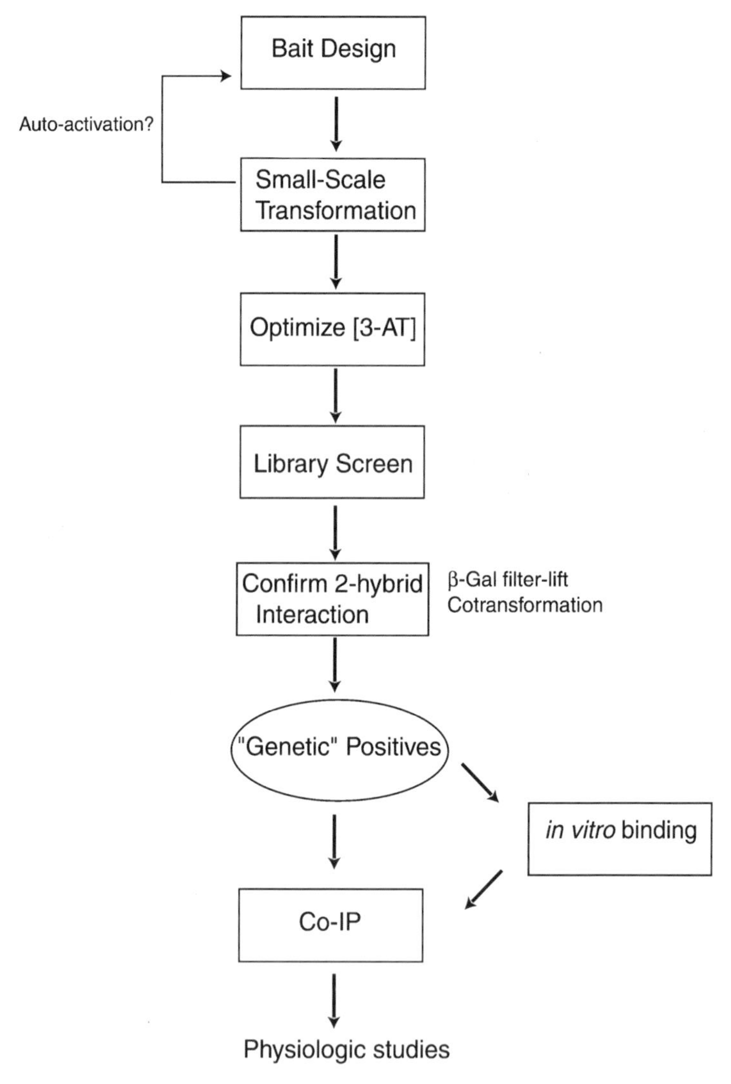

Fig. 1. Flow diagram of typical interaction cloning strategy.

two-hybrid screening is a powerful method to identify candidate interacting proteins but is generally not able to distinguish physiologically important protein interactions from spurious interactions, which reflect nonspecific binding or aberrant partnering of interacting domains outside of their normal cellular context. Testing candidate protein interactions using in vitro binding assays is a useful first step for confirming the biochemical specificity of the candidate protein interaction but does not provide information on whether this protein interaction occurs in vivo. Coimmunoprecipitation has proven to be a useful method for addressing this important question. Of course none of these methods are definitive by themselves and, ultimately, the significance of a newly identified receptor-interacting protein awaits demonstration of a physiological function. Nevertheless, yeast two-hybrid screening and coimmunoprecipitation are key methods used in strategies that have successfully identified a number of important protein interactions with specific GPCRs.

1.1. Yeast Two-Hybrid Screening

The yeast two-hybrid section describes yeast two-hybrid methods based on the GAL4 system *(1)*. The LexA-based yeast screens (interaction trap) are not discussed; however, the basic principles are similar. It is assumed that the necessary vectors, appropriate library, and yeast strain are available to the reader. All necessary components are widely available from commercial vendors. The two-hybrid screen exploits the modular nature of the yeast transcription factor GAL4. Many eukaryotic transcription factors contain distinct DNA binding and transcriptional activation domains. The DNA binding domain recognizes specific promoter sequences upstream of the target gene, whereas the activation domain activates transcription of the gene to which it has been brought into close proximity. A crucial feature is that the DNA binding and transcriptional activation domains do not require covalent attachment to each other to activate transcription of target genes *(2)*. The yeast two-hybrid approach takes advantage of these features of the yeast GAL4 transcription factor by generating an in-frame fusion of the GAL4 DNA binding domain (GAL4 BD) with a particular "bait" sequence to screen a library of GAL4 transcriptional-activation domain (GAL4 AD)–cDNA fusion proteins *(1)*. Transcription of reporter genes (typically *HIS3*, which controls production of an enzyme involved in histidine synthesis and *lacZ*, which controls production of β-galactosidase) is activated in yeast expressing DNA-binding domain and activation domain fusion proteins that interact with each other. The system must use an appropriately engineered yeast strain. In general, the screen is performed by transforming the yeast strain, either simultaneously or sequentially, with the bait construct and the plasmid library. The transformants are then spread onto media plates that lack the nutrients whose syntheses are conferred by the transformed plasmids and activation of reporters, followed by isolation and further testing of growing yeast colonies.

The first step in preparing for the screen involves bait selection and testing. There are various factors to consider, both generally and with respect to GPCRs. General considerations include the solubility of the protein sequence, the likelihood that it will exist in its native conformation in the hybrid protein, and the potential for it to activate transcription intrinsically. It is generally held that the larger the sequence the more likely that it will encompass a properly folded domain. Acidic amphipathic sequences can produce unwanted transcriptional activation *(3,4)*. Screens that have used GPCR-derived sequences have most often used portions of the C-terminal cytoplasmic tail *(5–7)*. The third intracellular loop has also been used successfully *(8)*. In principle, the extracellular loops could be screened by the two-hybrid method, although for cell surface interactions a different screening method might be preferred (e.g., phage display). In addition to factors that can lead to false-positive results, it should be kept in mind that there are various reasons why interactions may be missed. A poor or inappropriate library is chief among them and great care should be taken when making or obtaining the library to be screened. Other factors leading to false-negative results include poor expression or inefficient nuclear localization of the bait protein and improper folding or masking of the bait sequence.

1.2. Assessing Protein Interactions With GPCRs by Coimmunoprecipitation

Coimmunoprecipitation provides a useful means for investigating if a candidate interacting protein can bind to the full-length receptor, and for assessing whether this protein interaction is likely to occur in vivo. This method has been used by a number of investigators as a primary means for assessing protein interactions with GPCRs (e.g., **refs.** *5,6,9–11)*. Basically, coimmunoprecipitation involves purifying from a cell or tissue extract one of the components of the proposed protein complex, either the GPCR or candidate interacting protein, and then assaying for copurification of the other component. Although the basic principles of coimmunoprecipitation are simple and have been applied to many types of protein interaction, in practice this method is not always straightforward and can lead to aberrant or misleading results. Many aspects of this method are not unique to GPCRs and have been discussed in great detail elsewhere. However, several issues deserve particular attention when performing coimmunoprecipitation studies with GPCRs.

1.2.1. Initial Solubilization of the Receptor–Protein Complex

The conditions under which the initial cell or tissue extract is prepared is one of the most important experimental considerations. On the one hand, it is essential to solubilize membrane components completely to allow the highly hydrophobic GPCR to be efficiently isolated and to prevent the formation of nonspecific protein aggregates. On the other hand, one wants to maintain protein interactions with the GPCR and may need to solubilize receptors under nondenaturing conditions in which receptors retain their ability to bind ligands (this is essential if one wants to study the regulation of protein interactions by ligand). Finding the ideal balance is not always possible, which may be why some well known protein interactions with GPCRs (such as interactions with certain heterotrimeric G-proteins and arrestins) are notoriously difficult to detect by coimmunoprecipitation. In these cases one may choose to try chemical crosslinking of proteins prior to cell lysis, which can prevent protein partners from fully dissociating under harsher solubilization conditions (*see* **Subheading 3.2.2.**). A converse problem is that of generating a nonphysiological protein interaction during the solubilization process. A common means by which this occurs is via formation of aberrant intermolecular disulfide bonds during cell or tissue lysis. The cytoplasm is a highly reducing environment compared with typical extraction buffers, so many proteins will rapidly oxidize on cell lysis with attendant formation of intramolecular and intermolecular disulfide bonds. The degree to which this is a problem varies widely among proteins and extraction conditions but should always be considered. In mild cases it is often sufficient simply to include a small amount of reducing agent (e.g., 2 m*M* dithiothreitol [DTT]) in the extraction buffer. In more difficult cases one often uses fairly high concentrations of a suitable alkylating reagent (e.g., 25–100 m*M* *N*-ethylmaleimide or iodoacetamide). The idea here is to alkylate free sulhydryls rapidly before they form aberrant disulfide bonds, without disrupting normal disulfide bounds formed before cell lysis.

1.2.2. Obtaining Sufficient Signal-to-Noise Ratio and Controlling for Specificity

Most GPCRs are expressed in cells and tissues at low levels, so signals representing protein interactions with these receptors are often very small. This necessitates the use of highly sensitive detection systems (e.g., enzyme-linked chemiluminescence) and, consequently, levels of nonspecific background that are often negligible in a simple immunoprecipitation of a more abundant protein can be prohibitive in studies conducted with a GPCR. Thus, one must pay careful attention to the task of minimizing nonspecific binding in all stages of the analysis.

2. Materials

2.1. Yeast Materials

2.1.1. Media

1. DIFCO peptone.
2. Yeast extract.
3. 40% Dextrose, filter sterilized.
4. 0.2% Adenine hemisulfate, filter sterilized.
5. Yeast nitrogen base without amino acids.

6. Trp, Leu, Trp/Leu, and Trp/Leu/His 10X dropout solution, filter sterilized (*see* **Subheading 2.1.6.**).
7. 1 *M* 3-Amino triazolole (3-AT), filter sterilized.
8. Agar.
9. 10- and 15-cm media plates.

2.1.2. Transformations

1. 10X TE: 100 m*M* Tris-HCl, pH 7.5; 10 m*M* EDTA, sterile.
2. 1 *M* Lithium acetate (LiOAc) (10X) pH 7.5, sterile.
3. LiOAc–TE: 10 m*M* Tris-HCl, pH 7.5, 1 m*M* EDTA, 100 m*M* LiOAc; make fresh from 10X stocks.
4. 50% Polyethylene glycol (PEG) 4000, sterile.
5. Sterile 40% PEG 4000 in LiOAc–TE (made from 50% PEG 4000 and 10X LiOAc, TE stocks).
6. 100% Dimethyl sulfoxide (DMSO).
7. 2 mg/mL of high-quality single-stranded carrier DNA (*see* **Subheading 2.1.7.**).
8. 30°C and 42°C water baths.
9. Sterile bent glass rods or, preferably 5-mm glass beads for spreading cells on plates.

2.1.3. Yeast DNA Minipreps

1. 425- to 600-μm acid-washed glass beads.
2. Lysis buffer: 10 m*M* Tris-HCL, pH 8.0, 1 m*M* EDTA, 100 m*M* NaCl, 1% sodium dodecyl sulfate (SDS); 2% Triton X-100.
3. TE: 10 m*M* Tris-HCl, 1 m*M* EDTA, pH 8.0.
4. Saturated phenol, equilibrated to pH 8.0.
5. Chloroform.
6. 100% and 70% ethanol.

2.1.4. YPD (Yeast Extract, Peptone, Dextrose)

1. 10 g/L of yeast extract.
2. 20 g/L of DIFCO peptone.
3. 2% Dextrose.

Dissolve the yeast extract and peptone in 950 mL of water, adjust the pH to 6.5, and autoclave. After solution has cooled to 55°C add 50 mL of 40% sterile dextrose. To supplement with adenine (YPDA), add 10 mL of 0.3% adenine hemisulfate. For plates, include 20 g/L of agar before autoclaving.

2.1.5. Synthetic Dropout Medium (SD)

1. 6.7 g/L of yeast nitrogen base without amino acids.
2. 100 mL of appropriate 10X dropout solution/L.
3. 2% Dextrose.

Dissolve the nitrogen base in 850 mL of water, adjust the pH to 5.6, and autoclave. After the solution has cooled to ≤55°C, add 50 mL of 40% dextrose, 100 mL of 10X dropout solution, and 3-AT, if applicable. For plates, include 20 g/L of agar before autoclaving.

2.1.6. 10X Dropout Solution

Add all of the following nutrients except those that are to be excluded from the relevant dropout solution: 200 mg/L of L-adenine hemisulfate, 200 mg/L of L-arginine HCL, 200 mg/L of L-histidine, 300 mg/L of L-isoleucine, 1000 mg/L of L-leucine, 300 mg/L of L-lysine–HCl, 200 mg/L of L-methionine, 500 mg/L of L-phenylalanine, 2000 mg/L of L-threonine, 200 mg/L of L-tryptophan, 300 mg/L of L-tyrosine, 200 mg/L of L-uracil, and 1500 mg/L of L-valine.

Alternatively (and more conveniently), premixed dropout powders can be purchased from commercial vendors and used to make the 10X solutions.

2.1.7. Carrier DNA

Just before the first use, heat an aliquot to 95–100°C for 5 min, then immediately chill on ice. Repeat twice and keep on ice during use and before freezing. Repeat this procedure after several uses or if the tranformation efficiencies appear to be declining.

2.1.8. Z Buffer/X-Gal

2.1.8.1. Z BUFFER

16.1 g/L of $Na_2HPO_4 \cdot 7H_2O$, 5.5 g/L of $NaH_2PO_4 \cdot H_2O$, 0.75 g/L of KCl, and 0.246 g/L of $MgSO_4 \cdot 7H_2O$. Adjust the pH to 7.0 and sterilize.

2.1.8.2. Z BUFFER/X-GAL

Prepare fresh by adding 0.27 mL of β-mercaptoethanol and 1.67 mL of X-gal solution (20 mg/mL in dimethyl formamide) to 100 mL of Z buffer.

2.2. Immunoprecipitation

1. Isotonic phosphate-buffered saline solution (PBS).
2. Extraction buffer: 0.5% (v/v) Triton X-100, 50 m*M* Tris-HCl, pH 7.5, 120 m*M* NaCl, 25 m*M* KCl, 2 m*M* EDTA, 25 m*M* *N*-ethylmaleimide or iodoacetamide (added fresh), protease inhibitors (1 μg/mL of leupeptin, 1 μg/mL of pepstatin, 2 μg/mL of aprotinin, and 0.5 m*M* of phenylmethylsulfonyl fluoride), and DNase (optional, *see below*).
3. Wash buffer: same as extraction buffer, without *N*-ethylmaleimide or iodoacetamide.
4. Sucrose cushion buffer: wash buffer plus 1 *M* sucrose.

5. Antibody-recognizing receptor or epitope tag (e.g., 12CA5 anti-HA; Boehringer Manheim).
6. Antibody recognizing the candidate interacting protein or epitope tag (e.g., M2 anti-FLAG; Sigma-Aldrich).
7. Protein A/G resin, 50% suspension (Pierce Chemical Company).
8. Heat-killed, aldehyde-fixed *S. aureus* suspension (Pansorbin, Calbiochem).
9. Preimmune or nonimmune serum from same species as anti-receptor antibody.
10. Apparatus and solutions for standard SDS-polyacrylamide gel electrophoresis (SDS-PAGE) and immunoblotting.

2.3. Crosslinking

1. Disuccinimidyl suberate (DSS) noncleaveable.
2. Dithiobis(succinimidyl propronate) (DSP).

3. Methods

3.1. Yeast Methods

3.1.1. Yeast Transformation

To simplify method descriptions it is assumed that the bait plasmid carries the *Trp* nutritional marker, the library plasmid carries the *Leu* nutritional marker, and that reporter activation results in expression of the His and LacZ reporters. The relevant selection markers can be substituted in the protocols if a different system is being used.

The following protocol is based on the high-efficiency lithium acetate method developed by Geitz et al. ***(12)***. It may be scaled up or down as required. The desired transformation efficiency is on the order of 10^5 transformants/μg. The transformation efficiency becomes important when deciding the quantity of library to be used for the library transformation. More than one type of plasmid can be transformed at one time, as when performing secondary tests to verify that a library clone interacts with your bait. Be aware that the efficiency of cotransformation is generally an order of magnitude lower owing to the reduced probability that a yeast will take up both plasmids (the efficiency for the individual plasmids, however, should remain unchanged). When the bait strain is being used to make competent cells they should be grown in the appropriate SD media (Trp–) to maintain selection for the plasmid. During the final two doublings YPDA may be used with negligible plasmid loss.

3.1.1.1. Making Transformation-Competent Yeast

1. Transfer several large (>2 mm) fresh yeast colonies to 1–2 mL of YPDA or appropriate SD, vortex-mix to disperse clumps, and transfer to 50 mL of YPDA or SD for overnight growth at 30°C with shaking (~200 rpm).
2. Use the overnight culture to inoculate 400 mL of YPDA to OD_{600} approx 0.1–0.2 and grow at 30°C with shaking until the culture is in mid-log phase, OD_{600} =

0.4–0.8 (3–5 h). It is important to allow cells to complete at least two doublings before harvest.

3. Pellet cells at 2000*g* for 5 min at room temperature (RT), remove the supernatant, and resuspend in 50 mL of sterile water. Repellet cells and resuspend in 10 mL of LiOAc–TE.
4. Repellet cells and resuspend in 2 mL of LiOAc. For highest efficiency, competent cells should be used immediately. One hundred microliters is a sufficient amount of competent cells to be transformed with up to 2 μg of DNA. Transformation with higher DNA quantities increases the likelihood of introducing multiple plasmids into individual yeast—a situation that should be avoided when performing the library screen owing to the additional work and confusion that may be created when trying to identify the plasmid responsible for an identified positive interaction.

3.1.1.2. Small-Scale Transformation

The following small-scale transformation protocol can be used when first introducing the bait plasmid into yeast to create a bait strain for tests of auto-activation and for subsequent library transformation. Use the bait strain in small-scale transformations to assess the transformation efficiency obtained with library plasmid and to determine how much library can be used while remaining in the linear range of transformation. Exceeding this amount increases the chance of introducing multiple library plasmids into individual yeast.

1. To 100 μL of competent cells add 0.1–2.0 μg of DNA and 60 μg of single-stranded carrier DNA (*see* **Note 4**); mix by gentle vortexing.
2. Add 600 μL of sterile 40% PEG 4000 in LiOAc–TE, mix by gentle vortexing, and incubate at 30°C for 30 min.
3. Add DMSO to 10% (70 μL), mix by inversion, and heat shock at 42°C for 15 min. Place on ice for 2 min and return to RT. The addition of DMSO at this step can increase transformation efficiency two- to fivefold.
4. Pellet cells and resuspend in sterile 0.9% saline (can also use TE or water). Determine the transformation efficiency by making serial dilutions and plating 100 μL on 10-cm SD plates (either SD/Trp– or SD/Trp–/Leu–) using 5-mm glass beads *(5–7)* and shaking to spread the mixture evenly. If using bent glass rods use a minimum of strokes. To increase transformation efficiency, pellet cells after **step 3** and resuspend in YPDA for outgrowth at 30°C for 1–2 h before proceeding to **step 4**.

3.1.1.3. Library-Scale Transformation

For library-scale transformations either sequential or simultaneous transformations can be used. In the majority of cases it is preferable to do a sequential transformation, as it consumes significantly less library. If there is reason to believe, however, that the presence of the bait plasmid inhibits growth of the yeast or adversely affects the transformation efficiency, a simultaneous cotransformation should be considered. Use the small-scale transformation to determine the transformation efficiency of the library plasmid in the bait strain and calculate the amount of DNA and competent yeast necessary to achieve the desired number of transformants. To achieve good library coverage, plan to screen at least two to three times the number of independent clones in your library, usually 3–10 $\times$ 10^6 transformants. For example: an efficiency of 0.8 $\times$ 10^5 transformants/µg of library DNA was obtained and a library with 2 $\times$ 10^6 independent clones is to be used. Thus, 3 $\times$ 2 $\times$ 10^6 transformants/ 0.8 $\times$ 10^5 transformants/µg = 75 µg of library DNA will be required, and 75 µg/2 µg per transformation $\times$ 0.1 mL of competent cells per transformation = 3.75 mL of competent cells should be prepared.

1. Set up for the library screen by preparing the required amount of competent yeast and scaling up from the small-scale transformation. During incubations swirl every 5 min to ensure even heating of the larger volume mixture.
2. In addition, extend the incubation times to 40 min and 20 min for the 30°C and 42°C incubations, respectively.
3. At the end of the transformation, pellet and resuspend cells in 10–20 mL of 0.9% saline, TE, or water and plate 200 µL/15-cm plate (SD/Trp–/Leu–/His ± 3-AT) using 10–20 5-mm glass beads. Incubate 50–100 plates upside down at 30°C for 2–5 d or until colonies appear. Plate a small amount of the transformation mixture (1–10 µL in 100 µL of diluent) onto SD/Trp–/Leu 10-cm plate(s) to calculate the number of transformants screened.

3.1.2. Testing for Auto-Activation and Leaky His3 Expression

The success of a two-hybrid screen largely depends on the control of background false-positives. Two common sources of high background involve activation of the reporter gene(s) by the bait plasmid alone (auto-activation) and low-level expression of His3p in the absence of reporter activation (leaky expression). The problem of leaky His3p expression is overcome by including an optimal concentration of 3-AT, a competitive inhibitor of His3p, in the library screening plates ***(13)***. Weak auto-activation can often be minimized in the same way. In some cases, however, it may be necessary to modify the bait to eliminate sequences that are responsible for the autonomous transcriptional activation. Switching the bait to a lower expressing vector might also improve the situation. Follow the protocol below to determine the optimal concentration of 3-AT to include in the screen.

1. Using the small-scale transformation, transform the bait strain with an empty library plasmid or, ideally, with the library AD fused to an irrelevant protein sequence that does not interact with the bait.
2. Plate the transformation on a series of SD/Trp/Leu/His plates with 0, 2.5, 5, 10, 20, and 50 m*M* 3-AT. Plate an equal amount on SD/Trp/Leu plates for comparison. Aim to plate 500–1000 colonies/plate.

3. If possible, do a parallel transformation with a Gal4 AD-fusion that is known to interact with the bait or a cotransformation with a pair of positive two-hybrid interactors to serve as a positive control.
4. Use the lowest concentration of 3-AT at which only small pinpoint colonies appear after 1 wk of growth but still allows large colonies to grow with the positive control.

3.1.3. Filter-Lift Assay for β-Galactosidase

After growing the library plates to identify the initial positives it will be necessary to test the secondary reporter, usually β-galactosidase. Transfer the initial positives to a master plate and/or replica plate. Before performing the filter-lift assay described in the list that follows ***(14)*** make sure there are master or replica plates for backups, as this and other various manipulations carry the risk of introducing fungal and bacterial contamination.

1. Place a sterile clean dry Whatman filter onto the plate to be assayed and gently rub it evenly over the surface until it has been evenly wetted. Puncture holes into the filter and agar in three or more locations for orientation.
2. Gently lift the filter and transfer it to liquid nitrogen and submerge. Remove the filter from the liquid nitrogen and allow it thaw on plastic film. Repeat the freeze–thaw cycle one or two times.
3. Place the filter colony side up onto a second filter that has been presoaked in Z buffer/X-gal solution. Incubate the filters for 30°C until some colonies begin to turn blue (positive). Continue incubation as needed. Strong positives may turn blue within 30 min–2 h. Weak positives require longer times. Prolonged incubation will increase the number of false-positives.
4. Use the orientation holes to identify the positive colonies on the original plate and streak them onto a second master plate.

3.1.4. Rescuing the Library Plasmid from Yeast

Once the yeast containing the presumptive positive interaction are identified it will be necessary to "rescue" the plasmid via a yeast miniprep for further evaluation. This is not a trivial step, as yeast express low amounts of the plasmid and their cell walls require harsh treatments that result in the release and copurification of chromosomal DNA. We have found that the simplest and most reliable method of isolating the plasmid DNA is the "smash and grab" method of Rose et al. ***(15)***, followed by transformation into *E. coli*. Competent bacteria that are capable of ≥107 cfu/μg are required.

1. Grow 2–3 mL of overnight culture and pellet by centrifugation.
2. Add 0.3 g (~0.3 mL) of glass beads, 0.2 mL of lysis buffer, and 0.2 mL of a 1:1 mix of phenol–chloroform to the cell pellet in a 1.5-mL Eppendorf tube.
3. Vortex-mix at top speed for 2 min, add 0.2 mL of TE, pH 8, and vortex-mix again for 5 s.
4. Centrifuge the tube for 5 min at RT.

5. Transfer the upper aqueous phase (~0.38 mL) to a fresh Eppendorf tube and discard the tube with organic phase/beads.
6. Add two volumes of 100% ethanol, mix thoroughly, and centrifuge at top speed for 15 min.
7. Aspirate the supernatant and wash the pellet with 70% ethanol. Centrifuge for a few minutes to remove the supernatant and air-dry the pellet.
8. Resuspend the pellet in approx 15–20 μL of TE.
9. Transform into competent *E. coli*.

Some systems use separate antibiotic markers for the bait and library plasmids. In many systems, however, restriction digests must be performed to determine which of the bacterial colonies contain the library plasmid.

3.1.5. Confirmation of Two-Hybrid Interaction by Yeast Cotransformation

To test independently whether the isolated library clone interacts with the bait, cotransform yeast with the bait and library plasmids using the small-scale transformation. Perform parallel cotransformations using an empty bait vector as a negative control. Plate transformations on both SD/Trp/Leu/His and SD/Trp/Leu. True interactors should produce roughly the same number of colonies on the SD/Trp/Leu/His as on the SD/Trp/Leu plates, but show no growth on the SD/Trp/Leu/His plates with the negative bait control.

Once it has been verified that a library clone produces a specific two-hybrid interaction, proceed to additional tests such as in vitro binding, coimmunoprecipitation, colocalization, and so forth.

3.2. Immunoprecipitation Methods

3.2.1. A Generic Coimmunoprecipitation Protocol

3.2.1.1. Preparation of Cell Extract

1. Transfected HeLa cells expressing HA-tagged GPCR and FLAG-tagged candidate interacting protein are grown to confluence on 10-cm plastic dishes and treated at 37°C under the conditions of interest (e.g., ± agonist).
2. The dishes are chilled to 4°C in the cold room, and monolayers are washed three times in PBS and aspirated nearly dry. One milliliter of extraction buffer is added to each dish and dishes are incubated at 4°C with gentle rocking for 10 min. The extract and associated insoluble material are transferred to 1.5-mL microcentrifuge tubes and centrifuged at 14,000*g* for 15 min in the cold room.

3.2.1.2. Preclearing (Optional)

1. The supernatant from **Subheading 3.2.1.1., step 2** is removed using a transfer pipet, keeping clear of the insoluble pellet. Add 30 μL of Pansorbin and 5 μL of preimmune or nonimmune serum. Rotate 60 min at 4°C.

2. Centrifuge 14,000*g* for 5 min at 4°C, and transfer the clear supernatant to fresh tubes (*see* **Note 1**).

3.2.1.3. Immune Complex Formation

1. Add the appropriate amount of anti-receptor antibody to the supernatant from **step 2** of either **Subheading 3.2.1.1.** or **3.2.1.2.** dependent on whether preclearing has been performed. This amount will vary depending on the antibody and conditions used (for 12CA5 anti-HA we typically use 5 μg/mL). Rotate at 4°C for 60–120 min.

3.2.1.4. Precipitation and Washing of Immune Complexes

1. Following **step 1** in **Subheading 3.2.1.3.**, add 30–50 μL of protein A/G–Sepharose and rotate for an additional 60 min.
2. Centrifuge at 6000*g* for 2 min (Sepharose beads should sediment rapidly), discard the supernatant, and resuspend in 1 mL of wash buffer. Repeat five times, rotating for 5 min between each wash and removing as much liquid as possible using a vacuum aspirator attached to a 22- to 25-gage needle (to prevent aspiration of beads).

3.2.1.5. Detection of Candidate Protein in Immunoprecipitates

1. Add 75 μL of Laemmli gel loading buffer to each pellet from **Subheading 3.2.1.4., step 2** and incubate at 37°C for 15 min We typically do not boil samples because we find this promotes the formation of aberrant high molecular weight receptor aggregates.
2. Centrifuge 14,000*g* for 10 min at room temperature and load the supernatant on SDS-polyacrylamide gels suitable for resolving the candidate interacting protein.
3. Transfer to nitrocellulose or polyvinylidene fluoride (PVDF) membranes using standard methods and immunoblot for candidate interacting protein.

We typically use enzyme-linked chemiluminescence (ECL) for immunoblot detection, which is advantageous because of its high sensitivity. Consequently, significant signal can result from nonspecific binding or fluid "carryover" occurring during the immunoprecipitation. These artifactual sources of signal are of particular concern when using transfected cells that overexpress proteins of interest (*see* **Note 2**). It is often useful to estimate the amount of candidate interacting protein that is coimmunoprecipated relative to the amount of receptor protein that is directly precipitated by bound antibody (*see* **Note 3**). Another useful control is to assure that the protein complex isolated formed prior to cell lysis, and does not represent an artifact of protein binding or aggregation that can occur after cell disruption (*see* **Note 4**).

3.2.2. Use of Chemical Crosslinking Prior to Immune Complex Formation

A number of physiologically important protein interactions are difficult to detect by coimmunoprecipitation. In some cases this is because the proteins of interest dissociate during the initial solubilization of receptors from the cell or tissue preparation or during the extensive washing involved in isolating the immunoprecipitate. One approach to detect such interacting proteins is to crosslink them covalently to the GPCR prior to cell lysis. Popular reagents for performing this are amine-reactive succinimidyl esters, such as DSS and DSP, which are membrane-permeant bifunctional crosslinkers. DSS is cleavable using a reducing agent such as DTT or β-mercaptoethanol, whereas DSP forms crosslinks that are resistant to reducing agents. There are many other crosslinkers available and many variations on this method have been employed. As one might imagine, potential problems with nonspecific protein interactions are even greater using chemical crosslinking, necessitating very careful evaluation of specificity controls. A useful starting point for incorporating chemical crosslinking into a coimmunoprecipitation method is to substitute **step 2** in **Subheading 3.2.1.1.** with the following:

1. The dishes are washed three times with PBS. A 10 mg/mL solution of DSP (Pierce) is freshly prepared (in dry DMSO at room temperature) and diluted 1:1000 into ice-cold PBS immediately before use (DSP hydrolyzes within minutes in aqueous solution). Washed monolayers are overlayed with DSP-containing PBS (5 mL/dish) and incubated with gentle rocking at 4°C for 30 min. DSP-containing medium is removed and cells are washed another three times with PBS containing 10 m*M* glycine (to facilitate quenching of unreacted DSP). One milliliter of extraction buffer is added to each dish and dishes are incubated at 4°C with gentle rocking for 10 min. The extract and associated insoluble material are transferred to 1.5-mL microcentrifuge tubes and centrifuged at 14,000*g* for 15 min in the cold room, and proceed to **Subheading 3.2.1.2., step 1** or **3.2.1.3, step 1**.

4. Notes

1. At this point the extract solution should be completely clear and should not contain suspended materials that will sediment in subsequent centrifugation steps. If this is not the case, it is necessary to repeat the preceding steps above or modify the clarification method. In some cases (particularly with highly concentrated lysates) one observes a viscous lysate that is difficult to clarify. In this case one can add DNase to the extraction buffer or try mechanical shearing such as sonication or passage through a fine (21- or 23-gage) needle, to disrupt chromatin-containing material that will bind to many cellular proteins and contribute to nonspecific signals in the immunoprecipitation. In our hands the failure to clarify lysate at this initial stage of the experiment is a major source of nonspecific signal.
2. Nonspecific signals can arise from a number of sources (including nonspecific binding to the antibodies or surfaces of beads or plasticware, volume carryover caused by incomplete washing, or nonspecific signals mistaken for the protein of interest in the immmunoblotting procedure). A useful negative control that can be performed in studies using epitope-tagged receptors is to transfect and prepare

extract from cells exactly as done for the other samples, but to use a receptor construct in which the specific epitope tag sequence recognized by the precipitating antibody is not present. A positive signal in this control is a certain indication of nonspecific interaction. A more powerful control includes testing whether a distinct GPCR or mutation of the originally used receptor, which is not expected to interact with the protein of interest based on yeast two-hybrid or in vitro studies, is capable of generating the observed signal in the coimmunoprecipitation. Nonspecific controls can be more difficult if one is using antibodies recognizing endogenous receptor protein. On the other hand, even specific protein association observed in transfected cells may not represent physiologically meaningful associations, as it is not uncommon for irrelevant protein interactions to be "forced" by overexpression or to occur because the proteins of interest are expressed in an inappropriate cell type. Thus the evaluation of coimmunoprecipitation from a native cell or tissue can be very important. To evaluate specificity in a coimmunoprecipitation prepared from a native cell or tissue source, typically one tests preimmune and nonimmune serum and asks whether the coimmunoprecipitation can be obtained using more than one receptor antibody. Another useful approach is to examine whether the interaction can be detected using the converse order of isolation, that is, immunoprecipitating the protein of interest and looking for the presence of copurified receptor. In principle, gene knock-out animals provide ideal controls for specificity in such experiments (*see* Chapter 24), although the feasibility of this elegant control is dependent on the availability of the required mutant animals.

3. It is often useful to estimate how much of the total candidate protein present in the lysate is immunoprecipitated with the receptor of interest. To do this it is useful to load aliquots corresponding to 1% and 5% of the total cell extract on gel lanes, which are well separated (by blank lanes or molecular weight markers) from the experimental lanes to avoid false-positive signals caused by spillover between lanes. If one knows the relative levels of receptor and interacting protein expression in cells, it is then possible to estimate the stoichiometry of the coimmunoprecipitation obtained. Typically coimmunoprecipitations with GPCRs are obtained at low stoichiometry (<5%). A higher stoichiometry could indicate a remarkably stable protein interaction but should also lead one to reconsider the possibility of aberrant signal caused by nonspecific binding.
4. In the event that the experiment reveals a significant amount of coimmunoprecipitated protein relative to the negative control (*see* **Note 1**), it is useful to determine whether the protein association in question formed before or after cell lysis. Ideally one only wants to examine protein interactions that exist before cell lysis, as protein complexes formed after lysis are unlikely to be physiologically relevant. The formation of protein complexes postlysis may occur because proteins that are localized in different subcellular compartments in vivo can bind aberrantly when released from their normal environment on solubilization. This problem is likely to be most problematic in transfected cells, where protein expression levels may be unnaturally high, but may also occur with native proteins when

released from their normal cellular environment. To control for this, one can express the GPCR of interest by itself in one set of transfected cells and the candidate interacting protein by itself in another, and then combine the extracts prior to immunoprecipitation. A positive coimmunoprecipitation observed in this control suggests either that there is a prohibitively high nonspecific binding (such as to the tubes or beads) in the immunoprecipitation or that the detected protein complex formation has occurred after cell lysis.

References

1. Fields, S. and Song, O. (1989) A novel genetic system to detect protein–protein interactions. *Nature* **340,** 245–246.
2. Ma, J. and Ptashne, M. (1987) A new class of yeast transcriptional activators. *Cell* **51,** 113–119.
3. Ruden, D. M., Ma, J., Li, Y., Wood, K., and Ptashne, M. (1991) Generating yeast transcriptional activators containing no yeast protein sequences. *Nature* **350,** 250–252.
4. Ruden, D. M. (1992) Activating regions of yeast transcription factors must have both acidic and hydrophobic amino acids. *Chromosoma* **101,** 342–348.
5. Whistler, J. L., Enquist, J., Marley, A., et al. (2002) Modulation of postendocytic sorting of G protein-coupled receptors. *Science* **297,** 615–620.
6. Cong, M., Perry, S. J., Hu, L. A., Hanson, P. I., Claing, A., and Lefkowitz, R. J. (2001) Binding of the beta2 adrenergic receptor to *N*-ethylmaleimide-sensitive factor regulates receptor recycling. *J. Biol. Chem.* **276,** 45,145–45,152.
7. Hu, L. A., Tang, Y., Miller, W. E., et al. (2000) Beta 1-adrenergic receptor association with PSD-95. Inhibition of receptor internalization and facilitation of beta 1-adrenergic receptor interaction with *N*-methyl-D-aspartate receptors. *J. Biol. Chem.* **275,** 38,659–38,666.
8. Tang, Y., Hu, L. A., Miller, W. E., et al. (1999) Identification of the endophilins (SH3p4/p8/p13) as novel binding partners for the beta1-adrenergic receptor. *Proc. Natl. Acad. Sci. USA* **96,** 12,559–12,564.
9. Cao, T. T., Deacon, H. W., Reczek, D., Bretscher, A., and von Zastrow, M. (1999) A kinase-regulated PDZ-domain interaction controls endocytic sorting of the beta2-adrenergic receptor. *Nature* **401,** 286–290.
10. Hebert, T. E., Moffett, S., Morello, J. P., et al. (1996) A peptide derived from a beta2-adrenergic receptor transmembrane domain inhibits both receptor dimerization and activation. *J. Biol. Chem.* **271,** 16,384–16,392.
11. Klein, U., Ramirez, M. T., Kobilka, B. K., and von Zastrow, M. (1997) A novel interaction between adrenergic receptors and the alpha-subunit of eukaryotic initiation factor 2B. *J. Biol. Chem.* **272,** 19,099–19,102.
12. Geitz, D. M., St. Jean, A., Woods, R. A., and Schiestl, R. H. (1992) Improved method for high efficiency transformation of intact yeast cells. *Nucleic Acids Res.* **20,** 1425.
13. Fields, S. (1993) The two hybrid system to detect protein-protein interactions. *Methods Compan. Methods Enzymol.* **5,** 116–124.

14. Breedan, L. and Nasmyth, K. (1985) Regulation of the yeast HO gene. *Cold Spring Harbor Symp. Quant. Biol.* **50,** 643–650.
15. Rose, M. D., Winston, F., and Heiter, P. (1990) *Methods in Yeast Genetics: A Laboratory Course Manual*, Cold Spring Harbor Laboratory Press, Cold Spring Harbor, New York.

23

Study of G-Protein-Coupled Receptor–Protein Interactions Using Gel Overlay Assays and Glutathione-*S*-Transferase–Fusion Protein Pull-Downs

Ashley E. Brady, Qin Wang, and Lee E. Limbird

Summary

Numerous recent studies have suggested that the predicted cytosolic domains of G-protein-coupled receptors (GPCRs) represent a surface for association with proteins that may serve multiple roles in receptor localization, turnover, and signaling beyond the well-characterized interactions of these receptors with heterotrimeric G-proteins. This chapter describes two in vitro methods for ascertaining interactions between GPCRs and various binding partners: gel overlay strategies and GST-fusion protein pull-downs.

Key Words

Gel overlay, G-protein-coupled receptor, glutathione-*S*-transferase pull-down, protein–protein interaction.

1. Introduction

Gel overlays have been used historically to examine proteins whose folding is reachieved following sodium dodecyl sulfide-polyacrylamide gel electrophoresis (SDS-PAGE), a property that is determined empirically. For example, a variety of protein–protein interactions have been identified previously using gel overlay strategies, including interacting proteins for calmodulin *(1)* and for cAMP-dependent protein kinase (called A kinase anchoring proteins, or AKAPs) *(2)*. These approaches have been adapted for studying interactions with G-protein-coupled receptors (GPCRs) *(3)*.

The overall approach for gel overlay studies is to separate proteins in a mixture (tissue or cell lysates, membrane or cytosolic fractions), using SDS-PAGE, renature (or not) the polyacrylamide gel, transfer to polyvinylidene fluoride (PVDF) or nitrocellulose membranes, and probe with a radiolabeled ligand or

From: *Methods in Molecular Biology, vol. 259, Receptor Signal Transduction Protocols, 2nd ed.*
Edited by: G. B. Willars and R. A. J. Challiss

protein, typically created via metabolic labeling in bacteria followed by purification or by in vitro translation, as we describe in this chapter. The term "overlay" is a residual terminology, describing very early protocols in which the SDS-polyacrylamide gel was probed with radiolabeled ligands (including proteins) resuspended in agarose and "overlaid" (and thus immobilized) on the wet polyacrylamide gel.

The variable in current gel overlay protocols that likely needs to be optimized for each protein-interacting pair is the incubation of the SDS-polyacrylamide gel prior to electrophoretic transfer to PVDF or nitrocellulose membranes. Some investigators think that renaturation (e.g., in buffer containing urea or nondenaturing detergents) is essential before transfer; others think renaturation incubation should occur after transfer to the filter, and some include no incubations intended to renature proteins following SDS-PAGE. Naturally, optimization of these protocols depends on the proteins involved and the chemical and physical properties of their interacting domains.

Glutathione-*S*-transferase (GST)-fusion proteins provide a powerful in vitro tool for detecting direct interactions between GPCRs and other accessory or regulatory proteins ***(4–9)***. The principle of the methodology is that the water-soluble GPCR domain of interest (e.g., 3i loop or C-terminal tail) for examining protein–protein interactions is fused to GST, a dimeric enzyme. Design of "spacer" sequences between the cDNA encoding GST and that encoding the GPCR domain appears routinely to permit unperturbed folding of the GST and the independent fused protein domain. Known quantities of GST fusion protein are incubated with a protein mixture of interest (e.g., cellular extract) or, alternatively, coexpressed in a target cell of interest, and the fused protein is allowed to bind to possible interacting proteins. The incubations (or cells, after lysis) are then exposed to glutathione (GSH) conjugated to agarose. The GST enzyme binds its substrate, GSH, and proteins interacting with its fusion protein are similarly isolated. The appropriate control is to compare eluates of GSH-agarose bound to GST alone vs those bound to GST-fusion protein.

2. Materials

2.1. Gel Overlay

1. TnT rabbit reticulocyte in vitro transcription and translation kit (Promega).
2. [^{35}S]Methionine (1000 Ci/mmol, at 10 mCi/mL).
3. RNasin ribonuclease inhibitor (Promega).
4. cDNA template: example: 3i loop of the α_2-adrenergic receptor fused to Gen10, a methionine-rich viral coat protein ***(10)***, referred to as G_{10}-α_2-3i loop.
5. Nuclease-free water.
6. 12% SDS-PAGE minigel for analysis of ^{35}S-protein product.
7. Biological preparation of interest to separate on SDS-PAGE.

8. PVDF, such as Immobilon-P (0.45 μm) by Millipore (preferred), or nitrocellulose membranes.
9. Tris-buffered saline (TBS): 20 m*M* Tris-HCl, 137 m*M* NaCl, pH 7.6.
10. Blocking buffer: TBS containing 3% (v/v) Tween-20 and 5% (w/v) nonfat powdered milk.
11. Rinsing buffer: TBS containing 0.1% (v/v) Tween-20 and 5% (w/v) nonfat powdered milk.
12. Tris-glycine transfer buffer: 25 m*M* Tris-base, 192 m*M* glycine, 20% MeOH, pH 8.3.

2.2. GST-Fusion Protein Pull-Down

1. pGEMEX-2 or other expression vector encoding GST prepared for in-frame fusion with DNA sequence encoding the protein or domain of interest.
2. *E. coli* strain DH5α or BL21, or any strain that allows for expression of the gene of interest as a GST-fusion protein.
3. Luria broth (LB) (1 L): 10 g of bacto-tryptone, 5 g of Bacto-yeast extract, 10 g of NaCl, pH 7.0; autoclave for 20 min.
4. 50–100 μg/mL of ampicillin (or other appropriate antibiotic for selection).
5. Isopropyl-β-D-thiogalactopyranoside (IPTG).
6. Tris-Triton (TT) buffer: 100 m*M* Tris-HCl, pH 8.0, 1% Triton X-100, 100 μ*M* phenylmethylsulfonyl fluoride (PMSF), 1 μg/mL of soybean trypsin inhibitor (STI), 1 μg/mL of leupeptin, and 10 U/mL of aprotinin.
7. 1 mg/mL Lysozyme in TT buffer.
8. GSH-agarose (Pierce).
9. 333 m*M* NaCl in TT buffer.
10. 0.8 × 4 cm Poly-Prep column (Bio-Rad).
11. 10 m*M* GSH (free acid) in TT buffer.
12. Dulbecco's phosphate-buffered saline (PBS), sterile: 0.1 g of KCl, 0.1 g of KH_2PO_4, 4.0 g of NaCl, 1.08 g of Na_2HPO_4 in 500 mL of double-distilled water (ddH_2O), autoclaved.
13. Tris-Triton binding (TTB) buffer (*see* **Note 1**): 50 m*M* Tris-HCl, pH 7.4, 0.05% Triton X-100, 10% glycerol, 0.01% bovine serum albumin (BSA), 100 μ*M* PMSF.
14. Bio-Rad protein assay reagent (cat. no. 500-0006).
15. 1X Laemmli sample buffer: 62.5 m*M* Tris-HCl, pH 6.8, 700 m*M* β-mercaptoethanol, 2% SDS, 10% glycerol.
16. EN[^{3}H]ANCE (NEN).

3. Methods

3.1 Gel Overlay Assays

3.1.1. Generation of ^{35}S-Labeled Protein for Probing Transferred Proteins (Example: ^{35}S-3i Loop of the α_{2A}-Adrenergic Receptor)

1. Mix the following in an autoclaved microfuge tube: 25 μL of TnT lysate, 1 μL of amino acid mix (1 m*M*, minus methionine), 2 μL of TnT reaction buffer, 1 μL of TnT T7 RNA polymerase (all reagents provided in TnT kit).

2. Add 4 μL of [^{35}S]methionine (1000 Ci/mmol, at 10 mCi/mL) and 1 μL of RNasin ribonuclease inhibitor (40 U/μL).
3. Add 1 μg of the appropriate cDNA template (presented as the circular plasmid DNA in a vector possessing the T7 RNA polymerase recognition sequence).
4. Adjust the volume to 50 μL with nuclease-free water.
5. Incubate the mixture for 90 min at 30°C.
6. Stop the reaction by placing the tube on ice.
7. Analyze and quantitate the reaction product by separation of precisely 1 μL on a 12% SDS-PAGE minigel.
8. Dry the gel and visualize bands by autoradiography.
9. Cut the band corresponding to the estimated probe M_r out of the dried gel and count it in scintillation fluor in a β-counter. Determine cpm of total product per microliter which corresponds to cpm of ^{35}S-protein of the correct M_r.

3.1.2. Overlay Assay

1. Protein aliquots (2.5–3.0 mg/sample) of biological sample of interest are separated with a SDS-PAGE preparative gel (*see* **Note 2**) using 7.5–20% polyacrylamide gradients on 16-cm long and 1.5-mm thick gels.
2. Transfer the resolved proteins to PVDF at 4°C by electrophoresis overnight at 30 V in Tris-glycine transfer buffer.
3. Cut the membrane into vertical strips (2–4 mm wide) for gel overlay and Western blot analysis.
4. Block nonspecific binding to PVDF membranes by incubation for at least 1 h in blocking buffer at room temperature.
5. Wash PVDF membranes for 30 min in rinsing buffer.
6. Incubate PVDF membrane strips with 300,000 cpm of the appropriate amount of [^{35}S]Met-labeled G_{10}-α_2-3i loop structure diluted in 1 mL rinsing buffer for 4 h (to overnight) at 4°C with constant rocking (*see* **Note 3**).
7. Wash membranes three times for 15 min each with rinsing buffer, twice for 10 min with TBS, and air-dry before autoradiography (*see* **Note 4**).
8. Expose the pressed filter to a PhosphorImager and/or X-ray film to permit band identification and quantitation.

3.2. GST-Fusion Protein Pull-Downs

3.2.1. Synthesis of GST-Fusion Proteins

1. Subclone the gene of interest into a plasmid designed for GST-fusion protein expression (*see* **Note 5**).
2. Transform this plasmid into an *E. coli* strain such as DH5α or BL21.
3. Use bacteria transformed with this plasmid to inoculate 25 mL of LB medium containing the appropriate antibiotic (e.g., ampicillin at 50–100 μg/mL) to select for the GST-fusion plasmid.
4. Incubate this culture overnight (16–18 h) in a shaking incubator (250 rpm, 37°C).

5. Add 25 mL of the overnight culture to 250 mL (1:10 dilution) of LB without antibiotic.
6. Grow bacteria in the shaker (250 rpm, 37°C) until the OD_{600} = 0.6 (~1.5–2 h).
7. Add 1 m*M* IPTG to induce expression of the fusion protein.
8. Continue incubating for 2–6 h (determined empirically for optimal production of each GST-fusion protein) in the shaking incubator (250 rpm, 37°C).
9. Centrifuge the samples at 13,500*g* at 4°C for 15 min to pellet bacteria.
10. Discard the supernatant.
11. Resuspend the bacterial pellet in 20 mL of ice-cold TT buffer containing 1 mg/mL of lysozyme and transfer solution to a 50-mL centrifuge tube.
12. Probe sonicate the sample for a 30-s burst with the bacterial preparation suspended in an ice bath.
13. Freeze sample in a dry ice–EtOH bath.
14. Thaw in water bath at 37°C.
15. Repeat **steps 12**, **13**, and **14** twice more (three times total).
16. Centrifuge samples at 39,000*g* at 4°C for 15 min.
17. Collect supernatant and save (GST-fusion protein is in this supernatant).
18. Remove 20 µL of the sample for preliminary analysis of the GST-fusion protein product on SDS-PAGE.

3.2.2. Purification of GST-Fusion Proteins

1. Hydrate 300 mg of GSH-agarose powder in 45 mL of ddH_2O for 30 min with rotation at room temperature.
2. Centrifuge the tube at 1000*g* for 5 min to pellet the hydrated agarose resin.
3. Aspirate the ddH_2O, leaving approx 2 mm above the GSH-agarose.
4. Equilibrate agarose by adding 20 mL TT.
5. Centrifuge at 1000*g* for 5 min to pellet agarose.
6. Aspirate, leaving approx 2 mm of TT above the GSH-agarose.
7. Add enough TT to make a 1:1 agarose:TT slurry.
8. Add sample (up to 20 mL) to GSH-agarose and mix by inversion at room temperature (30 min) or 4°C (2 h).
9. Centrifuge at 1000*g* for 5 min.
10. Remove sample, leaving approx 2 mm of buffer above the GSH-agarose (retain the sample to assay for proteins *not* adsorbed to GST-fusion protein).
11. Wash the resin twice with 6 mL of TT.
12. Centrifuge and aspirate the sample, leaving approx 2 mm of buffer above the GSH-agarose.
13. Wash the resin with 3 mL of 333 m*M* NaCl in TT.
14. Centrifuge and aspirate the wash as described in **step 12**.
15. Wash the resin with 6 mL of TT.
16. Centrifuge and aspirate the wash as described in **step 12**.
17. Add 2 mL of TT.
18. Transfer the mixture to a 0.8 × 4 cm Poly-Prep column (*see* **Note 6**).

19. Add 2 mL of TT to collect residual resin and transfer this to the column as well.
20. Aspirate the wash buffer from the surface of the settled resin using a 27-gage needle (*see* **Note 7**).
21. To elute the fusion protein, add 3 mL of 10 m*M* GSH in TT buffer to the resin in the Poly-Prep column. Rotate at room temperature for 10 min and collect the flow-through in a sterile tube.
22. Separate the free GSH from the fusion protein by dialysis or by several cycles of concentration-dilution in an Amicon stirred cell with a PM10 (>10,000 mol wt) filter (Amicon) using sterile PBS or TTB buffer.
23. Resuspend GST-fusion protein to a final volume of approx 2 mL and assay protein concentration using the Bio-Rad protein assay reagent (according to the manufacturer's instructions).

3.2.3. GST Pull-Down Assay (Specific Example of Saturation Binding of ^{35}S-α_2-AR-3i Loop to GST-Spinophilin 151-444[3])

1. Add increasing amounts of GST (control) or GST-fusion protein (*see* **Note 8**) to 50 μL of GSH-agarose (1:1 slurry equilibrated with TTB buffer) in a total volume of 235 μL.
2. Incubate with rotation for 2 h at 4°C.
3. Add 15 μL of ^{35}S-labeled α_2-AR-3i loop (31,000 cpm; estimated as 2.3×10^{-11} *M*).
4. Rotate 2–16 h at 4°C via inversion in a 1.5-mL microcentrifuge tube.
5. Collect the resin by microcentrifugation.
6. Wash resin three times with 1 mL of TTB by resuspension and recollection of the resin by microcentrifugation.
7. Determine the amount of the ^{35}S-labeled α_2-AR-3i loop bound to GST vs GST-fusion protein by elution into 1X Laemmli sample buffer and separation of samples by SDS-PAGE (*see* **Note 9**).
8. Treat gels with EN[^{3}H]ANCE (NEN) according to the manufacturer's instructions prior to drying the gel to facilitate detection of ^{35}S-labeled bands by autoradiography.
9. Determine the quantity of bound protein by cutting the appropriate gel bands (e.g., corresponding to the ^{35}S-labeled α_2-AR-3i loop) from the dried gel and counting in scintillation fluor in a β-counter.

4. Notes

1. NaCl can be added to this buffer up to 200 m*M*. However, the actual NaCl concentration should be titrated for each specific protein interaction studied. Detection of the interaction between spinophilin and α_2-AR-3i loops was optimal in the absence of NaCl.
2. A preparative gel has two lanes. One narrow lane is at the far edge for running a molecular weight standard. The remainder of the gel consists of a wide lane that spans the entire length of the gel, so the sample is continuous over the entire membrane once transferred. Prestained molecular weight markers should be run in the narrow lane to permit estimation of the approximate molecular weights of the proteins identified by gel overlay analysis.

3. Based on the concentration of methionine contributed to the [^{35}S]Met-labeling reaction by the rabbit reticulocyte lysate (5 μM) and the specific activity of the [^{35}S]Met radiolabel, we estimate that 300,000 cpm of G_{10}-α_2-3i loop represents 5–10 pmol of probe.
4. The number of washes, temperature of wash, and duration of wash can be modified to enhance the signal and minimize background.
5. For example, the pGEMEX-2 vector from Promega.
6. Use a pipet tip with the end cut off so as not to damage the agarose.
7. The wash buffer can be thoroughly aspirated by briefly touching the surface of the resin with a 27-gage needle attached to a vacuum source. The resin will not be disturbed because of the small diameter of the needle.
8. A quantity of 0.00034–34 µg of the fusion protein represents 2.34×10^{-11}–1.77×10^{-6} M of this particular fusion protein in the incubation. Examining binding over a range of GST-fusion protein concentrations allows examination of the saturability of binding to the GST-fusion protein. Specificity is evaluated by "competition" studies with various "competitors" to determine their ability to bind ^{35}S-α_2-AR-3i loop, and thus diminish the quantity of radioligand available for interaction with the GST-fusion protein. To assess the relative affinity of these "competitors" with spinophilin for the 3i loop, spinophilin itself is included as one of the competitors. Incubate 5 µg of GST or GST-Sp151-444 with 20 µL of GSH-agarose slurry for 2 h at 4°C. Increasing amounts (0–16 µL) of in vitro-translated Sp151-444 or other "competitors" can be mixed with rabbit reticulocyte lysate to reach a total volume of 16 µL and are then added to each incubation together with 6000 cpm (estimated as 9.2×10^{-12} M) of ^{35}S-labeled 3i loop with a total reaction volume at 120 µL. After 2 h of incubation at 4°C, resins are washed and eluted. The ^{35}S-3i loop in the eluate is separated and quantitated as described in **Subheading 3.2.3., steps 7–9**. In these experiments, GST-fusion protein (the "receptor") is in excess of the ^{35}S-α_2-AR-3i loop (ligand) and thus the relative affinity of competitors can be compared, but true thermodynamic constants for the interaction cannot be obtained.
9. The percent acrylamide in the gel, with or without a gradient, is determined empirically for optimally resolving proteins in eluates; use 12% SDS-PAGE to identify ^{35}S-labeled α_2-AR-3i loops.

References

1. Sarkar, D., Erlichman, J., and Rubin, C. S. (1984) Identification of a calmodulin-binding protein that co-purifies with the regulatory subunit of brain protein kinase II. *J. Biol. Chem.* **259,** 9840–9846.
2. Lester, L. B., Coghlan, V. M., Nauert, B., and Scott, J. D. (1996) Cloning and characterization of a novel A-kinase anchoring protein. AKAP 220, association with testicular peroxisomes. *J. Biol. Chem.* **271,** 9460–9465.
3. Prezeau, L., Richman, J. G., Edwards, S. W., and Limbird, L. E. (1999) The ζ isoform of 14-3-3 proteins interacts with the third intracellular loop of different α_2-adrenergic receptor subtypes. *J. Biol. Chem.* **274,** 13,462–13,469.

4. Richman, J. G., Brady, A. E., Wang, Q., Hensel, J. L., Colbran, R. J., and Limbird, L. E. (2001) Agonist-regulated Interaction between α_2-adrenergic receptors and spinophilin. *J. Biol. Chem.* **276,** 15,003–15,008.
5. Becamel, C., Alonso, G., Galeotti, N., et al. (2002) Synaptic multiprotein complexes associated with 5-HT_{2C} receptors: a proteomic approach. *EMBO J.* **21,** 2332–2342.
6. Hall, R. A., Ostedgaard, L. S., Premont, R. T., et al. (1998) A C-terminal motif found in the β_2-adrenergic receptor, P2Y1 receptor and cystic fibrosis transmembrane conductance regulator determines binding to the Na^+/H^+ exchanger regulatory factor family of PDZ proteins. *Proc. Natl. Acad. Sci. USA* **95,** 8496–8501.
7. Smith, F. D., Oxford, G. S., and Milgram, S. L. (1999) Association of the D2 dopamine receptor third cytoplasmic loop with spinophilin, a protein phosphatase-1-interacting protein. *J. Biol. Chem.* **274,** 19,894–19,900.
8. DeGraff, J. L., Gurevich, V. V., and Benovic, J. L. (2002) The third intracellular loop of α_2-adrenergic receptors determines subtype specificity of arrestin interaction. *J. Biol. Chem.* **277,** 43,247–43,252.
9. Wang, Q. and Limbird, L. E. (2002) Regulated interactions of the α_{2A} adrenergic receptor with spinophilin, 14-3-3ζ, and arrestin 3. *J. Biol. Chem.* **277,** 50,589–50,596.
10. Shieh, B. H., Zhu, M. Y., Lee, J. K., Kelly, I. M., and Bahiraei, F. (1997) Association of INAD with NORPA is essential for controlled activation and deactivation of *Drosophila* phototransduction in vivo. *Proc. Natl. Acad. Sci. USA* **94,** 12,682–12,687.

24

Receptor Knock-Out and Knock-In Strategies

Ikuo Matsuda and Atsu Aiba

Summary

Accumulating examples have demonstrated that knock-out and knock-in mice of G-protein-coupled receptors (GPCRs) are useful in elucidating physiological functions of the receptor in vivo. GPCR knock-out and knock-in are achieved by either (1) manipulation of the endogenous locus of the receptor gene or (2) transgenic expression of the modified receptor. Historically speaking, the first generation knock-outs made the best use of homologous recombination in embryonic stem (ES) cells and their totipotency to introduce the desired mutation into the endogenous receptor locus. In the second-generation knock-outs using the Cre/loxP system, the disruption of the receptor gene is cell-type specific or region-specific but is irreversible in principle. In contrast, transgenic expression in the receptor knock-out mice of the wild-type receptor protein under a tissue- and stage-specific promoter (conditional "rescue" of the receptor knock-out) can be easily applied to create "reversible" or "inducible" knock-out of the receptor. This is called the third generation knock-out. In the following sections, we introduce examples of the materials and methods based on our in vivo analyses of the metabotropic glutamate receptor-subtype 1 (mGluR1).

Key Words

Cre/loxP system, embryonic stem cell, first-generation knock-out, homologous recombination, knock-in, tetracycline-inducible system, third-generation knock-out, transgenic mice.

1. Introduction

Accumulating examples have demonstrated that G-protein-coupled receptor (GPCR) knock-out and knock-in mice are useful in elucidating physiological functions of the GPCR in vivo. In this chapter, we first overview the general considerations on the receptor knock-out and knock-in strategies. We place emphasis on two aspects: the first one is the comparison between manipulation of the endogenous locus of the receptor gene and transgenic expression of the

From: *Methods in Molecular Biology, vol. 259, Receptor Signal Transduction Protocols, 2nd ed.*
Edited by: G. B. Willars and R. A. J. Challiss © Humana Press Inc., Totowa, NJ

receptor; the second one is the recent development from global (first-generation) knock-out to conditional (second- and third-generation) knock-out.

1.1. Manipulation of the Endogenous Locus of the Receptor Gene vs Transgenic Expression of the Receptor

Receptor knock-outs and knock-ins are achieved by either (1) manipulation of the endogenous locus of the receptor gene or (2) transgenic expression of the modified receptor. For example, the endogenous receptor gene can be knocked out by using homologous recombination in embryonic stem (ES) cells or transgenic expression of the mutant form of the receptor under an appropriate promoter. Likewise, any desired mutations or cDNAs such as human homologs of the receptor can be introduced in frame into the endogenous locus of the receptor of interest by homologous recombination in ES cells to achieve receptor knock-in.

The manipulation of the endogenous receptor locus and the transgenic expression of the receptor have their own advantages and disadvantages as described below.

First, it is possible to recapitulate faithfully the endogenous expression pattern of the receptor when manipulating the endogenous locus. However, in the transgenic expression of the receptor, the receptor transgene is integrated randomly into the host genome. As a result, the specificity and the strength of the selected promoter of the transgene depend on where in the chromosome the transgene is integrated, that is, a "position effect" *(1)*. In addition, some other genes may be disrupted by the random integration of the transgene. By these ways, the random integration of the receptor transgene gives rise to unexpected twists of the phenotypes of the transgenic mice. To obtain a transgenic mouse line with expected phenotypes, tremendously labor-intensive screening of a number of the transgenic mouse lines are required. To circumvent the position effect, bacterial artificial chromosome (BAC) transgenes *(2)* or insulator elements *(3)* are available. It should be noted that by taking the best advantage of the position effect, it is possible to achieve unexpectedly useful variations of specificity and strength of the transgene expression using the selected promoter *(4,5)*.

Second, the transgenic expression of the receptor is easily applicable to achieve reversible (ON/OFF) expression of the receptor (*see* **Subheading 1.2.**).

1.2. From Global (First-Generation) Knock-Out to Conditional (Second- and Third-Generation) Knock-Out

For simplicity, only the receptor knock-out is discussed in the following subsection. However, in general, a similar discussion is applicable to the receptor knock-in as well.

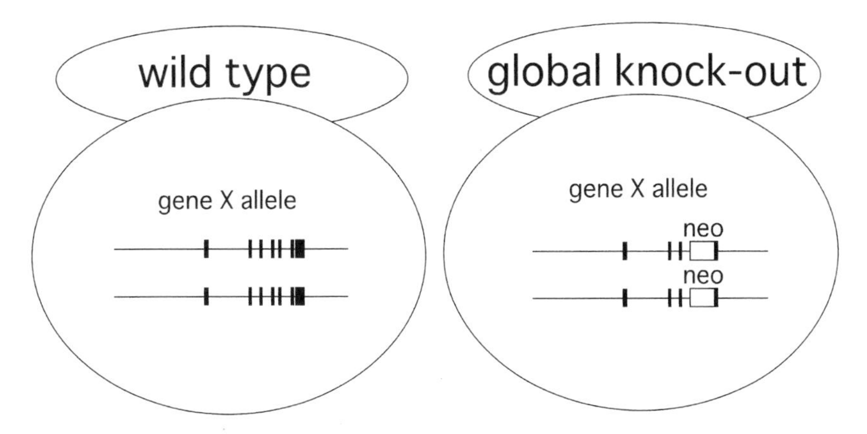

Fig. 1. Schematic representation of global knock-out. The disruption of the receptor gene X locus can be achieved by partial replacement of the gene X by an expression cassette for a positive selection marker such as neomycin-resistant gene *(neo)*.

Historically speaking, the first-generation knock-out made the best use of homologous recombination in ES cells and their totipotency to disrupt the function of receptor genes. This knock-out strategy has also been called "global knock-out," as the receptor gene locus is disrupted in all cells constituting the whole body from the initiation of development **(Fig. 1)**. Conventionally, the global knock-out of the receptor X gene is achieved by replacement of part of the gene X by an expression cassette for a positive selection marker such as a neomycin-resistant gene *(neo)* (**Fig. 1** and *see* **Notes**). Two problems in the global knock-out are as follows:

1. Except for cases in which the expression of the receptor is limited within a specific region and at a specific developmental stage, phenotypes of the receptor global knock-out are difficult to ascribe to a particular tissue or cell expressing the receptor at a particular stage. Because the receptor gene locus is disrupted from the initiation of the development, it is difficult to determine whether phenotypes of the receptor knock-out are the primary effects of receptor loss or the accumulation of secondary effects during development.
2. When the receptor knock-outs are embryonic lethal, it is impossible to analyze the in vivo function of the receptor protein in adults.

To circumvent these problems of global knock-outs, mice have been created in which conditional knock-out of the receptor can be achieved.

The second-generation conditional knock-out mice often involve the Cre/loxP system *(6)*. In this system, two lines of transgenic mice are prepared. One expresses the site-specific DNA recombinase of the bacteriophage P1, Cre, under a tissue- and stage-specific promoter. The second line has the gene locus of the receptor of interest flanked on both 5′- and 3′-sides by a loxP sequence, which is a target sequence of the Cre recombinase. The latter mouse can be created using homologous recombination in ES cells. Mating these two lines yields a Cre- and loxP-double positive line. In this line, the targeted receptor locus is disrupted by Cre/loxP-mediated deletion of the target chromosomal segment only in Cre-expressing tissues (or cells). The promoter selected to express the Cre recombinase in the first transgenic mice line directs where and when the disruption of the target receptor locus occurs in the whole body. Furthermore, embryonic lethality, if any, of global knock-outs can be circumvented by selecting a promoter that functions at a postnatal stage in the Cre-expressing transgenes.

The second-generation conditional knock-out mice can also be created by either (1) transgenic expression of the mutant form of the receptor in the wild-type mice under a tissue- and stage-specific promoter or (2) transgenic expression of the wild-type receptor protein under a tissue- and stage-specific promoter in the receptor knock-out mice (conditional "rescue" of the receptor knock-out) **(Fig. 2)**. We successfully applied the latter conditional rescue approach to the analyses of the metabotropic glutamate receptor-subtype 1 (mGluR1) function in cerebellar Purkinje cells as described in detail *(7)*.

In the second generation knock-out using the Cre/loxP system, the disruption of the target receptor locus is irreversible in principle. In contrast, transgenic expression of the dominant-negative form of the receptor or conditional rescue approach can be easily applied to create "reversible" or "inducible" knock-out of the receptor, which is called the third-generation knock-out **(Fig. 3)**. In the case of conditional rescue, the third generation knock-out mice can be created by introducing the tetracycline-inducible system as follows. Two transgene constructs are necessary. One transgene expresses the tetracycline-dependent transcription factor, tTA, under an appropriate region- or stage-specific promoter. The other transgene expresses a wild-type or modified receptor under the tetracycline responsive element (TRE). Introducing these two transgenes into the receptor knock-out mouse yields a tTA- and TRE-double transgenic mouse homozygous mutant for the endogenous receptor locus. Application of tetracycline or doxycycline to this line prevents tTA from binding to the TRE, resulting in no rescue. However, withdrawal of tetracycline or doxycycline from this line induces tTA binding to the TRE, leading to the expression of the receptor transgene and rescue of the receptor knock-outs. As described earlier, tTA could also be knocked-in to the endogenous receptor locus *(8)*.

In conclusion, the third-generation knock-out with conditional rescue is currently one of the most sophisticated knock-out strategies of the receptor. In the

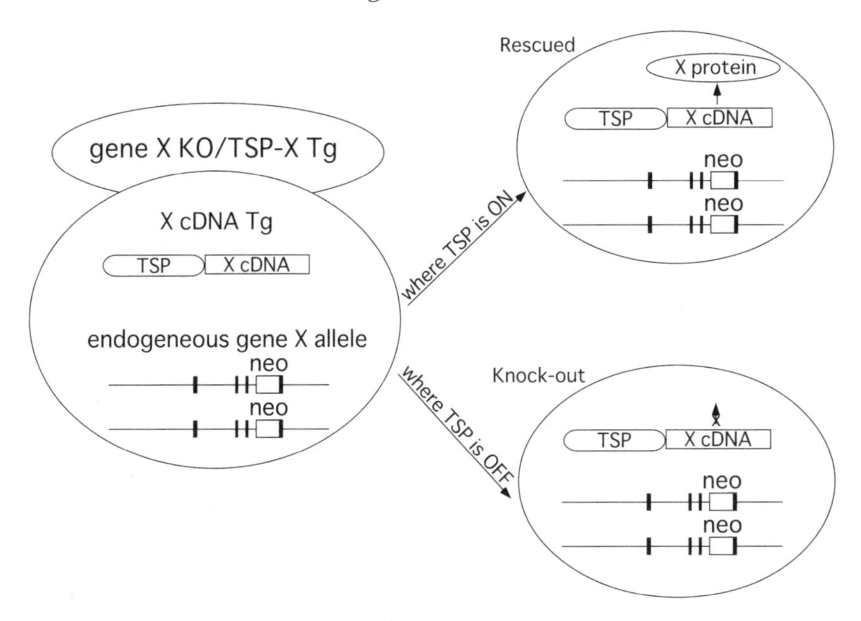

Fig. 2. Second-generation receptor knock-out: conditional "rescue" of the receptor knock-out. The transgene expressing a receptor X cDNA driven by a tissue-specific promoter (TSP) is introduced into the receptor gene X knock-out mouse, yielding the gene X KO/TSP-X Tg mouse. In this transgenic mouse line, the receptor X protein is expressed in the tissues where the TSP is active (ON), leading to "rescue" of the tissues from knock-out phenotype. However, the remaining tissues in which the TSP is inactive (OFF) are left "knocked-out."

following sections, we introduce examples of the materials and methods based on our in vivo analyses of the mGluR1.

2. Materials

2.1. Generation of the Receptor Global Knock-Out Mice (in Our Case, mGluR1 *Knock-Out Mice) (9)*

1. Source of the genomic fragment containing the receptor gene locus of interest (*see* **Note 1**). We used EMBL3 phage DNA library prepared from D3 ES cells ***(10)*** for screening and cloning an *mGluR1* genomic clone.
2. Molecular cloning reagents for restriction mapping (*see* **Note 2**). Please refer to standard molecular biology protocols ***(11)***.
3. ES cells (*see* **Note 3**). We used D3 ES cells.
4. Reagents for ES cell culture. Detailed discussions on ES cell culture are beyond the scope of this chapter. Please refer to ***(12)***. *See* **Note 3** for some basic points.
5. Electroporator. We used the Gene Pulser II (Bio-Rad).

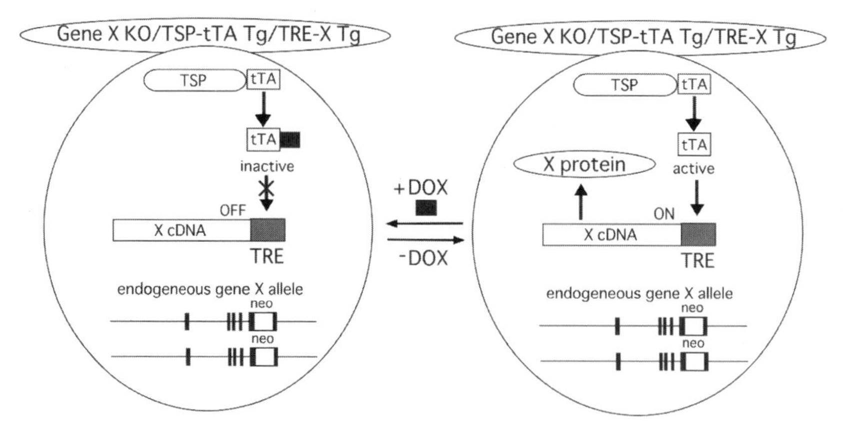

Fig. 3. Third-generation receptor knock-out: a combination of conditional "rescue" and the tetracycline-inducible system. Two transgenes, TSP-tTA and TRE-X, are introduced into the receptor gene X knock-out mouse to yield the gene X KO/TSP-tTA Tg/TRE-X Tg mouse. The TSP-tTA Tg expresses tTA under the TSP. The TRE-X Tg expresses the receptor X cDNA under the TRE. In this transgenic mouse line, application of doxycycline (DOX; the *closed black box*) inactivates tTA and suppresses the receptor expression from the TRE-X Tg. On the other hand, removal of DOX activates tTA in the tissues where the TSP is active (ON), restoring the receptor expression from the TRE-X Tg and leading to "rescue" from knock-out phenotype. TSP, tissue-specific promoter.

6. Drugs for positive selection of homologous recombinants. We used G418.
7. Microinjection. Detailed description of microinjection techniques is beyond the scope of this chapter. Please refer to ***(13)***.
8. Reagents for Southern, Northern, and Western blot analyses. Please refer to standard molecular biology protocols ***(11)***.

2.2. Generation of the Region- and Stage-Specific Conditional Rescue Mice (in Our Case, mGluR1*-Rescue Mice) (7)*

1. cDNA for the receptor of interest. We used the rat mGluR1a cDNA (a gift from Dr. Nakanishi).
2. Promoter of interest to express the receptor transgene for conditional rescue. We used the cerebellar Purkinje cell-specific *L7* promoter ***(14)***.
3. Microinjection. Detailed description of microinjection techniques is beyond the scope of this chapter. Please refer to ***(13)***.
4. Reagents for Western blot analysis. Please refer to standard molecular biology protocols ***(11)***.
5. Antibody that recognizes the receptor of interest. We used polyclonal antibodies to rat mGluR1 (Upstate Biotechnology).

3. Methods

3.1. Generation of the Receptor Global Knock-Out Mice (in Our Case, mGluR1 Knock-Out Mice) (9)

1. Obtaining the genomic fragment containing the receptor gene locus of interest (*see* **Note 1**). The genomic fragment containing exon 1 of *mGluR1* was isolated from a genomic EMBL3 phage DNA library prepared from D3 ES cells ***(10)*** after screening with a probe corresponding to amino acids 1–200 of the mouse mGluR1 protein, generated by the polymerase chain reaction (PCR).
2. Mapping and cloning of the receptor gene locus (for the mouse *mGluR1* gene; *see* **Note 2**). The exon 1 encodes amino acids 1–233 of the mouse mGluR1 protein. The amino acid sequence deduced from the DNA sequence *(unpublished data)* was identical to the published rat amino acid sequence in the corresponding region ***(15)***.
3. Construction of the targeting vector. The targeting vector was prepared in a quatrimolecular ligation reaction, using a 9.0-kb *Bam*HI–*Sal*I fragment located 5′ region of exon 1, a 1.8-kb *Sal*I–*Eco*RI fragment containing a *neo* gene driven by the *pgk1* promoter (a gift from Dr. Rudnicki), a 1.7-kb *Eco*RI-*Hin*dIII fragment located 3′ of exon 1, and the plasmid pBluescript digested with appropriate restriction enzymes. The targeting vector was designed to delete a 2.1-kb fragment from the *mGluR1* gene, including the 3′ part of exon 1 that encodes amino acids 1–233 of the mGluR1 protein.
4. Production of mutant mice (in our case, *mGluR1* mutant mice; *see* **Note 3**). D3 ES cells (gift from Dr. R. Kemler) were transfected with 50 μg of linearized targeting vector by electroporation (Bio-Rad gene pulser set at 800 V and 3 μF). G418 selection (150–175 μg/mL) was applied 24 h after transfection, and G418-resistant colonies were isolated during d 7–10 of selection. Genomic DNA from these clones was digested with *Bam*HI and *Xba*I and hybridized with a 600-bp DNA fragment 3′ to exon 1. Chimeric mice were generated as described ***(16)***. The contribution of ES cells to the germline of chimeric mice was determined by breeding with C57BL/6 mice and screening for agouti offspring. Germline transmission was confirmed by Southern blot analysis of tail DNA, and mice heterozygous for the mutation were interbred to homozygosity. Homozygous females and males are fertile, but are poor breeders. To generate litters from homozygous intercrosses, we transferred in vitro fertilized eggs to a pseudopregnant mother as described elsewhere ***(13)***.
5. Genotyping (*see* **Note 4**). Initial screening of mice was done by Southern blot analysis of the tail DNAs. Subsequently, mice were typed by PCR analysis with a set of *neo* primers and a set of primers from the deleted region of the mutant *mGluR1* allele.

3.2. Generation of the Region- and Stage-Specific Conditional Rescue Mice (in Our Case, Transgenic Mice That Expressed mGluR1a Under the Control of the Cerebellar Purkinje Cell-Specific L7 Promoter [mGluR1-Rescue Mice]) (7)

1. Construction of the transgenic vector. A 3.7-kb *Sac*II–*Fsp*I fragment of rat mGluR1a cDNA containing a 26-bp 5′ untranslated region, a 3597-bp coding region, and a 36-bp 3′ UTR was introduced into exon 4 of the *L7* gene cassette.

2. Generation of the transgenic mice (*see* **Note 5**). We obtained eight independent *L7-mGluR1* transgenic founder mice by microinjecting the transgene into the pronuclei of fertilized eggs derived from *mGluR1* heterozygous mutants. The mGluR1-rescue mice were obtained by breeding *mGluR1* heterozygous mutants with *L7–mGluR1* transgenic mice.
3. Confirmation of the transgene expression (*see* **Note 5**). The cerebellum-restricted expression of the transgene was examined by Western blotting with polyclonal antibodies against rat mGluR1 (Upstate Biotechnology). One line expressed the *L7–mGluR1* transgene in the cerebellum and not in the cerebral cortex.

4. Notes

1. The genomic fragments can be obtained by screening genomic libraries such as BAC libraries. Ideally, the mouse strain of the genomic libraries should be identical to that of ES cells to be used. Without published data, it is generally believed that mismatch of genetic backgrounds between the library and ES cells decreases the efficiency of homologous recombination between a targeting vector and the ES genomic DNA.

   Genomic fragments to be used should contain the exon(s) to be targeted and regions surrounding the exon(s) (at least several kilobases upstream and downstream of the exon[s]). The exon(s) to be targeted should be selected so that the disruption or deletion of the exon(s) makes the receptor protein nonfunctional. For example, an exon containing the initial methionine is a candidate. In the case of GPCRs whose coding regions are intronless, the whole coding exon could be deleted.

   To obtain the necessary genomic fragment, screening of phage or cosmid libraries with a radiolabeled probe containing a target exon has been a general method. However, screening of commercially available BAC libraries is a convenient alternative. First, the average insert size of BAC libraries (usually 100–150 kb) is larger than that of phage or cosmid libraries, which gives us more opportunities to obtain the desired genomic fragment to be used. Second, unlike phage libraries in which positive clones should be purified to be identified, positive BAC clones need not to be purified and are readily available by order after the library has been screened by hybridization.
2. Detailed discussions on targeting vector design are beyond the scope of this chapter ***(17)***. Here we have only comments on basic points. The simplest targeting vector should contain the exon(s) to be disrupted by a positive selection marker and homologous regions surrounding the exon(s) (at least several kilobases upstream and downstream of the exon[s]). As a positive selection marker, *neo* is frequently used and G418 as a selection drug against *neo*. Addition of a negative selection maker (e.g., HSV-thymidine kinase [TK]) to either terminus of the homologous regions may decrease the occurrence of nonhomologous recombination in ES cells. If coding sequences for some marker proteins, such as β-galactosidase or green fluorescent protein (GFP), are knocked-in in frame into the targeted exon of the receptor, expression of the marker proteins is expected to

recapitulate the endogenous receptor protein. In this case, disruption of the receptor gene locus and monitoring of its expression profile can be achieved simultaneously.

Generally, to construct a targeting vector, restriction mapping or sequencing of the genomic clone is necessary. However, recent progress in genome projects enables us to obtain these sequence information *in silico* without restriction mapping or sequencing.

The construction strategy of the targeting vector should include consideration of a Southern blot strategy to identify ES homologous recombinants. For definite confirmation of homologous recombinants, two outer probes (in the 5′ upstream or 3′ downstream of the targeted genomic region) and an internal probe (e.g., *neo* probe) should be used. It should be noted that CG methylase-sensitive restriction enzymes, for example, *Sma*I, *Cla*I, should not be used for the Southern blot analysis.

3. Detailed discussions on ES cell culture are beyond the scope of this chapter ***(12)***. Here we have only comments on basic points. The most important point in ES cell culture is to maintain the totipotency of ES cells. Feeder cells and leukemia inhibitory factor (LIF) are usually used to prevent ES cells from differentiating. Some ES cell lines can be cultured without feeder cells ***(18)***, which is both technically and economically advantageous.

   There are at least two points to consider on the genetic background of ES cells. First, the genetic background of genomic fragments to be used for targeting vectors is ideally identical to that of the ES cells. Second, the genetic background of ES cells to be chosen depends on what experimental purpose mutant mice derived from the ES cells are to be used for. For example, if the mutant mice will be used to examine their learning ability, C57BL/6 background may be better than a 129 background, because there is a large amount of data on learning tests involving C57BL/6 mice.

   Theoretically, after positive selection, colonies that appear are either homologous recombinants or random integrants. These two can be distinguished from each other by Southern blot analysis of their genomic DNA. Sometimes homologous recombinant colonies are contaminated with wild-type ES cells that evade exposure to positive selection drug. Therefore, in Southern blot analysis of ES genomic DNA, care should be taken in comparing the intensity of the band from the wild-type allele with that from the mutant allele.

4. Initial genotyping of mice should be done by Southern blot analysis of tail DNA; subsequently, for convenience, genotyping of mice can be determined by PCR analysis.

   In genotyping by PCR, it is important that an appropriate internal control should be included to assess the quality of the genomic DNA and PCR reaction. Also, in the transgenic expression of the receptor, the transgene receptor should be distinguished from the endogenous one (if any). For example, in our case with *mGluR1*-rescue mice ***(7)***, PCR genotyping was performed for *L7* transgene and *mGluR1* locus in the following way **(Fig. 4)**.

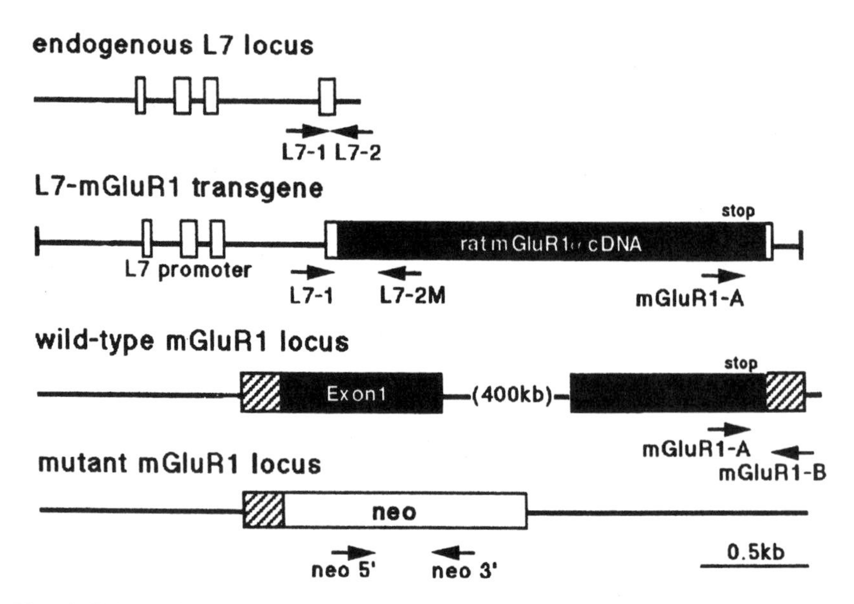

Fig. 4. Schematic representation of the strategy for genotyping by PCR: mGluR1-rescue mice. *Hatched boxes* are parts of the exons that are noncoding. The size of each PCR primer is arbitrary. For the *L7* transgene, PCR with primers L7-1 and L7-2 yields a 190-bp fragment from the endogenous *L7* locus, whereas that with primers L7-1 and L7M-2 yields a transgene-specific 480-bp fragment. For the *mGluR1* locus, PCR with primers mGluR1-A and mGluR1-B yields a 230-bp fragment from the endogenous *mGluR1* locus. The mGluR1-B primer anneals with the *mGluR1* noncoding region which is not contained in the *L7–mGluR1* transgene. In addition, PCR with primers *neo* 5′ and *neo* 3′ yields a 450-bp fragment from *neo* gene in the targeted allele.

For the *L7* transgene, PCR with primers L7-1 and L7-2 yields a 190-bp fragment from the endogenous *L7* locus, whereas that with primers L7-1 and L7M-2 yields a transgene-specific 480-bp fragment. These two PCR reactions are performed in a tube with tail genomic DNA and the above three primers. In this way, the endogenous 190-bp fragment can be used as an internal control for the quality of the genomic DNA and the PCR reaction.

For the *mGluR1* locus, PCR with primers mGluR1-A and mGluR1-B yields a 230-bp fragment from the endogenous *mGluR1* locus. The mGluR1-B primer anneals with the *mGluR1* noncoding region which is not contained in the *L7–mGluR1* transgene. This ensures that the 230-bp fragment is derived from the endogenous *mGluR1* locus. In addition, PCR with primers *neo* 5′ and *neo* 3′ yields a 450-bp fragment from the *neo* gene in the targeted allele.

5. It is critical to select a particular line of transgenic mice that are the most appropriate for research purposes. Focusing on the most appropriate line not only alle-

viates the financial and spatial burden of maintaining the transgenic mice but also helps speedy and efficient analyses of the mice. The efficiency of screening transgenic mice depends heavily on the selection criteria.

Primary screening of transgenic mice usually involves examination of the expression level of the receptor of interest in a particular tissue. This is usually done quantitatively by Northen blot analysis or Western blot analysis with a specific antibody against the receptor of interest. Sites of expression of the transgene are examined by *in situ* hybridization or immunohistochemistry

There are two caveats for screening by the expression level. First, the specificity of the antibody may sometimes be unreliable; strictly speaking, the specificity of the antibody should be confirmed using the knock-out mice as negative controls. This problem can be overcome by using a well-characterized peptide (or protein) tag, either by fusing the tag with the receptor transgene or by knocking-in the tag into the endogenous receptor locus. By using well-characterized tags, we can easily distinguish the endogenous receptor from the receptor transgene. Second, the expression level of the transgene may not correlate with expression of its function. For example, sometimes the transgenic mice exhibit apparently no phenotypes even though the transgene expression is detected either at mRNA or protein levels. In this case, it is difficult to determine whether the apparent absence of phenotypes is due to insufficiency of the transgene protein or whether the transgene expression is functionally unrelated to the phenotypes.

To overcome these caveats, screening criteria other than expression level are often crucial. It is not only convenient but also elegant that the other criteria are noninvasive to the mice, because they enable us to examine the expression level and the function of the transgene at the same time. For example, in our case with *mGluR1*-rescue mice *(7)*, rescue by the *L7–mGluR1* transgene of the ataxic phenotype of *mGluR1* knock-out mice reveal that the transgene is expressed at sufficient level for function. Likewise, such criteria as rescue of embryonic lethality *(5)* or change in coat color *(8)* enable us to examine the expression level and the function of the transgene at the same time.

In addition to the expression level of the transgene, germline transmission of the transgene should be confirmed by genomic Southern blot of transgenic mice to establish the transgenic mice line.

References

1. Wilson, C., Bellen, H. J., and Gehring, W. J. (1990) Position effects on eukaryotic gene expression. *Annu. Rev. Cell. Dev. Biol.* **6,** 679–714.
2. Heintz, N. (2001) BAC to the future: the use of bac transgenic mice for neuroscience research. *Nat. Rev. Neurosci.* **2,** 861–870.
3. Burgess-Beusse, B., Farrell, C., Gaszner, M., et al. (2002) The insulation of genes from external enhancers and silencing chromatin. *Proc. Natl. Acad. Sci. USA* **99,** 16,433–16,437.
4. Tsien, J. Z., Chen, D. F., Gerber, D., et al. (1996) Subregion- and cell type-restricted gene knockout in mouse brain. *Cell* **87,** 1317–1326.

5. Nakazawa, K., Quirk, M. C., Chitwood, R. A., et al. (2002) Requirement for hippocampal CA3 NMDA receptors in associative memory recall. *Science* **297,** 211–218.
6. Sauer, B. (1993) Manipulation of transgenes by site-specific recombination: use of Cre recombinase. *Methods Enzymol.* **225,** 890–900.
7. Ichise, T., Kano, M., Hashimoto, K., et al. (2000) mGluR1 in cerebellar Purkinje cells essential for long-term depression, synapse elimination, and motor coordination. *Science* **288,** 1832–1835.
8. Shin, M. K., Levorse, J. M., Ingram, R. S., and Tilghman, S. M. (1999) The temporal requirement for endothelin receptor-B signalling during neural crest development. *Nature* **402,** 496–501.
9. Aiba, A., Chen, C., Herrup, K., Rosenmund, C., Stevens, C. F., and Tonegawa, S. (1994) Reduced hippocampal long-term potentiation and context-specific deficit in associative learning in mGluR1 mutant mice. *Cell* **79,** 365–375.
10. Silva, A. J., Stevens, C. F., Tonegawa, S., and Wang, Y. (1992) Deficient hippocampal long-term potentiation in alpha-calcium-calmodulin kinase II mutant mice. *Science* **257,** 201–206.
11. Sambrook, J. and Russell, D. W. (2001) *Molecular Cloning: A Laboratory Manual*, Cold Spring Harbor Laboratory Press, Cold Spring Harbor, New York.
12. Wurst, W. and Joyner, A. L. (1993) Production of targeted embryonic stem cell clones, in *Gene Targeting: A Practical Approach* (Joyner, A. L., ed.), IRL Press, Oxford, pp. 33–62.
13. Hogan, B., Constantini, F., and Lacy, E. (1986) *Manipulating the Mouse Embryo*, Cold Spring Harbor Laboratory Press, Cold Spring Harbor, New York.
14. Oberdick, J., Smeyne, R. J., Mann, J. R., Zackson, S., and Morgan, J. I. (1990) A promoter that drives transgene expression in cerebellar Purkinje and retinal bipolar neurons. *Science* **248,** 223–226.
15. Masu, M., Tanabe, Y., Tsuchida, K., Shigemoto, R., and Nakanishi, S. (1991) Sequence and expression of a metabotropic glutamate receptor. *Nature* **349,** 760–765.
16. Bradley, A. (1987) Production and analysis of chimeric mice, in *Teratocarcinomas and Embryonic Stem Cells: A Practical Approach* (Robertson, E. J., ed.), IRL Press, Oxford, pp. 113–151.
17. Hasty, P. and Bradley, A. (1993) Gene targeting vectors for mammalian cells, in *Gene Targeting: A Practical Approach* (Joyner, A. L., ed.), IRL Press, Oxford, pp. 33–62.
18. Niwa, H., Miyazaki, J., and Smith, A. G. (2000) Quantitative expression of Oct-3/4 defines differentiation, dedifferentiation or self-renewal of ES cells. *Nat. Genet.* **24,** 372–376.

25

Statistical Methods in G-Protein-Coupled Receptor Research

Pat Freeman and Domenico Spina

Summary

In this chapter we provide an introduction to statistical methods appropriate in G-protein-coupled receptor research, including examples. Topics covered include the choice of appropriate averages and measures of dispersion to summarize data sets, and the choice of tests of significance, including *t*-tests and one- and two-way analysis of variance (ANOVA) plus posttests for normally distributed (Gaussian) data and their nonparametric equivalents. Techniques for transforming non-normally distributed data to more Gaussian distributions are discussed. Concepts of statistical power, errors, and the use of these in determining the optimal size of experiments are considered. Statistical aspects of linear and nonlinear regression are discussed, including tests for goodness-of-fit to the chosen model and methods for comparing fitted lines and curves.

Key Words

Correlation, power, probability, regression, significance, statistics.

1. Introduction

Experimental results of all kinds are prone to uncertainty or error, particularly in biology. This error can be minimized by applying the accepted criteria for good experimental design, such as making sure that data values are independent, assigning treatments at random to experimental units, making sure that sources of bias are eliminated, controlling for known sources of variability, and using an experimental approach with a wide range of applicability. It is not possible to eliminate error completely, however.

When performing an experiment to investigate a phenomenon or to measure some quantity it becomes important to have some measure of this error, which is where statistical methods come in. For the most part, biologists use statistics

From: *Methods in Molecular Biology, vol. 259, Receptor Signal Transduction Protocols, 2nd ed.*
Edited by: G. B. Willars and R. A. J. Challiss © Humana Press Inc., Totowa, NJ

to describe their data by calculating summary statistics (averages) and quoting quantities that describe variability in these data (measures of dispersion), to look for relationships between data values using regression and correlation techniques, and also to evaluate the effects of experimental manipulations, using tests of significance. It should not be forgotten, however, that these are tests of statistical and not scientific significance; they cannot conceal poor results and should not be used as a substitute for scientific argument *(1)*. They can help to interpret results of experiments but it is vital to choose the appropriate method for statistical analysis of results at the time of designing the experiment, rather than (as happens all too often) at the end of the project. Then the experimental protocol can be designed to make sure the appropriate data are collected to allow the chosen analysis.

It should be remembered that all statistical methods are based on the calculation of probabilities, rather than certainties. Thus, statistical results cannot prove that any scientific hypothesis is true, but only provide some justification for believing that it might not be false. However, it will be expected that most biologists seeking to publish their work will include statistical results to satisfy journal editors and reviewers. The purpose of this chapter is to discuss how to choose appropriate methods to analyze data, and how to use statistical techniques to improve experimental design.

2. Summary Statistics

Implicit in the quoting of an average or summary value for a quantity measured in an experiment is the idea that this represents the true value for this average—the population average. Because experiments can never measure the entire population they have to rely on measuring the average for a sample. Statistical methods assume that this sample is randomly chosen from the population, so when designing your experimental protocol you should consider whether or not you violate this assumption and try to avoid bias in both sample selection and measurement. Most biological data fit a Gaussian or normal (as in "pattern," not "usual") distribution, which has defined parameters with particular properties **(Fig. 1)**. Statistical techniques that assume that data fit a particular distribution are known as parametric techniques and are the most powerful and widely used. Nonparametric statistics, on the other hand, do not make any assumption about the shape of data distribution and are thus universally applicable.

2.1. Averages—Measures of Central Tendency or Position

Given the choice, most scientists calculate the arithmetic mean as an appropriate average value for any data set. This is normally appropriate as a summary as long as the data fit a normal distribution. Fortunately, most biological data do

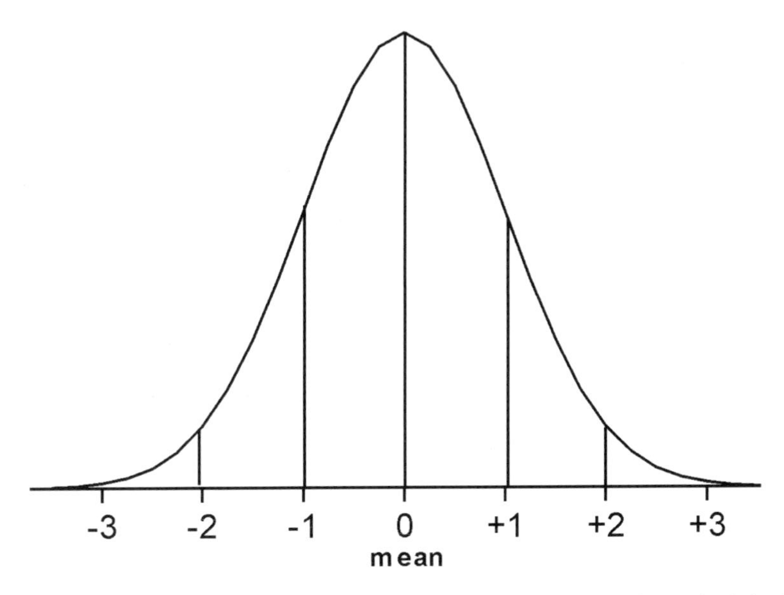

Fig. 1. The normal (Gaussian) distribution calibrated in terms of standard deviation. Sixty-eight percent of data values in such a distribution are expected to fall within ± 1 SD of the mean, 97% within ± 1.96 SD, and 99.7% within ± 3 SD.

Table 1
Some Useful Transformations

Type of data	Distribution	Normalizing transform
Count	Poisson distribution	$\sqrt{C}$
Proportion or %	Binomial distribution	Arcsine $\sqrt{P}$
Measurement	Log normal	Log (M)
Duration		$1/D$

not deviate too far from normality and the mean is usually a good average. If data are clearly not normally distributed, as has been shown, for example, for IC_{50} and K_i values from ligand-binding experiments or EC_{50} values from agonist concentration–effect curves ***(2,3)***, it is appropriate to transform them (in these cases using a logarithmic transformation) so that they fit the normal curve and then perform the statistical analysis on the transformed data. **Table 1** lists this and some other transformations that have been found to be useful in normalizing the distribution of different types of data before undertaking statistical analysis. You then need to transform data back to the original units to quote final results, of course.

Other data types commonly used in pharmacology that may present problems in analysis include data expressed as percentages, fractions, or multiples of control values. These are paired data values, but the differences between pairs will be larger as the control value increases, hence the usefulness of expressing the difference as a ratio. But ratios are asymmetric, so it makes more sense to analyze a logarithmic transformation of this type of data (*see* **ref. *4*** and **Subheading 3.1.1.**). Alternatively, percentages between 0 and 100% follow a binomial distribution and for data in this range the arcsine$\sqrt{P}$ transformation may be used to convert to a more Gaussian distrbution.

Transformation is not possible for certain other data types that are clearly not normally distributed—for example, discontinuous data, category data, or heavily skewed continuous data—and here too the arithmetic mean is not a good summary statistic. It is better in these cases to use the median (the middle value) of the data set, as this is less affected by outliers, or even the mode or modes (the most frequent value[s]) as an appropriate summary.

2.2. Measures of Dispersion

Having selected an average, it remains to choose a measure of dispersion to describe the data distribution further. For normally distributed data, it is possible to use an estimate of the population variance (the sum of squares of deviations from the mean, divided by the number of data values minus one) to describe the distribution. This is not very convenient, as the units of variance are the square of the data units. More conventionally we use the standard deviation (SD, the square root of the variance) to indicate the spread of values in a normally distributed data set. In fact, SD has a precise meaning defined by the Gaussian curve; 68% of data values in such a distribution are expected to fall within ± 1 SD of the mean, 97% within ± 1.96 SD, and 99.7% within ± 3 SD (**Fig. 1**). Thus, the larger the SD, the more scattered the data, so that the sample SD describes the variability in a particular set of data. As the size of the data set increases, the sample SD approaches the population SD (σ).

If you wish to estimate the true value of a particular parameter you may make repeated determinations in separate experiments and calculate a mean. To describe the distribution of means, you can then use an equivalent measure of dispersion, the standard error of the mean (SEM, calculated as the SD divided by the square root of n). As the size of the data set increases, the SEM approaches zero as the sample mean approaches the population mean. Thus a small SEM may not represent tight data, just a large sample size.

So, should you quote SD or SEM? After all, you can calculate both for any data set. Many experimenters prefer SEM because it is clearly smaller than the SD, but it should be remembered that it is describing a different property of the data set—the variability in the mean, not the spread of data values. As a general rule it is usually more appropriate to use mean ±SD when summarizing a single experiment, so that the reader knows how variable the values were, and mean ± SEM when presenting results from a series of repeated experiments because you are then estimating how close the sample mean is to the population mean.

Some (mainly clinical) journals prefer the use of confidence intervals (CI) as summary statistics rather than SDs and SEMs and probability *(p)* values *(5)*. These give a better idea of the (im)precision of the study estimates as generalized to the wider population. A 95% CI for a mean (the most frequently chosen value for CI) expresses the range of values between which you can be 95% certain that the population mean lies. Certainly it is easy for the reader to grasp an idea of variation in the data from the 95% CIs, which are expressed in the same units as the original data. For a population mean the 95% CI is calculated as the mean ± 1.96SEM but for experimental (i.e., real) data it is necessary to invoke the *t*-distribution to correct for small sample size (*see* **Example 1**):

$$95\%\ \text{CI} = \text{sample mean} \pm t_{(0.05,n)}\ \text{SEM}$$

Example 1.
Calculation of 95% confidence intervals.

Blood pressure measurements for a group of 41 randomly selected students yieled a mean of 123.4 mm Hg with SD of 14.93.
The *t*-value for this group size is 2.021; thus the 95% CI is 123.4 – (2.021 × 14.93/√41) to 123.4 + (2.021 × 14.93/√41) or 118.7–128.1 mm Hg.

This would be quoted as 123.4 (118.7–128.1) mm Hg.

A random sample of 10 students was drawn from this group and their blood pressures used to perform a similar calculation. The mean value was 110.5 mm Hg with SD of 16.76.
The *t*-value for $n = 10$ (9 df) is 2.262 and so the mean and 95% CI in this case are calculated as 110.5 (98.5–122.4) mm Hg, wider than that found from the larger group.

If you are using the median or mode as an average (i.e., implying that you do not consider the data to be Gaussian) you should not quote SD or SEM as measures of dispersion. The simplest such measure in these cases is the range of data values. If using the median, which is actually the location of the second quartile, you can calculate the first and third quartiles for your data and present your data as median (Q2) and interquartile range (Q1–Q3). Either of these approaches has the advantage of making the direction of any skew in the data values obvious to the reader. (*See* **Example 2**.)

Example 2.
Calculation of the median, range, and interquartile range for skewed data.

In an experiment measuring the time taken for control mice to respond to a standard noxious stimulus (being placed on a hot plate at 49°C) the following response time (s) were obtained: 29.4, 14.8, 60.0, 29.2, 17.5, 19.5, 29.0.

The median response time is 29.0 s, the range is 14.8–60.0 s, and the interquartile range is 17.5–29.4 s. In this case, the interquartile range is more indicative of the direction of skew, but the range clearly shows the extreme outlier.

An alternative to quoting these numbers would be to draw a box plot of the data where the median, Q1, Q3, and the range could be shown.

2.3. Are Your Data Normally Distributed?

If there is no published consensus on the matter, deciding if your data are normally distributed can be a problem. It may be helpful to inspect the distribution by eye using a dot plot or box and whisker diagram (there are rarely enough data to draw a meaningful histogram) and come to a conclusion. Also one can have a good idea of whether or not particular data types are expected to be normally distributed from their nature—continuous measurements are usually normally distributed whereas category data such as scores are not. The experimental protocol may lead to skewed data if there are prohibited values or cutoff points imposed by the experimental procedure. In **Example 2** above, a condition of the Home Office Project Licence was that all mice should be removed from the hot plate at 60 s, so that no value in excess of this was possible, even when some mice had not reacted by this time. In some experiments the range of possible values is dictated by physiological factors.

Statistical software may offer one or more tests for normality such as the Kolmogorov–Smirnov, Levene, or Wilk–Shapiro tests. However, a problem with these is that they give reliable results only when applied to large data sets (typically $n > 30$), and the small group sizes most commonly used in real experiments violate their assumptions *(3)*.

3. Tests of Significance

It is conventional to use tests of statistical significance to help assess the scientific significance of differences seen between treatment groups in experiments. The general procedure for such a test is to start by defining a null hypothesis—that there is no difference between treatment groups—and then perform the test to calculate the probability that a result as extreme as that observed experimentally could have arisen by chance. If that probability is small—conventionally one takes $p < 0.05$ as the critical probability (α)—then it can be assumed that the probability of the null hypothesis being true is also small (i.e., <5%) and that by implication there is a difference between groups. This is an entirely arbitrary level at which to decide to reject the null hypothesis, due originally to Fisher, and any other level may be chosen. Obviously, the smaller the value of p obtained in any case, the less likely it is that the null hypothesis is true, but it should be remembered that extreme values may occur in any experiment by chance and do not necessarily invalidate it. Statistical reasoning is all about probabilities, not certainties (*see* discussion of statistical power, and type 1 and type 2 errors in **Subheadings 5.1.** and **5.2.**).

Choosing a statistical test is fairly straightforward. If data values are normally distributed then tests assuming this may be used. In other cases it is safer to rely on results of tests that make no assumptions about the shape of the data's distribution; these are referred to as nonparametric tests or distribution-free methods. Several statistics texts provide flow charts or tables to guide you through selection of an appropriate test for the type of data you have (good examples are found in **refs.** ***6*** and ***7*** and the text by Dyson ***[8]*** is devoted entirely to this subject). Several statistics software packages provide some guidance; of these, GraphPad Instat™ is probably the most user-friendly. We discuss applications and limitations of the most widely used tests below.

3.1. Tests Based on the Normal Distribution

3.1.1. Tests for Two Groups

For the comparison of two groups of normally distributed data values, t-tests are appropriate. These were devised to be applicable to small group sizes, and different procedures are available for applying them to single sample data (comparing observed data with known values), to two independent treatment groups or to treatment groups where the values are paired (either because you have repeated measures data, or observations on paired individuals). The t-statistic is calculated as the ratio of the difference between group means to the pooled standard error of that difference, calculated by the method appropriate to the t-test procedure you are using. It assumes that data fit the normal distribution and that variances in the two groups are the same. A variant of the independent samples t-test (Welch's test) has been developed for cases in which the variances are not the same; this is offered by several statistics packages. In cases in which you have paired data expressed as a percent or ratio of control values (*see* **Subheading 2.1.**), Motulsky ***(4)*** suggests performing a t-test of the ratios by converting the data to logarithms and using a paired t-test to find the probability of obtaining a ratio as far from 1.0 (i.e., no difference) as observed in your experiment.

The probability of the calculated *t*-value arising by chance (*p*) depends on the degrees of freedom (df, best explained as the number of observations that are free to vary; $[n - 1]$ for a paired test and $[n_1 + n_2 - 2]$ for an unpaired test) and can be represented by the area under the tail of the curve for the appropriate value of *t*. Because the distribution is symmetrical about 0, there are both positive and negative values of *t* and hence two tails and two areas to be added together to obtain a value for *p*. The sign of *t* obtained by calculation depends on the direction of the difference between the means and is disregarded in assessing *p*. It is conventional in biology not to specify in advance the direction of this difference, that is, to perform a two-tailed test. This is regarded as the robust way to proceed. However, it is sometimes appropriate to predict in advance the direction of difference and perform a one-tailed test, in which case the area, and hence the *p* value, is halved.

3.1.2. Comparing More Than Two Groups—One-Way Analysis of Variance

Experiments comparing only two groups are not generally considered to be examples of good experimental design because they yield limited information. Designing your experiment to compare more than two groups at once allows you to examine the effects of several treatments (e.g., different drugs, doses of a single drug, time courses, etc.) under the same conditions using a single control group and is thus economical of time, effort, and materials. However, the analysis of such experiments becomes more complex, requiring use of the powerful technique of analysis of variance (ANOVA) rather than performing repeated *t*-tests. ANOVA avoids the problem of multiple comparisons, where the chance of making a type 1 error (erroneously rejecting the null hypothesis, *see* **Subheading 5.1.**) increases with the number of pairwise comparisons being made. ANOVA considers the overall variability in the data and compares intragroup variation with intergroup variation producing a ratio of variances, the *F*-ratio. If there is no difference between the groups the *F*-ratio will be close to 1. Where the intergroup variability exceeds the intragroup variability *F* is large and the *p*-value for this arising by chance may be determined, taking into account the df from both the number of groups and the group sizes. As with all tests of significance, conventionally $p < 0.05$ is taken as indicating a statistically significant result, allowing rejection of the null hypothesis. **Example 3** shows a typical one-way ANOVA results table.

This first stage of ANOVA does not distinguish which group means differ from each other; it is then necessary to retest the data to perform intergroup comparisons. Statisticians distinguish between comparisons planned before doing the experiment, that is, based on predictions from the scientific hypothesis (*a priori* tests) and tests to investigate differences shown up after the initial ANOVA (*post hoc* tests). There is much discussion in the literature as to the

Example 3.

Output from a one-way ANOVA to analyze the effects of two monoclonal antibodies on antigen-induced eosinophil accumulation in rabbit skin. The table below shows the number of eosinophils at each skin site under the four treatments. The null hypothesis is that the four groups come from populations with the same mean and variance.

Saline	Antigen	Antigen + MAb A	Antigen + MAb B
33	142	131	30
36	143	132	27
36	139	127	23
22	125	116	9
51	160	139	34

Testing this data with a one-way ANOVA gives the following standard output ANOVA table:

ANOVA

	Sum of squares	df	Mean square	*F*	Sig
Between group	56152.550	3	18717.517	175.422	<0.001
Within groups	1707.200	16	106.700		
Total	57859.750	19			

The value for *F*, the ratio between the between-group and within-group variances, is 175.422. The significance value is the probability of this value arising by chance if the null hypothesis is true (i.e., *F* is actually 1.0) and is <0.001. In this case the *F*-ratio suggests that there are differences between the groups.

most appropriate test procedure, based on the need to avoid type 1 errors by restricting the pairwise error rate, as opposed to detecting real differences between groups, that is, avoiding type 2 errors (*see*, e.g., discussions in ***3,6,8–11***). In general, our advice is to choose your comparison procedure based on the design of your experiment. So, if you start by wishing to compare only selected pairs of means (i.e., an *a priori* approach), the Bonferroni or Dunn–Sidiak tests are an appropriate choice (*see* **Example 4**). Both of these are conservative and restrict the pairwise error rate to control for type 1 errors. The least significant difference (LSD) procedure, another purely *a priori* test, does not restrict error rates and is consequently not considered to be as reliable.

If you want to restrict the number of comparisons by only comparing each group with a specified control group, as is sometimes appropriate in pharmacology, then Dunnett's posttest is designed to do this specifically.

To make all possible pairs of comparisons following a significant ANOVA result (the usual situation, you have $p < 0.05$ in your initial ANOVA and now need a *post hoc* test to identify which groups are different), the Bonferroni test will give an answer, although it is not advisable to use this if you have more than five groups, as it is so conservative and thus has low power to detect differences. Most texts recommend Tukey's or Student–Newman–Keuls tests as being more powerful than Bonferroni for large numbers of comparisons, or the Ryan–Einot–Gabriel–Welsch procedure ***(10)***. Although widely used in the literature, Duncan's multiple range test is not recommended for *post hoc* testing in statistics texts, because it has no means of controlling the pairwise error rate ***(10)***. If you have a very large number of complex comparisons or contrasts to make, then Sheffé's method is probably the best to use, although rather conservative.

If you expect differences between groups to follow a specific order (e.g., in a dose–response or a time course), it is possible to apply a posttest for trend, rather like the procedure for linear regression, rather than testing pairs of groups. This tests whether there is a trend for mean values to increase or decrease as you move through the groups.

3.1.3. More Complex Designs of ANOVA

One of the advantages of ANOVA as a statistical technique is its flexibility; it is possible to analyze complex experiments that make use of pairing, block-

Example 4.
Posttest results for eosinophil data.

Because there are only four groups and the experimenter was interested in relatively few specific comparisons, that is, between Saline and Antigen, Saline and MAb A, Saline and MAb B, and MAb A vs MAb B, a Bonferroni test was used. The relevant parts of the output (from SPSS) are shown below:

Multiple comparisons
Bonferroni

I	*J*	Mean diff (*I–J*)	Std. error	Sig
Saline	Antigen	106.2*	6.533	<0.001
Antigen/MAb A		93.4*	6.533	<0.001
Antigen/MAb B		11.0	6.533	0.670
Ag/MAb A	Antigen/MAb B	104.4*	6.533	<0.001

*SPSS helpfully identifies mean differences where $p < 0.05$ to help you make sense of the output. So we can see that antigen clearly increased eosinophil recruitment and that MAb A was not as effective as MAb B in inhibiting this response.

ing, and repeated measures designs to reduce variability, as well as designs examining the effects and possible interactions of two or more treatments (usually called factors in statistical texts), using two-way or multiway analyses, which can accommodate repeated measures designs. One problem with these procedures is the complexity of the output from statistics packages. **Example 5** attempts to demonstrate which parts of the output you need to look for to interpret the results.

Where you are comparing the effects of more than one factor in a single experiment the ANOVA can become more complex but can also reveal if one factor influences (or interacts with) another. These factorial designs are economical in terms of experimental units and can also feature repeated measures if your experimental design controls for intersubject variability, either by matching subjects or repeating measurements on the same subjects. If the results of two-way ANOVA indicate significant differences between groups, posttests can identify these as for one-way ANOVA. (*See* **Example 6.**)

Most statisticians regard *t*-tests and ANOVA as robust to slight deviations from normality, particularly for large group sizes. If the output gives significant results for tests of homogeneity of variances, sphericity. and so forth (i.e., indications that the assumptions of the ANOVA are violated), you have to make the decision as to whether to accept the results of the analysis or not, remembering that these tests give reliable results only with large group sizes, as mentioned earlier for tests of normality. In most laboratories group sizes are kept to a minimum, so you will have to make an individual decision based on your experience of similar experiments and on how much your results seem to violate the assumptions. The alternative to accepting the analysis results is to proceed to nonparametric analyses, which are less powerful procedures.

4. Nonparametric Methods

Where data clearly do not fit a Gaussian distribution there are several possibilities. You can use parametric tests on the basis that they are robust to deviations from normality. Alternatively, you can transform the data to fit a normal distribution so that parametric tests may be used **(Table 1)** or you can use distribution-free methods, so-called nonparametric statistics.

Nonparametric tests make no assumptions about data distributions. They are less powerful than the corresponding tests based on the normal distribution, so significant results obtained using them might in one sense be regarded as more reliable. The usual nonparametric equivalents to *t*-tests are the Mann–Whitney U test (for independent samples) and the Wilcoxon test (for paired data). (*See* **Example** 7.)

For multiple comparisons of independent samples the Kruskal–Wallis test (a nonparametric equivalent of one-way ANOVA, which is a generalization of

Example 5.
One-way ANOVA with repeated measures.

Contraction heights (mm) in response to 10^{-6} *M* histamine were measured in five sections of guinea pig ileum at four successive 15-min periods, giving four repeated observations in each tissue. The order of dosing was randomized, to avoid violating one of the assumptions of repeated measures ANOVA.

Tissue	Observation			
	1	2	3	4
1	33.0	42.0	31.0	30.0
2	36.0	43.0	32.0	27.0
3	36.0	39.0	27.0	23.0
4	22.0	25.0	16.0	9.0
5	51.0	60.0	39.0	34.0

Another assumption of this test is of homogeneity of covariance. Most statistics packages will test for this (SPSS uses Mauchly's test of sphericity). As long as the *p* value for this test is >0.05 (in this example *p* for Mauchly's test was 0.398), it is valid to proceed to the ANOVA. The resulting ANOVA table has more lines than for one-way but provides similar information, this time two *F* values:

Source	Sum of squares	df	mS	*F*	Sig
Tissue	1600	4	400	44.8	<0.001
Time	852.550	3	284.183	31.812	<0.001
Error	107.200	12	8.933		
Total	2559	19			

This analysis reveals significant differences between tissues as well as over time. A Bonferroni posttest of the time results reveals that results at time points 3 and 4 are significantly different from results at times 1 and 2—selected output below:

Multiple comparisons
Bonferroni

I (Time)	*J* (Time)	Mean diff (*I*–*J*)	Std. error	Sig
2	1	6.200	1.356	0.062
	3	12.800*	2.107	0.022
	4	17.200*	2.332	0.011

*Symbol inserted by SPSS to identify a difference where $p < 0.05$.

Example 6.
A two-way ANOVA.

The following results were obtained in a study to determine the effects of shock intensity and duration on the paw withdrawal response in rats (response time [s]):

	Intensity	
Duration	0.3 mA	1.0 mA
0.5 s	14, 19, 24, 16, 22, 29	16, 19, 14, 15, 12, 17
3 s	12, 16, 19, 14, 20, 19	9, 6, 12, 13, 4, 16

The two-way ANOVA assumes that the variances in all groups are equal. The results of a test for this are usually included in output from statistics packages (SPSS uses Levene's test of equality of error variances), and if these give $p >$ 0.05 (i.e., no significant differences) the ANOVA is valid. In the above case Levene's test gives $p = 0.168$, that is, variances are similar.

The ANOVA table has to account for five sources of variation; between groups for each factor (duration and intensity in this example) separately, interaction between the factors, total variation, and error variation. Thus, the table has more lines and gives three F-values of interest:

Test of between-subjects effects
Dependent variable—Response

Source	Sum of squares	df	MS	F	Sig
Intensity	210.042	1	210.042	12.571	0.002
Duration	135.375	1	135.375	8.102	0.01
Intensity × duration	3.375	1	3.375	0.202	0.658
Error	334.167	20	16.708		
Corrected total	682.958	23			

Thus, the table shows that shock intensity and duration both have significant effects on response time but that there is no indication of any interaction between the two ($p = 0.658$). Some statistics packages will plot the group means as a visual display of the results. In this case you would get two parallel lines denoting no interaction. Where there is evidence of interaction the lines intersect or approach each other.

the Mann–Whitney U test for more than two groups) may be used, with Dunn's test as a posttest (*see* **Example 8**). Not many packages offer posttests for non-parametric ANOVA, but the calculations are not too involved. Methods may be found in the classic text of Siegel and Castellan ***(12)*** or a more recent book by Neave and Worthington ***(13)***.

Example 7.
Mann–Whitney U test results for response times in mice.

In the experiment that yielded the control data presented in Example 2, a further group of mice treated with 9 mg/kg of a κ opioid agonist had the following response times (s) to the same stimulus: 60, 60, 48.8, 46.0, 60, 60, 60. Here median (range) = 60 (46.0–60) s, a very skewed data set, hence the decision to use a nonparametric method for analysis.

Comparison of the two groups using a Mann–Whitney test gave the following results (taken from SPSS output):

Response		*N*	Mean rank	Sum of ranks
	Control	7	5.29	37.0
	κ Agonist	7	9.71	68.0
	Total	14		

Mann–Whitney U 9.000
Two-tailed sig. 0.034

This test uses the ranks of data values to compute the test statistic, not the actual values. The final p value suggests that response times in the two groups were significantly different.

These tests rely on ranking data values and comparing ranks within groups to assess whether or not they differ. If you have category data or proportions, you may find χ^2 tests applicable. The usual applications for χ^2 tests are as tests for proportions or for goodness of fit. Where the assumptions of a χ^2 test for proportions are violated (usually because the group sizes are too small, making expected values <5) an alternative is Fisher's exact test, which can, as its name suggests, calculate exact probabilities.

5. What Is the Optimal Size of an Experiment?

If attempts have been made to improve sensitivity and reduce variability, then sample size and, by default, df available to estimate experimental error will determine the precision of the experiment. To illustrate this in graphical terms, **Fig. 2** represents the critical values of t at the 5% probability level as a function of df. At 10 df, the t-value is 2.23, which falls to 2.09 with 20 df.

As we can see from this figure, df greater than 10 will not yield a substantial reduction in the critical value of t. Thus, for comparisons between two samples, a minimum of six experimental units (independent data values) per group should be employed. If one can achieve a df for error (denominator) between 10 and 20, then one will obtain high power; any number above this value will

Example 8.
Kruskal–Wallis ANOVA applied to haematology data.

The following T-lymphocyte cell counts were obtained in an experiment comparing the responses of groups of mice to recombinant α-thymosin. Because cell counts are regarded as not being normally distributed, a nonparametric ANOVA was used.

Control	1 μg of thymosin	5 μg of thymosin	10 μg of thymosin
609	700	678	800
584	569	987	578
573	790	945	593
711	987	934	493
456	800	888	579

Kruskal–Wallis results (from SPSS):

Test statistic 8.730 df 3 Significance 0.0331

This indicates that the groups differ. GraphPad Prism™ gives identical results for the ANOVA and offers Dunn's posttest, with the following results (comparison of selected groups):

Control vs 1 μg of thymosin $p > 0.05$
Control vs 5 μg of thymosin $p < 0.05$
Control vs 10 μg of thymosin $p > 0.05$

Thus the 5-μg thymosin dose appeared to be the only one that altered T-cell counts in these mice.

lead to only a slight increase in precision at the expense of wasting experimental units, which is particularly unethical if the units are animals.

5.1. Power of Experiments and Type I and 2 Errors

Hypothesis testing involves making a decision as to whether two distributions are the same or different. We normally set the significance level at which the null hypothesis will be rejected at the small probability of 5%, and in this way avoid making exaggerated claims that a treatment has had a significant effect. Rejecting the null hypothesis when it is in fact true is known as a **type I error**. Type I errors are usually committed when multiple *t*-tests are applied between a control group and various treatment groups. If five "independent" comparisons are made, the probability of rejecting the null hypothesis is no longer 5% but increases to 23%. Thus, it is highly likely that a rejection of the null hypothesis will be made. To reduce the potential of a type I error in this example, the threshold to keep the overall risk of a type I error to 5% is now

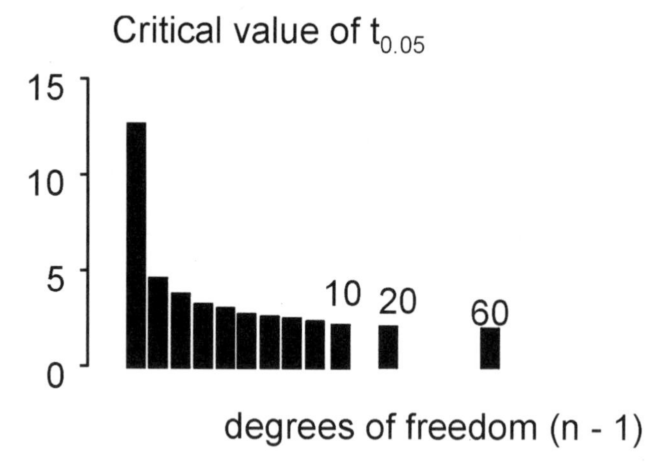

Fig. 2. Relationship between critical values of t and degrees of freedom.

0.0102. Thus, the null hypothesis could be rejected only if a probability of 1.02% or less was obtained. This is an example of a Bonferroni correction.

On the other hand, there is a chance that an effect beyond the significance level arose from the "control" distribution. This risk is denoted by α and the accepted level of significance is 5%. By accepting the null hypothesis, you are stating that you have not found a big enough difference *in your experiment* to reject the possibility that the difference arose simply by chance. So, if in your experiment you fail to find a significant difference, it could be that there really is no difference in the two distributions, or there is a difference, but you have missed it. It is this latter case that is commonly referred to as a **type 2 error** (i.e., failure to detect a difference when it in fact exists). This is denoted by β. In a perfect world, distributions would not overlap and we could avoid making both type I and type 2 errors, but in the real world we can only try to balance the two.

5.2. Power of a Test

To observe a statistical difference between treatments, your test requires sufficient **power:**

$$\text{Power of a test} = 1 - \beta$$

Statistical power is thus the complement to a type 2 error—the probability of detecting a given effect in a sample if it actually exists in the population. Techniques of power analysis are important in planning investigations, particularly

Example 9.
Using the resource equation to estimate an appropriate n for different experimental protocols.

In a comparison of cellular responses to drug treatment using one control and one treatment group (suitable for analysis by independent samples *t*-test) we have $B = 0$ and $T = 1$. So to obtain $E = 10$, we would need $N = 12 - 1 = 11$; that is, six experimental units (animals, 96-well plates, etc.) in each group. Any fewer, and *E* would be too small, whereas using more might be a waste of resources without providing any extra information.

If the experimental protocol was changed to one comparing three different treatments with the control (requiring one-way ANOVA analysis), $T = 3$ and, to obtain $E \geq 10$, *N* needs to be at least 13. You would probably elect to have a balanced design (equal numbers in each group) and so have $N = 15$; that is, four experimental units per group. So for the use of four extra experimental units (16 rather than 12) you have a protocol that provides three times the amount of information while still being of an appropriate size.

to help determine necessary sample sizes to detect particular effect magnitudes. By convention, an acceptable level of statistical power is set to 0.80 (80%), and may be expressed as:

$$\text{Power} \propto \frac{\text{Effect size} \times \alpha \times \sqrt{n}}{\sigma}$$

where α = significance level (conventionally set to 0.05), n = sample size, and σ = population standard deviation. The effect size depends on the nature of the experiment, that is, how small a difference between groups you would consider to be scientifically significant, or how small a difference your experimental system can actually detect. How these parameters are precisely linked to power is a function of the statistical test to be used; hence you must have decided on the method of analysis before you start! Because of the complexity of the calculations, it is usual to use proprietory software such as nQuery Advisor™ to calculate *n*, the minimum number of subjects or replicates for a particular effect size. This is mandatory when planning clinical trials, to avoid subjecting patients to needless risk by undertaking a study that is either too small to detect anything of interest or too large and thus a waste of resources. A similar position could become standard in laboratory experiments, given pressure on time and resources. Unfortunately, for most laboratory investigations, the population variance and hence the SD are not known, so that the calculation is difficult. Possible solutions might be to estimate the population SD based on pilot experiments, or to use values based on other published studies in the appropriate area (i.e., prior knowledge). Quinn and Keough ***(10)*** give examples and a comprehensive discussion of this topic.

Alternatively, Mead ***(14)*** suggested a rule of thumb resource equation, which is somewhat easier to apply, viz.:

$$E = N - T - B$$

where E = error degrees of freedom, N = total degrees of freedom (number of experimental units minus 1), T = treatment degrees of freedom, and B = block degrees of freedom (number of blocks minus 1).

This approach is valid only for data conforming to a Gaussian distribution capable of being analyzed by parametric methods such as t-tests or ANOVA. The desirable value for E is between 10 and 20, based on the distribution of critical values for t as shown in **Fig. 2**, where it is seen that increasing the df above 10 has very little effect on critical values of t. This approach does not take into account any consideration of effect size, unlike the full power analysis. (*See* **Example 9**.)

In most cases in research it may be of more use to use power analysis retrospectively after a preliminary experiment to answer the question, "Given the observed variability in the experiment, what is the smallest change that we could be expected to identify?" You could substitute the observed effect size and SD in the appropriate power equation to work out the power of your experiment. It is then possible to choose group sizes for follow-up studies to have a good chance of performing an experiment that will have 75–80% power. This "reversed" power analysis may be used to justify publication of negative results if it shows that the sample size used was sufficient to demonstrate a specified effect and yet none was detected.

6. Statistical Aspects of Line and Curve Fitting

6.1. Choosing a Model

Before the advent of personal computers, most models for experimental data used transformations so that the data could be fitted to straight lines, the easiest manipulation to carry out, using the least-squares method. These methods are now regarded as inappropriate because the transformed data often violate the assumptions of linear regression and hence lead to inaccurate values for parameters such as K_d, B_{max}, and so forth in radioligand-binding experiments ***(3,15)***. Software packages make it possible to use nonlinear regression to fit any data to a bewildering variety of curves with ease—the problem is not how to do it but rather how to choose an appropriate model and then to tell if the results indicate a good fit. Choosing a model for any particular data set must depend on custom and practice in that area and accepted methods. If your curve-fitting package does not contain the equation you require as a built-in function, you will have to add it as a user-defined option. To start the curve-fitting process, which is carried out by a process of iteration, you need to supply initial values for parameters of the curve. Most packages will automatically generate such values for their built-in equations based on the range of data provided, but you may need to inspect these and decide if you could improve on them before starting the curve-fitting process. If you use a user-defined equation, you may have to provide the initial values. The iteration process is automatic to converge on the best fit based on the model that you specify, but the software has no knowledge of the biology of the system you are modeling and will not identify any clearly impossible values. Consequently, if there are values that you want to fix (e.g., if modeling exponential decay, the curve must plateau at zero, or you have a set of results where you wish to fix the maximum response to 100%) you should do so at the outset.

For further discussion of appropriate methods for receptor studies, see Kenakin ***(15)*** or Motulsky ***(3)*** and earlier chapters in this book.

6.2. Telling If Data Fit Your Model and Comparing Data Sets

6.2.1. Linear Regression

The least-squares method for linear regression minimizes the squares of the vertical distances of each data point from the regression line and finds the slope and intercept of the straight line that best fits the data, assuming a linear relationship. Thus any data entered into regression software will generate values for these parameters. To tell if the fit is good, you need to examine other statistics from the regression. The Pearson correlation coefficient, r, is often quoted, but a better parameter is the square of this quantity, r^2. This is a measure of the fraction of variation in y that may be explained by its dependence on x. It is computed by comparing the ratio of the sums of squares (Ss) of distances of y values from the regression line with the sums of squares of deviations from the null hypothesis line ($y = 0$).

$$r^2 = 1 - \frac{\text{Ss(regr)}}{\text{Ss(total)}}$$

The better the data fit a straight line, the closer r^2 is to 1.0

An F-test for trend may be used to test the null hypothesis that the overall slope of the line is zero. A p-value < 0.05 for this test is additional evidence for a linear relationship between x and y. Most statistics packages will give this output as an ANOVA table in the form we have already seen for one-way ANOVA.

An alternative approach is to use t-tests to test the null hypotheses that the slope and the intercept are equal to zero (this would imply no linear relation-

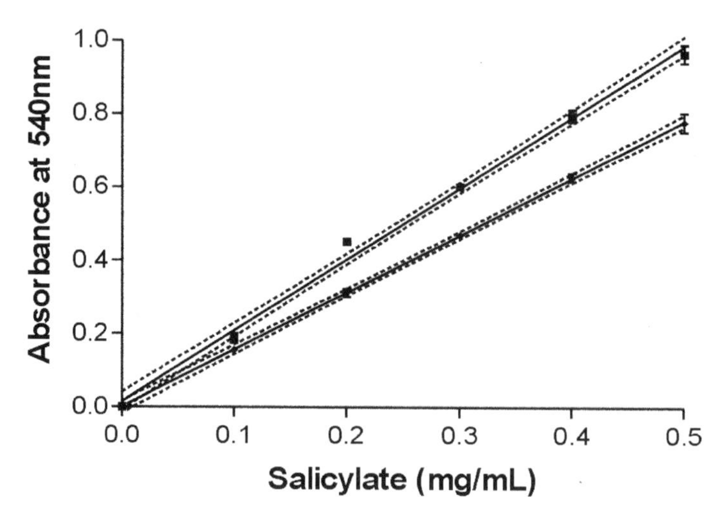

Fig. 3. The two regression lines (with 95% confidence intervals) from **Example 10**, showing the difference in slopes.

ship in the data). Again, this is often standard output from a statistics package and you have to look for p-values < 0.05 in these tests to confirm that the calculated coefficients are not zero. The p-value for the slope, of course, duplicates the information given by the F-test. A further confirmation is to inspect the 95% CIs for the coefficients, if given, as these will not span zero if the relationship is linear. The 95% confidence intervals for the regression line give limits within which the true line may lie. From a plot it may be seen that these are curved **(Fig. 3)**, demonstrating how this technique places most emphasis on data points nearest the mean values.

A plot of residuals (vertical distances for each data point from the regression line, which are easier to see from a graph than from a table of the figures) should show points randomly scattered about $y = 0$, not varying with x. There should be no clusters of adjacent points all above or below the line. A runs test (commonly available as an option for output) will highlight such clusters, by calculating the probability of observing as few or fewer runs or clusters of points as appear in the data. If the p-value for the runs test is < 0.05, one may conclude that the data do not fit a straight line.

6.2.1.1. Comparison of Several Regression Lines

Zar ***(12)*** gives a method for comparing the slopes and intercepts of regression lines using analysis of covariance (ANCOVA). This analysis is available as an option for linear regression in GraphPad Prism™. The starting point is to compare the slopes using an F-test. If the resulting p-value is < 0.05, the lines have different slopes and are not identical. Applying a suitable posttest (such as Tukey's test to compare pairs of lines, or Dunnett's test if one of the lines is a control and the other lines are to be compared with it) will allow you to identify where the differences are.

If, on the other hand, the slopes are not significantly different (i.e., $p > 0.05$) then it is possible to use a further F-test to compare the intercepts to decide whether the lines are identical (i.e., the intercepts are not significantly different). If p for this comparison is < 0.05 then one can conclude that the lines are parallel but separate and, as before, a posttest will identify the differences. (*See* **Example 10**.)

6.2.2. Goodness of Fit for Nonlinear Regression

As with linear regression, curve-fitting software will fit any data to any equation requested, so that assessing the appropriateness of the output relies on close inspection of the results of the analysis. First, the generated best-fit values for curve parameters should be within the expected range(s). Obviously impossible EC_{50} values and so forth (e.g., negative values or values off the scale of measurement) suggest that the data are not a good fit. The reason may be that you have too few points in the region of the curve closest to the EC_{50} and need to repeat the experiment to obtain better data.

Assuming the estimated parameters and the generated graph look reasonable, then values for r^2, calculated from the sums of squares of deviations from the curve as for linear regression and plots of residuals and runs test results can tell you if the fit was close to the chosen curve.

6.2.2.1. Comparison of Several Curves (*see* Note Added in Proof, p. 414)

Although it is possible to use two-way ANOVA with appropriate posttests to compare dose–response or binding curves, the results may not be easy to interpret because they ignore trends in the data and also depend on sample size. Instead, Motulsky *(4)* recommends using nonlinear regression to fit curves and compare the derived parameters. Just as with linear regression, you can use F-tests to determine whether slopes and derived parameters are the same. If any of these tests give a significant result ($p < 0.05$), a suitable posttest will identify the differences. It is also possible to use an F-test to compare the goodness of fit of two different models (e.g., one-site vs two-site binding).

7. Choosing a Statistics Package

Before embarking on any statistical analysis it is good practice to inspect your data. This usually means plotting it in some way and looking for trends, obvious deviations from expectation, anomalies, and so forth. Thus, using a graphics package to try out different forms of presentation is a good first step.

Example 10.
Linear regression and comparison of two lines.

An assay for salicylate gave the following results for absorbance at 540 nm:

Salicylate conc. (mg/mL)	Absorbance at 540 nm		
	1	2	3
0	0	0	0
0.1	0.175	0.189	0.201
0.2	0.456	0.448	0.455
0.3	0.607	0.605	0.596
0.4	0.799	0.811	0.801
0.5	0.990	0.968	0.941

The triplicate samples had been prepared individually rather than as repeated measurements on one sample, so were treated as independent determinations in the analysis.

The linear regreession gave the following results for coefficients (± SEM):

Slope = 1.945 ± 0.03912
Intercept = 0.01498 ± 0.01184
$r^2 = 0.9936$

The F-test for deviation of the slope from zero gave:

F (df 16.1) = 2472, $p < 0.001$

The runs test for deviations from linearity gave $p = 0.300$.

Thus the data (as we would hope for assay standards) fit a good straight line.

A second experiment performed later under the same conditions gave the following results:

Salicylate conc. (mg/mL)	Absorbance at 540 nm		
	1	2	3
0	0	0	0
0.1	0.158	0.162	0.146
0.2	0.324	0.323	0.291
0.3	0.465	0.483	0.461
0.4	0.640	0.641	0.606
0.5	0.760	0.831	0.748

Results as before:

Slope = 1.565 ± 0.02671
Intercept = –0.0001111 ± 0.008086
r^2 = 0.9954

The F-test for deviation of the slope from zero gave:

F(df 16,1) = 3432, $p < 0.0001$

The runs test for deviations from linearity gave $p = 0.700$.

The two lines are shown in **Fig. 3** with 95 CIs. Using ANCOVA to compare the slopes gave an F-value of 64.4669 (df 32,1) and for this, $p < 0.001$, that is, the slopes (and therefore the lines) are significantly different and there is no reason to compare the intercepts. Thus the results of this assay vary from day to day, and it will always be necessary to include standards.

Modern graphics software is mostly user-friendly, often includes statistical routines, and may be all that is needed for your data analysis. GraphPad Prism™ has the advantage that it was specifically designed for presentation and analysis of pharmacological results. It also has particularly good world wide web support for users (*see* Web Resources at end of References).

Many scientists make widespread use of spreadsheets for data manipulation and presentation. The graphics capabilities of these packages are generally of sufficient quality for scientific presentation. They also have some statistical routines built in, although these tend to be limited. One of the most widely used spreadsheets, Microsoft Excel™, comes with a data analysis add-in option that will do limited statistical analyses, including *t*-tests, correlation and linear regression, and one-way and two-way ANOVA (for balanced designs only) but does not include routines for posttests or nonparametric analyses. Various commercial packages developed as Excel add-ins will extend these capabilities.

Spreadsheets and graphics packages usually cannot cope with more complex analyses nor in some cases with unbalanced designs (unequal group sizes) and missing data values. On the other hand, dedicated statistics packages will perform more tests than most users will ever need and allow graphical representation of data, although this is rarely in a suitable form for publication. Most traditional statistics packages have evolved over time from mainframe versions and can be user-unfriendly (e.g. SPSS™; Minitab™ is slightly less unfriendly). Even packages written originally for PCs (Systat™, Statgraphics™, Origin™, etc.) can require some getting used to. Our best advice is to obtain one of the specialized texts dealing with your particular software, choosing one relevant to your discipline. These are often easier to understand than the official manual. You may well be constrained to use whatever package is supplied on your in-house network. If you need to do particular analyses routinely there may be specialized software available (e.g., nQuery Advisor™ for power calculations, Arcus ProStat Biomedical™ for analysis of clinical trial data) or you may choose a package that allows you to customize analyses. In the end, the choice of software is a personal one, balancing your needs with the availability of user support, budgetary constraints, and user-friendliness.

References

1. Huff, D. (1955) *How to Lie With Statistics.* Penguin Books, London.
2. Hancock, A. A., Bush, E. N., Stanisic, D., Kyncl, J. J., and Lin, C. T. (1988) Data normalization before statistical analysis: keeping the horse before the cart. *Trends Pharmacol. Sci.* **9,** 29–32.
3. Motulsky, H. J. (1995) *Intuitive Biostatistics.* Oxford University Press, New York.
4. Motulsky, H. J. (1999) *Analysing Data With GraphPad Prism.* GraphPad Software Inc., San Diego CA. www.graphpad.com.

5. Altman, D. G., Machin, D., Bryant, T. N., and Gardner, M. J. (2000) *Statistics With Confidence.* BMJ Books, London.
6. Sokhal, R. R. and Rolf, F. J. (1995) *Biometry.* W.H. Freeman & Co., New York.
7. Petrie, A. and Sabin, C. (2000) *Medical Statistics at a Glance.* Blackwell Science, Oxford.
8. Dyson, C. (2003) *Choosing and Using Statistics.* Blackwell Science, Oxford.
9. Wallenstein, S., Zucker, C., and Fleiss, J. L. (1980) Some statistical methods useful in circulation research. *Circ. Res.* **47,** 1–9.
10. Quinn, G. P. and Keough, M. J. (2002) *Experimental Design and Data Analysis for Biologists.* Cambridge University Press, Cambridge.
11. Zar, J. (1999) *Biostatistical Analysis*, 4th ed., Prentice-Hall, Englewood Cliffs, NJ.
12. Siegel, S. and Castellan, N. J. (1988) *Nonparametric Statistics for the Behavioural Sciences*, 2nd ed. McGraw-Hill, New York.
13. Neave, H. R. and Worthington, P. L. (1988) *Distribution-Free Tests.* Unwin Hyman Ltd., London.
14. Mead, R. (1988) *The Design of Experiments.* Cambridge University Press, Cambridge, Massachusetts.
15. Kenakin, T. (1996) *Molecular Pharmacology: A Short Course.* Blackwell Science, Oxford.

Web Resources

Software houses tend to have useful websites, with support for users. In particular, the GraphPad site (www.graphpad.com) offers advice, down-loadable manuals, and links to other sites of interest. Modern textbooks often have supplementary information available on the relevant publishers' websites.

Note Added in Proof

The most recent release of GraphPad Prism (version 4) now includes methods for comparison of curves fitted to exponential data by a process of global curve fitting.

Index

K

L

M

N

O

P